GEOMETRY AND ANALYSIS OF AUTOMORPHIC FORMS OF SEVERAL VARIABLES

Series on Number Theory and Its Applications ISSN 1793-3161

Published

Vol. 1 Arithmetic Geometry and Number Theory
edited by Lin Weng & Iku Nakamura

Vol. 2 Number Theory: Sailing on the Sea of Number Theory
edited by S. Kanemitsu & J.-Y. Liu

Vol. 4 Problems and Solutions in Real Analysis
by Masayoshi Hata

Vol. 5 Algebraic Geometry and Its Applications
edited by J. Chaumine, J. Hirschfeld & R. Rolland

Vol. 6 Number Theory: Dreaming in Dreams
edited by T. Aoki, S. Kanemitsu & J.-Y. Liu

Vol. 7 Geometry and Analysis of Automorphic Forms of Several Variables
Proceedings of the International Symposium in Honor of Takayuki Oda on the Occasion of His 60th Birthday
edited by Yoshinori Hamahata, Takashi Ichikawa, Atsushi Murase & Takashi Sugano

Series on Number Theory and Its Applications **Vol. 7**

GEOMETRY AND ANALYSIS OF AUTOMORPHIC FORMS OF SEVERAL VARIABLES

Proceedings of the International Symposium in Honor of Takayuki Oda on the Occasion of His 60th Birthday

Tokyo, Japan 14 – 17 September 2009

Editors

Yoshinori Hamahata *(Kansai University, Japan)*
Takashi Ichikawa *(Saga University, Japan)*
Atsushi Murase *(Kyoto Sangyo University, Japan)*
Takashi Sugano *(Kanazawa University, Japan)*

World Scientific

NEW JERSEY • LONDON • SINGAPORE • BEIJING • SHANGHAI • HONG KONG • TAIPEI • CHENNAI

Published by

World Scientific Publishing Co. Pte. Ltd.

5 Toh Tuck Link, Singapore 596224

USA office: 27 Warren Street, Suite 401-402, Hackensack, NJ 07601

UK office: 57 Shelton Street, Covent Garden, London WC2H 9HE

British Library Cataloguing-in-Publication Data
A catalogue record for this book is available from the British Library.

Series on Number Theory and Its Applications — Vol. 7
GEOMETRY AND ANALYSIS OF AUTOMORPHIC FORMS OF SEVERAL VARIABLES
Proceedings of the International Symposium in Honor of Takayuki Oda on the Occasion of His 60th Birthday

ISBN-13 978-981-4355-59-9
ISBN-10 981-4355-59-3

Printed in Singapore by World Scientific Printers.

Preface

This volume is the Proceedings of the International Symposium "Geometry and Analysis of Automorphic Forms of Several Variables" in honor of Professor Takayuki Oda's 60th birthday, which was held at the University of Tokyo in Japan during the period of 14-17 September 2009. During the symposium, seventeen invited lectures were presented, and 124 participants gathered from all over the world. We would like to thank all the speakers and participants for their vital activities throughout the symposium. We are also grateful to Toshiyuki Kobayashi and other organizers who made the symposium a great success. Further, we notice with gratitude that this symposium was partially supported by JSPS Grant-in-aid for Scientific Research A-21244001 (Tomoyoshi Ibukiyama), A-19204001 (Takayuki Oda), A-19204002 (Makoto Matsumoto), B-19340009 (Masanobu Kaneko), and by the fund of Saga University.

This proceedings consists of selected papers by the principal speakers of the symposium on recent trends and results in the area of automorphic forms of several variables. This volume presents both research articles and survey articles which are closely connected with main themes of Professor Oda's work including arithmetic of modular varieties, lifting of modular forms and representation theory of Lie groups. It would be our great pleasure if this volume would contribute to the development of the theory of automorphic forms.

Last but not the least, we would like to express our sincere gratitude to the contributors for their articles as well as to the anonymous referees for their careful review. Special thanks go to Miki Hirano, Tomonori Moriyama and Hiro-Aki Narita who gave us helpful advice throughout our editorial work.

Editors
Yoshinori Hamahata
Takashi Ichikawa
Atsushi Murase
Takashi Sugano

Professor Takayuki Oda

Program of symposium

Monday, September 14, 2009

14:00-15:00	Werner Hoffmann (Universität Bielefeld) Fourier transforms of weighted orbital integrals on $Sp(2, R)$
15:15-16:15	Takao Watanabe (Osaka University) On Voronoi's theorem and related problems
16:30-17:30	Yumiko Hironaka (Waseda University) p-adic spherical functions and singular series

Tuesday, September 15, 2009

10:00-11:00	Joachim Schwermer (Universität Wien) Geometric cycles, arithmetic groups and related automorphic forms
11:15-12:15	Shushi Harashita (Institute for the Physics and Mathematics of the Universe) The supremum of Newton polygons of p-divisible groups with a given p-kernel type
14:00-15:00	Fumihiro Sato (Rikkyo University) Representations of Clifford Algebras and local functional equations of non-prehomogeneous type
15:15-16:15	Atsushi Murase (Kyoto Sangyo University) Hecke symmetries for theta lifting and Borcherds products on orthogonal groups (a joint work with Bernhard Heim)
16:30-17:30	Tomoyoshi Ibukiyama (Osaka University) Differential operators on Siegel modular forms as special functions and applications

Wednesday, September 16, 2009

10:00-11:00	Henri Darmon (McGill University) Arithmetic of cycles on modular varieties
11:15-12:15	Makoto Matsumoto (Hiroshima University) Study of Galois representations via Teichmüller modular groups
14:00-15:00	Noriko HIRATA-Kohno (Nihon University) Exceptional classes of unit equations
15:15-16:15	Toshiyuki Kobayashi (University of Tokyo) Breaking Symmetries and Locally Symmetric Spaces
16:30-17:30	Takayuki Oda (University of Tokyo) Quest for more paths between analysis and geometry of automorphic forms on Hermitian domains

Thursday, September 17, 2009

10:00-11:00	T. N. Venkataramana (Tata Institute of Fundamental Research) Restriction of cohomology of locally symmetric spaces
11:15-12:15	Taku Ishii (Seikei University), Miki Hirano (Ehime University), Tadashi Miyazaki (University of Tokyo) Archimedean Whittaker functions on $GL(3)$
14:00-15:00	Masao Tsuzuki (Sophia University) Spectral square means for central values of L-series of Maass cusp forms
15:15-16:15	Marko Tadić (University of Zagreb) On automorphic duals of classical groups

Contents

The Birch and Swinnerton-Dyer conjecture for $\mathbb{Q}$-curves and Oda's period relations

Henri Darmon

Department of Mathematics and Statistics, McGill University,
805 Sherbooke Street West, Montreal, Canada H3A 2K6
E-mail: darmon@math.mcgill.ca

Victor Rotger*

Departament de Matemàtica Aplicada II, Universitat Politècnica de Catalunya,
C. Jordi Girona 1-3, 08034 Barcelona, Spain
E-mail: victor.rotger@upc.edu

Yu Zhao

Department of Mathematics and Statistics, McGill University,
805 Sherbooke Street West, Montreal, Canada H3A 2K6
E-mail: zhao@math.mcgill.ca

To Takayuki Oda on his 60th birthday

Let $N \equiv 1 (\mathrm{mod}\ 4)$ be a square-free positive integer, let ε be the primitive quadratic character of conductor N, and let $f \in S_2(\Gamma_0(N), \varepsilon)$ be a newform with fourier coefficients in a quadratic field. Shimura associates to f an elliptic curve E defined over the real quadratic field $F = \mathbb{Q}(\sqrt{N})$ which is isogenous over F to its Galois conjugate. Let M/F be a quadratic extension of F which is neither CM nor totally real, and denote by E_M the twist of E with respect to M/F. The first main result of this article is that, if $L'(E_M/F, 1) \neq 0$, then $E_M(F)$ has rank one and $Ш(E/F)$ is finite. The proof rests on a fundamental result[42] of Yuan, Zhang and Zhang and on an explicit Heegner point in $E_M(F)$ arising from the modular parametrization $X_1(N) \to E$. We formulate a conjecture relating this Heegner point to Stark-Heegner points[8] arising from ATR cycles of real dimension one on Hilbert modular surfaces, and present some numerical evidence for this conjecture.

Keywords: $\mathbb{Q}$-curves; Birch and Swinnerton-Dyer conjecture; Oda's period

*The research of the second author is financially supported by DGICYT Grant MTM2009-13060-C02-01 and the Grup de recerca consolidat de Catalunya 2009 SGR 1220.

relations; Heegner points; Stark-Heegner points; ATR cycles; Hilbert modular forms

1. Introduction

Let E be an elliptic curve over a number field F and let $L(E/F, s)$ denote its Hasse-Weil L-series. It is widely believed that the Shafarevich-Tate group $Ш(E/F)$ is finite and that $L(E/F, s)$ extends to an entire function of the complex variable s. The order of vanishing of this function at $s = 1$, denoted by $r_{\mathrm{an}}(E/F)$, is commonly referred to as the *analytic rank* of E over F, a terminology justified by the Birch and Swinnerton-Dyer conjecture which asserts that

$$\mathrm{rank}(E(F)) \stackrel{?}{=} r_{\mathrm{an}}(E/F). \tag{1}$$

The most convincing evidence for the Birch and Swinnerton-Dyer conjecture is the fact that it is proved when $F = \mathbb{Q}$ and $L(E, s) := L(E/\mathbb{Q}, s)$ has at most a simple zero at $s = 1$:

Theorem 1.1 (Gross-Zagier, Kolyvagin). *If* $r_{\mathrm{an}}(E/\mathbb{Q}) \leq 1$, *then* (1) *holds for* $E/\mathbb{Q}$, *and* $Ш(E/\mathbb{Q})$ *is finite.*

The proof of Theorem 1.1, which is briefly recalled in Section 2.1, rests on two key ingredients. The first is the modularity of E, in the strong geometric form which asserts that E is a quotient of the Jacobian of a modular curve over $\mathbb{Q}$. The second is the collection of Heegner points on this modular curve, which satisfies the axioms of an "Euler system" and provides a valuable bridge between the arithmetic of E and the analytic behaviour of its L-series.

Both these ingredients are available in greater generality, most notably when F is a *totally real field.* In this setting, a modular elliptic curve E over F is said to satisfy the *Jacquet-Langlands hypothesis* (JL) if either $[F : \mathbb{Q}]$ is odd, or there is at least one prime of F at which the automorphic form on $\mathrm{GL}_2(\mathbb{A}_F)$ attached to E is not in the principal series. Here, $\mathbb{A}_F$ stands for the ring of adèles of F. The meaning of condition (JL), which only fail to hold for certain elliptic curves of square conductor, is described more concretely in Section 2.1.

Most importantly for the proof of (1), the Jacquet-Langlands hypothesis implies that E is the quotient of the Jacobian of a suitable Shimura curve over F. Shimura curves are equipped with a plentiful supply of CM points, which have been parlayed into the proof of the following number field generalisation of Theorem 1.1.

Theorem 1.2 (Zhang). *Let E be a modular elliptic curve over a totally real field F satisfying (JL). If $r_{\mathrm{an}}(E/F) \leq 1$, then $Ш(E/F)$ is finite and (1) holds for E/F.*

Denote by $\mathfrak{N}$ the conductor of E/F. If either $[F:\mathbb{Q}]$ is odd or there exists a prime $\mathfrak{p}$ of F for which $\mathrm{ord}_{\mathfrak{p}}(\mathfrak{N}) = 1$, this is just [43, Theorem A]. The full result in which E is only assumed to satisfy (JL) follows from the subsequent strengthening of the Gross-Zagier formula[42] of Yuan, Zhang and Zhang. Both results are discussed further in Section 2.1.

In analytic rank zero, the Jacquet-Langlands hypothesis can be dispensed with:

Theorem 1.3 (Longo). *Let E be a modular elliptic curve over a totally real field F. If $L(E/F,1) \neq 0$, then $E(F)$ and $Ш(E/F)$ are finite.*

The proof[24] of Longo, extending to totally real fields an earlier treatment[2] of the case $F = \mathbb{Q}$, exploits the theory of congruences between modular forms to realise the Galois representation $E[p^n]$ in the Jacobian of a Shimura curve X_n whose level may (and indeed does) depend on n. The Euler system of CM points on X_n then gives rise to a collection of p^n-torsion cohomology classes which is used to bound the p^n-Selmer group of E over F independently of n, and thereby obtain the finiteness of $E(F)$ and $Ш(E/F)$.

The problem of removing the Jacquet-Langlands hypothesis from Theorem 1.2—or equivalently, of extending Theorem 1.3 to the case where $L(E/F,s)$ has a simple zero at $s=1$—is still very much open.

To better understand the difficulty which arises, it is instructive to examine the simplest setting where the Jacquet-Langlands hypothesis fails to hold. Assume for the rest of the introduction that F is a real quadratic field, and consider for now the case where E/F is an elliptic curve of conductor 1.

Assuming E is modular, the L-series $L(E/F,s)$ is known to have a functional equation relating its values at s and $2-s$, and the sign $w_E \in \{-1,1\}$ in this functional equation is always equal to 1 in this case.

Let M be any quadratic extension of F, let

$$\chi_M : G_F \longrightarrow \pm 1$$

be its associated Galois character, and denote by E_M the twist of E over F by χ_M, so that the L-series $L(E/M,s)$ factors as

$$L(E/M,s) = L(E/F,s)L(E/F,\chi_M,s) = L(E/F,s)L(E_M/F,s).$$

Since E has conductor 1, the sign w_{E_M} of the twisted L-series is controlled by the local signs attached to the archimedean places ∞_1 and ∞_2 of F, which are equal to $\chi_M(\infty_1)$ and $\chi_M(\infty_2)$ respectively. It follows that $w_{E_M} = 1$ if M is either totally real or CM. In particular, the elliptic curve E is always of *even* analytic rank over such M. Since an Euler system of Heegner points attached to a quadratic CM extension M/F is only expected to be available when E has odd analytic rank over M, this suggests that the mathematical objects so crucial in Kolyvagin's descent method may be unavailable for elliptic curves of conductor 1.

A similar expectation can be derived more generally for all elliptic curves which do not satisfy (JL). Indeed, if E/F is an elliptic curve of square conductor $\mathfrak{N}$ and M is a quadratic extension of F which is unramified at the primes dividing $\mathfrak{N}$, then the same analysis as above reveals that

$$r_{\mathrm{an}}(E/M) \equiv \begin{cases} 0 \pmod 2, & \text{if } M \text{ is CM or totally real;} \\ 1 \pmod 2 & \text{otherwise.} \end{cases} \tag{2}$$

A quadratic extension M of F which is neither CM nor totally real is called an *ATR extension* of F. An ATR extension of F thus has two real places and one complex place. (The acronym "ATR" stands for "Almost Totally Real" and is used more generally[8] to designate quadratic extensions of a totally real field having exactly one complex place.)

The present article is motivated by the following specific instance of the Birch and Swinnerton-Dyer conjecture which emerges naturally from the discussion above.

Conjecture 1.1. *Let E be a (modular) elliptic curve over a real quadratic field F of square conductor $\mathfrak{N}$ for which $w_E = 1$, and let M/F be an ATR extension of F of discriminant prime to $\mathfrak{N}$. If $L'(E_M/F, 1) \neq 0$, then $E_M(F)$ has rank one and $Ш(E_M/F)$ is finite.*

Although it seems tantalisingly close to the setting of Theorem 1.2, Conjecture 1.1 presents a real mystery and appears to lie beyond the reach of known methods. The difficulty is that, in the absence of the Jacquet-Langlands hypothesis, no natural "modular" method presents itself in general for constructing the point of infinite order on $E_M(F)$ whose existence is predicted by the Birch and Swinnerton–Dyer conjecture.

One of the original motivations for singling out Conjecture 1.1 for special study lies in the conjectural construction[8] of a so-called *Stark-Heegner point* $P_M^? \in E(M)$. This construction, which is recalled briefly in Section 2.2, involves the images under a complex Abel-Jacobi map attached to the

Hilbert modular form associated to E/F of certain "ATR cycles" indexed by ideals of M. The ATR cycles are null-homologous cycles of real dimension one on the corresponding Hilbert modular surface. It is conjectured[8] that the point $P_M^?$ is of infinite order precisely when $L'(E/M,1) \neq 0$, and that $P_M^?$ is part of a norm-coherent collection of points defined over abelian extensions of M satisfying Euler-System-like properties. However, progress on Conjecture 1.1 through the theory of ATR cycles is thwarted by our inability to provide much theoretical evidence for the algebraicity of ATR points.

The first aim of this note is to study Conjecture 1.1 for the class of elliptic curves E/F which are isogenous over F to their Galois conjugate. Following a terminology that was first introduced[32] by Ribet, these elliptic curves are called $\mathbb{Q}$-*curves*. Their basic properties are reviewed in Section 3. As explained in that section, the case of $\mathbb{Q}$-curves is ultimately made tractable by the existence of a classical elliptic cusp form f (with *non-trivial* nebentypus character in general) satisfying

$$L(E/F,s) = L(f,s)L(\bar{f},s),$$

leading to a modular parametrisation of E by a classical modular curve $X_1(N)$ for a suitable $N \geq 1$. The main theorem of Section 3 is a proof of Conjecture 1.1 for $\mathbb{Q}$-curves:

Theorem 1.4. *Let E/F be a $\mathbb{Q}$-curve of square conductor $\mathfrak{N}$, and let M/F be an ATR extension of F of discriminant prime to $\mathfrak{N}$. If $L'(E_M/F,1) \neq 0$, then $E_M(F)$ has rank one and $Ш(E_M/F)$ is finite.*

The key ingredients in the proof of Theorem 1.4 are a strikingly general recent extension of the theorem of Gross-Zagier obtained by Xinyi Yuan, Shouwu Zhang and Wei Zhang[42] covering cusp forms with possibly non-trivial nebentypus characters, and a strengthening of Kolyvagin's descent method to cover abelian variety quotients of $J_1(N)$, as worked out in the forthcoming book[39] of Ye Tian and Shouwu Zhang. Section 3 explains how Theorem 1.4 follows from these results and the *Artin formalism* for certain Rankin L-series.

The second part of the article focuses on the special case where the $\mathbb{Q}$-curve E is of conductor 1. Such elliptic curves, which were first systematically studied[35] by Shimura, are essentially in bijection with newforms f in $S_2(\Gamma_0(N),\varepsilon_N)$ with quadratic Fourier coefficients, where N is the discriminant of the real quadratic field F, and

$$\varepsilon_N : (\mathbb{Z}/N\mathbb{Z})^\times \longrightarrow \pm 1$$

is the corresponding even Dirichlet character.

Section 4 describes the explicit construction, for all quadratic ATR extensions M of F, of a canonical point $P_M \in E(M)$ arising from suitable CM divisors on $X_1(N)$. The trace to $E(F)$ of P_M is shown to vanish, so that P_M can also be viewed as an F-rational point on the twisted curve E_M.

After explaining how the points P_M can be computed complex analytically by integrating the elliptic modular form f, we tabulate these points for a few ATR extensions M of small discriminant. One expects that the height of the point P_M is related in a simple way to $L'(E_M/F, 1)$.

Finally, Conjecture 4.1 spells out a precise conjectural relationship between the *classical* Heegner point P_M and the Stark-Heegner point $P_M^?$ arising from ATR cycles on the Hilbert modular variety. This conjecture, which relates certain complex analytic invariants attached to an elliptic modular form f and its Doi-Naganuma lift, can be viewed as an analogue for Abel-Jacobi maps of Oda's period relations which are studied[28] in Oda's influential monograph. It is therefore a pleasure to dedicate this article to Takayuki Oda whose work on periods of Hilbert modular surfaces was a major source of inspiration for the conjectures[8] on ATR points.

It is also a pleasure to thank Xavier Guitart, Ariel Pacetti and David Rohrlich for their comments on a previous version of this manuscript.

2. Background

2.1. *The Birch and Swinnerton-Dyer conjecture in low analytic rank*

We begin by recalling in greater detail the main ideas behind the proofs of Theorems 1.1 and 1.2. We start with the assumption that F is a number field and write $\mathcal{O}_F$ for its ring of integers. Let $\mathfrak{N} \subset \mathcal{O}_F$ denote the conductor of the elliptic curve E/F.

The proofs of Theorems 1.1 and 1.2 can be broken up into five steps:

(i) *Modularity*: When $F = \mathbb{Q}$, the main results of Wiles[41] and Tatlor-Wiles,[38] as completed[1] by Breuil, Conrad, Diamond and Taylor, imply that there is a normalised newform f of weight 2 on $\Gamma_0(N)$ satisfying $L(E, s) = L(f, s)$. In particular, $L(E, s)$ has an analytic continuation to the left of the half plane $\mathrm{Re}(s) > 3/2$, and its order of vanishing at $s = 1$ is therefore well defined. For general F, the modularity of E/F is just the assertion that $L(E/F, s)$ is the L-series attached to an automorphic representation of $\mathrm{GL}_2(\mathbb{A}_F)$. Such a property is predicted to hold, as a (very special) case of the Langlands functoriality conjectures. In spite of the powerful ideas introduced into the subject building on Wiles' breakthrough, a proof in the general number field setting still seems a long way off. When F is totally real, modularity can be phrased in terms of modular forms much as in the case $F = \mathbb{Q}$. Namely, E/F is modular whenever there is a normalised Hilbert modular eigenform f of parallel weight 2 on the congruence group $\Gamma_0(\mathfrak{N}) \subseteq \mathrm{SL}_2(\mathcal{O}_F)$ satisfying $L(E/F, s) = L(f, s)$. The methods originating from Wiles' work seem well suited to yield a proof of modularity of all elliptic curves over totally real fields. (See for example the works of Skinner-Wiles,[37] Fujiwara,[10] Jarvis-Manoharmayum[19] and the references therein for an overview of the significant progress that has been achieved in this direction.) Currently, the case which offers most difficulties arises when the residual Galois representation at 3 is *reducible*.

(ii) *Geometric modularity*: Thanks to the geometric construction of Eichler-Shimura and to Faltings' proof of the Tate conjecture for abelian varieties over number fields, the modularity of E in the case where $F = \mathbb{Q}$ can be recast as the statement that E is a quotient of the jacobian $J_0(N)$ of the modular curve $X_0(N)$ over $\mathbb{Q}$, where $\mathfrak{N} = (N)$, $N \geq 1$. A non-constant morphism

$$\pi_E : J_0(N) \longrightarrow E \tag{3}$$

of abelian varieties over $\mathbb{Q}$ is called a *modular parametrisation* attached to E.

When F is a totally real field and E/F is known to be modular, the modular parametrisation arising from Eichler-Shimura theory admits no counterpart in general. However, such a modular parametrisation can be obtained when the Jacquet-Langlands hypothesis formulated in the introduction holds. More precisely, as it is explained in [43, §3], hypothesis (JL) implies that E is a quotient of the Jacobian of a suitable Shimura curve X attached to an order in a quaternion algebra over F which splits at exactly one archimedean place of F. That is, there is a non-constant map

$$\pi_E : J(X) \longrightarrow E \tag{4}$$

of abelian varieties over F generalising (3). The condition that the automorphic form $\pi = \otimes\pi_v$ attached to E be a principal series representation at a place v of F is satisfied precisely when E acquires good reduction over an abelian extension of F_v. For $v \nmid 2$, the meanings of various conditions on the local representations π_v in terms of the behaviour of E over F_v are summarised in the table below.

π_v	E/F_v	$\mathrm{ord}_v(\mathfrak{N})$
Unramified principal series	Good reduction over F_v	0
Principal series	Good reduction over an abelian extension of F_v	even
Steinberg	Potentially multiplicative reduction over F_v	1 or 2
Supercuspidal	Otherwise	≥ 2

(5)

We refer the reader to Gelbart [12, p. 73], Pacetti,[30] and Rohrich [33, Prop. 2], [34, Prop. 2 and 3] for proofs of these statements. (Note that, although in the latter article the ground field is assumed to be $F = \mathbb{Q}$, the results remain valid for arbitrary F as the questions at issue are purely local). See Pacetti[30] for the behaviour at places v above 2.

In particular, an elliptic curve which fails to satisfy hypothesis (JL) is necessarily of square conductor. The converse is not true, but note that it also follows from the table that all elliptic curves of conductor 1 over a totally real number field of even degree fail to satisfy (JL). We will often restrict our attention to elliptic curves of square conductor, thus encompassing all elliptic curves that do not satisfy hypothesis (JL).

(iii) *Heegner points and L-series*: Suppose first that $F = \mathbb{Q}$ and let K be an imaginary quadratic field of discriminant relatively prime to N and satisfying the

Heegner hypothesis: $\mathcal{O}_K$ has an ideal $\mathcal{N}$ of norm N satisfying $\mathcal{O}_K/\mathcal{N} \simeq \mathbb{Z}/N\mathbb{Z}$.

An ideal $\mathcal{N}$ of K with this property is sometimes called a *cyclic ideal* of norm N. When the Heegner hypothesis is satisfied, it can be shown that the L-function $L(E/K, s)$ has sign -1 in its functional equation, and therefore vanishes at $s = 1$. The CM points on $X_0(N)$ attached to the moduli of elliptic curves with complex multiplication by $\mathcal{O}_K$, and their images under π_E, can be used to construct a canonical point $P_K \in E(K)$: the so-called *Heegner point* on E attached to K. The main result of Gross-Zagier[15] expresses $L'(E/K, 1)$ as a multiple by a simple non-zero scalar of the Néron-Tate height of P_K. In particular, the point P_K is of infinite order if and only if $L'(E/K, 1) \neq 0$.

In the setting where F is a totally real field, the Shimura curve X is equipped with an infinite supply of CM points enjoying properties similar to their counterparts on modular curves. The auxiliary field K is now a totally complex quadratic extension of F satisfying a suitable Heegner hypothesis relative to X. The CM points attached to K can be used to construct a canonical point $P_K \in E(K)$ as in the case $F = \mathbb{Q}$. A general extension of the Gross-Zagier theorem [43, Theorem C] to this context relates the height of P_K to the derivative $L'(E/K, 1)$. In particular, the point P_K is of infinite order precisely when $L(E/K, s)$ has a simple zero at $s = 1$. We emphasise that this more general Heegner point construction relies crucially on E/F satisfying hypothesis (JL).

(iv) *The Euler system argument*: The Heegner point P_K does not come alone, but can be related to the norms of algebraic points on E defined over abelian extensions of K. Using this fact, it is shown[20] in the case $F = \mathbb{Q}$ that the point P_K, when it is of infinite order, necessarily generates $E(K)\otimes\mathbb{Q}$. Koyvagin's argument extends without essential difficulties to the context of Shimura curves over totally real fields (cf. Kolyvagin-Logachev,[22] Zhang [43, §7.2], or the forthcoming book[39] of Tian-Zhang).

(v) *Descending from K to F*: Assume first that $F = \mathbb{Q}$. If $\mathrm{ord}_{s=1}(L(E, s)) \leq 1$, the analytic non-vanishing results of Bump-Friedberg-Hoffstein[3] or Murty-Murty[27] produce an imaginary quadratic field K satisfying the Heegner hypothesis, and for which $L(E/K, s)$ has a simple zero at $s = 1$. By the Gross-Zagier theorem, the Heegner point P_K generates $E(K)$, and its trace therefore generates $E(\mathbb{Q})$. The known properties of the Heegner point P_K imply in particular that its trace to $\mathbb{Q}$ vanishes when $L(E, 1) \neq 0$, and is of infinite order when $L(E, 1) = 0$. Theorem 1.1 for $E/\mathbb{Q}$ follows from this. The proof of Theorem 1.2 is deduced similarly, by noting that the analytic non-vanishing results of Bump, Friedberg and Hoffstein[3] or Waldspurger[40] generalize to any number field and again produce a totally complex imaginary quadratic extension K/F satisfying the Heegner hypothesis for which $\mathrm{ord}_{s=1}(L(E/K, s)) = 1$.

2.2. *Oda's period relations and ATR points*

This section briefly recalls the main construction[8] which to any ATR extension M of F (satisfying a suitable Heegner condition) associates a point $P_M \in E(\mathbb{C})$ belonging conjecturally to $E(M)$. The points P_M arise by considering the images of certain *non-algebraic* cycles on Hilbert modular varieties under a map which is formally analogous to the Grffiths-Weil Abel-Jacobi maps on null-homologous algebraic cycles.

The general setting. We begin by treating a more general context where F is a totally real field of degree $r+1$. (This extra generality does not unduly complicate the notations, and may even clarify some of the key features of the construction.) Fix an ordering $v_0, v_1, \ldots, v_r$ of the $r+1$ distinct real embeddings of F. Let E be an elliptic curve over F, and let

$$E_j := E \otimes_{v_j} \mathbb{R} \qquad (0 \leq j \leq r)$$

be the $r+1$ elliptic curves over $\mathbb{R}$ obtained by taking the base change of E to $\mathbb{R}$ via the embedding v_j. To ease the exposition, we will make the following inessential assumptions:

(1) The field F has narrow class number one;
(2) the conductor of E/F is equal to 1 (i.e., E has everywhere good reduction).

(For a more general treatment where these assumptions are significantly relaxed, see for instance the forthcoming PhD thesis[11] of Jérôme Gärtner.)

The Hilbert modular form G on $\mathrm{SL}_2(\mathcal{O}_F)$ attached to E is a holomorphic function on the product $\mathcal{H}_0 \times \mathcal{H}_1 \times \cdots \times \mathcal{H}_r$ of $r+1$ copies of the complex upper half plane, which is of parallel weight $(2, 2, \ldots, 2)$ under the action of the Hilbert modular group $\mathrm{SL}_2(\mathcal{O}_F)$. The latter group acts discretely on $\mathcal{H}_0 \times \cdots \times \mathcal{H}_r$ by Möbius transformations via the embedding

$$(v_0, \ldots, v_r) : \mathrm{SL}_2(\mathcal{O}_F) \longrightarrow \mathrm{SL}_2(\mathbb{R})^{r+1}.$$

Because of this transformation property, the Hilbert modular form G can be interpreted geometrically as a holomorphic differential $(r+1)$-form on the complex analytic quotient

$$X(\mathbb{C}) := \mathrm{SL}_2(\mathcal{O}_F) \backslash (\mathcal{H}_0 \times \mathcal{H}_1 \times \cdots \times \mathcal{H}_r), \tag{6}$$

by setting

$$\omega_G^{\mathrm{hol}} := (2\pi i)^{r+1} G(\tau_0, \ldots, \tau_r) d\tau_0 \cdots d\tau_r.$$

It is important to replace ω_G^{hol} by a closed, but non-holomorphic differential $(r+1)$-form ω_G on $X(\mathbb{C})$. When $r = 1$, the differential ω_G is defined by choosing a unit $\epsilon \in \mathcal{O}_F^\times$ of norm -1 satisfying

$$\epsilon_0 := v_0(\epsilon) > 0, \qquad \epsilon_1 := v_1(\epsilon) < 0,$$

and setting

$$\omega_G = (2\pi i)^2 \left(G(\tau_0, \tau_1) d\tau_0 d\tau_1 - G(\epsilon_0 \tau_0, \epsilon_1 \bar{\tau}_1) d\tau_0 d\bar{\tau}_1 \right).$$

For general r, one defines ω_G similarly, but this time summing over the subgroup of $\mathcal{O}_F^\times/(\mathcal{O}_F^+)^\times$ of cardinality 2^r consisting of units ϵ with $v_0(\epsilon) > 0$. Note that the closed $(r+1)$-form ω_G is holomorphic in τ_0, but only harmonic in the remaining variables $\tau_1, \ldots, \tau_r$. The justification for working with ω_G rather than ω_G^{hol} lies in the following statement which is a reformulation of a conjecture of Oda[28] in the special case of modular forms with rational fourier coefficients:

Conjecture 2.1 (Oda). *Let*

$$\Lambda_G := \left\{ \int_\gamma \omega_G, \quad \gamma \in H_{r+1}(X(\mathbb{C}), \mathbb{Z}) \right\}.$$

Then Λ_G is a lattice in $\mathbb{C}$ and the elliptic curve $\mathbb{C}/\Lambda_G$ is isogenous to E_0.

This conjecture is shown to hold[28] for Hilbert modular forms which are base change lifts of classical elliptic modular forms, which corresponds to the case where the associated elliptic curve E is a $\mathbb{Q}$-curve. But it should be emphasised that no $\mathbb{Q}$-curve hypothesis on E is necessary in Conjecture 2.1.

Let

$$\mathcal{Z}_r(X(\mathbb{C})) := \left\{ \begin{array}{c} \text{Null-homologous cycles} \\ \text{of real dimension } r \\ \text{on } X(\mathbb{C}) \end{array} \right\}.$$

Conjecture 2.1 makes it possible to define an "Abel-Jacobi map"

$$\mathrm{AJ}_G : \mathcal{Z}_r(X(\mathbb{C})) \longrightarrow E_0(\mathbb{C}), \tag{7}$$

by choosing an isogeny $\iota : \mathbb{C}/\Lambda_G \longrightarrow E_0(\mathbb{C})$, and setting

$$\mathrm{AJ}_G(\Delta) := \iota\left(\int_{\tilde{\Delta}} \omega_G \right), \qquad (\text{for any } \tilde{\Delta} \text{ with } \partial\tilde{\Delta} = \Delta). \tag{8}$$

Note that the domain $\mathcal{Z}_r(X(\mathbb{C}))$ of AJ_G has no natural algebraic structure, and that the map AJ_G bears no simple relation (beyond an analogy in its definition) with the Griffiths-Weil Abel-Jacobi map on the Hilbert modular variety X.

ATR Cycles. Generalising slightly the definitions given in the Introduction to the case $r > 1$, a quadratic extension M of F is called an ATR extension if

$$M \otimes_{F,v_0} \mathbb{R} \simeq \mathbb{C}, \qquad M \otimes_{F,v_j} \mathbb{R} \simeq \mathbb{R} \oplus \mathbb{R}, \quad (1 \le j \le r).$$

The acronym ATR stands for "Almost Totally Real", since an ATR extension of F is "as far as possible" from being a CM extension, without being totally real.

Fix an ATR extension M of F, and let $\Psi : M \longrightarrow M_2(F)$ be an F-algebra embedding. Then

(1) Since $M \otimes_{F,v_0} \mathbb{R} \simeq \mathbb{C}$, the torus $\Psi(M^\times)$ has a unique fixed point $\tau_0 \in \mathcal{H}_0$.
(2) For each $1 \le j \le r$, the fact that $M \otimes_{F,v_j} \mathbb{R} \simeq \mathbb{R} \oplus \mathbb{R}$ shows that $\Psi(M^\times)$ has two fixed points τ_j and τ_j' on the boundary of $\mathcal{H}_j$. Let $\Upsilon_j \subset \mathcal{H}_j$ be the hyperbolic geodesic joining τ_j to τ_j'.

An embedding $\Psi : M \longrightarrow M_2(F)$ has a *conductor*, which is defined to be the $\mathcal{O}_F$-ideal c_Ψ for which

$$\Psi(M) \cap M_2(\mathcal{O}_F) = \Psi(\mathcal{O}_F + c_\Psi \mathcal{O}_M).$$

The $\mathcal{O}_F$-order $\mathcal{O}_\Psi := \mathcal{O}_F + c_\Psi \mathcal{O}_M$ is called the *order associated to* Ψ. It can be shown that there are finitely many distinct $\mathrm{SL}_2(\mathcal{O}_F)$-conjugacy classes of embeddings of M into $M_2(F)$ associated to a fixed order $\mathcal{O} \subset \mathcal{O}_M$, and that the Picard group (in a narrow sense) of $\mathcal{O}$ acts simply transitively on the set of such conjugacy classes of embeddings.

By the Dirichlet unit theorem, the group

$$\Gamma_\Psi := \Psi((\mathcal{O}_\Psi^+)^\times) \subset \mathrm{SL}_2(\mathcal{O}_F)$$

is of rank r and preserves the region

$$R_\Psi := \{\tau_0\} \times \Upsilon_1 \times \cdots \times \Upsilon_r.$$

The ATR cycle associated to the embedding Ψ is defined to be the quotient

$$\Delta_\Psi := \Gamma_\Psi \backslash R_\Psi.$$

It is a closed cycle on $X(\mathbb{C})$ which is topologically isomorphic to an r-dimensional real torus. In many cases, one can show that Δ_Ψ is null-homologous, at least after tensoring with $\mathbb{Q}$ to avoid the delicate issues arising from the possible presence of torsion in integral homology. (The homological triviality of Δ_Ψ always holds, for instance, when $r = 1$, and follows from the fact that the group cohomology $H^1(\mathrm{SL}_2(\mathcal{O}_F), \mathbb{C})$ is trivial.) Assume from now on that Δ_Ψ is homologically trivial, and therefore that it belongs to $\mathcal{Z}_r(X(\mathbb{C}))$.

The following conjecture lends arithmetic meaning to the Abel-Jacobi map AJ_G and to the ATR cycles Δ_Ψ.

Conjecture 2.2. *Let $\Psi : M \longrightarrow M_2(F)$ be an F-algebra embedding of an ATR extension M of F. Then the complex point $\mathrm{AJ}_G(\Delta_\Psi) \in E_0(\mathbb{C})$ is algebraic. More precisely, the isogeny ι in the definition (8) of AJ_G can be chosen so that, for all Ψ,*

$$\mathrm{AJ}_G(\Delta_\Psi) \textit{ belongs to } E(H_{c_\Psi}),$$

where H_{c_Ψ} is the ring class field of M of conductor c_Ψ. Furthermore, if $\Psi_1, \ldots, \Psi_h$ is a complete system of representatives for the $\mathrm{SL}_2(\mathcal{O}_F)$-conjugacy classes of embeddings of M in $M_2(\mathcal{O}_F)$ of a given conductor c, then the Galois group $\mathrm{Gal}(H_c/M)$ acts (transitively) on the set $\{\mathrm{AJ}_G(\Delta_{\Psi_1}), \ldots, \mathrm{AJ}_G(\Delta_{\psi_h})\}$.

Conjecture 2.2 is poorly understood at present. For instance, it is not clear whether the Tate conjecture sheds any light on it. On the positive side, the ATR points that are produced by Conjecture 2.2 are "genuinely new" and go beyond what can be obtained using only CM points on Shimura curves. Indeed, the former are defined over abelian extensions of ATR extensions of totally real fields, while the latter are defined over abelian extensions of CM fields.

Most germane to the concerns of this paper, Conjecture 2.2 can be used as a basis for the construction of a point $P_M^? \overset{?}{\in} E(M)$, by letting $\Psi_1, \ldots, \Psi_h$ be a complete system of representatives for the $\mathrm{SL}_2(\mathcal{O}_F)$-conjugacy classes of embeddings of M in $M_2(\mathcal{O}_F)$ of conductor 1 and setting

$$P_M^? := \mathrm{AJ}_G(\Delta_{\Psi_1}) + \cdots + \mathrm{AJ}_G(\Delta_{\Psi_h}). \tag{9}$$

3. The Birch and Swinnerton-Dyer conjecture for $\mathbb{Q}$-curves

3.1. *Review of $\mathbb{Q}$-curves*

The first goal of the present work is to study Conjecture 2.2 for $\mathbb{Q}$-curves, which are defined as follows:

Definition 3.1. Let F be a number field and fix an algebraic closure $\bar{\mathbb{Q}}$ of $\mathbb{Q}$ containing F. We say that an elliptic curve E/F is a $\mathbb{Q}$-*curve* if it is *isogenous over* F to all its Galois conjugates over $\mathbb{Q}$.

In the literature, these curves are sometimes referred as $\mathbb{Q}$-curves *completely defined over* F, reserving the term $\mathbb{Q}$-curve for the wider class of elliptic curves over F which are isogenous *over* $\bar{\mathbb{Q}}$ to all their Galois conjugates over $\mathbb{Q}$.

$\mathbb{Q}$-curves are known to be modular, thanks to the work of Ellenberg and Skinner [ES] (who proved (geometric) modularity of $\mathbb{Q}$-curves under local conditions at 3), now vastly superseded by Khare-Wintenberger[23] which implies modularity of all $\mathbb{Q}$-curves as a very particular case. Combining this with the older work of Ribet (cf. the survey[32]), it follows that $\mathbb{Q}$-curves E/F are arithmetically uniformisable *over* $\bar{\mathbb{Q}}$ by the classical modular curves $X_1(N)$. By this, we mean that there exists a non-constant morphism of curves

$$\pi_E : X_1(N)_{\bar{\mathbb{Q}}} \longrightarrow E_{\bar{\mathbb{Q}}} \tag{10}$$

over $\bar{\mathbb{Q}}$, for some $N \geq 1$.

For simplicity, in this article we shall restrict our attention to $\mathbb{Q}$-curves over a *quadratic* field F, which represents the simplest non-trivial scenario. However, we believe that the ideas present in this note should allow, with some more effort, to treat more general cases; see the forthcoming Ph.D thesis[45] of the third author.

Let F be a quadratic field with ring of integers $\mathcal{O}_F$ and write $\mathrm{Gal}\,(F/\mathbb{Q}) = \{1, \tau\}$. Let E be a $\mathbb{Q}$-curve over F of conductor $\mathfrak{N} \subset \mathcal{O}_F$.

Given a Dirichlet character ε of conductor N, let

$$\Gamma_\varepsilon(N) = \{ \left(\begin{smallmatrix} a & b \\ c & d \end{smallmatrix}\right) \mid N|c,\ \varepsilon(a) = 1\} \subseteq \mathrm{SL}_2(\mathbb{Z})$$

and let $X_\varepsilon(N)$ be the modular curve associated to this congruence subgroup. The curve $X_\varepsilon(N)$ admits a canonical model over $\mathbb{Q}$, and coarsely represents the moduli problem of parametrizing triples (A, C, t) where A is a generalised elliptic curve, C is a cyclic subgroup of order N of $A(\bar{\mathbb{Q}})$ and t is an orbit in $C \setminus \{O\}$ for the action of $\ker(\varepsilon) \subset (\mathbb{Z}/N\mathbb{Z})^\times$. Note that the group $(\mathbb{Z}/N\mathbb{Z})^\times$ acts on $X_\varepsilon(N)$ via the diamond operators, and that the subgroup $\ker(\varepsilon)$ fixes it.

As discussed for example in Guitart-Quer[14] or Ribet,[32] the modular parametrisation (10) is particularly well-behaved when F is quadratic. More precisely, there exists a positive integer $N \geq 1$, an even Dirichlet character $\varepsilon : (\mathbb{Z}/N\mathbb{Z})^\times \to \{\pm 1\} \subset \mathbb{C}^\times$, and a pair $f_E, f_E' \in S_2(\Gamma_\varepsilon(N)) \subseteq S_2(\Gamma_0(N), \varepsilon)$ of normalised *newforms* of weight 2, level N and nebentypus ε, such that

$$L(E/F, s) = L(f_E, s) \cdot L(f_E', s). \tag{11}$$

In this case, the uniformisation in (10) factors through a modular parametrisation

$$\pi_E : X_\varepsilon(N)_F \longrightarrow E_F \tag{12}$$

defined over F.

Let K_f denote the field generated by the Fourier coefficients of f_E. It is either $\mathbb{Q}$ or a quadratic field. When $K_f = \mathbb{Q}$, the elliptic curve E is in fact isogenous to the base change of an elliptic curve defined over $\mathbb{Q}$ and question (1) can rather be tackled with the classical techniques reviewed in §2; we assume throughout that this is not the case. Hence $[K_f : \mathbb{Q}] = 2$ and, letting σ denote the single nontrivial automorphism of K_f, we have

$$f'_E = {}^{\sigma}f_E.$$

Weil's restriction of scalars $A := \mathrm{Res}_{F/\mathbb{Q}}(E)$ is an abelian surface of GL_2-type over $\mathbb{Q}$ such that

$$\mathrm{End}_{\mathbb{Q}}(A) \otimes \mathbb{Q} \simeq K_f \quad \text{(and thus is simple over } \mathbb{Q}), \tag{13}$$

$$A_{/F} \simeq E \times {}^{\tau}E, \text{ and} \tag{14}$$

$$L(A/\mathbb{Q}, s) = L(E/F, s) = L(f_E, s) \cdot L(f'_E, s). \tag{15}$$

Moreover, for any field extension $L/\mathbb{Q}$, there is a canonical isomorphism

$$A(L) \simeq E(F \otimes_{\mathbb{Q}} L) \tag{16}$$

and in particular $A(\mathbb{Q}) \simeq E(F)$. As shown by Carayol, the conductor of A over $\mathbb{Q}$ is N^2, and it follows from [25, Prop. 1] (see also [13, Remark 9] for a more detailed discussion) that the conductor of E/F is $\mathfrak{N} = N_0 \cdot \mathcal{O}_F$, where $N_0 \in \mathbb{Z}$ satisfies

$$N = N_0 \cdot |\mathrm{disc}(F)|. \tag{17}$$

As we shall now explain, when F is imaginary the problem can be reduced to the classical setting considered by Gross-Zagier and Kolyvagin-Logachev, and presents no mysteries. It is the case of F real that deserves more attention, and to which the main bulk of this note will be devoted.

If ε is trivial, then K_f is real and F can be either real or imaginary (and indeed both cases occur in examples). As a direct consequence of (11), (16) and the generalization[21] of the work of Kolyvagin to higher dimensional quotients of $J_0(N)$ over $\mathbb{Q}$, (1) also holds for E/F provided $\mathrm{ord}_{s=1} L(f_E, s) \leq 1$.

Assume $\varepsilon \neq 1$ for the rest of this article. Now K_f is an imaginary quadratic field. Besides, it follows from an observation of Serre (cf. [32, Proposition 7.2]) that F is necessarily *real.* In fact, F can be computed explicitly from f_E as $F = \bar{\mathbb{Q}}^{\ker(\varepsilon)}$. In particular it follows that ε is the quadratic Dirichlet character associated with F.

Let ω_N denote the Fricke involution of $X_\varepsilon(N)_F$ defined on the underlying Riemann surface by the rule $\tau \mapsto -\frac{1}{N\tau}$. It induces an involution on the jacobian $J_\varepsilon(N)_F$ of $X_\varepsilon(N)_F$ which leaves A_F stable. We have

$$A_F \sim (1+\omega_N)A_F \times (1-\omega_N)A_F, \tag{18}$$

where both factors on the right have dimension 1, are isogenous over F and conjugate one each other over $\mathbb{Q}$ (cf. [6, §5]). By replacing E by its conjugate if necessary, we shall assume throughout that $E = (1+\omega_N)A_F$.

It then follows that (12) factors through the following commutative diagram:

$$\begin{array}{ccccc} X_\varepsilon(N)_F & \longrightarrow & J_\varepsilon(N)_F & \longrightarrow & A_F \\ \downarrow \pi_N & & \downarrow & & \downarrow \\ X_\varepsilon^+(N) & \longrightarrow & J_\varepsilon^+(N) & \xrightarrow{\varphi_E} & E, \end{array} \tag{19}$$

where we set $X_\varepsilon^+(N) := X_\varepsilon(N)_F/\langle\omega_N\rangle$ and $J_\varepsilon^+(N) := (1+\omega_N)J_\varepsilon(N)_F$. The reader should keep in mind that both are varieties over F, not over $\mathbb{Q}$.

3.2. *The main result*

The goal of this section is to prove Theorem 1.4 of the Introduction. Let M be a quadratic ATR extension of F. Since M has two real places and one complex place, it is not Galois over $\mathbb{Q}$. Let M' denote its Galois conjugate over $\mathbb{Q}$, and let $\mathcal{M}$ be the Galois closure of M over $\mathbb{Q}$. It is not hard to see that $\mathcal{M}$ is the compositum over F of M and M' and that $\mathrm{Gal}(\mathcal{M}/\mathbb{Q})$ is isomorphic to the dihedral group of order 8. The subgroup $V_F := \mathrm{Gal}(\mathcal{M}/F)$ is isomorphic to a Klein 4-group. The dihedral group of order 8 contains two distinct, non-conjugate subgroups which are isomorphic to the Klein 4-group. This is most easily seen by viewing D_8 as the symmetry group of a square, as in the figure below, in which V_F is identified with the subgroup generated by the reflections about the two diagonals. These two reflections can be labeled as τ_M and τ_M', in such a way that

$$\mathcal{M}^{V_F} = F, \qquad \mathcal{M}^{\tau_M} = M, \qquad \mathcal{M}^{\tau_M'} = M'.$$

The second Klein four-group, which shall be denoted V_K, is generated by the reflections about the vertical and horizontal axes of symmetry of the square. We label these reflections as τ_L and τ_L', as shown in the figure below.

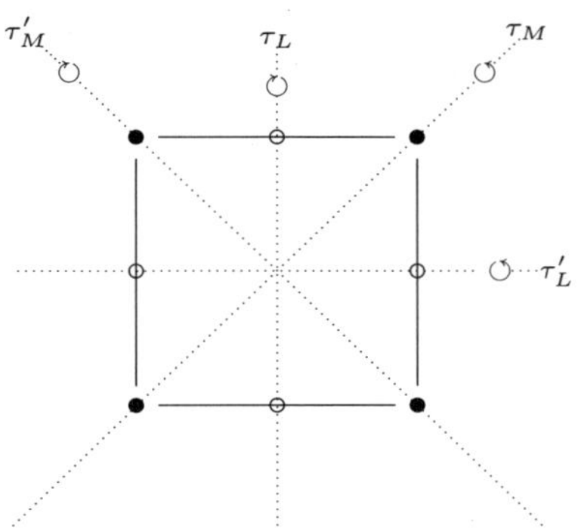

(20)

Now let

$$K := \mathcal{M}^{V_K}, \qquad L := \mathcal{M}^{\tau_L}, \qquad L' = \mathcal{M}^{\tau_L'}.$$

These fields fit into the following diagram of field extensions, where each unbroken line indicates an extension of degree 2:

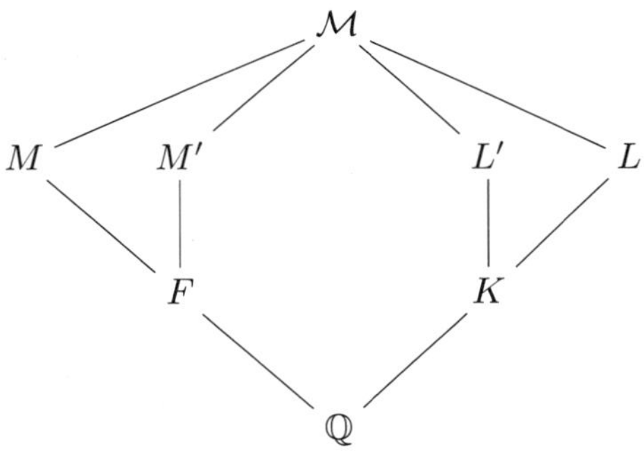

(21)

Let

$$\chi_M, \chi_M' : G_F \longrightarrow \{\pm 1\}$$

denote the Galois characters of the real quadratic field F which cut out the extensions M and M', and let

$$\chi_L, \chi_L' : G_K \longrightarrow \{\pm 1\}$$

be the quadratic characters of the imaginary quadratic field K which cut out the extensions L and L'. We will often view these characters as idèle class characters defined on $\mathbb{A}_F^\times$ and on $\mathbb{A}_K^\times$ respectively. Finally, let ε_F and

ε_K denote the quadratic Dirichlet characters attached to F and K, and let $\mathrm{N}^F_{\mathbb{Q}} : \mathbb{A}_F^\times \longrightarrow \mathbb{A}_{\mathbb{Q}}^\times$ and $\mathrm{N}^K_{\mathbb{Q}} : \mathbb{A}_K^\times \longrightarrow \mathbb{A}_{\mathbb{Q}}^\times$ denote the norms on adèles.

Proposition 3.1.

(1) The field K is a quadratic imaginary field.

(2) The characters χ_M, χ'_M, χ_L and χ'_L, viewed as idèle class characters of F and K respectively, satisfy

$$\chi_M \chi'_M = \varepsilon_K \circ N^F_{\mathbb{Q}}; \qquad \chi_L \chi'_L = \varepsilon_F \circ N^K_{\mathbb{Q}}.$$

(3) The central character of χ_M and χ'_M is ε_K, and the central character of χ_L and χ'_L is ε_F.

(4) The following two-dimensional representations of $G_{\mathbb{Q}}$ are isomorphic:

$$Ind^{\mathbb{Q}}_F \chi_M = Ind^{\mathbb{Q}}_F \chi'_M = Ind^{\mathbb{Q}}_K \chi_L = Ind^{\mathbb{Q}}_K \chi'_L.$$

Proof. The quadratic field K is of the form $\mathbb{Q}(\sqrt{-d})$, where $-d$ is defined (modulo squares in $\mathbb{Q}^\times$) by

$$-d = \mathrm{N}_{F/\mathbb{Q}}(\alpha), \qquad \text{with} \qquad M = F(\sqrt{\alpha}).$$

The fact that M is ATR implies that $-d$ is a negative rational number, and therefore that K is an imaginary quadratic field. The second part follows directly from the field diagram (21) above. To prove the third part, note that part (2) implies that the central character of L restricted to the group of norms from K is equal to ε_F. (This is because $\chi_L(x) = \chi'_L(\bar{x})$, where $x \mapsto \bar{x}$ is complex conjugation). Class field theory implies that this central character differs from ε_F by a power of ε_K. But the central character of χ_L cannot be $\varepsilon_F \varepsilon_K$ since this is an odd Dirichlet character and the central character of a finite order Hecke character of an imaginary quadratic field is necessarily even, because the map from the group of components of $\mathbb{R}^\times$ to the group of components of $\mathbb{C}^\times$ is trivial. Finally, the proof of part (4) is a simple exercise in representation theory: the four representations that are listed in (4) are all isomorphic to the unique irreducible two-dimensional representation of $\mathrm{Gal}(\mathcal{M}/\mathbb{Q})$. □

As at the end of the previous section §3.1, let E be a $\mathbb{Q}$-curve over a real quadratic field F and $A = \mathrm{Res}_{F/\mathbb{Q}}(E)$, let $f = f_E \in S_2(\Gamma_\varepsilon(N))$ denote the modular form associated to it and let K_f denote the imaginary quadratic field generated over $\mathbb{Q}$ by the Fourier coefficients of f.

Theorem 3.1 (Tian, Yuan, Zhang and Zhang). *Let K be a quadratic imaginary field satisfying the Heegner hypothesis, and let $\chi : \mathbb{A}_K^\times \longrightarrow \mathbb{C}^\times$ be a finite order Hecke character of K satisfying*

$$\chi|_{\mathbb{A}_\mathbb{Q}^\times} = \varepsilon_f^{-1}. \tag{22}$$

Then

(i) The L-function $L(f/K, \chi, s)$ vanishes to odd order at $s = 1$;
(ii) If $L'(f/K, \chi, 1) \neq 0$, then $(A(K^{\mathrm{ab}}) \otimes \mathbb{C})^\chi$ has rank one over $K_f \otimes_\mathbb{Q} \mathbb{C}$, and $Ш(A/K^{\mathrm{ab}})^\chi$ is finite.

Proof. The modular form f gives rise to a cuspidal automorphic representation π of $\mathrm{GL}_2(\mathbb{A}_\mathbb{Q})$ whose central character is $\omega_\pi = \varepsilon_f$, the nebentypus of f. Condition (22) ensures that the tensor product of the motives attached to π and χ is self-dual, and therefore the L-function

$$L(f/K, \chi, s) = L(\pi, \chi, s - \frac{1}{2})$$

satisfies a functional equation whose central critical point is $s = 1$; since the discriminant of K is relatively prime to N, the sign of this functional equation is $(-1)^{\sharp\Sigma}$, where

$$\Sigma = \{\text{primes } \ell \text{ inert in } K \text{ such that } \mathrm{ord}_\ell(N) \text{ is odd}\} \cup \{\infty\}.$$

The Heegner hypothesis satisfied by K says that $\Sigma = \{\infty\}$ and thus the sign is -1; this implies (i).

As a consequence of (13), the complex vector space $(A(K^{\mathrm{ab}}) \otimes \mathbb{C})^\chi$ is naturally a $K_f \otimes_\mathbb{Q} \mathbb{C}$-module. Part (ii) is a theorem of Tian-Zhang[39] which follows as a corollary of [42, Theorem 1.3.1] by applying Kolyvagin's method. Since the reference[39] is not currently available, the reader may consult [42, Theorem 1.4.1] and, for the precise statement quoted here, [44, Theorem 4.3.1]. □

We are now ready to prove Theorem 1.4 of the introduction.

Theorem 3.2. *Let E be a $\mathbb{Q}$-curve over F of square conductor $\mathfrak{N}$, and let M/F be an ATR extension of F of discriminant prime to $\mathfrak{N}$. If $L'(E_M/F, 1) \neq 0$, then $E_M(F)$ has rank one and $Ш(E_M/F)$ is finite.*

Proof. By (11) and the Artin formalism for L-series,

$$L(E_M/F, s) = L(E, \chi_M, s) = L(f \otimes \chi_M/F, s) = L(f \otimes \mathrm{Ind}_F^\mathbb{Q} \chi_M, s). \tag{23}$$

By part 4 of Proposition 3.1,

$$L(f \otimes \mathrm{Ind}_F^{\mathbb{Q}} \chi_M, s) = L(f \otimes \mathrm{Ind}_K^{\mathbb{Q}} \chi_L, s) = L(f \otimes \chi_L/K, s). \tag{24}$$

It follows from (23) and (24) that

$$L'(E_M/F, 1) = L'(f \otimes \chi_L/K, 1) = L'(f^\sigma \otimes \chi_L/K, 1). \tag{25}$$

Therefore the two rightmost expressions in (25) are non-zero by assumption, so that the product

$$L(f \otimes \chi_L/K, s) L(f^\sigma \otimes \chi_L/K, s) = L(A/K, \chi_L, s)$$

vanishes to order exactly $2 = [K_f : \mathbb{Q}]$ at $s = 1$. By Theorem 3.1, it follows that $A(L)^-$ is of rank two, where $A(L)^-$ denotes the subgroup of the Mordell-Weil group of $A(L)$ of points whose trace to K is trivial. In particular, the Galois representation $\mathrm{Ind}_K^{\mathbb{Q}} \chi_L$ occurs in $A(\overline{\mathbb{Q}}) \otimes \mathbb{C}$ with multiplicity 2. Hence, invoking once again part 4 of Proposition 3.1, we find that

$$\mathrm{rank}(A(M)^-) = 2.$$

But since M contains F and since A is isogenous over F to $E \times E$, it follows that

$$\mathrm{rank}(E(M)^-) = \mathrm{rank}(E_M(F)) = 1.$$

The result about the ranks follows. The result about the finiteness of $Ш(E_M/F)$ follows in the same way from part (ii) of Theorem 3.1. □

4. Heegner points on Shimura's elliptic curves

Implicit in the proof of Theorem 3.2 (via the use that is made of it in the proof of Theorem 3.1) is the construction of a Heegner point $P_M \in E_M(F)$ arising from the image of certain CM divisors on $X_\varepsilon(M)$ via the modular parametrisation (10). We now wish to make this construction explicit in the case where the $\mathbb{Q}$-curve E has everywhere good reduction over the real quadratic field F. The $\mathbb{Q}$-curves with this property are sometimes called *Shimura elliptic curves* because they were first systematically considered by Shimura. More precisely, it is shown in Shimura[35] how to associate a Shimura elliptic curve over $F = \mathbb{Q}(\sqrt{N})$ to any classical elliptic modular form $f \in S_2(\Gamma_0(N), \varepsilon_N)$ with quadratic fourier coefficients. (Cf. also (17).)

It will be assumed throughout this chapter that E/F is a Shimura elliptic curve, and that f is the corresponding elliptic modular form. We also assume for simplicity that N *is odd*, and thus square-free.

Remark 4.1. According to calculations performed by the third author using PARI[29] (extending the data gathered in [6, §6], Pinch[31] in the range $N \leq 1000$), there exists an eigenform $f \in S_2(\Gamma_0(N), \varepsilon_N)$ of prime level $1 < N < 5000$, with fourier coefficients in a quadratic imaginary extension $K_f = \mathbb{Q}(\sqrt{-d})$ for precisely the values of N and d listed in the following table.

N	29	37	41	109	157	229	257	337	349
d	5	1	2	3	1	5	2	2	5
N	373	397	421	461	509	877	881	997	1069
d	1	1	7	5	5	1	2	3	1
N	1709	1861	2657	4481	4597				
d	5	5	2	11	1				

Associated to each such eigenform there is a Shimura elliptic curve over $F = \mathbb{Q}(\sqrt{N})$. Furthermore, according to computations due to Cremona, Dembelé, Elkies and Pinch, there are only four primes N in the range $[1, 1000]$ for which there exists an elliptic curve with good reduction everywhere over $F = \mathbb{Q}(\sqrt{N})$ which is *not* a $\mathbb{Q}$-curve, namely, $N = 509$, 853, 929 and 997. It is hard to predict whether the preponderance of Shimura elliptic curves among elliptic curves of conductor one will persist or is merely an artefact of the relatively low ranges in which numerical data has been gathered. Note that it is not even known whether there exist infinitely many Shimura elliptic curves over real quadratic fields, while it is a theorem of S. Comalada[5] that there are infinitely many elliptic curves over real quadratic fields with good reduction everywhere.

4.1. *An explicit Heegner point construction*

Let us recall the diagram of field extensions introduced in (21):

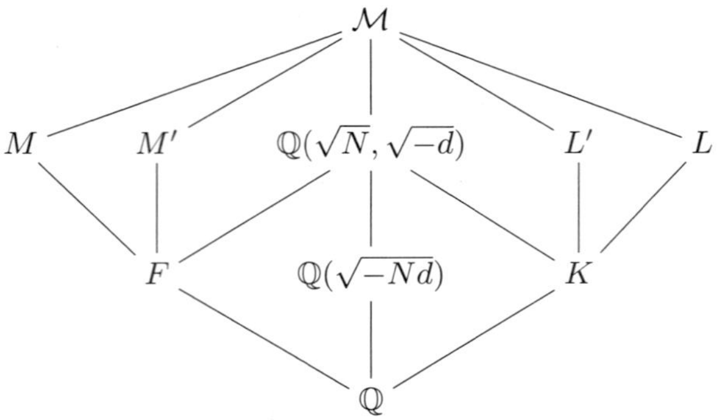

(26)

The following lemma is crucial in constructing the point $P_M \in E(M)^- = E_M(F)$ explicitly.

Lemma 4.1. *Let M be an ATR extension of F and let K be the quadratic imaginary field attached to M as in the diagram* (26)*. Then K has a (canonical) ideal $\mathcal{N}$ of K of norm N. In particular, all the prime divisors of N are either split or ramified in K.*

Proof. The conductor-discriminant formula combined with part 4 of Proposition 3.1 show that

$$\operatorname{disc}(F)\operatorname{Nm}_{F/\mathbb{Q}}(\operatorname{disc}(M/F)) = \operatorname{disc}(K)\operatorname{Nm}_{K/\mathbb{Q}}(\operatorname{disc}(L/K)).$$

Therefore, after setting

$$N_{\mathrm{ram}} = \gcd(N, \operatorname{disc}(K)), \qquad N_{\mathrm{split}} = N/N_{\mathrm{ram}},$$
$$\mathcal{N}_{\mathrm{ram}} = (N_{\mathrm{ram}}, \sqrt{\operatorname{disc}(K)}), \qquad \mathcal{N}_{\mathrm{split}} = (N_{\mathrm{split}}, \operatorname{disc}(L/K)),$$

we find that $\mathcal{N} := \mathcal{N}_{\mathrm{ram}}\mathcal{N}_{\mathrm{split}}$ gives the desired ideal of norm N. $\square$

Let $\mathbb{A}_K$ denote the ring of adèles of K, and let

$$\hat{\mathcal{O}}_K^\times := \prod_v \mathcal{O}_v^\times$$

denote the maximal compact subgroup of the group $\mathbb{A}_{K,\mathrm{fin}}^\times$ of finite idèles of K. Given a rational integer $c \geq 1$, $(c, N) = 1$, we define

$$U_c = \hat{\mathbb{Z}}^\times(1 + c\hat{\mathcal{O}}_K)\mathbb{C}^\times \subset \mathbb{A}_K^\times.$$

By class field theory, the quotient $G_c := \mathbb{A}_K^\times/(K^\times U_c)$ is identified with $\operatorname{Gal}(H_c/K)$, where H_c is the *ring class field* of K of conductor c.

As a piece of notation, we shall write H_c for the ring class field attached to the order in K of conductor $c \geq 1$ and write $K_{\mathfrak{a}}$ for the ray class field of conductor $\mathfrak{a}$.

Define

$$U_c^+ = \{\beta \in U_c \text{ such that } (\beta)_{\mathcal{N}} \in \ker(\varepsilon) \subset (\mathbb{Z}/N\mathbb{Z})^\times\},$$

$$U_c^- = \{\beta \in U_c \text{ such that } (\beta)_{\overline{\mathcal{N}}} \in \ker(\varepsilon) \subset (\mathbb{Z}/N\mathbb{Z})^\times\},$$

and $\tilde{U}_c = U_c^+ \cap U_c^-$. Here $(\beta)_{\mathcal{N}}$ denotes the image of the local term of the idèle β at $\mathcal{N}$ in the quotient $\mathcal{O}_{\mathcal{N}}^\times/(1 + \mathcal{N} \cdot \mathcal{O}_{\mathcal{N}}) \simeq (\mathbb{Z}/N\mathbb{Z})^\times$. Similarly for $\overline{\mathcal{N}}$. This way we can regard the character ε as having source either $\mathcal{O}_{\mathcal{N}}^\times$ or $\mathcal{O}_{\overline{\mathcal{N}}}^\times$.

Set

$$\tilde{G}_c := \mathbb{A}_K^\times/(K^\times \tilde{U}_c) = \mathrm{Gal}(\tilde{H}_c/K),$$

where $\tilde{H}_c$ is a biquadratic extension of the ring class field H_c. It can be written as $\tilde{H}_c = L_c L_c'$, where L_c (resp. L_c') is the class field attached to U_c^+ (resp. U_c^-).

Proposition 4.1. *The relative discriminant of L/K factors as $d(L/K) = c \cdot \mathcal{N}$, where c is a positive integer such that $L \subset L_c$ and $L' \subset L_c'$ and thus $\mathcal{M} \subset \tilde{H}_c$.*

More precisely, we have $c = 2^t \cdot c_0$ where $0 \leq t \leq 3$ and c_0 is odd and square-free. The proof of this proposition is an exercise in class field theory, which we relegate to §4.4 for the convenience of the reader.

We now explain how to construct a degree zero divisor on $X_\varepsilon(N)$ defined over $\tilde{H}_c$. To do this, let A_c be an elliptic curve satisfying

$$\mathrm{End}(A_c) = \mathcal{O}_c,$$

where $\mathcal{O}_c := \mathbb{Z} + c\mathcal{O}_K$ is the order in K of conductor c. Such a curve, along with its endomorphisms, may be defined over the ring class field H_c. The module $A_c[\mathcal{N}]$ of $\mathcal{N}$-torsion points is therefore defined over H_c, yielding a point $P_c := [A_c, A_c[\mathcal{N}]] \in X_0(N)(H_c)$.

The action of $G_{H_c} := \mathrm{Gal}(\overline{\mathbb{Q}}/H_c)$ on the points of this group scheme gives a Galois representation

$$\rho_{\mathcal{N}} : G_{H_c} \longrightarrow (\mathbb{Z}/N\mathbb{Z})^\times.$$

The composition of $\rho_{\mathcal{N}}$ with the nebentypus character ε is a quadratic character of G_{H_c}, which cuts out the quadratic extension L_c of H_c. The point P_c lifts to two points P_c^+ and P_c^- in $X_\varepsilon(N)(L_c)$ which are interchanged by the action of $\mathrm{Gal}(L_c/H_c)$; we do not specify the order in which these points are to be taken. Similarly, we can replace the module $A_c[\mathcal{N}]$ by $A_c[\overline{\mathcal{N}}]$, mimic the above construction and obtain points $P_c'^+$ and $P_c'^-$ defined over L_c'.

Definition 4.1. Let

$$\mathrm{CM}(c) = \bigcup \{P_c^+, P_c^-, P_c'^+, P_c'^-\} \subset X_\varepsilon(N)(\tilde{H}_c)$$

be the set of Heegner points on $X_\varepsilon(N)$ obtained by letting A_c run over all isomorphism classes of elliptic curves with CM by $\mathcal{O}_c$.

If we let $h(O_c)$ denote the cardinality of the group $\mathrm{Pic}(\mathcal{O}_c)$ of classes of locally free ideals of $\mathcal{O}_c$, the cardinality of $\mathrm{CM}(c)$ is $4h(\mathcal{O}_c)$. In fact, $\mathrm{CM}(c)$ is naturally the disjoint union of the two subsets $CM(c) \cap X_\varepsilon(N)(L_c)$ and $CM(c) \cap X_\varepsilon(N)(L_c')$, each of cardinality $2h(\mathcal{O}_c)$.

A Heegner point $P \in \mathrm{CM}(c)$ of conductor c may be described by a triple $([\mathfrak{a}], \mathfrak{n}, t)$, where

- $[\mathfrak{a}] \in \mathrm{Pic}(\mathcal{O}_c)$ is the class of an invertible $\mathcal{O}_c$-module of K,
- $\mathfrak{n}$ is an integral ideal of $\mathcal{O}_c$ such that the quotient $\mathcal{O}_c/\mathfrak{n}$ is cyclic of order N,
- t is an orbit for the action of $\ker(\varepsilon)$ of an element of order N in $\mathfrak{a}\mathfrak{n}^{-1}/\mathfrak{a} \cong \mathbb{Z}/N\mathbb{Z}$.

Let $\tilde{C}$ be the quotient of the ray class group of K of conductor $c\mathcal{N}$ for which Artin's reciprocity map of global class field theory furnishes a canonical isomorphism

$$\mathrm{rec} : \tilde{C} \xrightarrow{\sim} \mathrm{Gal}(\tilde{H}_c/K).$$

Let $\mathcal{O} = \mathcal{O}_c$ denote the order of conductor c in K. There are natural exact sequences, sitting in the commutative diagram

$$\begin{array}{ccccccccc} 1 & \longrightarrow & \mathrm{Gal}(\tilde{H}_c/H_c) & \longrightarrow & \mathrm{Gal}(\tilde{H}_c/K) & \xrightarrow{\mathrm{res}_{\tilde{H}_c/H_c}} & \mathrm{Gal}(H_c/K) & \longrightarrow & 1 \\ & & \downarrow{\scriptstyle \mathrm{rec}} & & \downarrow{\scriptstyle \mathrm{rec}} & & \downarrow{\scriptstyle \mathrm{rec}} & & \\ 1 & \longrightarrow & \langle [\beta_0], [\beta_0'] \rangle & \longrightarrow & \tilde{C} & \longrightarrow & \mathrm{Pic}(\mathcal{O}) & \longrightarrow & 1, \end{array}$$

where the vertical arrows are isomorphisms. Here, $\beta_0 \in \mathcal{O}_{\mathcal{N}}^\times$ and $\beta_0' \in \mathcal{O}_{\overline{\mathcal{N}}}^\times$ are elements such that $\varepsilon(\beta_0) = -1$ and $\varepsilon(\beta_0') = -1$. Artin's reciprocity map induces an isomorphism

$$\mathrm{Gal}(\tilde{H}_c/H_c) \simeq \mathcal{O}_{\mathcal{N}}^{\times}/\ker(\varepsilon) \times \mathcal{O}_{\overline{\mathcal{N}}}^{\times}/\ker(\varepsilon) \simeq \mathbb{Z}/2\mathbb{Z} \times \mathbb{Z}/2\mathbb{Z}.$$

We thus can formally write elements of $\tilde{C}$ as classes of *enhanced ideals*, which are defined as elements of the form $\underline{\mathfrak{b}} := \beta_{\mathcal{N}}\beta_{\overline{\mathcal{N}}}\prod_{\wp\nmid N}\wp^{n_\wp}$, taken up to principal ideals (b) with $b \in K^{\times}$. Here $\beta_{\mathcal{N}}$ and $\beta_{\overline{\mathcal{N}}}$ belong to $K_{\mathcal{N}}^{\times}/\ker(\varepsilon)$ and $K_{\overline{\mathcal{N}}}^{\times}/\ker(\varepsilon)$ respectively, $\wp$ runs over all prime invertible ideals of $\mathcal{O}$ not dividing N, and the exponents $n_\wp$ are integers which are almost all zero. We say an enhanced ideal is *integral* if $\beta_{\mathcal{N}}$ and $\beta_{\overline{\mathcal{N}}}$ have representatives in $\mathcal{O}_{\mathcal{N}}^{\times}$ and $\mathcal{O}_{\overline{\mathcal{N}}}^{\times}$ respectively, and $n_\wp \geq 0$ for all $\wp$. The image of the class $\underline{\mathfrak{b}}$ in $\mathrm{Pic}(\mathcal{O})$ is simply the class of the ideal $\mathfrak{b} = \mathcal{N}^{\mathrm{ord}_{\mathcal{N}}(\beta_{\mathcal{N}})}\overline{\mathcal{N}}^{\mathrm{ord}_{\overline{\mathcal{N}}}(\beta_{\overline{\mathcal{N}}})}\prod_{\wp\nmid N}\wp^{n_\wp}$ generated by it.

By Shimura's reciprocity law,

$$\mathrm{rec}(\underline{\mathfrak{b}})(D) = \underline{\mathfrak{b}}^{-1} \star D \tag{27}$$

for all $\underline{\mathfrak{b}} \in \tilde{C}$ and all divisors $D \in J_\varepsilon(N)(\tilde{H}_c)$ supported on $\mathrm{CM}(c)$.

On the left hand side we make use of the natural Galois action of $\mathrm{Gal}(\tilde{H}_c/K)$ on $J_\varepsilon(N)(\tilde{H}_c)$, via Artin's reciprocity isomorphism. On the right hand side, a class $[\underline{\mathfrak{b}}] \in \tilde{C}$ acts on $\mathrm{CM}(c)$ by the rule

$$\underline{\mathfrak{b}} \star P = ([\mathfrak{a}\mathfrak{b}^{-1}], \mathfrak{n}, \varphi_{\mathfrak{b}}(\beta_{\mathcal{N}}t)), \tag{28}$$

where $P = ([\mathfrak{a}], \mathfrak{n}, t) \in \mathrm{CM}(c)$, $\underline{\mathfrak{b}} = \beta_{\mathcal{N}}\beta_{\overline{\mathcal{N}}}\prod_{\wp\nmid N}\wp^{n_\wp}$ is an integral representative of its class and $\varphi_{\mathfrak{b}} : \mathbb{C}/\mathfrak{a} \to \mathbb{C}/\mathfrak{a}\mathfrak{b}^{-1}$ is the natural projection map. Writing $P = [\tau] \in X_\varepsilon(N)(\mathbb{C})$ for some $\tau \in \mathcal{H}$, let $\gamma_{\underline{\mathfrak{b}}} \in \mathrm{GL}_2^+(\mathbb{Q})$ be such that $\underline{\mathfrak{b}} \star P = [\gamma_{\underline{\mathfrak{b}}}\tau]$.

Besides this action, there is also the diamond involution W_ε, acting on $P = [\tau] \in X_\varepsilon(N)(\mathbb{C})$ as $W_\varepsilon([\tau]) = [\gamma_\epsilon\tau]$ and on $P = ([\mathfrak{a}], \mathfrak{n}, t) \in \mathrm{CM}(c)$ as

$$W_\varepsilon(P) = ([\mathfrak{a}], \mathfrak{n}, dt), \qquad \text{for } \gamma_\epsilon = \left(\begin{smallmatrix} a & b \\ Nc & d \end{smallmatrix}\right) \in \Gamma_0(N) \setminus \Gamma_\varepsilon(N). \tag{29}$$

The cardinality of $\mathrm{CM}(c)$ is $4h(\mathcal{O})$ and it is acted on freely and transitively by the group $\langle W_N, W_\varepsilon\rangle \times \tilde{C}_{\mathcal{M}}$, where we let $\tilde{C}_{\mathcal{M}} := \mathrm{rec}^{-1}(\mathrm{Gal}(\tilde{H}_c/\mathcal{M})) \subset \tilde{C}_K$. Note that the restriction map $\mathrm{res}_{\tilde{H}_c/H_c}$ induces an isomorphism $\tilde{C}_{\mathcal{M}} \cong \mathrm{Pic}(\mathcal{O}) \cong \mathrm{Gal}(H_c/K)$.

It is our aim now to define a point $P_M \in E(M)$ (and thus also, by conjugation over F, a point $P_{M'} \in E(M')$) on the elliptic curve E, rational over the ATR extension M/F. We shall construct P_M as a suitable linear combination of certain points $P_L \in A(L)$ and $P_{L'} \in A(L')$ on the abelian surface $A = \mathrm{Res}_{F/\mathbb{Q}}(E)$. These points are defined as the trace to L of the projection of $P_c^+ \in X_\varepsilon(N)(L_c)$ (respectively of $P_c^{'+} \in X_\varepsilon(N)(L_c'))$ on A.

Before doing so, we first observe that choosing $P_c^- = W_\varepsilon(P_c^+)$ instead of P_c^+ (and similarly $P_c'^-$ instead of $P_c'^+$) is unimportant for our construction, as the next lemma shows that both lead to the same point on A up to sign and torsion. Recall the canonical projection $\pi_f : J_\varepsilon(N) \to A$ defined over $\mathbb{Q}$ and reviewed in (19), which can be composed with the natural embedding of $X_\varepsilon(N)$ into its jacobian $J_\varepsilon(N)$ given by the map $P \mapsto P - i\infty$. By an abuse of notation, we continue to denote by π_f this composition.

Lemma 4.2. *For any $P \in X_\varepsilon(N)(\overline{\mathbb{Q}})$, the point $\pi_f(P)+\pi_f(W_\varepsilon(P))$ belongs to $A(F)_{\mathrm{tors}}$.*

Proof. There is a natural decomposition $S_2(\Gamma_\varepsilon(N)) = S_2(\Gamma_0(N)) \oplus S_2(\Gamma_0(N), \varepsilon)$ corresponding to the eigenspaces of eigenvalue ± 1 with respect to the action of the involution W_ε. The rule $f(z) \mapsto f(z)dz$ yields an identification of $S_2(\Gamma_\varepsilon(N))$ with the space of holomorphic differentials on $X_\varepsilon(N)_{\mathbb{C}}$. Via this isomorphism, $\pi_f^* H^0(\Omega_A^1)$ is contained in $S_2(\Gamma_0(N), \varepsilon)$. Consequently, $\pi_f(P - i\infty) = -\pi_f(W_\varepsilon(P - i\infty))$ and

$$\begin{aligned}
&\pi_f(P) + \pi_f(W_\varepsilon(P)) = \pi_f(P - i\infty) + \pi_f(W_\varepsilon(P) - i\infty) \\
&= \pi_f(P - i\infty) + \pi_f(W_\varepsilon(P) - W_\varepsilon(i\infty)) + \pi_f(W_\varepsilon(i\infty) - i\infty) \\
&= \pi_f(W_\varepsilon(i\infty) - i\infty).
\end{aligned}$$

This last expression is a torsion point on $A(F)$ by the Manin-Drinfeld theorem which asserts that degree zero cuspidal divisors on a modular curve give rise to torsion elements in its Jacobian. □

We now set

$$P_L = \mathrm{Tr}_{L_c/L}(\pi_f(P_c^+)) \in A(L).$$

Note that $\tau_M(P_L^+)$ is either equal to $\mathrm{Tr}_{L'_c/L'}(\pi_f(P_c'^+))$ or to $\mathrm{Tr}_{L'_c/L'}(\pi_f(P_c'^-))$. Without loss of generality, assume that $\tau_M(P_L^+) = \mathrm{Tr}_{L'_c/L'}(\pi_f(P_c'^+))$ and denote it by P'_L.

Set

$$u = \begin{cases} 2 & \text{if } K = \mathbb{Q}(\sqrt{-1}) \text{ and } c = 1; \\ 3 & \text{if } K = \mathbb{Q}(\sqrt{-3}) \text{ and } c = 1; \\ 1 & \text{otherwise,} \end{cases}$$

and define

$$P_{A,M} := \frac{1}{u}(P_L + P'_L), \qquad P_{A,M'} := \frac{1}{u}(P_L - P'_L).$$

The construction of the point $P_{A,M}$ is illustrated in the figure below.

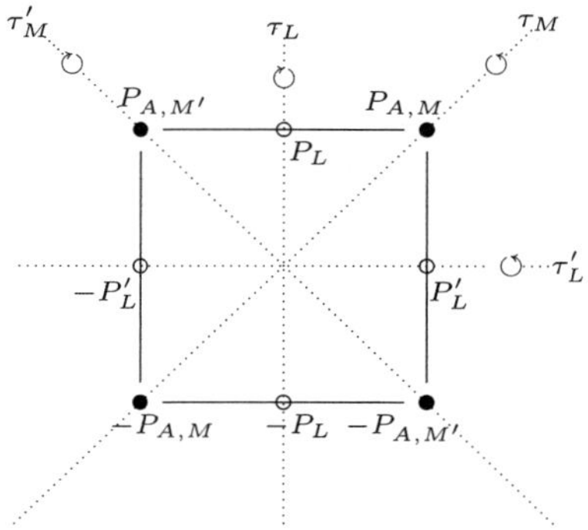

This figure suggests–and it is indeed easy to check–that

$$P_{A,M} \in A(M), \qquad P_{A,M'} \in A(M').$$

Recall that the morphism $\varphi_F : A_F \longrightarrow E$ introduced in (19) is defined over $F \subset M$, and therefore that the point

$$P_M := \varphi_F(P_{A,M})$$

belongs to $E(M)$. As a by-product of our explicit construction we obtain the following analytic formula for calculating the point P_M.

Theorem 4.1. *Let $\tau_c, \tau'_c \in \mathcal{H}$ be elements representing the Heegner points $P_c^+, P_c'^+ \in X_\varepsilon(N)(\tilde{H}_c)$. Set*

$$z_M = \sum_{\underline{\mathfrak{b}} \in \tilde{C}_M} \left[\int_{i\infty}^{\gamma_{\underline{\mathfrak{b}}}\tau_c} (f_E(\tau) + f_E|_{W_N}(\tau))\, d\tau + \int_{i\infty}^{\gamma_{\underline{\mathfrak{b}}}\tau'_c} (f_E(\tau) + f_E|_{W_N}(\tau))\, d\tau \right]. \tag{30}$$

Then $P_M = \eta(z_M)$ where η is the Weierstrass parametrization

$$\eta : \mathbb{C}/\Lambda_E \longrightarrow E(\mathbb{C}), \quad \eta(z) = (\wp(z), \wp'(z)). \tag{31}$$

Here, $\wp$ is the Weierstrass function associated with the lattice of periods

$$\Lambda_E := \left\{ \int_\delta (f_E + f_E|_{W_N}) d\tau \right\}$$

where $\delta \in H_1(X_\varepsilon(\mathbb{C}), \mathbb{Z})$ runs over the cycles of $X_\varepsilon(\mathbb{C})$ such that $\int_\delta (f_E - f_E|_{W_N}) d\tau = 0$.

4.2. *Heegner points and ATR cycles*

The main conjecture that will be formulated in this section relates the Heegner point P_M with the Stark Heegner point $P_M^?$ arising from ATR cycles. Recall that $\mathrm{Gal}(F/\mathbb{Q}) = \{1, \tau\}$ and $D_F = \mathrm{disc}(F)$. Let also $c_{E/F}$ (resp. $c_{E^\tau/F}$) denote either the real period or twice the real period of $E/\mathbb{R}$ (resp. of $E^\tau/\mathbb{R}$) depending on whether $E(\mathbb{R})$ (resp. $E^\tau(\mathbb{R})$) is connected or not.

Conjecture 4.1. *The ATR point $P_M^?$ is of infinite order if and only if P_M is of infinite order and $L(E/F,1) \neq 0$. More precisely,*

$$P_M^? = 2^s \ell \cdot P_M \tag{32}$$

where $\ell \in \mathbb{Q}^\times$, which depends only on (E, F) and not on M, satisfies

$$\ell^2 = \frac{L(E/F,1)}{\Omega_{E/F}}, \qquad \text{with } \Omega_{E/F} = \frac{c_{E/F} \cdot c_{E^\tau/F}}{D_F^{1/2} \cdot |E_{\mathrm{tor}}(F)|^2},$$

and $s \in \mathbb{Z}$ depends on M.

Below we collect some numerical data in support of Conjecture 4.1. Besides the numerical evidence, Conjecture 4.1 is also motivated by the equality

$$L'(E/M,1) = L(E/F,1)L'(E/F,\chi_M,1)$$

as recalled in the proof of Theorem 3.2, and by the facts that

(1) the Néron-Tate height of the Stark-Heegner point $P_M^?$ is expected[8] to be related in a simple way to $L'(E/M,1)$.
(2) the extension of the Gross-Zagier formula[42] proved by Yuan, Zhang and Zhang should in principle lead to an analogous relationship between the Heegner point P_M and the derivative $L'(f/K,\chi_L,1) = L'(f/F,\chi_M,1) = L'(E/F,\chi_M,1)$.

It would be interesting to formulate a precise recipe predicting the power of 2 that arises as a fudge factor in (32). The authors have not made a serious attempt to do this.

4.3. *Numerical examples*

For $N = 29, 37, 41$ it is known (cf. Shiota[36] and [6, § 6]) that there is a unique Shimura elliptic curve defined over $F = \mathbb{Q}(\sqrt{N})$ up to isogeny over F.

The aim of this section is to provide numerical evidence for conjecture 4.1, which we have gathered by explicitly computing the points P_M and $P_M^?$

for several ATR extensions M/F on each of these three elliptic curves. The computation of the Heegner point P_M was performed with the software package PARI[29] by exploiting formula (30) and the material in Shiota[36] and [6, § 6] to produce a complex invariant $z_M \in \mathbb{C}/\Lambda_E$ mapping to P_M under the Weierstrass uniformisation. Similarly, the ATR point $P_M^?$ was computed by following the method explained in [8, Section 3].

In fact, for our experiments it was sufficient to compute the element $z_M^? \in \mathbb{C}/\Lambda_E$ mapping to $P_M^?$ under the Weierstrass uniformisation. For several values of M, the invariants z_M and $z_M^?$ were calculated to roughly 50 digits of decimal accuracy, and the constants s and ℓ in (32) could then be obtained by picking a basis (e_1, e_2) for Λ_E and searching for a linear dependence relation with small integer coefficients between the four complex numbers z_M, $z_M^?$, e_1 and e_2, using Pari's `lindep` command.

This approach represents a dramatic improvement over the one that had to be followed in [8, Section 3] in which only the point $P_M^?$ was computed. In practice, recognizing $P_M^?$ as an algebraic point using standard rational recognition programs is difficult once the height of $P_M^?$ becomes large. The strategy followed in [8, Section 3] consists in performing an independent search for a generator of $E(M)$–a computationally difficult and time-consuming task–in order to check that $P_M^?$ indeed agreed with a point of small height on $E(M)$ to within the calculated decimal accuracy. The new approach based on the Heegner point P_M makes the experimental verifications of the conjectures of [8, §2] much more systematic and efficient, and allows them to be carried out for much further ranges.

In the tables below, we have followed almost the same notations as in [8, §3]. In particular, we have written $M = F(\beta)$ with $\beta^2 \in F$, and denoted by $D_M = \mathrm{Nm}_{F/\mathbb{Q}}(\mathrm{disc}(M/F))$ the absolute discriminant. (Note that K is used in Darmon-Logan[8] to denote the field that we call M in the present work.) As before, L/K denotes the quadratic extension sitting in the Galois closure of M as in the field diagram (26), and we denote $D_K = \mathrm{disc}(K/\mathbb{Q})$ and $D_L = \mathrm{Nm}_{K/\mathbb{Q}}(\mathrm{disc}(L/K))$ and $\ell^2 \in \mathbb{Q}^\times$ is the value we found numerically for the constant alluded to in Conjecture 4.1. Finally note that ℓ and s uniquely determine P_M up to sign and $E(M)_{\mathrm{tor}}$.

The case $N = 29$. Let $\delta = 2 + \omega = (5 + \sqrt{29})/2$. Shiota's Weierstrass equation for E_N is given by

$$E_{29} : y^2 + xy + \delta^2 y = x^3,$$

whose discriminant is $\Delta_{29} = -\delta^{10}$. Our calculations convincingly suggest that

$$\ell^2 = \frac{L(E_{29}/F,1)}{\Omega_{E_{29}/F}} = 1$$

and that the point $P_M^?$ and s are given in the following table. The table suggests that $s = -2$ in all cases that have been calculated for this particular curve.

The case $N = 37$. Letting $\omega = \frac{1+\sqrt{37}}{2}$, Shiota's Weierstrass equation for E_{37} is given by

$$E_{37} : y^2 + y = x^3 + 2x^2 - (19+8\omega)x + (28+11\omega),$$

and its discriminant is $\Delta_{37} = (5+2\omega)^6$. Note that $5+2\omega$ is a fundamental unit of F. Our calculations are consistent with the fact that

$$\ell^2 = \frac{L(E_{37}/F,1)}{\Omega_{E_{37}/F}} = 1.$$

More precisely, the Stark-Heegner point $P_M^?$ and s are given in the tables below.

The case $N = 41$. Shiota's Weierstrass equation for E_{41} is

$$E_{41} : y^2 = x^3 - \frac{17}{48}x + \left(-\frac{5}{32} + \frac{1}{27}\sqrt{41}\right)$$

The computations in [8, Section 3] used instead curve $E'_{41} : y^2 + xy = x^3 - (32 + 5\sqrt{41})x$. This Weierstrass equation was first found by Oort, and there is an explicit isogeny $\psi : E'_{41} \to E_{41}$ of degree 2. Following the approach in [8, Section 3], points $P_M^?$ listed below are points on E'_{41}. Since the isogeny ψ is explicit, it is an easy task to transfer them to points on E_{41}, and this is what we did in order to compare the Heegner points $P_M \in E_{41}(M)$ with the Stark-Heegner points $\psi(P_M^?) \in E_{41}(\mathbb{C})$. In this case, calculations suggest once again that

$$\ell^2 = \frac{L(E_{41}/F,1)}{\Omega_{E_{41}/F}} = 1,$$

while the values of the exponent s also appear in Table 3.

4.4. *Proof of Proposition 4.1*

The aim of this section is proving Proposition 4.1, which was left unproved in §4.1 and asserts that the relative discriminant of L/K factors

$D_M = D_K \cdot c^2$	β^2	D_L	$\lvert\mathrm{Pic}(\mathcal{O}_c)\rvert$	$P_M^?$	s
$-7 = -7 \cdot 1$	$-1+\omega$	29	1	$(\beta^2+3,\ -\frac{5}{2}\beta^3 - 3\beta^2 - 8\beta - \frac{19}{2})$	-2
$-16 = -4 \cdot 2^2$	$2+\omega$	$2^2 \cdot 29$	1	$(\frac{\beta^2}{2},\ -\frac{5}{4}\beta^3 - \frac{11}{4}\beta^2 - \frac{\beta}{4} - \frac{1}{2})$	-2
$-23 = -23 \cdot 1$	$17+8\omega$	29	3	$(\frac{1}{8}(11\beta^2+5),\ -\frac{13}{8}\beta^3 - \beta^2 - \frac{7}{8}\beta - \frac{1}{2})$	-2
$-35_1 = -35 \cdot 1$	$19+9\omega$	29	2	$(\frac{1}{5}(2\beta^2+1),\ -\frac{59}{225}\beta^3 - \frac{43}{90}\beta^2 - \frac{89}{450}\beta - \frac{29}{90})$	-2
$-35_2 = -35 \cdot 1$	$4+3\omega$	29	2	$(-\frac{1}{15}(4\beta^2+11),\ -\frac{1}{150}(17\beta^3+105\beta^2+43\beta+270))$	-2
$-59 = -59 \cdot 1$	$61+28\omega$	29	3	$(-\frac{1}{9},\ -\frac{11}{1512}\beta^3 - \frac{5}{56}\beta^2 - \frac{1}{1512}\beta + \frac{1}{504})$	-2
$-63 = -7 \cdot 3^2$	3ω	$3^2 \cdot 29$	4	$(\frac{7}{9}\beta^2+5,\ \frac{26}{27}\beta^3 - \frac{11}{9}\beta^2 + \frac{57}{9}\beta - 8)$	-2
$-64 = -4 \cdot 4^2$	$4+2\omega$	$2^4 \cdot 29$	2	$(-\frac{1}{4},\ -\frac{3}{8}\beta^3 - \frac{5}{4}\beta^2 - \frac{\beta}{4} - \frac{3}{8})$	-2
$-80 = -20 \cdot 2^2$	$1+\omega$	$2^2 \cdot 29$	4	$(\frac{1}{10}(43\beta^2+51),\ -\frac{517}{50}\beta^3 - \frac{93}{20}\beta^2 - \frac{1233}{100}\beta - \frac{111}{20})$	-2
$-91 = -91 \cdot 1$	$7+5\omega$	29	2	$(\frac{1}{13}(98\beta^2+387),\ -\frac{18939}{845}\beta^3 - \frac{111}{26}\beta^2 - \frac{150109}{1690}\beta - \frac{439}{26})$	-2
$-175 = -7 \cdot 5^2$	$-5+5\omega$	$5^2 \cdot 29$	6	$(-\frac{6}{50}\beta^2 - 2,\ \frac{1}{10}\beta^3 - \frac{11}{25}\beta^2 + \frac{98}{100}\beta - \frac{45}{10})$	-2

Table 1: ATR extensions of $\mathbb{Q}(\sqrt{29})$ and ATR points on E_{29}

$D_M = D_K \cdot c^2$	β^2	D_L	$\lvert\mathrm{Pic}(\mathcal{O}_c)\rvert$	$P_M^?$	s
$-3 = -3 \cdot 1$	$-3 + \omega$	37	1	$(-\frac{2}{3}\beta - \frac{13}{3}, -\frac{61}{18}\beta^3 - \frac{169}{9}\beta - \frac{1}{2})$	-1
$-7 = -7 \cdot 1$	$1 + \omega$	37	1	$(\frac{2}{7}\beta - \frac{3}{7}, -\frac{57}{98}\beta^3 - \frac{44}{49}\beta - \frac{1}{2})$	-1
$-11 = -11 \cdot 1$	$38 + 15\omega$	37	1	$(-\frac{2}{165}\beta^2 - \frac{104}{165}, -\frac{17}{1210}\beta^3 - \frac{2}{605}\beta - \frac{1}{2})$	-1
$-16 = -4 \cdot 2^2$	$5 + 2\omega$	$2^2 \cdot 37$	1	$(\frac{\beta^2}{8} - \frac{5}{8}, \frac{\beta^3}{8} - \frac{1}{2})$	-2
$-48 = -3 \cdot 4^2$	$2 + \omega$	$4^2 \cdot 37$	3	$(\frac{115}{588}\beta^2 - \frac{80}{147}, -\frac{11225}{24696}\beta^3 - \frac{1529}{6174}\beta - \frac{1}{2})$	-1
$-64 = -4 \cdot 4^2$	$10 + 4\omega$	$4^2 \cdot 37$	2	$(-\frac{\beta^2}{8} - \frac{3}{4}, -\frac{\beta^3}{8} - \frac{1}{2})$	-2
$-67 = -67 \cdot 1$	$193 + 76\omega$	67	1	$(-1, -\frac{1}{2} + \frac{1}{2}\beta)$	-2
$-75 = -3 \cdot 5^2$	$-15 + 5\omega$	$5^2 \cdot 37$	3	$(\frac{196}{775}\beta^2 + \frac{136}{27}, -\frac{1559}{12150}\beta^3 - \frac{25732}{6075}\beta - 1/2)$	-1
$-192 = -3 \cdot 8^2$	$18 + 8\omega$	$8^2 \cdot 37$	6	$(\frac{7}{3} + \frac{7}{6}\omega, -\frac{1}{2} + \frac{1}{36}(\frac{85}{3} + \frac{14}{3}\sqrt{37})\beta)$	-2
$-275 = -11 \cdot 5^2$	$445 + 180\omega$	$5^2 \cdot 37$	4	$(\frac{2}{11} + \frac{4}{11}\omega, -\frac{1}{2} + \frac{1}{242}(\frac{62}{7} + \frac{9}{7}\sqrt{37})\beta)$	-2
$-448 = -7 \cdot 8^2$	$2 + 2\omega$	$8^2 \cdot 37$	4	$(\frac{45}{7} + \frac{39}{14}\omega, -\frac{1}{2} + \frac{1}{196}(\frac{689}{2}\sqrt{37} + \frac{4191}{2})\beta)$	-2

Table 2: ATR extensions of $\mathbb{Q}(\sqrt{37})$ and ATR points on E_{37}

$D_M = D_K \cdot c^2$	β^2	D_L	$\lvert \mathrm{Pic}(\mathcal{O}_c) \rvert$	$P_M^?$	s
$-4 = -4 \cdot 1$	$27 + 10\omega$	41	1	$(-\frac{1}{4},\ -\frac{\beta}{2} + \frac{1}{8})$	1
$-8 = -8 \cdot 1$	$-248 + 67\omega$	41	1	$\left(-\frac{1}{268}(3\beta^2 + 1481),\ \frac{1}{536}(-254\beta^3 + 3\beta^2 - 108954\beta + 1481)\right)$	0
$-20 = -20 \cdot 1$	$697 + 258\omega$	41	2	$(\frac{1}{43}(\beta^2 - 9),\ \frac{1}{258}(-\beta^3 - 3\beta^2 + 181\beta + 27))$	0
$-23 = -23 \cdot 1$	$398 + 144\omega$	41	3	$\left(\frac{-71027\beta^2 - 1271153}{9884736},\ \frac{-1095348\beta^3 + 9304537\beta^2 + 16459332\beta + 166521043}{2589800832}\right)$	0
$-32 = -8 \cdot 2^2$	$1 + \omega$	$2^2 \cdot 41$	2	$\left(\frac{29\beta^2 + 49}{4},\ \frac{1}{16}(-359\beta^3 - 58\beta^2 - 611\beta - 98)\right)$	0
$-36 = -4 \cdot 3^2$	$6 + 3\omega$	$3^2 \cdot 41$	4	$\left(-8 + 2\omega, (\frac{7}{2} - \frac{1}{2}\sqrt{41})(1 + 5\beta)\right)$	-1
$-40 = -40 \cdot 1$	$35 + 13\omega$	41	2	$\left(9 + \frac{27}{8}\omega, -\frac{171}{32} - \frac{27}{32}\sqrt{41} + \frac{3}{32}(\frac{109}{2} + \frac{17}{2}\sqrt{41})\beta\right)$	-1
$-100 = -4 \cdot 5^2$	$10 + 5\omega$	$5^2 \cdot 41$	2	$\left(\frac{9}{2} + \frac{7}{4}\omega, -\frac{43}{16} - \frac{7}{16}\sqrt{41} + (\frac{3}{8}\sqrt{41} + \frac{19}{8})\beta\right)$	-2
$-115 = -115 \cdot 1$	$177 + 68\omega$	41	2	$\left(-\frac{31}{9} - \frac{11}{9}\omega, \frac{73}{36} + \frac{11}{36}\sqrt{41} + \frac{1}{108}(\frac{59}{5} + \frac{9}{5}\sqrt{41})\beta\right)$	-1
$-160 = -40 \cdot 2^2$	4ω	$2^2 \cdot 41$	4	$\left(32 + 12\omega, -19 - 3\sqrt{41} + (\frac{173}{2} + \frac{27}{2}\sqrt{41})\beta\right)$	-2
$-368 = -23 \cdot 4^2$	$43 + 16\omega$	$4^2 \cdot 41$	6	$\left(\frac{29}{4} + \frac{11}{4}\omega, -\frac{69}{16} - \frac{11}{16}\sqrt{41} + (\frac{13}{4} + \frac{1}{2}\sqrt{41})\beta\right)$	-2

Table 3: ATR extensions of $\mathbb{Q}(\sqrt{41})$ and ATR points on E'_{41}

as $d(L/K) = c \cdot \mathcal{N}$, where c is a positive integer such that $L \subset L_c$ (and similarly $L' \subset L'_c$).

Recall our assumption on $N = \mathrm{disc}(F)$ to be odd, and thus square-free. Here we shall assume for notational simplicity that $K \neq \mathbb{Q}(\sqrt{-1})$ and $\mathbb{Q}(\sqrt{-3})$, so that $\mathcal{O}_K^\times = \{\pm 1\}$; we leave to the reader the task of filling the details for the two excluded fields.

Let us recall first the following classical lemma ([16, §38-39]; cf. also Daberkow[7]) on Kummer extensions of local fields, which applies in particular to our quadratic extension L/K.

Lemma 4.3. *Let k be a local field containing all p-th roots of unity for some prime p and let $v_k : k^\times \to \mathbb{Z}$ denote the valuation map of k, normalized so that $v_k(k^\times) = \mathbb{Z}$. Let K/k be a Kummer extension of degree p with discriminant $\mathfrak{d}_{K/k}$. Then $K = k(\sqrt[p]{\vartheta})$ for some $\vartheta \in k$ such that $v_k(\vartheta) \in \{0, 1\}$. Moreover,*

(i) If $v_k(\vartheta) = 1$, $v_k(\mathfrak{d}_{K/k}) = pv_k(p) + (p-1)$.

(ii) Assume $v_k(\vartheta) = 0$. If $v_k(p) = 0$, then $v_k(\mathfrak{d}_{K/k}) = 0$. Otherwise, write $\mathfrak{p}_k$ for the unique maximal ideal in k. We have:

- *(a) If equation $x^p \equiv \vartheta (\mathrm{mod}\ \mathfrak{p}_k^{pv_k(p)/(p-1)})$ can be solved in k, then $v_k(\mathfrak{d}_{K/k}) = 0$.*
- *(b) If not, $v_k(\mathfrak{d}_{K/k}) = pv_k(p) + (p-1)(1-\eta)$, where $\eta = \max_\ell\{0 \leq \ell < pv_k(p)/(p-1) \mid x^p \equiv \vartheta (\mathrm{mod}\ \mathfrak{p}_k^\ell)$ can be solved in $\mathcal{O}_k\}$.*

We use the above result in order to deduce several lemmas which shall allow us to reduce the proof of Proposition 4.1 to the case in which L/K is unramified at dyadic primes.

Lemma 4.4. *Let $p \nmid \mathrm{disc}(K)$ be a prime and put $p^* = 8$ if $p = 2$, $p^* = p$ if $p \equiv 1 (\mathrm{mod}\ 4)$ and $p^* = -p$ if $p \equiv -1 (\mathrm{mod}\ 4)$. Then $K(\sqrt{p^*})$ is contained in the* ring class field *H_c of K associated to the order $\mathcal{O}_c$ of conductor $c = |p^*|$.*

Proof. Suppose first that p is split in K and fix a prime $\mathfrak{p}|p$ in K. Let

$$U = K_\mathfrak{p}^\times \bigcap K^\times \prod_v \mathcal{O}_{c,v}^\times,$$

where the intersection is computed by regarding $K_\mathfrak{p}^\times$ as a subgroup of $\prod_v K_v^\times$ by means of the usual embedding $x_\mathfrak{p} \mapsto (1, ..., 1, x_\mathfrak{p}, 1, ..., 1)$.

Since the map $K_\mathfrak{p}^\times / U \to \mathbb{I}_K / \left(K^\times \prod_v \mathcal{O}_{c,v}^\times\right)$ is injective by [26, p. 173, Prop. 5.2], it follows that $U \subset K_\mathfrak{p}^\times \simeq \mathbb{Q}_p^\times$ corresponds to $H_{c,\mathfrak{P}}/K_\mathfrak{p}$ by local class field theory for any prime $\mathfrak{P}$ of H_c above $\mathfrak{p}$.

Write $c = p^r$ with $r = 3$ if $p = 2$, $r = 1$ if p is odd. Since $1 + p^r\mathbb{Z}_p \subseteq U$, $1 + p^{r-1}\mathbb{Z}_p \not\subset U$ by [4, p. 197], an easy calculation shows that

$$U = \{\frac{\alpha}{\overline{\alpha}} \mid \alpha \in V\} \cdot (1 + \mathfrak{p}^r),$$

where $V = \{\alpha \in K^\times \mid \mathrm{ord}_v(\alpha) = 0 \, \forall v \neq \mathfrak{p}\}$. Note that $V = \{\pm\alpha_0^n, n \in \mathbb{Z}\}$ for some $\alpha_0 \in K^\times$ such that $\mathrm{ord}_v(\alpha_0) = 0$ for all $v \neq \mathfrak{p}$ and $\mathrm{ord}_{\mathfrak{p}}(\alpha_0) = n_0 \geq 1$ is minimal. With this notation we have

$$U = \{(\frac{\alpha_0}{\overline{\alpha}_0})^n, n \in \mathbb{Z}\} \cdot (1 + \mathfrak{p}^r). \tag{33}$$

Suppose now that p remains inert in $\mathcal{O}_K$. Arguing similarly as before we obtain that the open subgroup $U \subset K_p^\times$ corresponding to $H_{p^r,\mathfrak{P}}/K_p$ by local class field theory is $U = K_p^\times \cap \big(K^\times \prod_v \mathcal{O}_{p^r,v}^\times\big)$, i.e.

$$\begin{aligned} U &= \{\alpha \mid \alpha \in K^\times, \, \mathrm{ord}_v(\alpha) = 0, \, \forall v \neq p\} \cdot (1 + p^r\mathcal{O}_{K_p}) \\ &= \{(\pm\alpha_0)^n \mid n \in \mathbb{Z}\} \cdot (1 + p^r\mathcal{O}_{K_p}), \end{aligned} \tag{34}$$

where $\alpha_0 \in K^\times$ is chosen such that $\mathrm{ord}_v(\alpha_0) = 0$ for all $v \neq p$ and $\mathrm{ord}_p(\alpha_0) \geq 1$ is minimal. We can thus take $\alpha_0 = p^{n_0}$ for some $n_0 \geq 1$.

Put $K' = K(\sqrt{p^*})$. Any prime $\mathfrak{p}$ in K above p ramifies in K'. Fix one such prime $\mathfrak{p}$ and put $\mathfrak{p} = \wp^2$ in K' so that $K'_\wp = K_{\mathfrak{p}}(\sqrt{p^*})$. By class field theory, in order to prove that $K' \subset H_c$ it is enough to show that $U \subset \mathrm{Nm}_{K'_\mathfrak{p}/K_\mathfrak{p}}(K_\mathfrak{p}^{'\times})$. Since $d(K'/K) = p^*$ by Lemma 4.3, K' is contained in the *ray class field* K_c of conductor c of K and it thus suffices to verify that $\alpha_0\overline{\alpha}_0$ (resp. $\pm\alpha_0$) lies in $\mathrm{Nm}_{K'_\mathfrak{p}/K_\mathfrak{p}}(K_\mathfrak{p}^{'\times})$ if p splits (resp. remains inert) in K.

Assume $p = 2$. Then $\pm 1, \pm 2 \in \mathrm{Nm}_{K'_\wp/K_\mathfrak{p}}(K_\wp^{'\times})$ because $-1 = \mathrm{Nm}(1 + \sqrt{2})$ and $-2 = \mathrm{Nm}(\sqrt{2})$. The lemma thus follows automatically if 2 is inert in K, while if 2 splits, it follows because $\alpha_0\overline{\alpha}_0$ is a power of 2, hence $\alpha_0\overline{\alpha}_0$ lies in either $\pm\mathbb{Q}_2^{\times^2}$ or $\pm 2\mathbb{Q}_2^{\times^2}$.

Assume p is odd. Then $-p^* = \mathrm{Nm}(\sqrt{p^*}) \in \mathrm{Nm}_{K'_\mathfrak{p}/K_\mathfrak{p}}(K_\mathfrak{p}^{'\times})$. Suppose first p splits in K: as before, it is enough to show that $p \in \mathrm{Nm}_{K'_\mathfrak{p}/K_\mathfrak{p}}(K^{'\times})$, which we already did if $p^* = -p$. That the same holds when $p^* = p$ follows because $p \equiv 1 (\mathrm{mod}\ 4)$ implies that $-1 \in \mathrm{Nm}_{K'_\mathfrak{p}/K_\mathfrak{p}}(K_\mathfrak{p}^{'\times})$. Suppose now p remains inert in K; we must show that $\pm p \in \mathrm{Nm}_{K'_\mathfrak{p}/K_\mathfrak{p}}(K_\mathfrak{p}^{'\times})$. If $p^* = p$ this follows by the same reason as above; if $p^* = -p$, then $p \equiv 3 \, (\mathrm{mod}\ 4)$, $K_p = \mathbb{Q}_p(\sqrt{-1})$ and thus $-1 \in K_p^{\times^2}$, which allows us to conclude. □

Note that a direct consequence of the previous lemma is that for any odd square free integer m relatively coprime with $\mathrm{disc}(K)$ either $K(\sqrt{m})$ or $K(\sqrt{-m})$ is contained in H_m.

Lemma 4.5. *$d(L/K) = 2^t c_0\mathcal{N}$ for some integer $0 \le t \le 3$ and some positive integer $c_0 \ge 1$ relatively coprime to 2 and $\mathcal{N}$. If further 2 is ramified in K, $0 \le t \le 2$.*

Proof. Write $K = \mathbb{Q}(\sqrt{-d_0})$ for some square free integer $d_0 > 0$ and $L = K(\sqrt{\beta})$ for some $\beta \in \mathbb{Z}+\mathbb{Z}\sqrt{-d_0}$ and square free in K. Without loss of generality, N can be written as $\mathcal{N}\overline{\mathcal{N}}$ where $\mathcal{N}$ divides the square free part $\mathfrak{B}$ of (β) in K and $\overline{\mathcal{N}}$ is relatively coprime to $\mathfrak{B}$.

Write $\mathfrak{B}_2$ for the largest ideal which divides $\mathfrak{B}$ and is relatively coprime to any prime of K above 2. Since $v_{K_{\mathfrak{p}'}}(2) = 0$ and $v_{K_{\mathfrak{p}'}}(\mathfrak{B}_2) = 1$ for any prime $\mathfrak{p}' \mid \mathfrak{B}_2$, Lemma 4.3 shows that $v_{K_{\mathfrak{p}'}}(\mathfrak{d}_{L_{\mathfrak{P}'}/K_{\mathfrak{p}'}}) = 1$, where $\mathfrak{P}'$ is the prime in L above $\mathfrak{p}'$, thus the prime-to-2 part of $d(L/K)$ is $\mathfrak{B}_2$. Besides, $\mathfrak{B}_2 = \mathcal{N}\cdot\mathfrak{C}$ with $(\mathfrak{C},\mathcal{N}) = 1$. Since $\mathrm{Nm}_{K/\mathbb{Q}}(\beta)/N$ is a perfect square in $\mathbb{Z}$, $\mathfrak{C}$ is principal and can be written as $\mathfrak{C} = (c_0)$ for some integer $c_0 > 0$. Hence $\mathcal{N}c_0 \mid d(L/K) \mid 2^t\mathcal{N}c_0$ for some integer $t \ge 0$.

If 2 is unramified in K we have $v_{K_{\mathfrak{p}}}(2) = 1$ for any prime $\mathfrak{p} \mid 2$ in K and it follows from Lemma 4.3 that $d(L/K) = \mathcal{N}c_0 2^t$ with $0 \le t \le 3$.

Suppose now that 2 ramifies in K with $(2) = \mathfrak{p}^2$ in K. Then, since $v_{K_{\mathfrak{p}}}(2) = 2$ and $\mathrm{Nm}_{K/\mathbb{Q}}(\beta)/N$ is a perfect square in $\mathbb{Z}$, we fall into case (ii) of Lemma 4.3: for any prime $\mathfrak{P}$ in L above $\mathfrak{p}$, $L_{\mathfrak{P}}$ can be written as $K_{\mathfrak{p}}(\sqrt{\vartheta})$ for some $\vartheta \in K_{\mathfrak{p}}^{\times}$ such that $v_{k_{\mathfrak{p}}}(\vartheta) = 0$. Suppose $v_{K_{\mathfrak{p}}}(\mathfrak{d}_{L_{\mathfrak{P}}/K_{\mathfrak{p}}}) \neq 0$. Then Lemma 4.3 (b) asserts that

$$v_{K_{\mathfrak{p}}}(\mathfrak{d}_{L_{\mathfrak{P}}/K_{\mathfrak{p}}}) = 5 - \eta,$$

where

$$\eta = \max\{0 \le \ell < 4 \,\big|\, \exists \iota \in \mathcal{O}_{K_{\mathfrak{p}}},\ \iota^2 \equiv \vartheta (\mathrm{mod}\ \mathfrak{p}^{\ell})\}.$$

A classical result[18,17] of Hilbert implies that $v_{K_{\mathfrak{p}}}(\mathfrak{d}_{L_{\mathfrak{P}}/K_{\mathfrak{p}}})$ is even. Hence $d(L/K) = \mathcal{N}2^t c_0$ with $0 \le t \le 2$. □

Lemma 4.6. *It is enough to prove Proposition 4.1 when $d(L/K) = 2^t\mathcal{N}$ and $0 \le t \le 2$.*

Proof. Lemma 4.5 shows in general $d(L/K) = 2^t c_0\mathcal{N}$, where $0 \le t \le 3$ and $c_0 \ge 1$. Suppose first that $t = 3$, then 2 is unramified in K by the same lemma. Let $\mathfrak{P}$ and $\mathfrak{p}$ be prime ideals in L and K respectively such that $\mathfrak{P}|\mathfrak{p}|2$. Then $L_{\mathfrak{P}}$ can be written as $K_{\mathfrak{p}}(\sqrt{\vartheta})$ for some $\vartheta \in K$ with

$v_{K_\mathfrak{p}}(\vartheta) \in \{0,1\}$. Define $L' = K(\sqrt{\vartheta'})$, where ϑ' is defined as

$$\vartheta' = \begin{cases} \vartheta/2 & \text{if } v_{K_\mathfrak{p}}(\vartheta) = 1; \\ \vartheta & \text{if } v_{K_\mathfrak{p}}(\vartheta) = 0. \end{cases}$$

Hence $v_{K_\mathfrak{p}}(\vartheta') = 0$. Let $\mathfrak{P}'$ be a prime in L' above $\mathfrak{p}$. Then either case (a) or (b) of Lemma 4.3 applies. In case (a), $v_{K_\mathfrak{p}}(\mathfrak{d}_{L'_{\mathfrak{P}'}/K_\mathfrak{p}}) = 0$. In case (b), $v_{K_\mathfrak{p}}(\mathfrak{d}_{L'_{\mathfrak{P}'}/K_\mathfrak{p}}) = 3 - \eta$, where $0 \leq \eta \leq 2$, hence $\mathfrak{p}$ is ramified in L', so residue field of $L'_{\mathfrak{P}'}$ is equal to that of $K_\mathfrak{p}$ and consequently $\eta \geq 1$. We conclude that $v_{K_\mathfrak{p}}(\mathfrak{d}_{L'_{\mathfrak{P}'}/K_\mathfrak{p}}) \leq 2$. By Lemma 4.4, $L \subset K(\sqrt{2})L' \subset H_8 L'$ with $d(L'/K) = 2^{t'} c_0 \mathcal{N}$ for some integer $0 \leq t' \leq 2$.

Suppose now $c_0 > 1$. Setting $L'' = K(\sqrt{\delta\vartheta'/c_0})$ we have $d(L''/K) = 2^{t'}\mathcal{N}$, where $\delta \in \{\pm 1\}$ such that $K(\sqrt{\delta c_0}) \subset H_{c_0}$ as described in Lemma 4.4. By the same lemma, $L' \subset K(\sqrt{\delta c_0})L'' \subset H_{c_0} L''$.

So $L \subset H_8 L' \subset H_8 H_{c_0} L'' = H_{8c_0} L''$ such that L''/K is a quadratic extension and $d(L''/K) = 2^{t'}\mathcal{N}$ for some integer $0 \leq t' \leq 2$. This justifies the claim that we only need to prove proposition 4.1 when $0 \leq t \leq 2$ and $d(L/K) = 2^t \mathcal{N}$. □

Thanks to Lemma 4.6 we can assume in what follows that $c_0 = 1$ and $0 \leq t \leq 2$.

Lemma 4.7. *There is a unique quadratic extension $\mathcal{L}_{2^t}/K_{2^t}$ contained in $K_{2^t\mathcal{N}}$ such that the set of primes in K_{2^t} which ramify in $\mathcal{L}_{2^t}$ is the set of primes above $\mathcal{N}$. We have $L \subset \mathcal{L}_{2^t}$.*

Proof. Assume first $t = 0$ or 1. Then $\mathrm{Gal}(K_{2^t\mathcal{N}}/K_{2^t}) \cong (\prod_{\mathfrak{p}|\mathcal{N}} (\mathcal{O}_K/\mathfrak{p})^\times)/\{\pm 1\} \cong (\mathbb{Z}/N\mathbb{Z})^\times/\{\pm 1\}$. This is obvious for $t = 0$, and holds for $t = 1$ because $K_2 K_\mathcal{N} = K_{2\mathcal{N}}$. Extension $\mathcal{L}_{2^t}/K_{2^t}$ corresponds by Galois theory to the unique primitive even quadratic Dirichlet character ε of conductor N.

Suppose now $t = 2$. Then

$$\mathrm{Gal}(K_{2^t\mathcal{N}}/K_{2^t}) \cong G := \big(\{\pm 1\} \times (\mathbb{Z}/N\mathbb{Z})^\times\big)/\{\pm \mathbb{1}\},$$

where $\mathbb{1} = (1, \mathbf{1})$ is the identity element of $\{\pm 1\} \times (\mathbb{Z}/N\mathbb{Z})^\times$. Again, any extension $\mathcal{L}_{2^t}/K_{2^t}$ as in the statement corresponds to a non-trivial character $\varepsilon' : G \to \{\pm 1\}$ which is trivial on $\{\pm 1\} \times \{\mathbf{1}\}$ and is even and primitive on $\{1\} \times (\mathbb{Z}/N\mathbb{Z})^\times$. As above, the only such character is $\varepsilon' = 1 \times \varepsilon$.

Finally, note that LK_{2^t}/K_{2^t} is a quadratic extension contained in $K_{2^t\mathcal{N}}$. Since $\mathrm{disc}(LK_{2^t}/K_{2^t}) = \mathcal{N}$ it follows that $\mathcal{L}_{2^t} = LK_{2^t}$ and thus $L \subset \mathcal{L}_{2^t}$. □

Recall the quadratic extension L_c of the ring class field H_c introduced in §4.1, over which the Heegner points P_c^+ and $P_c^- \in \mathrm{CM}(c)$ are rational. Lemma 4.7 reduces the proof of Proposition 4.1 to showing that $L_c = \mathcal{L}_c$. Since L_c was defined as the quadratic extension of H_c cut out by the kernel of the single even primitive character ε of conductor N, it suffices to show that $H_{2^t} = K_{2^t}$ for $0 \leq t \leq 2$.

When $t = 0$ and we obviously have $H_1 = K_1$. If $t = 1$ or 2, the ratio of the ray class number h_{2^t} by the ring class number $h(\mathcal{O}_{2^t})$ is (cf. [26, p.154] for this and the remaining notations):

$$\frac{h_{2^t}}{h(\mathcal{O}_{2^t})} = \frac{[U : U_{2^t,1}]^{-1}\mathrm{Nm}(2^t)\prod_{\mathfrak{p}|2^t}(1 - \frac{1}{\mathrm{Nm}(\mathfrak{p})})}{\frac{2^t}{[\mathcal{O}_K^\times : \mathcal{O}_{2^t}^\times]}\prod_{p|2^t}(1 - (\frac{d_K}{p})\frac{1}{p})}$$
$$= \begin{cases} \frac{[\mathcal{O}_K^\times:\mathcal{O}_{2^t}^\times]}{[U:U_{2^t,1}]} \cdot \frac{2^{2t}\left(1-\frac{1}{4}\right)}{2^t\left(1-\left(\frac{d_K}{2}\right)\frac{1}{2}\right)} & \text{if 2 is inert in } K, \\ \frac{[\mathcal{O}_K^\times:\mathcal{O}_{2^t}^\times]}{[U:U_{2^t,1}]} \cdot \frac{2^{2t}\left(1-\frac{1}{2}\right)^2}{2^t\left(1-\left(\frac{d_K}{2}\right)\frac{1}{2}\right)} & \text{if 2 is split in } K, \\ \frac{[\mathcal{O}_K^\times:\mathcal{O}_{2^t}^\times]}{[U:U_{2^t,1}]} \cdot \frac{2^{2t}\left(1-\frac{1}{2}\right)}{2^t} & \text{if 2 is ramified in } K. \end{cases} \qquad (35)$$
$$= \frac{[\mathcal{O}_K^\times : \mathcal{O}_{2^t}^\times]}{[U : U_{2^t,1}]} \cdot 2^{t-1}.$$

Since $K \neq \mathbb{Q}(\sqrt{-1})$ and $\mathbb{Q}(\sqrt{-3})$, $[\mathcal{O}_K^\times : \mathcal{O}_{2^t}^\times] = 1$. If $t = 1$, then $U = U_{2^t,1}$, so $K_2 = H_2$. If $t = 2$, then $[U : U_{2^t,1}] = 2$, and therefore $K_{2^2} = H_{2^2}$.

References

1. C. Breuil, B. Conrad, F. Diamond, and R. Taylor. *On the modularity of elliptic curves over* **Q***: wild 3-adic exercises.* J. Amer. Math. Soc. **14** (2001), no. 4, 843–939.
2. M. Bertolini and H. Darmon. *Iwasawa's main conjecture for elliptic curves over anticyclotomic* $\mathbb{Z}_p$*-extensions.* Ann. of Math. (2) **162** (2005), no. 1, 1–64.
3. D. Bump, S. Friedberg, and J. Hoffstein. *Nonvanishing theorems for L-functions of modular forms and their derivatives.* Invent. Math. **102** (1990), no. 3, 543–618.
4. D. A. Cox. *Primes of the form* $x^2 + ny^2$*: Fermat, class field theory, and complex multiplication.* John Wiley & Sons (1989).
5. S. Comalada, *Elliptic curves with trivial conductor over quadratic fields*, Pacific J. Math. **144** (1990), 237-258.
6. J. E. Cremona. *Modular symbols for* $\Gamma_1(N)$ *and elliptic curves with everywhere good reduction.* Math. Proc. Camb. Phil. Soc. (1992), **111**, 199–218.
7. M. Daberkow. *On computations in Kummer extensions.* J. Symbolic Computation **31** (2001), 113–131.

8. H. Darmon and A. Logan. *Periods of Hilbert modular forms and rational points on elliptic curves.* International Math. Res. Not. **40** (2003), 2153–2180.
9. J. S. Ellenberg and C. Skinner. *On the modularity of* $\mathbb{Q}$*-curves.* Duke Math. J. **109** (2001), no. 1, 97–122.
10. K. Fujiwara. *Deformation rings and Hecke algebras in the totally real case.* Preprint 1996, revised 2004, 2006.
11. J. Gärtner. *Points de Darmon et variétés de Shimura.* Thèse de Doctorat, Université Paris 7 (Jussieu) (2010).
12. S. Gelbart. *Automorphic forms on adèle groups.* Ann. of Math. Studies **83**, Princeton Univ. Press, Princeton, NJ (1975).
13. E. González and X. Guitart. *On the modularity level of modular abelian varieties over number fields.* J. Number Theory **130** (2010), no. 7, 1560-1570.
14. X. Guitart and J. Quer. *Modular abelian varieties over number fields.* Preprint available at http://arxiv.org/abs/0905.2550.
15. B.H. Gross and D.B. Zagier. *Heegner points and derivatives of L-series.* Invent. Math. **84** (1986), no. 2, 225–320.
16. E. Hecke. *Lectures on the theory of algebraic numbers.* Translated from the German by George U. Brauer, Jay R. Goldman and R. Kotzen. Graduate Texts in Mathematics, 77. Springer-Verlag, New York-Berlin (1981).
17. D. Hilbert. *Über die Theories des relativquadratischen Zahlkörpers.* Math. Ann. **51** (1899), 1–127.
18. J.G. Huard, B.K. Spearman, and K.S. Williams. *Integral bases for quartic fields with quadratic subfields.* J. Number Theory **51** (1995), no. 1, 87–102.
19. F. Jarvis and J. Manoharmayum. *On the modularity of supersingular elliptic curves over certain totally real number fields.* J. Number Theory **128** (2008), no. 3, 589–618.
20. V.A. Kolyvagin. *Finiteness of* $E(\mathbf{Q})$ *and* $Ш(E, \mathbf{Q})$ *for a subclass of Weil curves.* Izv. Akad. Nauk SSSR Ser. Mat. 52 (1988), no. 3, 670–671; translation in Math. USSR-Izv. 32 (1989), no. 3, 523–541.
21. V. A. Kolyvagin and D. Yu. Logachev. *Finiteness of the Shafarevich-Tate group and the group of rational points for some modular abelian varieties.* Leningrad Math. J., **1** (1990), no. 5, 1229–1253.
22. V. A. Kolyvagin and D. Yu. Logachev. *Finiteness of* $Ш$ *over totally real fields.* Math. USSR Izvestiya **39** (1992), no. 1, 829–853.
23. C. Khare and J.-P. Wintenberger. *Serre's modularity conjecture I, II.* Invent. Math. **178** (2009), no. 3, 485–504 and 505–586.
24. M. Longo. *On the Birch and Swinnerton-Dyer conjecture for modular elliptic curves over totally real fields.* Ann. Inst. Fourier (Grenoble) **56** (2006), no. 3, 689–733.
25. J. S. Milne. *On the arithmetic of abelian varieties.* Invent. Math. **17** (1972), 177–190.
26. J. S. Milne. *Class field theory.* Available at http://www.jmilne.org/math.
27. M.R. Murty and V.K. Murty. *Mean values of derivatives of modular L-series.* Ann. of Math. (2) **133** (1991), no. 3, 447–475.
28. T. Oda. *Periods of Hilbert modular surfaces.* Progress in Mathematics, **19**. Birkhäuser, Boston, Mass. (1982).

29. PARI/GP, version `2.3.4`, Bordeaux (2008). Available at http://pari.math.u-bordeaux.fr/.
30. A. Pacetti, On the change of root number under twisting and applications, available at arXiv:1010.3781.
31. R.G.E. Pinch. *Elliptic curves over number fields.* D. Phil thesis, Oxford University (1982).
32. K. A. Ribet. *Abelian varieties over* $\mathbb{Q}$ *and modular forms*, in *Modular curves and abelian varieties.* J. Cremona, J.-C. Lario, J. Quer, K. Ribet (eds.), Progress in Mathematics **224**, Birkhäuser (2004), 241–261.
33. D.E. Rohrlich. *Nonvanishing of L-functions and structure of Mordell-Weil groups.* Journal für di Reine und Angewand. Math. **417** (1991), 1–26.
34. D.E. Rohrlich. *Variation of the root number in families of elliptic curves.* Compositio Math. **87** (1993) 119–151.
35. G. Shimura. *Introduction to the arithmetic theory of automorphic functions.* Reprint of the 1971 original. Publications of the Mathematical Society of Japan, 11. Kanô Memorial Lectures, 1. Princeton Univ. Press, Princeton, NJ (1994).
36. K. Shiota. *On the explicit models of Shimura's elliptic curves.* J. Math. Soc. Japan **38** (1986), no. 4, 649–659.
37. C. Skinner and A. Wiles. *Nearly ordinary deformations of irreducible residual representations.* Ann. Fac. Sci. Toulouse Math. (6) **10** (2001), 185–215.
38. R. Taylor and A. Wiles. *Ring-theoretic properties of certain Hecke algebras.* Ann. of Math. (2) **141** (1995), no. 3, 553–572.
39. Y. Tian and S. Zhang. Book, in progress.
40. J.-L. Waldspurger. *Sur les valeurs de certaines fonctions L automorphes en leur centre de symmétrie.* Compositio Math. **54** (1985), 173–242.
41. A. Wiles. *Modular elliptic curves and Fermat's last theorem.* Ann. of Math. (2) **141** (1995), no. 3, 443–551.
42. X. Yuan, W. Zhang, and S. Zhang. *Heights of CM points I: Gross–Zagier formula.* Preprint.
43. S. Zhang. *Heights of Heegner points on Shimura curves.* Ann. of Math. (2) **153** (2001), no. 1, 27–147.
44. S. Zhang. *Arithmetic of Shimura curves.* Science China Mathematics **53** (2010), 573–592.
45. Y. Zhao. McGill Univ. Ph. D Thesis, 2011.

The supremum of Newton polygons of p-divisible groups with a given p-kernel type

Shushi Harashita

Graduate School of Environment and Information Sciences, Yokohama National University, 79-7 Tokiwadai, Hodogaya-ku, Yokohama 240-8501 Japan
E-mail: harasita@ynu.ac.jp

Dedicated to Professor Takayuki Oda on his 60th birthday

In this paper we show that there exists the supremum of Newton polygons of p-divisible groups with a given p-kernel type, and provide an algorithm determining it.

Keywords: p-divisible groups; finite group schemes; Newton polygons

1. Introduction

Let k be an algebraically closed field of characteristic $p > 0$. We are concerned with estimating the isogeny type (=Newton polygon, cf. [11]) of a p-divisible group X over k from its p-kernel $X[p]$. In this paper we give an optimal estimation.

We fix once for all, non-negative integers c and d with $r := c + d > 0$. Let $\mathrm{W}(= \mathrm{W}_r)$ be the Weyl group of the general linear group GL_r. In the usual sense, we identify W and $\mathrm{Aut}(\{1, \ldots, r\})$. Let $s_i \in \mathrm{W}$ be the simple reflection $(i, i+1)$ for $i = 1, \ldots, r-1$. Let $\mathrm{S} = \{s_1, \ldots, s_{r-1}\}$ and set $\mathrm{J} := \mathrm{S} \setminus \{s_d\}$. Let W_J be the subgroup of W generated by elements of J. We denote by ${}^\mathrm{J}\mathrm{W}$ the set of $(\mathrm{J}, \emptyset)$-reduced elements of W (cf. [1], Chap. IV, Ex. §1, 3), which are the shortest representatives of $\mathrm{W}_\mathrm{J} \setminus \mathrm{W}$. A classification theory of BT_1's by Kraft, Oort, Moonen and Wedhorn says that the set of the isomorphism classes of BT_1's with tangent-dimension d and length r is bijective to the set ${}^\mathrm{J}\mathrm{W}$. Note that ${}^\mathrm{J}\mathrm{W}$ has a natural ordering $\subset$ introduced and investigated by He [7], also see Wedhorn [22] (we shall give a short review: Theorem 3.6).

Let us explain our main results: Theorem 1.1 combined with Corollary

2.2. Let w be any element of ${}^{\mathrm{J}}\mathrm{W}$. In Corollary 2.2 we show that there exists the supremum $\xi(w)$ of Newton polygons of p-divisible groups with p-kernel type w:

- every p-divisible group whose p-kernel is of type w has Newton polygon $\prec \xi(w)$;
- there exists a p-divisible group X such that $X[p]$ is of type w and the Newton polygon of X equals $\xi(w)$.

Theorem 1.1 below gives us a combinatorial algorithm determining $\xi(w)$, see Remark 5.1. For a Newton polygon ζ, let $\mu(\zeta) \in {}^{\mathrm{J}}\mathrm{W}$ denote the type of the p-kernel of the minimal p-divisible group $H(\zeta)$ having Newton polygon ζ (cf. [15] and also a review [4], §3).

Theorem 1.1. *$\xi(w)$ is the biggest one of the Newton polygons ζ with $\mu(\zeta) \subset w$.*

We shall see in §5 that this theorem follows from Proposition 5.2. The last two sections are devoted to the proof of Proposition 5.2. The theorem is an unpolarized analogue of [6], Corollary II. For a more effective algorithm determining the first/last slope of $\xi(w)$, see [3], Theorem 4.1 for the polarized case and [4], Corollary 1.3 for the unpolarized case. In the polarized case, the existence of the supremum $\xi(w)$ follows from the fact that any Ekedahl-Oort stratum on the moduli space $\mathcal{A}_g$ of principally polarized abelian varieties is irreducible if it is not contained in the supersingular locus ([2], Theorem 11.5). An obstruction in the unpolarized case has been the absence of a good moduli space like $\mathcal{A}_g$. However using Vasiu's $\mathbb{T}_m$-action instead, we have (Lemma 2.1) that there exists an irreducible catalogue of p-divisible groups with a fixed p-kernel type; this clearly shows the existence of $\xi(w)$. Then Theorem 1.1 can be shown by a similar argument as in [6] (which is relatively easier than the polarized case). Finally we mention a different approach announced by Viehmann [21], who seems to have generalized our results in terms of the loop groups of split reductive groups, making use of results on affine Deligne-Lusztig varieties.

Terminology

We naturally identify the category of affine schemes with the opposite category to the category of commutative rings. We fix once for all a rational prime p. In this paper we freely use a part of Zink's result [23], Theorem 9, which says that for a commutative ring R of finite type over a field of characteristic p, there exists a categorical equivalence from the category of formal p-divisible groups over R to that of nilpotent displays over R, where we follow the terminology of [10] for displays and nilpotent displays.

2. A catalogue of p-divisible groups with a given type

Let k be an algebraically closed field of characteristic p. Let $(P, Q, F, \dot{F})$ be a display over k, and $P = L \oplus T$ be a normal decomposition ([23], Introduction), where L and T are free $W(k)$-modules. Let c and d be the ranks of L and T respectively. Let $\mathrm{G} = \mathrm{GL}(P)$ be the general linear group over $W(k)$ of degree $r = c + d$. Let H be the parahoric subgroup of G stabilizing Q, which is a connected smooth affine group scheme over $W(k)$. Let $\mathcal{D}_m$ and $\mathcal{H}_m$ be connected smooth affine group schemes over k such that $\mathcal{D}_m(R) = \mathrm{G}(W_m(R))$ and $\mathcal{H}_m(R) = \mathrm{H}(W_m(R))$ respectively, see [19], 2.1.4 for more details. For any truncated Barsotti-Tate group of level m with codimension c and dimension d, there exists a $g \in \mathcal{D}_m$ such that its Dieudonné module is isomorphic to the $W_m(k)$-module P/p^mP with the Frobenius and the Verschiebung defined by gF and Vg^{-1} respectively. Vasiu introduced an action:

$$\mathbb{T}_m : \quad \mathcal{H}_m \times_k \mathcal{D}_m \longrightarrow \mathcal{D}_m, \tag{1}$$

and showed in [19], 2.2.2 that the set of $\mathbb{T}_m$-orbits is naturally bijective to the set of isomorphism classes of truncated Barsotti-Tate groups of level m over k with codimension c and dimension d. Let $\mathbf{BT}_m(k)$ be the set of isomorphism classes of truncated Barsotti-Tate groups of level m over k with codimension c and dimension d. We have

Lemma 2.1. *For any $u \in \mathbf{BT}_m(k)$, there exists an irreducible catalogue of p-divisible groups with p^m-kernel type u, i.e., there exists a family $\mathcal{X} \to S$ of p-divisible groups such that*

(1) *for any geometric point $s \in S$, the p^m-kernel of the fiber $\mathcal{X}_s$ is of type u;*
(2) *For any p-divisible group X over k with p^m-kernel type u, there exists an $s \in S(k)$ such that $X \simeq \mathcal{X}_s$;*
(3) *S is irreducible and of finite type over k.*

Proof. It suffices to consider the case that u has no étale part, since every (truncated) Barsotti-Tate group over k is the direct sum of its local part and its étale part and the decomposition is compatible with truncations. Let N be an integer $\geq m$ so that $X[p^N] \simeq Y[p^N]$ implies $X \simeq Y$ for any p-divisible groups X and Y over k (cf. [16], 1.7 and [20]). Let π be the natural map $\mathcal{D}_N \to \mathcal{D}_m$. Let $\mathcal{D}$ be the (group) scheme over k such that $\mathcal{D}(R) = \mathrm{GL}(W(R)$ for any k-algebra R, and let τ be a section of $\mathcal{D} \to \mathcal{D}_N$ as a morphism of schemes. Let $\mathbb{O}_u$ be the $\mathbb{T}_m$-orbit associated to u. Since $\mathcal{H}_m$ is irreducible, $\mathbb{O}_u$ is irreducible. Since π is smooth with connected fibers, $\pi^{-1}(\mathbb{O}_u)$ is also irreducible. Let S be the image of $\pi^{-1}(\mathbb{O}_u)$ by τ. Then S is irreducible and of finite type over k. By [23], Theorem 9, we have a p-divisible group $\mathcal{X}$ over S. Clearly $\mathcal{X}$ satisfies the required properties. □

Corollary 2.2. *There exists the supremum of Newton polygons of p-divisible groups with the given p^m-kernel type.*

Proof. Let $\mathcal{X} \to S$ be the family as in the lemma above. Let η be the generic point of S. It follows from Grothendieck and Katz ([8], Th. 2.3.1 on p. 143) that the Newton polygon of $\mathcal{X}_\eta$ is the supremum of Newton polygons of p-divisible groups with a given p^m-kernel type. □

3. Preliminaries on F-zips

In this section, we collect some basic facts on F-zips which we shall use later on.

We first recall the definition ([13], (1.5)) of F-zip in a particular case. Let S be a scheme of characteristic p. Let σ denote the absolute Frobenius on S. For any $\mathcal{O}_S$-module M we write $M^{(p)} = \mathcal{O}_S \otimes_{\sigma,\mathcal{O}_S} M$.

Definition 3.1. An *F-zip* over S is a quintuple $Z = (N, C, D, \varphi, \dot{\varphi})$ consisting of locally free $\mathcal{O}_S$-module N and $\mathcal{O}_S$-submodules C, D of N which are locally direct summands of N, and σ-linear homomorphisms $\varphi : N/C \to D$ and $\dot{\varphi} : C \to N/D$ whose $\mathcal{O}_S$-linearizations $\varphi^\sharp : (N/C)^{(p)} \to D$ and $\dot{\varphi}^\sharp : C^{(p)} \to N/D$ are isomorphisms. If S is connected, we define the *height* of Z to be the rank of N and the *type* of Z to be a map from $\{0, 1\}$ to $\mathbb{Z}_{\geq 0}$ sending 0 to $\mathrm{rk}\, D$ and 1 to $\mathrm{rk}\, C$; we will simply write the type as $(\mathrm{rk}\, D, \mathrm{rk}\, C)$.

If S is the spectrum of a perfect field K, then the (covariant) Dieudonné functor $\mathbb{D}$ makes an equivalence from the category of BT_1's over K to that of F-zips over K. The F-zip $(N, C, D, \varphi, \dot{\varphi})$ associated to a BT_1-group G is given by $N = \mathbb{D}(G)$ with $C = VN$ and $D = FN$, and φ and $\dot{\varphi}^{-1}$ are naturally induced by F and V respectively.

Let k be an algebraically closed field of characteristic p. Let c, d, r and ${}^J W$ be as in Introduction. As shown by Kraft, Oort, Moonen and Wedhorn, there exists a bijection from $\mathbf{BT}_1(k)$ to ${}^{\mathrm{J}}\mathrm{W}$ (this statement is due to [13]; also see [9], [12] and [17]). This classification is based on the fact that for any BT_1-group G over k, there uniquely exists $w \in {}^{\mathrm{J}}\mathrm{W}$ such that G is isomorphic to G_w defined below. To a $w \in {}^{\mathrm{J}}\mathrm{W}$ we associate a pair (B, δ), called a *final type*, cf. [4], Definition 2.6, where B is a totally ordered set $\{b_1 < \ldots < b_r\}$ and δ is a map $B \to \{0, 1\}$ defined by $\delta(b_i) = 1 \Leftrightarrow w(i) \leq d$. There uniquely exists an automorphism $\pi = \pi_\delta$ of B such that $\pi(b') > \pi(b) \Leftrightarrow \delta(b') > \delta(b)$ for any $b' < b$. We define G_w so that its F-zip $Z_w = (N, C, D, \varphi, \dot{\varphi})$ is given by $N = \bigoplus_{b\in B} kb$ (i.e., the k-vector space with basis indexed by B) and $C = \bigoplus_{\delta(b)=1} kb$, and $D = \bigoplus_{\delta(b)=0} k\pi(b)$ and φ, $\dot{\varphi}$ are defined by $\varphi(b) = \pi(b)$ for b with $\delta(b) = 0$ and $\dot{\varphi}(b) = \pi(b)$ for b with $\delta(b) = 1$.

Next we review a description of homomorphisms of F-zips over k for the reader's convenience (cf. [14], §2 and [12], §4 and also see [4], § 4.3), and show some facts used later on. Let Z_1 and Z_2 be two F-zips over k. Let w_1 and w_2 be the types of Z_1 and Z_2 respectively and let $\mathcal{B}_1 = (B_1, \delta_1)$ and $\mathcal{B}_2 = (B_2, \delta_2)$ be their final types. Set $\pi_1 = \pi_{\delta_1}$ and $\pi_2 = \pi_{\delta_2}$. A *finite slice* ω is a subset of $B_1 \times B_2$ of the form $\omega = \{(\pi_1^i(s_1), \pi_2^i(s_2)) \mid 1 \leq i \leq \ell\}$ with $|\omega| = \ell$ for $s_1 \in B_1$ and $s_2 \in B_2$ satisfying **(a)** $\delta_1(s_1) = 1$ and $\delta_2(s_2) = 0$, **(b)** $\delta_1(\pi_1^i(s_1)) = \delta_2(\pi_2^i(s_2))$ for all $1 \leq i < \ell$ and **(c)** $\delta_1(\pi_1^\ell(s_1)) = 0$ and $\delta_2(\pi_2^\ell(s_2)) = 1$. We denote by $\Omega_o = \Omega_o(\mathcal{B}_1, \mathcal{B}_2)$ the set of finite slices of $\mathcal{B}_1$ and $\mathcal{B}_2$. An *infinite slice* ω is a subset of $B_1 \times B_2$ of the form $\omega = \{(\pi_1^i(s_1), \pi_2^i(s_2)) \mid 1 \leq i \leq \ell\}$ with $|\omega| = \ell$ for $s_1 \in B_1$ and $s_2 \in B_2$ satisfying **(a)** $s_1 = \pi_1^\ell(s_1)$ and $s_2 = \pi_2^\ell(s_2)$, **(b)** $\delta_1(\pi_1^i(s_1)) = \delta_2(\pi_2^i(s_2))$ for all $1 \leq i < \ell$. We denote by $\Omega_\infty = \Omega_\infty(\mathcal{B}_1, \mathcal{B}_2)$ the set of infinite slices of $\mathcal{B}_1$ and $\mathcal{B}_2$. Set $\Omega = \Omega(\mathcal{B}_1, \mathcal{B}_2) := \Omega_o \sqcup \Omega_\infty$. For each slice ω, we define a group scheme $\mathbb{K}_\omega$ over k to be the additive group $\mathbb{G}_a$ over k if $\omega \in \Omega_o$ and to be $\mathrm{Ker}(F^{|\omega|} - \mathrm{id} : \mathbb{G}_a \to \mathbb{G}_a)$ if $\omega \in \Omega_\infty$. Let S be a k-scheme. Let $\omega = \{(\pi_1^i(s_1), \pi_2^i(s_2)) \mid 1 \leq i \leq \ell\}$ be a slice with $|\omega| = \ell$. For an element $r \in \omega$, we denote by $\varepsilon(r)$ $(= \varepsilon_\omega(r))$ the integer ε with $0 \leq \varepsilon < \ell$ satisfying $r = (\pi_1^{\varepsilon+1}(s_1), \pi_2^{\varepsilon+1}(s_2))$. For $a \in \mathbb{K}_\omega(S)$, we define a map

$$\mathrm{st}_{\omega,a} : \quad B_1 \times B_2 \longrightarrow \mathbb{K}_\omega(S) \tag{2}$$

by sending $r \in \omega$ to $a^{p^{\varepsilon(r)}}$ and $r \notin \omega$ to 0. The functor, from the category of k-schemes to the category of commutative groups, sending S to $\mathit{Hom}_S(Z_{1,S}, Z_{2,S})$ is represented by a group scheme $\mathrm{Hom}(Z_1, Z_2)$ over k; moreover there is an isomorphism as group schemes over k:

$$\Lambda : \quad \bigoplus_{\omega\in\Omega} \mathbb{K}_\omega \xrightarrow{\ \sim\ } \mathrm{Hom}(Z_1, Z_2). \tag{3}$$

Indeed, for $* = 1, 2$, we write $B_* = \{b_1^{(*)} < \cdots < b_{r_*}^{(*)}\}$ and also write $Z_* = (N_*, C_*, D_*, \varphi_*, \dot{\varphi}_*)$ with $N_* = \bigoplus_{i=1}^{r_*} kb_i^{(*)}$. Let S be any k-scheme. An $\mathcal{O}_S$-homomorphism $\mu : N_{1,S} \to N_{2,S}$, say $\mu(b_i^{(1)}) = \sum_j r_{ij} b_j^{(2)}$ with $r_{ij} \in \Gamma(S, \mathcal{O}_S)$ gives an element of $\mathcal{H}om_S(Z_{1,S}, Z_{2,S})$ if and only if r_{ij} is of the form $\sum_{\omega \in \Omega} \mathrm{st}_{\omega,a}(b_{ij})$ for a certain $a \in \mathbb{K}_\omega(S)$, where $b_{ij} = (b_i^{(1)}, b_j^{(2)}) \in B_1 \times B_2$. From now on we identify $\mathrm{Hom}(Z_1, Z_2)$ with $\bigoplus_{\omega \in \Omega} \mathbb{K}_\omega$. The connected component of zero in a commutative group scheme H will be denoted by H_o. Then $\mathrm{Hom}(Z_1, Z_2)_o$ is the product of $\mathbb{K}_\omega$ for $\omega \in \Omega_o$. We write $\mathrm{Hom}(Z_1, Z_2)_\infty$ for $\bigoplus_{\omega \in \Omega_\infty} \mathbb{K}_\omega$.

It is straightforward to prove

Lemma 3.2. *Let Z_1, Z_2, Z_3 be F-zips over k. The composition map*

$$\mathrm{Hom}(Z_1, Z_2) \times \mathrm{Hom}(Z_2, Z_3) \longrightarrow \mathrm{Hom}(Z_1, Z_3)$$

sends the pair of (ω_1, a_1) and (ω_2, a_2) (i.e., $\omega_i \in \Omega(\mathcal{B}_i, \mathcal{B}_{i+1})$ and $a_i \in \mathbb{K}_{\omega_i}$ for $i = 1, 2$) to $\sum_{\tilde{\omega}} (\omega, a_1^{p^e} a_2^{p^f})$, where the sum is over $\pi_1 \times \pi_2 \times \pi_3$-orbits $\tilde{\omega}$ in $\omega_1 \times_{B_2} \omega_2$ and $\omega = \mathrm{pr}_{13}(\tilde{\omega})$ and e is the minimal element of $\varepsilon_{\omega_1}(\mathrm{pr}_{12}(\tilde{\omega}))$ and f is the minimal element of $\varepsilon_{\omega_2}(\mathrm{pr}_{23}(\tilde{\omega}))$. Here we denote by pr_{ij} the projections $B_1 \times B_2 \times B_3 \to B_i \times B_j$ for $1 \le i < j \le 3$.

The next lemma shows that the ring scheme $\mathrm{End}(Z)_o$ consists of nilpotent endomorphisms.

Lemma 3.3. *Let $\omega \in \Omega_o(\mathcal{B}, \mathcal{B})$. Let (b, b') be an element of ω. Then we have $b > b'$.*

Proof. By the definition of finite slice, $\nu(b) := \sum_{l \in \mathbb{N}} \delta(\pi^{-l}(b)) 2^{-l}$ is greater than $\nu(b') := \sum_{l \in \mathbb{N}} \delta(\pi^{-l}(b')) 2^{-l}$. Then [5], Proposition 4.7 shows $b > b'$. □

For later use, we look at the action of $\mathrm{End}(Z)_o$ on $\mathrm{Hom}(Z, Z_1)$. Let $\Omega^i(\mathcal{B}, \mathcal{B}_1)$ be the subset of $\Omega(\mathcal{B}, \mathcal{B}_1)$ consisting of $\omega \in \Omega(\mathcal{B}, \mathcal{B}_1)$ with $\mathrm{pr}(\omega) \subset \{b_i, \ldots, b_r\}$, where pr is the projection $B \times B_1 \to B$. We define a subgroup scheme $\mathrm{Fil}^i \mathrm{Hom}(Z, Z_1)$ of $\mathrm{Hom}(Z, Z_1)$ by

$$\mathrm{Fil}^i \mathrm{Hom}(Z, Z_1) = \bigoplus_{\omega \in \Omega^i(\mathcal{B}, \mathcal{B}_1)} \mathbb{K}_\omega. \tag{4}$$

From the lemmas above, we have

Corollary 3.4. *The composition map induces*

$$\mathrm{End}(Z)_o \times \mathrm{Fil}^i \mathrm{Hom}(Z, Z_1) \longrightarrow \mathrm{Fil}^{i+1} \mathrm{Hom}(Z, Z_1).$$

At the end of this section, we recall Wedhorn's result ([22]) on specializations of F-zips.

Definition 3.5. Let w and w' be elements of ${}^{\mathrm{J}}\mathrm{W}$. We say $w \subset w'$ if there exists an F-zip over an irreducible scheme of characteristic p such that the generic fiber is of type w' and there exists a closed point such that the fiber over that point is of type w.

Let $x = w_0^{\mathrm{J}} : \mathrm{W} \to \mathrm{W}$ be the map sending i to $i + c$ if $i \le d$ and i to $i - d$ if $i > d$. Define $\delta : \mathrm{W} \to \mathrm{W}$ by $\delta(u) = x \cdot u \cdot x^{-1}$.

Theorem 3.6 ([22]). *Let $w, w' \in {}^{\mathrm{J}}\mathrm{W}$. We have $w \subset w'$ if and only if there exists $u \in \mathrm{W}_{\mathrm{J}}$ such that $u^{-1}w\delta(u)$ is less than or equal to w' with respect to the Bruhat order.*

4. Lifting of F-zips

Let R be a commutative ring of characteristic p. Let ^{F} and ^{V} denote the Frobenius and Verschiebung on $W(R)$. Write $I_R := {}^V W(R)$. Let $M = (P, Q, F, \dot{F})$ be a display over R. One can associate to M an F-zip $M/I_R M$, which is defined as follows. Let $P = L \oplus T$ be a normal decomposition of P with $Q = L \oplus I_R T$ (cf. [23], Introduction). We define $M/I_R M$ to be $(N, C, D, \varphi, \dot{\varphi})$ where $N = P/I_R P$ and $C = Q/I_R P \simeq L/I_R L$, and D is the submodule of N generated by the image of $F : T \to P \to N$, and φ and $\dot{\varphi}$ are canonically induced by F and $\dot{F}$ respectively.

Lemma 4.1. *Let Z be an F-zip over S. Let s be any closed point of S. Let M be a display over s. There is an open affine subscheme $U = \mathrm{Spec}(R)$ of S with $s \in U$ and a display $\mathcal{M}$ over R such that $\mathcal{M}/I_R\mathcal{M} \simeq Z_R$ and $\mathcal{M}_s \simeq M$.*

Proof. Write $Z = (N, C, D, \varphi, \dot{\varphi})$. Let U be an affine open subscheme of S containing s over which C and D are direct summands of N, say $N = C \oplus E$, and C, D and E are free. Write $U = \mathrm{Spec}(R)$ and $s = \mathrm{Spec}(R/\mathfrak{m})$. We replace S by U. We have an ^{F}-linear homomorphism

$$\phi : \quad C \oplus E \xrightarrow{\ \sim\ } C \oplus N/C \xrightarrow{\dot{\varphi} \oplus \varphi} N/D \oplus D \xrightarrow{\ \sim\ } N.$$

Let $\mathcal{L}$ and $\mathcal{T}$ be free $W(R)$-modules such that $\mathcal{L}/I_R\mathcal{L} \simeq C$ and $\mathcal{T}/I_R\mathcal{T} \simeq E$. Put $\mathcal{P} = \mathcal{L} \oplus \mathcal{T}$ and $\mathcal{Q} = \mathcal{L} \oplus I_R\mathcal{T}$. Let $M = (P, Q, F, \dot{F})$ and $L \oplus T$ a normal decomposition of P, and let Φ_0 be $\dot{F} \oplus F : L \oplus T \to P$ obtained in [23], Lemma 9; one can identify L and T with $\mathcal{L}/W(\mathfrak{m})\mathcal{L}$ and $\mathcal{T}/W(\mathfrak{m})\mathcal{T}$ respectively.

Since the canonical map from $\mathrm{GL}_r(W(R))$ to the fiber product of $\mathrm{GL}_r(R) \to \mathrm{GL}_r(R/\mathfrak{m})$ and $\mathrm{GL}_r(W(R/\mathfrak{m})) \to \mathrm{GL}_r(R/\mathfrak{m})$ is clearly surjective, there exists an ^{F}-linear homomorphism

$$\Phi : \quad \mathcal{L} \oplus \mathcal{T} \longrightarrow \mathcal{P}$$

such that $(\Phi \mod I_R) = \phi$ and $(\Phi \mod W(\mathfrak{m})) = \Phi_0$. Set $\mathcal{F} = \Phi \circ ({}^V1 \oplus \mathrm{id}) : \mathcal{L} \oplus \mathcal{T} \to \mathcal{P}$ and define $\dot{\mathcal{F}} : \mathcal{L} \oplus I_R\mathcal{T} \to \mathcal{P}$ by sending $l + {}^Vwt$ to $\Phi(l) + w\Phi(t)$ for every $l \in \mathcal{L}$, $t \in \mathcal{T}$ and $w \in W(R)$. Then we have a display $\mathcal{M} = (\mathcal{P}, \mathcal{Q}, \mathcal{F}, \dot{\mathcal{F}})$, which satisfies the required properties. □

For $v \in {}^J\mathrm{W}$, let $\xi(v)$ be the Newton polygon introduced in §1.

Corollary 4.2. *Let w and w' be elements of ${}^J\mathrm{W}$. If $w \subset w'$, then we have $\xi(w) \prec \xi(w')$.*

Proof. Assume $w \subset w'$, i.e., there exists an F-zip Z over an irreducible scheme S such that the type of the fiber of the generic point η is w' and the type of the fiber of a special point s is w. By the definition of $\xi(w)$, there exists a display M over s such that the Newton polygon of M is $\xi(w)$. Applying Lemma 4.1 to Z and M, there exist an open affine subscheme $U = \mathrm{Spec}(R)$ of S containing s and a display $\mathcal{M}$ over R such that $\mathcal{M}/I_R\mathcal{M} \simeq Z_R$ and $\mathcal{M}_s \simeq M$. It follows from Grothendieck-Katz ([8], Th. 2.3.1 on p. 143) that $\xi(w)$ is less than or equal to the Newton polygon, say ζ, of $\mathcal{M}_\eta$. By the definition of $\xi(w')$, we have $\zeta \prec \xi(w')$. □

5. A reduction of the problem

Let w be any element of ${}^J\mathrm{W}$. The purpose of this paper is to prove Theorem 1.1:

$$\xi(w) = \max_{\prec}\{\zeta \mid \mu(\zeta) \subset w\}, \tag{5}$$

where ζ is over Newton polygons $\sum(m_i, n_i)$ with $\sum m_i = d$ and $\sum n_i = c$ (see §1 for the definitions of $\xi(w)$ and $\mu(\zeta)$).

Remark 5.1. This gives, thanks to Theorem 3.6, a purely combinatorial algorithm determining $\xi(w)$ for a given w. See [5], Corollary 4.8 for a way to compute $\mu(\zeta)$.

We first prove that Theorem 1.1 follows from the next proposition. The subsequent sections are devoted to the proof of this proposition.

Proposition 5.2. *Assume that w is not minimal. Then there exist a scheme S of finite type over k with $\dim S \geq 1$, a p-divisible group $\mathcal{X}$ over S and a non-constant family of isogenies*

$$H(\xi(w))_S \longrightarrow \mathcal{X} \tag{6}$$

over S such that the isomorphism type of $\mathcal{X}_s[p]$ is w for every geometric point s of S.

Proof of (Proposition 5.2 $\Rightarrow$ Theorem 1.1). We first claim that Theorem 1.1 is equivalent to

$$\mu(\xi(w)) \subset w. \tag{7}$$

Clearly Theorem 1.1 implies (7). Suppose (7). Put $\Xi = \{\zeta \mid \mu(\zeta) \subset w\}$. We want to show that $\xi(w)$ is the biggest element of Ξ. Clearly (7) says $\xi(w) \in \Xi$. Let ζ be any element of Ξ. Then we have $\xi(\mu(\zeta)) \prec \xi(w)$ by Corollary 4.2. Note that we have $\xi(\mu(\zeta)) = \zeta$ by [15], (1.2) Theorem. Thus we have $\zeta \prec \xi(w)$.

Let us prove (7) under the assumption that Proposition 5.2 holds. We first consider the case that w is minimal, say $w = \mu(\zeta)$ the type of $H(\zeta)[p]$. Then we have $\xi(w) = \zeta$ by [15] (1.2), and therefore we have $\mu(\xi(w)) = w$; hence (7) holds in this case. Assume that w is not minimal. Let $\mathcal{M}$ be the moduli space of quasi-isogenies $H(\xi(w)) \to Y$ of p-divisible groups, see [18], Chapter 2. Let $\mathcal{I}$ be an irreducible component of $\mathcal{M}_{\mathrm{red}}$ containing the generic point of the family (6). Note that $\mathcal{I}$ is projective ([18], Proposition 2.32). Let $\mathcal{S}_w(\mathcal{I})$ be the locally closed subvariety consisting of isogenies $H(\xi(w)) \to X$ where $X[p]$ is of type w. It is known that $\mathcal{S}_w(\mathcal{I})$ is quasi-affine (cf. [19], 1.2 (g)). By the assumption, we have $\dim \mathcal{S}_w(\mathcal{I}) \geq 1$. Hence there exists $w' \in {}^{\mathrm{J}}\mathrm{W}$ such that $w' \subsetneq w$ and $\xi(w') = \xi(w)$. This shows in particular that any final type v with $\xi(v) = \xi(w)$ which is "minimal w.r.t. $\subset$" is minimal. We use induction on w with respect to $\subset$. We assume $\mu(\xi(w')) \subset w'$. Then we have $\mu(\xi(w)) = \mu(\xi(w')) \subset w' \subset w$. □

6. Extensions by a minimal p-divisible group

Let ξ be a Newton polygon. Let $\varrho = (m_1, n_1)$ be a segment of ξ, i.e., $m_1/(m_1+n_1)$ is a slope of ξ and $\gcd(m_1, n_1) = 1$. Let ξ' be the Newton polygon such that $\xi = \xi' + \varrho$. Let $Z_1 = Z(\varrho)_S$, where $Z(\varrho)$ is the F-zip of $H(\varrho)[p]$ over k. Let

$$0 \longrightarrow Z' \longrightarrow Z \xrightarrow{\ f\ } Z_1 \longrightarrow 0 \tag{8}$$

be a short exact sequence of F-zips over a reduced k-scheme S. Write $Z = (N, C, D, \varphi, \dot{\varphi})$ and $Z_1 = (N_1, C_1, D_1, \varphi_1, \dot{\varphi}_1)$ and so on. That f is surjective means that $f : N \to N_1$ and $f : C \to C_1$ are surjective, and also the injectivity is the dual notion of this surjectivity. Let $M' = (P', Q', F', \dot{F}')$ be any display lifting Z' with an isogeny $\rho' : M(\xi')_S \to M'$, where $M(\xi')$ is the display of $H(\xi')$. Let $P' = L' \oplus T'$ be a normal decomposition.

Proposition 6.1. *For any closed point $s \in S$, there exist an open affine subscheme U of S with $s \in U$ and a finite surjective morphism* $\mathrm{Spec}(R) \to U$ *such that there exist a display $\mathcal{M}$ over R with an isogeny $\rho : M(\xi)_R \to \mathcal{M}$ and a surjective homomorphism $\phi : \mathcal{M} \to (M_1)_R := M(\varrho)_R$ with kernel M'_R and an isomorphism $\theta : \mathcal{M}/I_R\mathcal{M} \to Z_R$ such that we have the commutative diagrams*

$$\begin{array}{ccccccccc} 0 & \longrightarrow & M(\xi')_R & \longrightarrow & M(\xi)_R & \xrightarrow{\mathrm{pr}} & M(\varrho)_R & \longrightarrow & 0 \\ & & {\scriptstyle\rho'}\big\downarrow & & {\scriptstyle\rho}\big\downarrow & & \big\| & & \\ 0 & \longrightarrow & M'_R & \longrightarrow & \mathcal{M} & \xrightarrow{\phi} & (M_1)_R & \longrightarrow & 0 \end{array} \tag{9}$$

and

$$\begin{array}{ccccccccc} 0 & \longrightarrow & M'_R/I_RM'_R & \longrightarrow & \mathcal{M}/I_R\mathcal{M} & \xrightarrow{\overline{\phi}} & (M_1)_R/I_R(M_1)_R & \longrightarrow & 0 \\ & & \big\| & & {\scriptstyle\theta}\big\downarrow{\scriptstyle\simeq} & & \big\| & & \\ 0 & \longrightarrow & Z'_R & \longrightarrow & Z_R & \xrightarrow{f_R} & (Z_1)_R & \longrightarrow & 0. \end{array} \tag{10}$$

Proof. Let $u = \min\{m_1, n_1\}$. Recall [4], Lemma 3.3 that the Dieudonné module $M(\varrho)$ is generated over the Dieudonné ring by X_i ($i \in \mathbb{Z}/u\mathbb{Z}$) and all relations are generated by $F^{\alpha_i}X_i - V^{\beta_{i+1}}X_{i+1} = 0$ for some non-negative integers α_i, β_i. Put $x_i := V^{\beta_i}X_i$. Set $\beta = \max\{\beta_i\}$.

Let s be any closed point of S. Let $U = \mathrm{Spec}(R)$ be an affine open subscheme of S containing s. We may replace S by U. We choose a lift $\overline{Y}_i \in N$ of $\overline{X}_i \in N_1$ for each $i \in \mathbb{Z}/u\mathbb{Z}$. Let ψ be the composition of $N \to N/D$ and $(\dot{\varphi}^\sharp)^{-1} : N/D \to C^{(p)}$. After replacing R with R' such that $(R')^{p^\beta} = R$, we can find $\overline{y}_{i,j} \in C$ lifting $V^j\overline{X}_i$ for $0 \leq j \leq \beta_i$ such that the composition $\psi^{(p^{j-1})} \circ \cdots \circ \psi^{(p)} \circ \psi$ sends $\overline{Y}_i$ to $1 \otimes \overline{y}_{i,j} \in R \otimes_{F^j, R} C$. We put $\overline{y}_i := \overline{y}_{i,\beta_i}$. After replacing R by an open affine subscheme over which N is free, we can find a section of $N \to N/D$, defining a lift $\tilde{\varphi} : C \to N$ of $\dot{\varphi} : C \to N/D$, such that $\tilde{\varphi}^{\beta_i - j}(\overline{y}_i) = \overline{y}_{i,j}$. It follows from the exact sequence (8) that N is generated by elements of N' and $\tilde{\varphi}^s\overline{y}_i$ ($0 \leq s < \beta_i$)

and $\varphi^r\tilde{\varphi}^{\beta_i}\overline{y}_i$ $(0 \leq r < \alpha_i)$ with relations

$$\varphi^{\alpha_i}\tilde{\varphi}^{\beta_i}\overline{y}_i - \overline{y}_{i+1} = \overline{v}_i \tag{11}$$

for some $\overline{v}_i \in N'$, where C is generated over $W(R)$ by elements of C' and $\tilde{\varphi}^s\overline{y}_i$ $(0 \leq s < \beta_i)$, and D is generated over $W(R)$ by elements of D' and $\varphi^r\tilde{\varphi}^{\beta_i}\overline{y}_i$ $(1 \leq r \leq \alpha_i)$.

Put $W_{\mathbb{Q}}(R) = \mathbb{Q} \otimes W(R)$. Write $\xi' = \sum_{l=2}^{t}(m_l, n_l)$. Then the isogeny ρ' induces an isomorphism

$$W_{\mathbb{Q}}(R) \otimes M' \xrightarrow{\ \sim\ } \bigoplus_{l=2}^{t} W_{\mathbb{Q}}(R) \otimes M((m_l, n_l)). \tag{12}$$

Let $e_l \in W_{\mathbb{Q}}(R) \otimes M'$ be the highest element of $M((m_l, n_l))$, see [4], Section 3.1. Recall that the ring of endomorphisms of H_{m_l,n_l} is described as $E_l := W(\mathbb{F}_{p^{m_l+n_l}})[\theta_l]/(\theta_l^{m_l+n_l} - p)$ for a uniformizer θ_l of $\mathrm{End}(H_{m_l,n_l})$. Let $E_l(R)$ be the $W(R)$-module $W(R) \otimes E_l$ and set $E_{l,\mathbb{Q}}(R) := \mathbb{Q} \otimes E_l(R)$. We extend the action of the Frobenius σ on $W(R)$ to that on $E_{l,\mathbb{Q}}(R)$ by the rule $\theta_l^{\sigma} = \theta_l$. Note the $W_{\mathbb{Q}}(R)$-homomorphism

$$E_{l,\mathbb{Q}}(R) \longrightarrow W_{\mathbb{Q}}(R) \otimes M((m_l, n_l)) \tag{13}$$

defined by sending $f(\theta_l)$ to $f(\theta_l)e_l$ is an isomorphism.

We have to define $\mathcal{M} = (\mathcal{P}, \mathcal{Q}, \mathcal{F}, \dot{\mathcal{F}})$. Note that $\mathcal{P}$ should have a normal decomposition $\mathcal{L} \oplus \mathcal{T}$. We will define $\mathcal{L}$ to be the $W(R)$-submodule of $W_{\mathbb{Q}}(R) \otimes (P_1 \oplus P')$ generated by elements of L'_R and $F^r\dot{F}^{\beta_i}y_i$ $(0 \leq r < \alpha_i)$ and $\mathcal{T}$ will be defined to be the $W(R)$-submodule of $(P_1 \oplus P') \otimes W_{\mathbb{Q}}(R)$ generated by elements of T'_R and $\dot{F}^s y_i$ $(0 \leq s < \beta_i)$ for $i = 1, 2, \ldots, n$ where $y_i \in (P_1 \oplus P') \otimes W_{\mathbb{Q}}(R)$ is of the form:

$$y_i = x_i + \sum_{l=2}^{t} a_{il}e_l \tag{14}$$

for some $a_{il} \in E_{l,\mathbb{Q}}(R)$, which will be chosen later so that $\mathcal{M}$ has the required properties. Here $\mathcal{M}$ is defined by $\mathcal{P} = \mathcal{L} \oplus \mathcal{T}$ and $\mathcal{Q} = \mathcal{L} \oplus I_R\mathcal{T}$ with $\mathcal{F}$ and $\dot{\mathcal{F}}$ naturally extending F and $\dot{F}$ on $M(\xi)_R$. Since M'_R contains $M(\xi')_R$, it suffices to find $a_{il} \in E_{l,\mathbb{Q}}(R)$ modulo $I_{R,\mu}E_l(R)$ for a sufficiently large natural number μ $(\geq \max\{m_i\beta_i; i \in \mathbb{Z}/u\mathbb{Z}\})$.

Let $v_i \in P'$ be a lift of $\overline{v}_i$. We define $b_{il} \in E_{l,\mathbb{Q}}(R)$ by $\sum_{l=2}^{t} b_{il}e_l = v_i$. It suffices to show that there exists a solution $\{a_{il}\}$ $(i \in \mathbb{Z}/u\mathbb{Z},\ 2 \leq l \leq t)$ satisfying

$$F^{\alpha_i}\dot{F}^{\beta_i}y_i - y_{i+1} \equiv \sum_{l=2}^{t} b_{il}e_l \qquad (\mathrm{mod}\ I_{R,\mu}M(\xi')_R). \tag{15}$$

Comparing the coefficients of e_l of the both sides of (15), we obtain

$$a_{i,l}^{\sigma^{\alpha_i+\beta_i}}\theta_l^{n_l\alpha_i-m_l\beta_i}-a_{i+1,l}\equiv b_{il}\qquad(\mathrm{mod}\ I_{R,\mu}E_l(R))\tag{16}$$

for $i\in\mathbb{Z}/u\mathbb{Z}$. Since l is the same in each equation, it suffices to solve the simultaneous equations for each l. Writing a_i, b_i, n, m and θ for a_{il}, b_{il}, n_l, m_l and θ_l respectively, we have

$$a_1^{\sigma^{\sum_{i=1}^u(\alpha_i+\beta_i)}}\theta^{\sum_{i=1}^u(n\alpha_i-m\beta_i)}-a_1\equiv r\qquad(\mathrm{mod}\ I_{R,\mu}E_l(R))\tag{17}$$

for some $r\in E_{l,\mathbb{Q}}(R)$. It suffices to show that this has a solution $a_1\in E_{l,\mathbb{Q}}(R)$ for a finite cover $\mathrm{Spec}(R)$ of S; then we get a required solution $\{a_i\}_{i=1}^u$ from (16).

Write $z:=a_1$ and $\varrho:=\sigma^{\sum_{i=1}^u(\alpha_i+\beta_i)}$. Note $\varrho\neq1$ by $\alpha_i,\beta_i>0$. We also put $\epsilon:=\sum_{i=1}^u(n\alpha_i-m\beta_i)$. Then (17) is written as $z^\varrho\theta^\epsilon-z\equiv r\ (\mathrm{mod}\ I_{R,\mu}E_l(R))$. If $\epsilon>0$, we have a solution $z=\sum_{\ell=0}^\infty(-r)^{\varrho^\ell}\theta^{\ell\epsilon}$. Also if $\epsilon<0$, let c be a sufficient large integer such that $\theta^{-c\epsilon}\in I_{R,\mu}E_l(R)$, and we replace R by R' so that $(R')^{p^c}=R$; then we have a solution $z=\sum_{\ell=1}^{c-1}r^{\varrho^{-\ell}}\theta^{-\ell\epsilon}$. Finally we consider the case $\epsilon=0$. Write $z=\sum_{i=0}^{m+n-1}z_i\theta^i$ and $r=\sum_{i=0}^{m+n-1}r_i\theta^i$ with $z_i,r_i\in W_{\mathbb{Q}}(R)$. It suffices to solve $z_i^\varrho-z_i\equiv r_i\ (\mathrm{mod}\ I_{R,\mu})$ for each $0\leq i<m+n$. Let ν_i be the biggest non-negative integer ν such that $r_i\in p^\nu W(R)$. We replace R with R' so that we have $(R')^{p^\nu}=R$. Then there exist elements t_j of R for all integers $j\geq\nu_i$ such that $z_i=\sum_{j=\nu_i}^\infty{}^{V^j}[t_j]$ is a solution. Indeed, putting $z_{ij}:=\sum_{j'<j}{}^{V^{j'}}[t_{j'}]$, we can find $t_{j'}$ successively so that $z_{ij}^\varrho-z_{ij}\equiv r_i\ (\mathrm{mod}\,I_{R,j})$. Let $j\geq\nu_i$ and suppose that we have already got such $t_{j'}$ for $j'<j$. Since $\varrho\neq1$, after replacing R by a finite cover R', there exists a solution $t_j\in R$ of the Artin-Schreier equation $t_j^\varrho-t_j=({}^{V^{-j}}(r_i-z_{ij}^\varrho+z_{ij})\ \mathrm{mod}\,I_R)$. Then clearly $z_i:=\sum_{j=\nu_i}^{\mu-1}{}^{V^j}[t_j]$ is a solution of $z_i^\varrho-z_i\equiv r_i\ (\mathrm{mod}\ I_{R,\mu})$. □

7. Proof of Proposition 5.2

Let $w\in{}^J\mathrm{W}$. Let (m_1,n_1) be the first segment of $\xi(w)$. By the definition of $\xi(w)$, there exists a p-divisible group X over an algebraically closed field k of characteristic p such that $X[p]$ is of type w and the Newton polygon of X is $\xi(w)$. Write $M=\mathbb{D}(X)$. Choose an embedding $\imath:M\to M(\xi(w))$ and let $\jmath:M(\xi(w))\to M_{m_1,n_1}$ be the natural projection. Put $M_1=\jmath\circ\imath(M)$. Let $f_0:X\to X_1$ be the homomorphism of p-divisible groups corresponding to $M\to M_1$. Let X_0' be the kernel of f_0. Note X_0' is a p-divisible group. Thus we have an exact sequence of p-divisible groups

$$0\longrightarrow X_0'\longrightarrow X\xrightarrow{f_0}X_1\longrightarrow0.\tag{18}$$

Lemma 7.1. *X_1 is minimal, i.e., $X_1 \simeq H_{m_1,n_1}$.*

Proof. Recall [4], Corollary 5.4, whose dual is as follows. Let λ_v be the optimal lower bound of the first Newton slopes of p-divisible groups with p-kernel type v for each $v \in {}^{\mathrm{J}}\mathrm{W}$; then we have

$$\lambda_v = \min\{m/(m+n) \mid G_{v,\Omega} \overset{\exists}{\twoheadrightarrow} H_{m,n}[p]_\Omega \text{ for some alg. closed field } \Omega\}. \tag{19}$$

Note that λ_v is equal to the first slope of $\xi(v)$.

Let w and w_1 be the final types of $X[p]$ and $X_1[p]$ respectively. Since $X[p] \twoheadrightarrow X_1[p]$, i.e., $G_{w,k} \twoheadrightarrow G_{w_1,k}$, we have $\lambda_w \leq \lambda_{w_1}$ by (19). By the construction of X_1, the (first) Newton slope of X_1 is λ_w; hence we have $\lambda_w \geq \lambda_{w_1}$. Thus $\lambda_w = \lambda_{w_1}$. Then (19) implies that there exists a surjective homomorphism $H_{m_1,n_1}[p]_\Omega \twoheadrightarrow G_{w_1,\Omega}$ for some $\Omega = \overline{\Omega}$. This is an isomorphism, since $H_{m_1,n_1}[p]$ and G_{w_1} have the same rank $(= m_1 + n_1)$. □

We use induction on the rank of w to prove Proposition 5.2. Assume that w is not minimal. It suffices to show the case that

(*) $\quad G_w$ has no direct factor which is isomorphic to $H_{m_1,n_1}[p]$.

Indeed if $G_w = G_v \oplus H_{m_1,n_1}[p]$, then v is not minimal and our problem can be reduced to the case v. Hence we assume (*) from now on. Let Z and Z_1 be F-zips of $X[p]$ and $X_1[p]$ respectively. Let $\mathcal{B}$ and $\mathcal{B}_1$ be the final types of Z and Z_1 respectively. Now (*) implies that $\Omega_\infty(\mathcal{B},\mathcal{B}_1) = \emptyset$. Consider the space $\Sigma := \mathrm{Hom}(Z, Z_1)$, which is isomorphic to $\prod_{\omega\in\Omega_o(\mathcal{B},\mathcal{B}_1)} \mathbb{K}_\omega$; hence Σ is irreducible. Let f be the universal homomorphism $Z_\Sigma \to (Z_1)_\Sigma$.

Lemma 7.2. *Let $T \to \Sigma$ be any dominant morphism of k-schemes. Then f_T is not "constant up to $\mathrm{Aut}(Z_T)$". Here we say that f_T is constant up to $\mathrm{Aut}(Z_T)$ if there exists a section $x = \mathrm{Spec}(k) \to T$ such that $f_T = (f_x)_T \circ \kappa$ for an automorphism κ of Z_T.*

Proof. Let $x = \mathrm{Spec}(k) \to T$ be any section and κ any automorphism of Z_T. Let i be the largest integer such that $\mathrm{Fil}^i \mathrm{Hom}(Z, Z_1) = \mathrm{Hom}(Z, Z_1)$. Since $\Omega(\mathcal{B},\mathcal{B}_1)$ consists of finite slices, we have $\dim \mathrm{Fil}^{i+1} \mathrm{Hom}(Z, Z_1) < \dim \mathrm{Hom}(Z, Z_1)$. We write $\kappa = \kappa_o + \kappa_\infty$ with $\kappa_o \in \mathrm{End}(Z)_o(T)$ and $\kappa_\infty \in \mathrm{End}(Z)_\infty(T)$. It follows from Corollary 3.4 that $(f_x)_T \circ \kappa$ is in $\mathrm{Fil}^{i+1} \mathrm{Hom}(Z, Z_1)(T) + (f_x)_T \circ \kappa_\infty$. Since $\mathrm{End}(Z)_\infty$ is discrete, κ_∞ factors through $\mathrm{Spec}(k)$. Hence the dimension of the scheme-theoretic image of $(f_x)_T \circ \kappa : T \to \mathrm{Hom}(Z, Z_1)$ is less than or equal to $\dim \mathrm{Fil}^{i+1} \mathrm{Hom}(Z, Z_1)$.

On the other hand, the morphism $f_T : T \to \mathrm{Hom}(Z, Z_1)$ is dominant. Hence we have $f_T \neq (f_x)_T \circ \kappa$. □

Let η denote the generic point of Σ and let w' be the type of the kernel of f_η. Let U be the open subvariety of Σ consisting of $u \in U$ such that f_u is surjective and the kernel of f_u is of type w'. Choose a finite surjective morphism $S \to U$ which trivializes $Z' := \ker \rho_U$, i.e., $Z'_S \simeq (Z_{w'})_S$.

$$0 \longrightarrow (Z_{w'})_S \longrightarrow Z_S \xrightarrow{f_S} (Z_1)_S \longrightarrow 0. \tag{20}$$

By Corollary 2.2, there exists a display M' over k such that $M'/IM' \simeq Z_{w'}$ and the Newton polygon of M' is $\xi(w')$. Choose an isogeny $M(\xi(w')) \to M'$. Put

$$\zeta := \xi(w') + (m_1, n_1). \tag{21}$$

Applying the result of §6 to (20), for a finite surjective morphism $\mathrm{Spec}(R) \to S$, we obtain an isogeny

$$\rho : \quad M(\zeta)_R \longrightarrow \mathcal{M} \tag{22}$$

with $\phi : \mathcal{M} \to (M_1)_R$ and $\theta : \mathcal{M}/I_R\mathcal{M} \simeq Z_R$ satisfying the commutative diagrams (9) and (10).

Lemma 7.3. *We have $\zeta = \xi(w)$.*

Proof. Since $\mathcal{M}$ has Newton polygon ζ and p-kernel type w, we have $\zeta \prec \xi(w)$ by the definition of $\xi(w)$. Let w'_0 be the type of $X'_0[p]$. Since $w'_0 \subset w'$, we have $\xi(w'_0) \prec \xi(w')$ by Corollary 4.2. Note the Newton polygon $\xi(w)$ of X is equal to $\xi(w'_0)+(m_1, n_1)$. This is less than or equal to $\xi(w')+(m_1, n_1) = \zeta$. □

It remains to show

Lemma 7.4. *ρ is not constant.*

Proof. Applying Lemma 7.2 to $T := \mathrm{Spec}(R) \to \Sigma$, we have that f_R is not constant up to $\mathrm{Aut}(Z_R)$. It follows from the diagram (10) that $\overline{\phi}$ is non-constant; hence so is ϕ. Then we have the lemma by the diagram (9). □

References

1. N. Bourbaki: *Lie groups and Lie algebras. Chapters 4–6.* Translated from the 1968 French original by Andrew Pressley. Elements of Mathematics (Berlin). Springer-Verlag, Berlin, 2002.

2. T. Ekedahl and G. van der Geer: *Cycle classes of the E-O stratification on the moduli of abelian varieties.* Algebra, arithmetic and geometry: in honor of Yu. I. Manin. Vol. I, Progr. Math., **269**, Birkhäuser, Boston, 2009; pp. 567–636.
3. S. Harashita: *Ekedahl-Oort strata and the first Newton slope strata.* J. Algebraic Geom. **16** (2007), no. 1, 171–199.
4. S. Harashita: *Successive extensions of minimal p-divisible groups.* J. Pure Appl. Algebra **213** (2009), 1823-1834.
5. S. Harashita: *Configuration of the central streams in the moduli of abelian varieties.* Asian J. Math. **13** (2009), no. 2, 215–250.
6. S. Harashita: *Generic Newton polygons of Ekedahl-Oort strata: Oort's conjecture.* Ann. Inst. Fourier (Grenoble) **60** (2010), no. 5, 1787–1830.
7. X. He: *The G-stable pieces of the wonderful compactification.* Trans. Amer. Math. Soc. **359** (2007), no. 7, 3005–3024.
8. N. M. Katz: *Slope filtration of F-crystals.* Géom. Alg. Rennes, Vol. I, Astérisque **63** (1979), Soc. Math. France, pp.113-164.
9. H. Kraft: *Kommutative algebraische p-Gruppen (mit Anwendungen auf p-divisible Gruppen und abelsche Varietäten).* Sonderforschungsbereich. Bonn, September 1975. Ms. 86 pp.
10. E. Lau: *Displays and formal p-divisible groups.* Invent. Math. **171** (2008), 617–628.
11. Yu. I. Manin, *Theory of commutative formal groups over fields of finite characteristic.* Uspehi Mat. Nauk **18** no. 6 (114) (1963), pp. 3-90.
12. B. Moonen: *Group schemes with additional structures and Weyl group cosets.* Moduli of abelian varieties (Ed. C. Faber, G. van der Geer, F. Oort), Progr. Math., **195**, Birkhäuser, Basel, 2001; pp. 255–298.
13. B. Moonen and T. Wedhorn: *Discrete invariants of varieties in positive characteristic.* Int. Math. Res. Not. 2004, no. **72**, 3855–3903.
14. F. Oort: *Simple p-kernels of p-divisible groups.* Adv. Math. **198** (2005), no. 1, 275–310.
15. F. Oort: *Minimal p-divisible groups.* Ann. of Math. (2) **161** (2005), no. 2, 1021–1036.
16. F. Oort: *Foliations in moduli spaces of abelian varieties.* J. Amer. Math. Soc. **17** (2004), no. 2, 267–296.
17. F. Oort: *A stratification of a moduli space of abelian varieties.* Progr. Math., **195**, pp. 345–416, Birkhäuser, Basel, 2001.
18. M. Rapoport and Th. Zink: *Period spaces for p-divisible groups.* Annals of Mathematics Studies **141** (1996), Princeton University Press.
19. A. Vasiu: *Level m stratifications of versal deformations of p-divisible groups.* J. Algebraic Geom. **17** (2008), no. 4, 599–641.
20. A. Vasiu: *Reconstructing p-divisible groups from their truncations of small level.* Comment. Math. Helv. **85** (2010), no. 1, 165–202.
21. E. Viehmann: *Truncations of level 1 of elements in the loop group of a reductive group.* arXiv:0907.2331v1
22. T. Wedhorn: *Specialization of F-zips.* arXiv: 0507175v1
23. Th. Zink: *The display of a formal p-divisible group.* Cohomologies p-adiques et applications arithmétiques, I. Astérisque No. **278** (2002), 127–248.

Borcherds lifts on $\mathrm{Sp}_2(\mathbb{Z})$

Bernhard Heim

German University of Technology in Oman,
Way No. 36, Building No. 331, North Ghubrah, Muscat, Sultanate of Oman
E-mail: bernhard.heim@gutech.edu.om

Atsushi Murase

Faculty of Science, Kyoto Sangyo University,
Motoyama, Kamigamo, Kita-ku, 603-8555 Kyoto, Japan
E-mail: murase@cc.kyoto-su.ac.jp

Dedicated to Professor Takayuki Oda on the occasion of his 60th birthday

In this paper, we give several necessary conditions for a holomorphic Siegel modular form on $\mathrm{Sp}_2(\mathbb{Z})$ to be obtained as a Borcherds lift. As an application, we show that Siegel Eisenstein series are not Borcherds lifts. We also give a condition satisfied by the weight of a Borcherds lift.

Keywords: Siegel modular forms; Borcherds lifts

1. Introduction and the main results

1.1. *Introduction*

The lifting of elliptic modular forms to Siegel modular forms of degree two was, suggested by a conjecture by Saito and Kurokawa, introduced by a pioneering work [20] by Maass (for an overview of this theory, see Zagier's article [26]). Using theta liftings, Oda [22] and Rallis-Schiffmann [23], independently, constructed a lifting of elliptic modular forms to automorphic forms on $O(2, m)$ generalizing Maass's lifting as well as Shimura correspondence ([24]). Later a *multiplicative* analogue of the Oda-Rallis-Schiffmann lifting was discovered by Borcherds ([2,3]). In our previous paper [17], we showed that Borcherds lifts satisfy certain symmetries. The object of the paper is to study Borcherds lifts on $\Gamma_2 := \mathrm{Sp}_2(\mathbb{Z})$ (namely the case of $O(2, 3)$) by using these symmetries.

A Borcherds lift on Γ_2 is a meromorphic automorphic form F on Γ_2 (with a multiplier system of finite order) whose divisor is of the form $\sum_d A(d)H_d$, where d runs over the positive integers congruent to 0 or 1 modulo 4, $A(d) \in \mathbb{Z}$ ($A(d) = 0$ except for a finite number of d) and H_d is the Humbert surface of discriminant d (for the precise definition of H_d, see **2.2**; for Borcherds lifts, we refer to [2,3,5]; see also [1,15]). Since every Borcherds lift is a quotient of two holomorphic Borcherds lifts, we are mainly concerned with holomorphic Borcherds lifts in this paper.

From the fundamental results of Borcherds ([2–4]), it follows that the weight of F is expressed in terms of $A(d)$ and Cohen numbers ([7]). By using this weight formula and certain properties of Cohen numbers, we determine the holomorphic Borcherds lifts of small weights (Proposition 4.4).

To discuss the arbitrary weight case, we employ our previous result on the multiplicative symmetries for Borcherds lifts ([17]). We first characterize the powers of the modular discriminant Δ in terms of multiplicative Hecke operators (Proposition 3.1). We then show that the image of a holomorphic Borcherds lift F on Γ_2 under the Siegel operator is proportional to a power of Δ by using the multiplicative symmetry for F and the above characterization. This immediately implies that the Siegel Eisenstein series are never Borcherds lifts. We also show that a holomorphic Borcherds lift on Γ_2 with trivial character is proportional to $\chi_{10}^a \chi_{35}^b F'$, where χ_{10} and χ_{35} are Borcherds lifts of weight 10 and 35 respectively, $a \in \mathbb{Z}_{\geq 0}, b \in \{0, 1\}$ and F' is a Borcherds lift of weight divisible by 12 such that the image of F' under the Witt operator is nonzero (Corollary 1.3).

1.2. *Siegel modular forms*

To explain our results more precisely, let

$$\mathrm{Sp}_n(\mathbb{R}) := \left\{ g \in \mathrm{GL}_{2n}(\mathbb{R}) \mid {}^t g \begin{pmatrix} 0_n & 1_n \\ -1_n & 0_n \end{pmatrix} g = \begin{pmatrix} 0_n & 1_n \\ -1_n & 0_n \end{pmatrix} \right\}$$

be the symplectic group of degree n, where 0_n (respectively 1_n) denotes the zero (respectively identity) matrix of degree n. Let $\Gamma_n := \mathrm{Sp}_n(\mathbb{R}) \cap \mathrm{GL}_{2n}(\mathbb{Z})$ be the Siegel modular group of degree n and $\mathfrak{H}_n := \{Z \in \mathrm{M}_n(\mathbb{C}) \mid {}^t Z = Z, \mathrm{Im}(Z) > 0\}$ the Siegel upper half space of degree n,

Let $M_k(\Gamma_n)$ be the space of holomorphic automorphic forms of weight k on Γ_n. By definition, $F \in M_k(\Gamma_n)$ is a holomorphic function on $\mathfrak{H}_n$ such that $F(\gamma\langle Z \rangle) = j(\gamma, Z)^k F(Z)$ holds for any $\gamma \in \Gamma_n$ and $Z \in \mathfrak{H}_n$, and that

F is holomorphic at ∞ if $n = 1$. Here

$$\gamma\langle Z\rangle := (AZ+B)(CZ+D)^{-1},\ j(\gamma, Z) := \det(CZ+D)$$
$$\left(\gamma = \begin{pmatrix} A & B \\ C & D \end{pmatrix} \in \Gamma_n, Z \in \mathfrak{H}_n\right).$$

In what follows, we are mainly concerned with the case of $n = 2$ and often write (τ_1, z, τ_2) for a point

$$\begin{pmatrix} \tau_1 & z \\ z & \tau_2 \end{pmatrix} \in \mathfrak{H}_2.$$

For $F \in M_k(\Gamma_2)$, we put

$$\Phi(F)(\tau) := \lim_{y\to\infty} F(\tau, 0, iy) \qquad (\tau \in \mathfrak{H}_1),$$
$$\mathcal{W}(F)(\tau_1, \tau_2) := F(\tau_1, 0, \tau_2) \qquad (\tau_1, \tau_2 \in \mathfrak{H}_1).$$

Then $\Phi(F) \in M_k(\Gamma_1)$ and $\mathcal{W}(F) \in \mathrm{Sym}^2(M_k(\Gamma_1))$, where $\mathrm{Sym}^2(M_k(\Gamma_1))$ is the space of functions on $\mathfrak{H}_1 \times \mathfrak{H}_1$ spanned by $f(\tau_1)f'(\tau_2) + f'(\tau_1)f(\tau_2)$ with $f, f' \in M_k(\Gamma_1)$. The operator Φ (respectively $\mathcal{W}$) is called the *Siegel* (respectively *Witt*) operator. Then $S_k(\Gamma_2) := \{F \in M_k(\Gamma_2) \mid \Phi(F) = 0\}$ is the space of cusp forms. Note that $\mathcal{W}(F) \neq 0$ if $\Phi(F) \neq 0$. A Siegel modular form $F \in M_k(\Gamma_2)$ admits the Fourier expansion

$$F(\tau_1, z, \tau_2) = \sum_{n,r,m\in\mathbb{Z}} A_F(n, r, m)\, \mathbf{e}(n\tau_1 + rz + m\tau_2),$$

where we put $\mathbf{e}(z) := \exp(2\pi i z)$ for $z \in \mathbb{C}$. Note that $A_F(n, r, m) = 0$ unless $n, m, 4nm - r^2 \geq 0$.

For $k \in 2\mathbb{Z}$ with $k \geq 4$, we define the Siegel Eisenstein series on Γ_2 of weight k by

$$E_k(Z) := \sum_{\gamma\in\Gamma_2^\infty\backslash\Gamma_2} j(\gamma, Z)^{-k} \qquad (Z \in \mathfrak{H}_2),$$

where

$$\Gamma_2^\infty := \left\{\begin{pmatrix} * & * \\ 0_2 & * \end{pmatrix} \in \Gamma_2\right\}.$$

Then $E_k \in M_k(\Gamma_2)$ and $\Phi(E_k) = e_k$, where

$$\begin{aligned} e_k(\tau) &:= \frac{1}{2} \sum_{c,d\in\mathbb{Z},\ (c,d)=1} (c\tau + d)^{-k} \\ &= 1 - \frac{2k}{B_k} \sum_{n=1}^{\infty} \sigma_{k-1}(n) q^n \qquad (\tau \in \mathfrak{H}_1) \end{aligned}$$

is the elliptic Eisenstein series of weight k. Here B_k denotes the k-th Bernoulli number, $q := \mathbf{e}(\tau)$ and

$$\sigma_{k-1}(n) := \sum_{0<d|n} d^{k-1}.$$

We define the modular discriminant by

$$\begin{aligned}\Delta(\tau) &:= q \prod_{n\geq 1}(1-q^n)^{24} \\ &= \frac{1}{1728}\left(e_4^3(\tau) - e_6^2(\tau)\right) \in S_{12}(\Gamma_1) \qquad (\tau \in \mathfrak{H}_1),\end{aligned}$$

where $S_k(\Gamma_1)$ denotes the space of cusp forms on Γ_1 of weight k.

Due to Igusa ([18]), the graded ring $\bigoplus_{k\geq 0} M_k(\Gamma_2)$ is generated by $E_4, E_6, \chi_{10}, \chi_{12}$ and χ_{35}, where

$$\chi_{10} := -43867 \cdot 2^{-12} \cdot 3^{-5} \cdot 5^{-2} \cdot 7^{-1} \cdot 53^{-1}(E_4E_6 - E_{10}) \in S_{10}(\Gamma_2),$$

$$\begin{aligned}\chi_{12} := {}& 131 \cdot 593 \cdot 2^{-13} \cdot 3^{-7} \cdot 5^{-3} \cdot 7^{-2} \cdot 337^{-1} \\ &\times \left(3^2 \cdot 7^2\, E_4^3 + 2 \cdot 5^3\, E_6^2 - 691\, E_{12}\right) \in S_{12}(\Gamma_2)\end{aligned}$$

and χ_{35} is a nonzero element of $S_{35}(\Gamma_2)$ (unique up to constant multiples). For an explicit form of χ_{35}, see **2.3**. Note that we here follow Igusa's normalizations of χ_{10} and χ_{12} so that

$$\begin{aligned}A_{\chi_{10}}(1,1,1) &= -1/4, \\ A_{\chi_{12}}(1,1,1) &= 1/12.\end{aligned}$$

We also recall that van der Geer ([10]) defined a Siegel modular form

$$\begin{aligned}G_{24} := {}& \left(\chi_{12} - 2^{-12} \cdot 3^{-6}(E_6^2 + E_4^3)\right)^2 \\ & - E_4\left(2 \cdot 3^{-1}\chi_{10} - 2^{-11} \cdot 3^{-6}E_4E_6\right)^2\end{aligned}$$

of weight 24 on Γ_2 whose divisor is the Humbert surface of discriminant 5. It is known that χ_{10}, χ_{35} and G_{24} are Borcherds lifts (see [13,14]), though χ_{12} is not a Borcherds lift (see [17]).

In this paper, employing our previous result on the multiplicative symmetries for Borcherds lifts ([17]; see Theorem 3.1), we give several necessary conditions for $F \in M_k(\Gamma_2)$ to be a Borcherds lift.

Theorem 1.1. *Assume that* $F \in M_k(\Gamma_2)$ *is a Borcherds lift. Then* $\Phi(F)$ *is proportional to a power* Δ^r *of the modular discriminant* Δ *with* $r \geq 0$.

As direct consequences of Theorem 1.1, we have

Corollary 1.1. *If $F \in M_k(\Gamma_2)$ is a Borcherds lift with $\Phi(F) \neq 0$, then the weight k is divisible by* 12.

Corollary 1.2. *The Siegel Eisenstein series E_k is not a Borcherds lift.*

Theorem 1.2. *Let $F \in M_k(\Gamma_2)$ be a Borcherds lift with $\mathcal{W}(F) \neq 0$. Then the following holds.*

(1) The weight k is divisible by 12.
(2) We have $k > 12$.

Corollary 1.3. *Let $F \in M_k(\Gamma_2)$ be a Borcherds lift. We put $b = 0$ if k is even and $b = 1$ otherwise. Let a be the nonnegative integer such that the multiplicity of H_1 in the divisor of F is equal to $2a + b$. Then F is proportional to $\chi_{10}^a \chi_{35}^b F'$. Here either $F' = 1$ or $F' \in M_{12c}(\Gamma_2)$ is a Borcherds lift with $\mathcal{W}(F') \neq 0$ (note that $c \geq 2$ in this case). In particular, the weight k of F is of the form $10a + 35b + 12c$ $(a \in \mathbb{Z}_{\geq 0}, b \in \{0,1\}, c \in \mathbb{Z}_{\geq 0}, c \neq 1)$.*

1.3. *The organization of the paper*

The paper is organized as follows.

In Section 2, we recall the definition of Borcherds lifts (**2.4**) attached to Jacobi forms of weight 0 and index 1 (**2.1**) after recalling the definition of Humbert surfaces (**2.2**). We also recall the construction of several known Borcherds lifts $\chi_5, \chi_{10}, \chi_{35}$ and G_{24} (**2.5**).

Theorem 1.1 and Theorem 1.2 (i) are proved in Section 3 in the following way: First observe that a Borcherds lift F satisfies a certain symmetry $(\mathrm{MS})_p$ for any prime number p (see Theorem 3.1). Then $f = \Phi(F)$ satisfies, for any p,

$$(*)_p \qquad f(p\tau) \prod_{a=0}^{p-1} f\left(\frac{\tau + a}{p}\right) = \epsilon_p f(\tau)^{p+1}$$

with some $\epsilon_p \in \mathbb{C}, |\epsilon_p| = 1$ (see Proposition 3.3). The key fact is that, if $f \in M_k(\Gamma_1)$ satisfies $(*)_2$, f is a constant multiple of a power of the modular discriminant Δ (Proposition 3.1 and Remark 3.1). Theorem 1.1 is then immediately deduced. We also show that the image φ of F under the Witt operator satisfies a similar symmetry $(\mathrm{ms})_2$ (Proposition 3.3) and that the weight of such a φ is divisible by 12 if $\varphi \neq 0$ (Proposition 3.2), which completes the proof of Theorem 1.2 (i).

In Section 4, after recalling the definition and several properties of Cohen numbers (**4.1**), we give a weight formula for Borcherds lifts (Theorem 4.1 (iv)) essentially due to Borcherds. Using the weight formula together with a lower bound of Cohen numbers (Lemma 4.2), we determine the holomorphic Borcherds lifts of weight less than or equal to 60 (Proposition 4.4). As a by-product, we obtain a proof of the second part of Theorem 1.2 (see Remark 4.4).

1.4. *Notation*

We let $\mathbb{C}^1 = \{z \in \mathbb{C} \mid |z| = 1\}$. For a nonnegative integer n, we write $n = \square$ if n is a square and $n \neq \square$ otherwise. Let $\mathrm{M}_{m,n}$ denote the space of $m \times n$ matrices and write M_m for $\mathrm{M}_{m,m}$. The transpose of a matrix X is denoted by ${}^tX$. We denote by $\mathrm{Sym}_m := \{X \in \mathrm{M}_m \mid {}^tX = X\}$ the space of symmetric matrices of degree m. For $S \in \mathrm{Sym}_m(\mathbb{C})$, we put $S(X,Y) = {}^tXSY$ and $S[X] = {}^tXSX$ for $X, Y \in \mathbb{C}^m$. For $x \in \mathbb{R}$, denote by $[x]$ the integer satisfying $0 \leq x - [x] < 1$. For a condition P, we put $\delta(P) = 1$ if P holds and $\delta(P) = 0$ otherwise. An integer D is called a *fundamental discriminant* either if $D = 1$ or if D is the discriminant of a quadratic field. Denote by $\mathcal{D}$ the set of positive integers congruent to 0 or 1 modulo 4. Note that, for $d \in \mathcal{D}$, there uniquely exist $f \in \mathbb{Z}_{>0}$ and a fundamental discriminant D such that $d = f^2D$.

2. Borcherds lifts

2.1. *Jacobi forms*

For $k \in \mathbb{Z}$, let $\mathcal{J}_{k,1}$ denote the space of holomorphic functions on $\mathfrak{H}_1 \times \mathbb{C}$ satisfying the following conditions:

(1) For $\begin{pmatrix} a & b \\ c & d \end{pmatrix} \in \Gamma_1$ and $(\tau, z) \in \mathfrak{H}_1 \times \mathbb{C}$, we have

$$\phi\left(\frac{a\tau + b}{c\tau + d}, \frac{z}{c\tau + d}\right) = (c\tau + d)^k \mathbf{e}\left(\frac{cz^2}{c\tau + d}\right) \phi(\tau, z).$$

(2) For $\lambda, \mu \in \mathbb{Z}$, we have

$$\phi(\tau, z + \lambda\tau + \mu) = \mathbf{e}(-\lambda^2\tau - 2\lambda z)\, \phi(\tau, z).$$

(3) Let $\phi(\tau, z) = \sum_{n,r \in \mathbb{Z}} a_\phi(n, r)\mathbf{e}(n\tau + rz)$ be the Fourier expansion of ϕ. Then $a_\phi(n, r) = 0$ if $4n - r^2$ is sufficiently small.

The Fourier coefficient $a_\phi(n,r)$ depends only on $N = 4n - r^2$ and is often denoted by $a_\phi(N)$. We put $a_\phi(N) = 0$ if $N \equiv 1$ or $2 \pmod 4$. We then have

$$\phi(\tau, z) = \sum_{N\in\mathbb{Z}} a_\phi(N) \sum_{r\in\mathbb{Z},\, r^2\equiv -N \bmod 4} \mathbf{e}\left(\frac{N+r^2}{4}\tau + rz\right).$$

Let $J_{k,1} := \{\phi \in \mathcal{J}_{k,1} \mid a_\phi(N) = 0 \text{ if } N < 0\}$ be the space of holomorphic Jacobi forms of weight k and index 1, and let $J_{k,1}^{\mathrm{cusp}} := \{\phi \in J_{k,1} \mid a_\phi(0) = 0\}$ be the space of cusp forms in $J_{k,1}$.

In what follows, we are mainly concerned with the weight 0 case. For $\phi \in \mathcal{J}_{0,1}$, we call $\{a_\phi(N) \mid N < 0\}$ the *principal part* of ϕ, which determines ϕ since $J_{0,1} = \{0\}$.

Remark 2.1. The vanishing of $J_{0,1}$ is proved as follows. Suppose that ϕ is a nonzero element of $J_{0,1}$. Then $\Delta\phi \in J_{12,1}^{\mathrm{cusp}}$. Since $J_{12,1}^{\mathrm{cusp}}$ is one-dimensional and spanned by ϕ' with $a_{\phi'}(3) = 1$ (cf. [9]), we have $\Delta\phi = c\,\phi'$ with $c \in \mathbb{C}^\times$ and hence $\phi = c\,\Delta^{-1}\phi'$. This implies that $a_\phi(-1) = c$, which contradicts to the holomorphy of ϕ.

2.2. *Humbert surfaces*

Let $O(Q) := \{g \in GL_5 \mid {}^t gQg = Q\}$ be the orthogonal group of

$$Q := \begin{pmatrix} & & & & 1 \\ & & & 1 & \\ & & -2 & & \\ & 1 & & & \\ 1 & & & & \end{pmatrix}.$$

For $Z = (\tau_1, z, \tau_2) \in \mathfrak{H}_2$, we put $i(Z) := {}^t(-\tau_1\tau_2 + z^2, \tau_1, z, \tau_2, 1) \in \mathbb{C}^5$. Note that $Q[i(Z)] = 0$ and $Q(i(Z), \overline{i(Z)}) = 4\det(\mathrm{Im}(Z)) > 0$. There exists a homomorphism $\iota\colon \mathrm{Sp}_2(\mathbb{R}) \to O(Q, \mathbb{R})$ such that $i(g\langle Z\rangle) = j(g,Z)^{-1}\iota(g)i(Z)$ for $g \in \mathrm{Sp}_2(\mathbb{R})$ and $Z \in \mathfrak{H}_2$.

Let

$$L := \mathbb{Z}^5,\ L^* := Q^{-1}L,$$
$$L^*_{\mathrm{prim}} := \{\lambda \in L^* \mid n^{-1}\lambda \notin L^* \text{ for any integer } n > 1\}.$$

For an integer d, let

$$\mathcal{H}_d := \sum_{X\in\mathcal{L}_d} \{Z \in \mathfrak{H}_2 \mid Q(X, i(Z)) = 0\}$$

be a divisor of $\mathfrak{H}_2$, where $\mathcal{L}_d := \{X \in L^*_{\text{prim}} \mid Q[X] = -d/2\}$. Note that $\mathcal{H}_d = 0$ unless $d \in \mathcal{D}$. Since $\mathcal{L}_d$ is $\iota(\Gamma_2)$-invariant, $\mathcal{H}_d$ is Γ_2-invariant. Denote by H_d the image of $\mathcal{H}_d$ in $\Gamma_2\backslash\mathfrak{H}_2$ by the natural projection $\mathfrak{H}_2 \to \Gamma_2\backslash\mathfrak{H}_2$. The divisor H_d of $\Gamma_2\backslash\mathfrak{H}_2$ is called the *Humbert surface* of discriminant d. It is known that H_d is nonzero and irreducible if $d \in \mathcal{D}$ (see [11], page 212, Theorem 2.4; see also [12], Section 3). Note that

$$\mathcal{H}_1 = \bigcup_{\gamma\in\Gamma_2} \gamma\{(\tau_1, 0, \tau_2) \in \mathfrak{H}_2 \mid \tau_1, \tau_2 \in \mathfrak{H}_1\},$$
$$\mathcal{H}_4 = \bigcup_{\gamma\in\Gamma_2} \gamma\{(\tau, z, \tau) \in \mathfrak{H}_2 \mid \tau \in \mathfrak{H}_1, z \in \mathbb{C}\}.$$

2.3. *Siegel modular forms with a nontrivial character*

Recall that the character group of Γ_2 consists of two elements, the trivial character and the nontrivial one υ given by

$$\upsilon\begin{pmatrix} 0_2 & 1_2 \\ -1_2 & 0_2 \end{pmatrix} = 1,$$
$$\upsilon(n(B)) = (-1)^{b_1+b_2+b_3},$$
$$\upsilon(d(A)) = (-1)^{(1+a_1+a_4)(1+a_2+a_3)+a_1a_4},$$

where

$$n(B) := \begin{pmatrix} 1_2 & B \\ 0_2 & 1_2 \end{pmatrix} \qquad \left(B = \begin{pmatrix} b_1 & b_2 \\ b_2 & b_3 \end{pmatrix} \in \mathrm{Sym}_2(\mathbb{Z})\right),$$
$$d(A) := \begin{pmatrix} A & 0_2 \\ 0_2 & {}^tA^{-1} \end{pmatrix}, \qquad \left(A = \begin{pmatrix} a_1 & a_2 \\ a_3 & a_4 \end{pmatrix} \in \mathrm{GL}_2(\mathbb{Z})\right)$$

(cf. [21], page 135). Let $M_k(\Gamma_2, \upsilon)$ be the space of holomorphic functions F on $\mathfrak{H}_2$ satisfying $F(\gamma Z) = \upsilon(\gamma)j(\gamma, Z)^k F(Z)$ for any $\gamma \in \Gamma_2$.

Let

$$\theta_\epsilon(Z) := \sum_{l\in\mathbb{Z}^2} \mathbf{e}\left(\frac{1}{2}{}^t\!\left(l + \frac{1}{2}\epsilon'\right) Z \left(l + \frac{1}{2}\epsilon'\right) + \frac{1}{2}{}^t\!\left(l + \frac{1}{2}\epsilon'\right)\epsilon''\right) \qquad (Z \in \mathfrak{H}_2)$$

be the theta constant associated with a characteristic $\epsilon = (\epsilon', \epsilon'') \in (\mathbb{Z}/2\mathbb{Z})^2 \times (\mathbb{Z}/2\mathbb{Z})^2$. Then $\theta_\epsilon \neq 0$ if and only if ϵ is even, that is ${}^t\epsilon'\epsilon'' \equiv 0 \pmod 2$. There are 10 even characteristics. Set $\chi_5 := 2^{-8}\prod_{\epsilon \text{ even}} \theta_\epsilon$. It is known that $\chi_5 \in M_5(\Gamma_2, \upsilon)$, $\mathrm{div}(\chi_5) = H_1$ and $\chi_{10} = \chi_5^2$ (see [19], IV. 9

for example). Put

$$\begin{aligned}\chi_{35}(\tau_1,z,\tau_2) &:= \chi_5(\tau_1,z,\tau_2)^{-8}\chi_5(2\tau_1,2z,2\tau_2)\\ &\times \prod_{a,b,c \bmod 2} \chi_5\left(\frac{\tau_1+a}{2},\frac{z+b}{2},\frac{\tau_2+c}{2}\right)\\ &\times \prod_{a \bmod 2} \chi_5\left(\frac{\tau_1+a}{2},z,2\tau_2\right)\chi_5\left(2\tau_1,z,\frac{\tau_2+a}{2}\right)\\ &\times \prod_{b \bmod 2} \chi_5\left(2\tau_1,-\tau_1+z,\frac{\tau_1-2z+\tau_2+b}{2}\right).\end{aligned}$$

Then $\chi_{35} \in S_{35}(\Gamma_2)$ and $\mathrm{div}(\chi_{35}) = H_1 + H_4$ (cf. [13]). It follows that $\chi_{30} := \chi_5^{-1}\chi_{35} \in M_{30}(\Gamma_2, v)$ and $\mathrm{div}(\chi_{30}) = H_4$. The following fact is easily proved.

Lemma 2.1. *Let $F \in M_k(\Gamma_2, v)$. If k is odd, $\chi_5^{-1}F \in M_{k-5}(\Gamma_2)$. If k is even, $\chi_{30}^{-1}F \in M_{k-30}(\Gamma_2)$.*

2.4. *Borcherds lifts on* Γ_2

As a special case of Borcherds theory ([2,3]; see also [15], §2.1), we have the following result:

Theorem 2.1. *Let $\phi \in \mathcal{J}_{0,1}$ and assume that $a_\phi(N) \in \mathbb{Z}$ if $N < 0$.*

(1) Let

$$\lambda_\phi := \frac{1}{24}\sum_{r\in\mathbb{Z}} a_\phi(-r^2), \tag{1}$$

$$\rho_\phi := \frac{1}{2}\sum_{r\in\mathbb{Z}_{>0}} a_\phi(-r^2)r, \tag{2}$$

$$\mu_\phi := \frac{1}{2}\sum_{r\in\mathbb{Z}_{>0}} a_\phi(-r^2)r^2 \tag{3}$$

and

$$\begin{aligned}\Lambda &:= \left\{(m,r,n)\in\mathbb{Z}^3 \mid m>0\right\} \cup \left\{(0,r,n)\in\mathbb{Z}^3 \mid n>0\right\}\\ &\quad \cup \left\{(0,r,0)\in\mathbb{Z}^3 \mid r>0\right\}.\end{aligned}$$

Set

$$\begin{aligned}\Psi_\phi(\tau_1,z,\tau_2) &:= \mathbf{e}\left(\mu_\phi\tau_1 - \rho_\phi z + \lambda_\phi\tau_2\right)\\ &\quad\times \prod_{(m,r,n)\in\Lambda} \left(1-\mathbf{e}(m\tau_1+rz+n\tau_2)\right)^{a_\phi(4mn-r^2)}.\end{aligned}$$

Then $\Psi_\phi(Z)$ *converges absolutely if* $\det(\mathrm{Im}(Z))$ *is sufficiently large, and is meromorphically continued to* $\mathfrak{H}_2$.

(2) *The function* Ψ_ϕ *is a meromorphic modular form on* Γ_2 *of weight* $k_\phi = a_\phi(0)/2$ *and character* v^{α_ϕ} $(\alpha_\phi \in \{0,1\})$.

(3) *The divisor of* Ψ_ϕ *is*

$$\sum_{d\in\mathcal{D}} \left(\sum_{f=1}^{\infty} a_\phi(-f^2 d) \right) H_d.$$

We call Ψ_ϕ the *Borcherds lift* of ϕ.

Remark 2.2. The Borcherds lift Ψ_ϕ is holomorphic if and only if

$$\sum_{f=1}^{\infty} a_\phi(-f^2 d) \geq 0$$

holds for any $d \in \mathcal{D}$.

Remark 2.3. As we will see in Remark 4.2, the weight k_ϕ of Ψ_ϕ is always an integer.

In view of Lemma 2.1, we are mainly concerned with Borcherds lifts of trivial character in this paper.

2.5. *Examples of Borcherds lifts*

For an even integer $k \geq 4$, put

$$\begin{aligned} &e_{k,1}(\tau, z) \\ &:= \frac{1}{2} \sum_{c,d\in\mathbb{Z},(c,d)=1} (c\tau+d)^{-k} \sum_{\lambda\in\mathbb{Z}} \mathbf{e}\left(\lambda^2 \frac{a\tau+b}{c\tau+d} + 2\lambda \frac{z}{c\tau+d} - \frac{cz^2}{c\tau+d} \right), \end{aligned}$$

where $a, b \in \mathbb{Z}$ are chosen to be $ad - bc = 1$. The Jacobi Eisenstein series $e_{k,1}$ belongs to $J_{k,1}$. Let

$$\begin{aligned} \phi^{(1)}(\tau, z) &:= \frac{e_4(\tau)^2 e_{4,1}(\tau, z) - e_6(\tau) e_{6,1}(\tau, z)}{144\Delta(\tau)} \\ &= \left(\zeta + 10 + \zeta^{-1}\right) + q\left(10\zeta^2 - 64\zeta + 108 - 64\zeta^{-1} + 10\zeta^{-2}\right) + \cdots, \end{aligned}$$

where $q := \mathbf{e}(\tau), \zeta := \mathbf{e}(z)$ and $e_k(\tau)$ is the elliptic Eisenstein series of weight k (cf. **1.2**). It is easily verified that $\phi^{(1)}$ is an element of $\mathcal{J}_{0,1}$ with principal part

$$a(N) = \begin{cases} 1 & \text{if } N = -1, \\ 0 & \text{if } N < -1 \end{cases}$$

and that χ_5 (respectively χ_{10}) is proportional to the Borcherds lift of $\phi^{(1)}$ (respectively $2\phi^{(1)}$).

Let

$$\phi^{(2)}(\tau,z) = q^{-1} + \left(\zeta^2 - \zeta + 60 - \zeta^{-1} + \zeta^{-2}\right) + O(q)$$

be the element of $\mathcal{J}_{0,1}$ with principal part

$$a(N) = \begin{cases} 1 & \text{if } N = -4, \\ -1 & \text{if } N = -1, \\ 0 & \text{if } N < 0, N \neq -4, N \neq -1. \end{cases}$$

Then χ_{30} is proportional to the Borcherds lift of $\phi^{(2)}$.

Similarly we let

$$\phi^{(3)}(\tau,z) = q^{-1}(\zeta + \zeta^{-1}) + 48 + O(q)$$

be the element of $\mathcal{J}_{0,1}$ with principal part

$$a(N) = \begin{cases} 1 & \text{if } N = -5, \\ 0 & \text{if } N < 0, N \neq -5. \end{cases}$$

Then G_{24} is proportional to the Borcherds lift of $\phi^{(3)}$.

3. Proof of the main results

3.1. *The multiplicative symmetry*

Let $F \in M_k(\Gamma_2)$ and let p be a prime number. Put

$$\left(F|\mathcal{T}_p^{\uparrow}\right)(\tau_1,z,\tau_2) := F(p\tau_1,pz,\tau_2)\prod_{a=0}^{p-1} F\left(\frac{\tau_1+a}{p},z,\tau_2\right),$$

$$\left(F|\mathcal{T}_p^{\downarrow}\right)(\tau_1,z,\tau_2) := F(\tau_1,pz,p\tau_2)\prod_{a=0}^{p-1} F\left(\tau_1,z,\frac{\tau_2+a}{p}\right).$$

We say that F satisfies the *multiplicative symmetry* for p if the condition

$$(\text{MS})_p \qquad F|\mathcal{T}_p^{\uparrow} = \epsilon_p(F)\, F|\mathcal{T}_p^{\downarrow}$$

holds with $\epsilon_p(F) \in \mathbb{C}^1$. The following result is a special case of [17].

Theorem 3.1. *Suppose that $F \in M_k(\Gamma_2)$ is a Borcherds lift. Then F satisfies the multiplicative symmetry for any prime number p.*

3.2. *A characterization of powers of the modular discriminant*

Recall that $S_{12}(\Gamma_1) = \mathbb{C} \cdot \Delta$ and that Δ has no zeros in $\mathfrak{H}_1$.

For $f \in M_k(\Gamma_1)$ and a prime number p, we define the *multiplicative Hecke operator* by

$$(f|\mathcal{T}_p)(\tau) := f(p\tau) \prod_{a=0}^{p-1} f\left(\frac{\tau + a}{p}\right).$$

Then $f|\mathcal{T}_p \in M_{(p+1)k}(\Gamma_1)$. The following elementary fact plays a crucial role in the proof of Theorem 1.1.

Proposition 3.1. *A nonzero element* f *of* $M_k(\Gamma_1)$ *satisfies*

$$(*)_p \qquad (f|\mathcal{T}_p)(\tau) = \epsilon_p(f) f(\tau)^{p+1} \qquad (\tau \in \mathfrak{H}_1)$$

for any prime number p *with* $\epsilon_p(f) \in \mathbb{C}^1$ *if and only if* $f = c\Delta^r$ *with* $c \in \mathbb{C}^\times$ *and* $r \in \mathbb{Z}_{\geq 0}$.

Proof. We first show the "if" part. It suffices to prove $(*)_p$ for $f = \Delta$. For a prime number p, we have

$$\begin{aligned}\frac{(\Delta|\mathcal{T}_p)(\tau)}{\Delta(\tau)^{p+1}} &= \mathbf{e}\left(p\tau + \sum_{a=0}^{p-1} \frac{\tau + a}{p} - (p+1)\tau\right) \\ &\quad \times \prod_{n \geq 1} \left(\frac{(1 - \mathbf{e}(pn\tau)) \prod_{a=0}^{p-1} \left(1 - \mathbf{e}\left(n \frac{\tau + a}{p}\right)\right)}{(1 - \mathbf{e}(n\tau))^{p+1}}\right)^{24}.\end{aligned}$$

Since

$$\prod_{a=0}^{p-1} \left(1 - \mathbf{e}\left(n\frac{\tau + a}{p}\right)\right) = \begin{cases} (1 - \mathbf{e}(p^{-1}n\tau))^p & \text{if } p|n, \\ 1 - \mathbf{e}(n\tau) & \text{if } p \nmid n, \end{cases}$$

we have

$$\begin{aligned}\frac{(\Delta|\mathcal{T}_p)(\tau)}{\Delta(\tau)^{p+1}} &= \mathbf{e}\left(\frac{p-1}{2}\right) \\ &\quad \times \left(\frac{\prod_{n \geq 1}(1 - \mathbf{e}(pn\tau)) \prod_{n \geq 1,\, p|n}(1 - \mathbf{e}(p^{-1}n\tau))^p \prod_{n \geq 1,\, p\nmid n}(1 - \mathbf{e}(n\tau))}{\prod_{n \geq 1}(1 - \mathbf{e}(n\tau))^{p+1}}\right)^{24} \\ &= (-1)^{p-1},\end{aligned}$$

which proves the claim. This also shows that $\epsilon_p(\Delta^r) = (-1)^{(p-1)r}$. We next show the "only if" part. Suppose that f satisfies $(*)_p$ for $p = 2$. Let

$f(\tau) = a_r q^r + O(q^{r+1})$ with $r \in \mathbb{Z}_{\geq 0}$ and $a_r \neq 0$ ($q = \mathbf{e}(\tau)$). We may assume $a_r = 1$ without loss of generality. Then $g := f\Delta^{-r} = 1 + O(q)$ is in $M_{k-12r}(\Gamma_1)$ and we have

$$g(2\tau)g\left(\frac{\tau}{2}\right)g\left(\frac{\tau+1}{2}\right) = \epsilon\, g(\tau)^3 \tag{4}$$

with $\epsilon = (-1)^r \epsilon_2(f) \in \mathbb{C}^1$. Suppose that g has a zero τ_0 in $\mathfrak{H}_1$. Let $\mathcal{F} = \{\tau \in \mathfrak{H}_1 \mid |\tau| \geq 1, -1/2 \leq \mathrm{Re}(\tau) \leq 1/2\}$. Since $\mathcal{F}$ is the closure of a fundamental domain of $\Gamma_1\backslash\mathfrak{H}_1$, we may suppose that $\tau_0 \in \mathcal{F}$. Note that $\mathrm{Im}(2\tau_0) \geq \sqrt{3} > 1$. Since $g(2\tau_0)^3 = \epsilon^{-1}g(4\tau_0)g(\tau_0)g(\tau_0 + 1/2) = 0$, we have $g(2\tau_0) = 0$. Take a point τ_1 in $\mathcal{F}$ such that $\tau_1 - 2\tau_0 \in \mathbb{Z}$. Then $g(\tau_1) = 0$ and $\tau_1 \neq \tau_0$. Repeating this procedure, we see that g has infinitely many zeros in $\mathcal{F}$, a contradiction. Thus g has no zeros in $\mathfrak{H}_1$ and hence $g^{-1} \in M_{-k+12r}(\Gamma_1)$. This implies that $k = 12r$ and $g = 1$. This completes the proof of the proposition. □

Remark 3.1. The proof of the proposition shows that f is a constant multiple of Δ^r if $f \in M_k(\Gamma_1)$ satisfies $(*)_2$.

3.3. *The multiplicative symmetry for* $\mathrm{Sym}^2(M_k(\Gamma_1))$

Recall that $\mathrm{Sym}^2(M_k(\Gamma_1))$ is the space of functions on $\mathfrak{H}_1 \times \mathfrak{H}_1$ spanned by $f(\tau_1)f'(\tau_2) + f'(\tau_1)f(\tau_2)$ with $f, f' \in M_k(\Gamma_1)$. Let $\mathrm{Sym}^2(S_k(\Gamma_1))$ be the subspace of $\mathrm{Sym}^2(M_k(\Gamma_1))$ spanned by $f(\tau_1)f'(\tau_2) + f'(\tau_1)f(\tau_2)$ with $f, f' \in S_k(\Gamma_1)$. For $\varphi \in \mathrm{Sym}^2(M_k(\Gamma_1))$, we put

$$\Phi'(\varphi)(\tau) = \lim_{y\to\infty} \varphi(\tau, iy).$$

Then $\Phi'(\varphi) \in M_k(\Gamma_1)$, and we have $\Phi'(\varphi) = 0$ if and only if $\varphi \in \mathrm{Sym}^2(S_k(\Gamma_1))$.

For $\varphi \in \mathrm{Sym}^2(M_k(\Gamma_1))$ and a prime number p, put

$$(\varphi|\mathcal{T}_p^{\uparrow})(\tau_1,\tau_2) := \varphi(p\tau_1,\tau_2)\prod_{a=0}^{p-1}\varphi\left(\frac{\tau_1+a}{p},\tau_2\right),$$
$$(\varphi|\mathcal{T}_p^{\downarrow})(\tau_1,\tau_2) := \varphi(\tau_1,p\tau_2)\prod_{a=0}^{p-1}\varphi\left(\tau_1,\frac{\tau_2+a}{p}\right) \qquad (\tau_1,\tau_2 \in \mathfrak{H}_1).$$

Note that $\varphi|\mathcal{T}_p^{\uparrow}(\tau_2,\tau_1) = \varphi|\mathcal{T}_p^{\downarrow}(\tau_1,\tau_2)$. We say that φ satisfies the *multiplicative symmetry* for p if there exists an $\epsilon_p(\varphi) \in \mathbb{C}^1$ such that

$$(\mathrm{ms})_p \qquad \varphi|\mathcal{T}_p^{\uparrow} = \epsilon_p(\varphi)\varphi|\mathcal{T}_p^{\downarrow}$$

holds. The following fact is easily verified.

Lemma 3.1. *If* $\varphi \in \mathrm{Sym}^2(M_k(\Gamma_1))$ *satisfies* $(\mathrm{ms})_2$ *and* $f := \Phi'(\varphi)$ *is not identically equal to zero, then* f *satisfies* $(*)_2$. *In particular, we have* $f = c\Delta^r$ *with* $c \in \mathbb{C}^\times, r \in \mathbb{Z}_{\geq 0}$ *and* k *is divisible by* 12.

Proposition 3.2. *If* $\varphi \in \mathrm{Sym}^2(M_k(\Gamma_1)) \setminus \{0\}$ *satisfies* $(\mathrm{ms})_2$, k *is divisible by* 12.

Proof. If $\Phi'(\varphi) \neq 0$, the assertion follows from Lemma 3.1. Suppose that $\Phi'(\varphi) = 0$. Since $S_k(\Gamma_1) = \Delta\, M_{k-12}(\Gamma_1)$, there exist a positive integer r and $\psi \in \mathrm{Sym}^2(M_{k-12r}(\Gamma_1))$ such that $\varphi(\tau_1, \tau_2) = (\Delta(\tau_1)\Delta(\tau_2))^r\, \psi(\tau_1, \tau_2)$ and $\Phi'(\psi) \neq 0$. Since ψ satisfies $(\mathrm{ms})_2$, we have $12|(k - 12r)$ by Lemma 3.1 and hence k is divisible by 12. □

The following is easily verified.

Proposition 3.3. *Suppose that* $F \in M_k(\Gamma_2)$ *satisfies* $(\mathrm{MS})_p$ *for a prime* p. *Put* $f := \Phi(F)$ *and* $\varphi := \mathcal{W}(F)$. *Then* f *(respectively* φ*) satisfies* $(*)_p$ *(respectively* $(\mathrm{ms})_p$*) and* $\epsilon_p(F) = \epsilon_p(f) = \epsilon_p(\varphi)$.

3.4. *Proofs of Theorem 1.1 and Theorem 1.2 (i)*

Proposition 3.4. *Assume that* $F \in M_k(\Gamma_2)$ *satisfies* $(\mathrm{MS})_2$ *and that* $f := \Phi(F)$ *is not identically equal to zero. Then* $f = c\,\Delta^r \;\; (c \in \mathbb{C}^\times, r \in \mathbb{Z}_{\geq 0})$. *In particular, the weight* k *is divisible by* 12.

Proof. The proposition follows from Proposition 3.3 and Remark 3.1. □

Theorem 1.1 is a direct consequence of Theorem 3.1 and Proposition 3.4. Theorem 1.2 (i) also follows from Theorem 3.1, Proposition 3.3 and Proposition 3.2.

4. The weight formula

4.1. *Cohen numbers*

To express the weight k_ϕ of Ψ_ϕ in terms of the principal part of ϕ, we recall the definitions of Cohen numbers $c(r, N)$ and $C(r, N)$.

Let $L(s,\chi)$ denote the Dirichlet L-function attached to a Dirichlet character χ modulo N. For $r, N \in \mathbb{Z}_{>0}$, define

$$c(r,N) := \begin{cases} (-1)^{[r/2]}(r-1)!\,N^{r-1/2}2^{1-r}\pi^{-r}L(r,\chi_{(-1)^r N}) \\ \qquad\qquad\qquad \text{if } (-1)^r N \equiv 0 \text{ or } 1 \pmod 4, \\ 0 \\ \qquad\qquad\qquad \text{if } (-1)^r N \equiv 2 \text{ or } 3 \pmod 4, \end{cases} \tag{5}$$

where $\chi_M(a) := \left(\dfrac{M}{a}\right)$ for $M \in \mathbb{Z}$ with $M \equiv 0$ or $1 \pmod 4$. We put $c(r,0) := \zeta(1-2r)$ for $r \in \mathbb{Z}_{>0}$. For $r \in \mathbb{Z}_{>0}$ and $N \in \mathbb{Z}_{\geq 0}$, we define

$$C(r,N) := \begin{cases} \sum_{f>0,\, f^2|N} c(r, f^{-2}N) & \text{if } (-1)^r N \equiv 0 \text{ or } 1 \pmod 4 \text{ and } N > 0, \\ \zeta(1-2r) & \text{if } N = 0, \\ 0 & \text{otherwise.} \end{cases} \tag{6}$$

We set

$$C^*(N) := -60\,C(2,N),$$
$$c^*(N) := -60\,c(2,N)$$

for $N \geq 0$. If $N > 0$, we have

$$C^*(N) = \sum_{f>0,\, f^2|N} c^*\left(f^{-2}N\right) \tag{7}$$

and

$$c^*(N) = \sum_{f>0,\, f^2|N} \mu(f) C^*\left(f^{-2}N\right), \tag{8}$$

where μ denotes the Möbius function. Note that $c^*(D) = C^*(D)$ for a fundamental discriminant D.

Remark 4.1. We give a table of $c^*(N)$ and $C^*(N)$ for $N \leq 20$ with $N \equiv 0$ or $1 \pmod 4$.

N	0	1	4	5	8	9	12	13	16	17	20
$c^*(N)$	$-1/2$	5	30	24	60	120	120	120	240	240	240
$C^*(N)$	$-1/2$	5	35	24	60	125	120	120	275	240	264

The following result is proved in [7] (Propositition 4.1).

Proposition 4.1. *For a nonnegative integer* N, *we have*

$$C^*(N) = 12 \sum_{s \in \mathbb{Z}} \sigma_1\left(\frac{N - s^2}{4}\right) + \begin{cases} 6N & \text{if } N = \square, \\ 0 & \text{otherwise,} \end{cases}$$

where

$$\sigma_1(m) = \begin{cases} \sum_{0<d|m} d & \text{if } m \in \mathbb{Z}_{>0}, \\ -1/24 & \text{if } m = 0, \\ 0 & \text{otherwise.} \end{cases} \tag{9}$$

In particular, $C^*(N)$ *is a positive integer if* $N > 0$.

For $N \in \mathcal{D}$, we set

$$B(N) := g^3 \prod_{p|g} \left(1 - \frac{\chi_D(p)}{p^2}\right),$$

where $g \in \mathbb{Z}_{>0}$ and D is a fundamental discriminant such that $N = g^2 D$. It is easy to see that $B(N)$ is a positive integer. The following is easily verified.

Lemma 4.1.

(1) For a fundamental discriminant D, *we have* $B(D) = 1$.
(2) Let N, g, D *be as above. Then we have* $c^*(N) = B(N)C^*(D)$ *and* $B(N) \geq g^2$. *In particular* $c^*(N)$ *is a positive integer.*

Lemma 4.2. *For* $N \in \mathcal{D}$, *we have*

$$C^*(N) \geq c^*(N) \geq 3N.$$

Proof. By (7), we have $C^*(N) \geq c^*(N)$. By Proposition 4.1, we have $C^*(N) \geq 3N$. Then Lemma 4.1 implies that

$$c^*(N) \geq g^2 C^*(D) \geq 3g^2 D = 3N. \qquad \square$$

Lemma 4.3. *If* $D > 1$ *is a fundamental discriminant with* $D \neq 5$ *and* $D \neq 8$, *then* $C^*(D)$ *is divisible by* 120.

Proof. The lemma immediately follows from Proposition 4.1 and [8] Corollary 10.3.9. $\square$

Lemma 4.4. *If* $N \geq 9$, $c^*(N)$ *is divisible by* 120.

Proof. If $N \equiv 2$ or $3 \pmod 4$, we have $c^*(N) = 0$. Thus we may suppose that $N = g^2 D$ with $g \in \mathbb{Z}_{>0}$ and D a fundamental discriminant. By Lemma 4.1, we have $c^*(N) = B(N)\, C^*(D)$. It suffices to consider the case where $D = 1$, $D = 5$ or $D = 8$ in view of Lemma 4.1. Since $C^*(1) = 5, C^*(5) = 24$ and $C^*(8) = 60$, the proof of the lemma is reduced to the following elementary fact, whose proof we omit.

(a) If $D = 1$ and $g \geq 3$, $B(N)$ is divisible by 24.
(b) If $D = 5$ and $g \geq 2$, $B(N)$ is divisible by 5.
(c) If $D = 8$ and $g \geq 2$, $B(N)$ is divisible by 2. □

4.2. *The weight formula for Borcherds lifts*

Proposition 4.2.

(1) For each $d \in \mathcal{D}$, there uniquely exists an element ϕ_d of $\mathcal{J}_{0,1}$ with principal part

$$a_{\phi_d}(N) = \begin{cases} 1 & \text{if } N = -d, \\ 0 & \text{if } N < 0 \text{ and } N \neq -d. \end{cases}$$

(2) For $\phi \in \mathcal{J}_{0,1}$, we have

$$\sum_{N \geq 0} a_\phi(-N) C^*(N) = 0$$

and hence

$$\frac{1}{2} a_\phi(0) = \sum_{N > 0} a_\phi(-N) C^*(N).$$

Proof. This fact is essentially due to Borcherds ([4]; see also [6], page 1721). For completeness, we give a sketch of the proof. For $k \in \mathbb{Z}$, denote by $J^+_{k,1}$ and $J^{+,\,\mathrm{cusp}}_{k,1}$ the space of skew-holomorphic Jacobi forms of weight k and index 1 and its subspace consisting of cusp forms respectively (for the definition, see [25]). For an odd integer k, we also denote by $M^+_{k/2}(\Gamma_0(4))$ and $S^+_{k/2}(\Gamma_0(4))$ the Kohnen plus space on $\Gamma_0(4)$ of weight $k/2$ and its subspace consisting of cusp forms respectively. It is known that $J^+_{k,1} \cong M^+_{k-1/2}(\Gamma_0(4))$ and $J^{+,\,\mathrm{cusp}}_{k,1} \cong S^+_{k-1/2}(\Gamma_0(4))$ if k is odd (see [25]; see also [16]). By [4], the obstruction space for $\mathcal{J}_{0,1}$ is $J^{+,\,\mathrm{cusp}}_{3,1}$. The first assertion of the proposition follows from the fact $J^{+,\,\mathrm{cusp}}_{3,1} \cong S^+_{5/2}(\Gamma_0(4)) \cong S_4(\mathrm{SL}_2(\mathbb{Z})) = \{0\}$. Next observe that $J^+_{3,1} \cong M^+_{5/2}(\Gamma_0(4))$ and that the latter space is one-dimensional and spanned by Cohen's Eisenstein series

$$\sum_{N\geq 0} C(2,N)\mathbf{e}(N\tau).$$

The second assertion follows from this. $\square$

Remark 4.2. Proposition 4.1 and Proposition 4.2 imply that, if $\phi \in \mathcal{J}_{0,1}$ satisfies the assumption of Theorem 2.1, the weight k_ϕ of the Borcherds lift Ψ_ϕ is an integer.

Theorem 4.1.

(1) For each $d \in \mathcal{D}$*, there exists an* $F_d \in M_{k_d}(\Gamma_2, v^{\alpha_d})$ *satisfying* $\mathrm{div}(F_d) = H_d$*, where*

$$\alpha_d = \begin{cases} 1 & \text{if } d = 1 \text{ or } 4, \\ 0 & \text{if } d > 4. \end{cases}$$

(2) We have $k_d = c^*(d)$*.*

(3) A Borcherds lift $F \in M_k(\Gamma_2, v^\alpha)$ $(\alpha \in \{0,1\})$ *is a constant multiple of* $\prod_{d\in\mathcal{D}} F_d^{A(d)}$*, where* $A(d)$ *is a nonnegative integer (*$A(d) = 0$ *except for a finite number of* d*) satisfying* $A(1)+A(4) \equiv \alpha \pmod 2$*. Furthermore we have*

$$k = \sum_{d\in\mathcal{D}} A(d)c^*(d).$$

Proof. Let F_d be the Borcherds lift of

$$\psi_d := \sum_{f>0,\, f^2|d} \mu(f)\,\phi_{f^{-2}d} \in \mathcal{J}_{0,1}.$$

By Theorem 2.1 (iii), we have $\mathrm{div}(F_d) = H_d$ and hence F_d is holomorphic. The weight of F_d is equal to

$$\sum_{f>0,\, f^2|d} \mu(f)C^*(f^{-2}d) = c^*(d)$$

by (8). By Lemma 2.1, the character of F_d is equal to 1 (respectively v) if $d > 4$ (respectively if $d = 1$ or $d = 4$). The remaining assertion of the theorem is easily verified. $\square$

Remark 4.3. We have $F_1 = c\chi_5, F_4 = c'\chi_{30}$ and $F_5 = c''G_{24}$ with $c, c', c'' \in \mathbb{C}^\times$.

Corollary 4.1. *Let $\phi \in \mathcal{J}_{0,1}$ and assume that $a_\phi(N) \in \mathbb{Z}$ for $N < 0$. Let $\alpha_\phi \in \{0,1\}$ be as in Theorem 2.1 (ii). Then we have $\alpha_\phi \equiv 2\mu_\phi \pmod 2$ (for the definition of μ_ϕ, see (3)).*

Proof. Since $\phi_d = \sum_{f>0,\, f^2|d} \psi_{f^{-2}d}$, we have

$$\phi = \sum_{d\in\mathcal{D}} a_\phi(-d)\phi_d = \sum_{d\in\mathcal{D}}\sum_{f=1}^{\infty} a_\phi(-f^2d)\psi_d.$$

Thus Ψ_d is a constant multiple of $\prod_{d\in\mathcal{D}} F_d^{A_\phi(d)}$ with $A_\phi(d) := \sum_{f=1}^{\infty} a_\phi(-f^2d)$. It follows that

$$\begin{aligned}
\alpha_\phi &\equiv A_\phi(1) + A_\phi(4) \pmod 2 \\
&= \sum_{f=1}^{\infty} \{a_\phi(-f^2) + a_\phi(-4f^2)\} \\
&\equiv \sum_{f\geq 1,\, f \text{ odd}} a_\phi(-f^2) \pmod 2.
\end{aligned}$$

On the other hand, we have

$$\begin{aligned}
2\mu_\phi &= \sum_{f=1}^{\infty} a_\phi(-f^2)f^2 \\
&\equiv \sum_{f\geq 1,\, f \text{ odd}} a_\phi(-f^2) \pmod 2,
\end{aligned}$$

which shows that $\alpha_\phi \equiv 2\mu_\phi \pmod 2$. □

Proposition 4.3. *The weight k_d of F_d is divisible by 24 if and only if $d > 4$ and $d \neq 8$. If $d \geq 9$, k_d is divisible by 120.*

Proof. The proposition is a direct consequence of Theorem 4.1, Lemma 4.4 and Remark 4.1. □

Proposition 4.4. *The Borcherds lifts in $M_k(\Gamma_2)$ with $k \leq 60$ are listed as follows:*

Borcherds lift	*weight*	*divisor*
F_1^{2a} $(1 \le a \le 6)$	$10a$	$2aH_1$
$F_1^{2a+1}F_4$ $(1 \le a \le 2)$	$10a + 35$	$(2a+1)H_1 + H_4$
$F_1^{2a}F_5$ $(1 \le a \le 3)$	$10a + 24$	$2aH_1 + H_5$
F_4^2	60	$2H_4$
F_5^2	48	$2H_5$
F_8	60	H_8

Remark 4.4. The above table shows that every Borcherds lift of weight less than or equal to 60 is a monomial of F_1, F_4, F_5 and F_8. We also see that there is no holomorphic Borcherds lift of weight 12, which proves the second assertion of Theorem 1.2. This also gives another proof of the fact that χ_{12} is not a Borcherds lift, which was proved in [17] in a different way.

Acknowledgement

This work started when we stayed at Mathematisches Forschungsinstitut Oberwolfach supported by the Research in Pairs program in 2008, and was completed during our stay at Max-Planck-Institut für Mathematik in Bonn in 2010. We would like to thank both institutes for their hospitality and financial support. The first author is grateful to GUtech (Oman) for their financial support. The second author was partially supported by Grants-in-Aids from JSPS (20540031) and Kyoto Sangyo University Research Grants.

References

1. H. Aoki and T. Ibukiyama, *Simple graded rings of Siegel modular forms, differential operators and Borcherds products*, International J. Math. **16** (2005), 249–279.
2. R. E. Borcherds, *Automorphic forms on* $O_{s+2,2}(\mathbb{R})$ *and infinite products*, Invent. Math. **120** (1995), 161–213.
3. R. E. Borcherds, *Automorphic forms with singularities on Grassmannians*, Invent. Math. **132** (1998), 491–562.
4. R. E. Borcherds, *The Gross-Kohnen-Zagier theorem in higher dimensions*, Duke Math. J. **97** (1999), 219–233.
5. J. H. Bruinier, *Borcherds Products on* $O(2,l)$ *and Chern Classes of Heegner Divisors*, Lecture Notes in Math. **1780** (2002), Springer Verlag.
6. J. H. Bruinier and U. Kühn, *Integrals of automorphic Green's functions associated to Heegner divisors*, IMRN **31** (2003), 1687–1729.
7. H. Cohen, *Sums involoving the values at negative integers of L-functions of quadratic characters* Math. Ann. **217** (1975), 271–285.

8. H. Cohen, *Number Theory, Volume II: Analytic and Modern Tools*, Springer Verlag 2007.
9. M. Eichler and D. Zagier, *Theory of Jacobi fomrs*, Progress in Math. **55**, Birkhäuser 1985.
10. G. van der Geer, *On the geometry of a Siegel modular threefold*, Math. Ann. **260** (1982), 317–350.
11. G. van der Geer, *Hilbert modular surfaces*, Springer Verlag 1987.
12. V. A. Gritsenko and K. Hulek, *Minimal Siegel modular threefolds*, Math. Proc. Camb. Phil. Soc. **123** (1998), 461–485.
13. V. A. Gritsenko and V. V. Nikulin, *The Igusa modular forms and "the simplest" Lorentzian Kac-Moody algebra*, Matem. Sbornik **187** (1996), 1601–1643.
14. V. A. Gritsenko and V. V. Nikulin, *Siegel automorphic form correction of some Lorentzian Kac-Moody Lie algebras*, Amer. J. Math. **119** (1997), 181–224.
15. V. A. Gritsenko and V. V. Nikulin,, *Automorphic forms and Lorentzian Kac-Moody algebra, part II*, Int. J. Math. **9** (1998), 201–275.
16. S. Hayashida, *Skew-holomorphic jacobi forms of index* 1 *and Siegel modular forms of half-integral weight*, J. Number Theory **106** (2004), 200–218.
17. B. Heim and A. Murase, *Symmetries for the Siegel theta functions, Borcherds lifts and automorphic Green functions*, arXiv: 1003.2248.
18. J. Igusa, *On Siegel modular forms of genus two (II)*, Amer. J. Math. **86** (1964), 392–412.
19. H. Klingen, *Introductory lectures on Siegel modular forms*, Cambridge Univ. Press, 1990.
20. H. Maass, *Über eine Spezialschar von Modulformen zweiten Grades I; II; III*, Invent. Math. **52** (1979), 95–104; **53** (1979), 249–253; **53** (1979), 255–265.
21. H. Maass, *Die Multiplikatorsystetme zur Siegelschen Modulgruppe*, Nachr. der Akad. der Wiss. in Göttingen, II. Math.-phys. Klasse **11** (1964), 125-135.
22. T. Oda, On modular forms associated with indefinite quadratic forms of signature $(2, n-2)$, *Math. Ann.* **231** (1977), 97–144.
23. S. Rallis and G. Schiffmann, Automorphic forms constructed from the Weil representation: Holomorphic case, *Amer. J. Math.* **100** (1978), 1049–1122.
24. G. Shimura, *On modular forms of half integral weight*, Ann. Math. **97** (1973), 440–481.
25. N.-P. Skoruppa, *Explicit formulas for the Fourier coefficients of Jacobi and elliptic modular forms*, Invent. Math. **102** (1990), 501–520.
26. D. Zagier, *Sur la conjecture de Saito-Kurokawa (d'aprè H. Maaß)*, Séminaire Delange-Pisot-Poitou 1979–1980, Progress in Math. **12**, 371–394, Birkhäuser 1980.

The archimedean Whittaker functions on $GL(3)$

Miki Hirano*

Department of Mathematics, Faculty of Science, Ehime University,
2-5, Bunkyo-cho, Matsuyama, Ehime, 790-8577, Japan
E-mail: hirano@math.sci.ehime-u.ac.jp

Taku Ishii†

Faculty of Science and Technology, Seikei University,
3-3-1 Kichijoji-Kitamachi, Musashino, Tokyo, 180-8633, Japan
E-mail: ishii@st.seikei.ac.jp

Tadashi Miyazaki

Department of Mathematics, Rikkyo University,
3-34-1, Nishi-Ikebukuro, Toshima, Tokyo, 171-8501, Japan
E-mail: miyaza@ms.u-tokyo.ac.jp

Dedicated to Professor Takayuki Oda on the occasion of his 60th birthday

We introduce the explicit formulas of archimedean Whittaker functions on $GL(3)$ and their application to archimedean zeta integrals.

Keywords: Whittaker functions; automorphic forms; zeta integrals

1. Introduction

The explicit Fourier(-Whittaker) expansions of the classical automorphic forms make essentials for their theory. This is also true in the case of higher rank groups, even though there are many difficulties in their concrete harmonic analysis. Following Oda's philosophy which asserts the explicit results of harmonic analysis on algebraic groups are indispensable

*The first author is supported by JSPS Research Fellowship for Young Scientists (No. 21740020).

†The second author is supported by JSPS Research Fellowship for Young Scientists (No. 21740028).

for further research of automorphic forms, we have studied various models of admissible representations of real reductive groups, which may not be spherical in general. In our former papers [MIO], [Mi] and [HO], we discussed archimedean Whittaker functions for generalized principal series representations of $GL(3)$. Spherical Whittaker functions are characterized only by the action of the center of the universal enveloping algebra. But, in the study of non-spherical Whittaker functions, we need more precise information of the $(\mathfrak{g}_{\mathbf{C}}, K)$-module structures of generalized principal series representations. In those papers, we obtained a holonomic system of partial differential equations for Whittaker functions, explicit formulas of the primary and the secondary Whittaker functions at the minimal K-types.

The theory of the Rankin-Selberg method is important for the study of automorphic L-functions, and this theory for $GL(n)$ was developed by Jacquet, Piatetski-Shapiro and Shalika (*cf.* [Co]). The essential step of this method is the analysis of the local zeta integrals, because of the uniqueness of Whittaker models. However, for example, we have few results of the evaluations of the archimedean zeta integrals for $GL(n)$ via explicit formulas of Whittaker functions such as Bump [Bu1], Hoffstein-Murty [HM] and Stade [St1], [St2], and these are only for the spherical case. Besides, the non-spherical computation for $GL(2, \mathbf{C})$ is done by Popa [Po], very recently.

Our project is to give an explicit computation for the archimedean zeta integrals for $GL(3)$ by using the explicit formulas of Whittaker functions. In this paper, we treat mainly the case of $GL(3) \times GL(1)$. To do this, our former results are not enough, because the zeta integral at the minimal K-type may vanish in some cases. Here, we give more explicit formulas of Whittaker functions in Sections 3 and 4, and compute the zeta integrals explicitly using them in Section 6.

2. Preliminaries

2.1. *Notation*

We denote by $\mathbf{Z}$, $\mathbf{Q}$, $\mathbf{R}$ and $\mathbf{C}$ the ring of rational integers, the rational number field, the real number field and the complex number field, respectively. Let $\mathbf{Z}_{\geq l}$ be the set of integers which are not less than $l \in \mathbf{Z}$ and $\mathbf{R}_+$ the multiplicative group of positive real numbers. For $z \in \mathbf{C}$, we denote the real part and the imaginary part of z by $\mathrm{Re}(z)$ and $\mathrm{Im}(z)$, respectively.

Let F be $\mathbf{R}$ or $\mathbf{C}$. For $c \in F$, we define the unitary character $\psi_F^{(c)}$ by

$$\psi_F^{(c)}(\xi) = \begin{cases} \exp(2\pi\sqrt{-1}c\xi) & \text{if } F = \mathbf{R}, \\ \exp(2\pi\sqrt{-1}(c\xi + \overline{c\xi})) & \text{if } F = \mathbf{C}, \end{cases} \qquad (\xi \in F).$$

Here $\overline{\xi}$ means the complex conjugate of ξ. We denote $\psi_F^{(1)}$ simply by ψ_F. We identify the additive group F with its dual group via the isomorphism defined by $c \mapsto \psi_F^{(c)}$. Then we denote by $d_F^+\xi$ the self-dual additive Haar measure on F, that is, $d_{\mathbf{R}}^+\xi = d\xi$ and $d_{\mathbf{C}}^+\xi = 2d\xi_1 d\xi_2$ ($\xi = \xi_1 + \sqrt{-1}\xi_2$). Here $d\xi$ is the ordinary Lebesgue measure on $\mathbf{R}$. For $c \in F$, we denote by $|c|_F$ the module of the endomorphism $\xi \mapsto c\xi$ of F, namely, the factor by which this transformation multiplies the additive Haar measure. Then $|c|_{\mathbf{R}} = |c|$ and $|c|_{\mathbf{C}} = |c|^2$ where $|\cdot|$ is the ordinary absolute value. We fix the multiplicative Haar measure $d_F^\times\xi$ on $F^\times$ by

$$d_F^\times\xi = m_F \frac{d_F^+\xi}{|\xi|_F}, \qquad m_F = \begin{cases} 2^{-1} & \text{if } F = \mathbf{R}, \\ (2\pi)^{-1} & \text{if } F = \mathbf{C}. \end{cases}$$

Moreover, it is convenient to define the measures $d_F\xi$ on $\mathbf{R}$ by $d_{\mathbf{R}}\xi = d\xi$ and $d_{\mathbf{C}}\xi = 2d\xi$. We set $\Gamma_{\mathbf{R}}(s) = \pi^{-s/2}\Gamma(s/2)$ and $\Gamma_{\mathbf{C}}(s) = 2(2\pi)^{-s}\Gamma(s)$ where $\Gamma(s)$ is the Gamma function.

Let $M_{m,n}(F)$ and $M_n(F)$ be the spaces of $m \times n$ and $n \times n$ matrices over F, respectively. Let 1_n be the unit matrix of degree n and E_{ij} the matrix unit in $M_3(F)$ with 1 at (i,j)-th entry and 0 at other entries.

2.2. *Basic objects*

Let $G_n = GL(n,F)$ be the general linear group of degree n over F. In this article, we view G_n as a real reductive Lie group even if $F = \mathbf{C}$. Let N_n be the group of upper triangular matrices in G_n with diagonal entries equal to 1, and let A_n be the group of diagonal matrices with diagonal entries in $\mathbf{R}_+$. Moreover, we fix a maximal compact subgroup K_n of G_n by

$$K_n = \begin{cases} O(n) \text{ if } F = \mathbf{R}, \\ U(n) \text{ if } F = \mathbf{C}, \end{cases}$$

where $O(n)$ is the orthogonal group of degree n and $U(n)$ is the unitary group of degree n. Then $G_n = N_n A_n K_n$ is an Iwasawa decomposition of G_n. It is convenient to take the coordinate on N_n and A_n as follows:

$$x = \begin{pmatrix} 1 & x_{1,2} & x_{1,3} & \cdots & x_{1,n} \\ 0 & 1 & x_{2,3} & \cdots & x_{2,n} \\ \vdots & & \ddots & & \vdots \\ 0 & & & 1 & x_{n-1,n} \\ 0 & 0 & \cdots & 0 & 1 \end{pmatrix} \in N_n,$$

$$y = \mathrm{diag}(y_1 y_2 \cdots y_n,\ y_2 \cdots y_n,\ \cdots,\ y_n) \in A_n,$$

where $x_{i,j} \in F$ $(1 \leq i < j \leq n)$ and $y_k \in \mathbf{R}_+$ $(1 \leq k \leq n)$. We denote the Lie algebras of G_n, N_n, A_n and K_n by $\mathfrak{g}_n, \mathfrak{n}_n, \mathfrak{a}_n$ and $\mathfrak{k}_n$, respectively.

For a Lie algebra $\mathfrak{l}$, we denote the complexification and the universal enveloping algebra of $\mathfrak{l}$ by $\mathfrak{l}_{\mathbf{C}} = \mathfrak{l} \otimes_{\mathbf{R}} \mathbf{C}$ and $U(\mathfrak{l})$, respectively. Let $C^\infty(G_n)$ be the space of smooth functions on G_n and denote the right regular action of G_n on $C^\infty(G_n)$ by R, that is,

$$R(g)f(x) = f(xg) \qquad (x, g \in G_n,\ f \in C^\infty(G_n)).$$

We denote the action of $U(\mathfrak{g}_{n\mathbf{C}})$ on $C^\infty(G_n)$ determined from the differential of R again by R.

2.3. *Whittaker functions on G_n*

For $c \in F^\times$, let $C^\infty(N_n\backslash G_n; \psi_F^{(c)})$ be the space of all C^∞-functions $W\colon G_n \to \mathbf{C}$ satisfying

$$W(xg) = \psi_F^{(c)}(x_{1,2} + x_{2,3} + \cdots + x_{n-1,n})W(g), \qquad (x, g) \in N_n \times G_n.$$

Here G_n acts on this space by the right regular action R, and we topologize this space by the usual Fréchet topology.

For an irreducible smooth admissible Fréchet representation (Π, H_Π) of G_n, let $\mathrm{Hom}_{G_n}(H_\Pi, C^\infty(N_n\backslash G_n; \psi_F^{(c)}))$ be the space of continuous G_n-homomorphism from H_Π to $C^\infty(N_n\backslash G_n; \psi_F^{(c)})$. When $\mathrm{Hom}_{G_n}(H_\Pi, C^\infty(N_n\backslash G_n; \psi_F^{(c)}))$ has a non-zero element, we denote its image by $\mathcal{W}(\Pi, \psi_F^{(c)})$. The space $\mathcal{W}(\Pi, \psi_F^{(c)})$ is called the Whittaker model for Π and functions in $\mathcal{W}(\Pi, \psi_F^{(c)})$ are called (primary) Whittaker functions for Π. Shalika [Sha] shows that the space $\mathrm{Hom}_{G_n}(H_\Pi, C^\infty(N_n\backslash G_n; \psi_F^{(c)}))$ is at most one dimensional. In other words, for fixed $c \in F^\times$, the Whittaker model for Π is unique if it exists.

In [Ja] Lemma 2.5, it is shown that an irreducible smooth admissible Fréchet representation (Π, H_Π) of G_n has the Whittaker model if and only if Π is isomorphic to an irreducible generalized principal series representation defined in §2.6 and §2.7. The purpose of this article is to study the explicit formulas of Whittaker functions for all generalized principal series representations of G_3 and their application to the archimedean zeta integrals.

Remark 2.1. For $c \in F^\times$, we set $M_c = \mathrm{diag}(c^{n-1}, c^{n-2}, \cdots, c, 1) \in G_n$. Then

$$C^\infty(N_n\backslash G_n; \psi_F) \ni W(g) \mapsto W^{(c)}(g) = W(M_c g) \in C^\infty(N_n\backslash G_n; \psi_F^{(c)})$$

is a G_n-isomorphism. Hence, in this article, we only consider the case of $c = 1$.

2.4. *The radial parts of Whittaker functions*

For a K_n-type (τ, V_τ) of Π and a K_n-embedding

$$V_\tau \ni v \mapsto W(v; g) \in \mathcal{W}(\Pi, \psi_F^{(c)}),$$

we have

$$W(v; xgk) = \psi_F^{(c)}(x_{1,2} + \cdots + x_{n-1,n}) W(\tau(k)v; g) \tag{1}$$

for $(x, g, k) \in N_n \times G_n \times K_n$. Therefore, because of the Iwasawa decomposition $G_n = N_n A_n K_n$, we note that a K_n-embedding $v \mapsto W(v; g)$ $(g \in G)$ is characterized by its restriction $v \mapsto W(v; y)$ $(y \in A_n)$ to A_n. We call $v \mapsto W(v; y)$ the radial part of $v \mapsto W(v; g)$.

2.5. *The contragradient Whittaker functions*

Let (Π, H_Π) be a smooth admissible Fréchet representation of G_n which has the Whittaker model $\mathcal{W}(\Pi, \psi_F^{(c)})$. We define the representation $(\Pi^\vee, H_{\Pi^\vee})$ by $H_{\Pi^\vee} = H_\Pi$ and $\Pi^\vee(g) = \Pi({}^t g^{-1})$ $(g \in G_n)$. Then $\Pi^\vee$ is isomorphic to the contragradient representation of Π and the Whittaker model $\mathcal{W}(\Pi^\vee, \psi_F^{(-c)})$ for $\Pi^\vee$ is given by

$$\mathcal{W}(\Pi^\vee, \psi_F^{(-c)}) = \{W^\vee(g) = W(w_n {}^t g^{-1}) \mid W \in \mathcal{W}(\Pi, \psi_F^{(c)})\}$$

where

$$w_n = \begin{pmatrix} 0 & 0 & \cdots & 0 & 1 \\ 0 & & & 1 & 0 \\ \vdots & & \iddots & & \vdots \\ 0 & 1 & & & 0 \\ 1 & 0 & \cdots & 0 & 0 \end{pmatrix} \in K_n.$$

2.6. *The generalized principal series representations of* $G_n = GL(n, \mathbf{R})$

In this subsection, we set $F = \mathbf{R}$. We shall specify certain irreducible representations of G_1 and G_2 as follows.

(1) For $\nu \in \mathbf{C}$ and $\delta \in \{0, 1\}$, we define the character $\chi_{(\nu,\delta)} \colon G_1 \to \mathbf{C}^\times$ by

$$\chi_{(\nu,\delta)}(t) = \left(\frac{t}{|t|}\right)^\delta |t|_\mathbf{R}^\nu \qquad (t \in G_1 = \mathbf{R}^\times).$$

(2) Let $D_\kappa^\pm$ ($\kappa \in \mathbf{Z}_{\geq 2}$) be the discrete series representations of $SL(2,\mathbf{R})$ with the minimal $SO(2)$-type

$$SO(2) \ni k_\theta = \begin{pmatrix} \cos\theta & \sin\theta \\ -\sin\theta & \cos\theta \end{pmatrix} \mapsto \exp(\pm\sqrt{-1}\kappa\theta) \in \mathbf{C}^\times.$$

For each $\nu \in \mathbf{C}$ and $\kappa \in \mathbf{Z}_{\geq 2}$, we denote by $D_{(\nu,\kappa)}$ the irreducible smooth admissible Fréchet representation of G_2 characterized by $D_{(\nu,\kappa)}(t1_2) = t^{2\nu}$ ($t \in \mathbf{R}_+$) and $D_{(\nu,\kappa)}|_{SL(2,\mathbf{R})} \simeq D_\kappa^+ \oplus D_\kappa^-$ as $(\mathfrak{sl}(2,\mathbf{R})_\mathbf{C}, SO(2))$-modules.

Let $(n_1, n_2, \cdots, n_r)$ be a partition of n into 1's and 2's, that is, each n_i is equal to 1 or 2 and $\sum_{j=1}^r n_j = n$. For a partition $(n_1, n_2, \cdots, n_r)$, we associate the block diagonal subgroup

$$M = \left\{ \begin{pmatrix} g_1 & & & \\ & g_2 & & \\ & & \ddots & \\ & & & g_r \end{pmatrix} \in G_n \;\middle|\; \begin{matrix} g_i \in G_{n_i} \\ (1 \leq i \leq r) \end{matrix} \right\} \simeq G_{n_1} \times G_{n_2} \times \cdots \times G_{n_r},$$

and the block strictly upper triangular subgroup U. For $1 \leq i \leq r$, let $\sigma_i = \chi_{(\nu_i,\delta_i)}$ if $n_i = 1$ and $\sigma_i = D_{(\nu_i,\kappa_i)}$ if $n_i = 2$. Then $(\sigma_1, \sigma_2, \cdots, \sigma_r)$ defines a representation (σ, V_σ) of M by the tensor product, and we extend this representation to the corresponding block upper triangular subgroup $P = UM$ by

$$\sigma(um) = \sigma(m) \qquad (u \in U,\ m \in M).$$

We define the function ρ on P by

$$\rho(p) = |\det(\mathrm{Ad}_\mathfrak{u}(p))|^{\frac{1}{2}} \qquad (p \in P),$$

where $\mathrm{Ad}_\mathfrak{u}$ means the adjoint action on the Lie algebra $\mathfrak{u}$ of U. Then the generalized principal series representation $I_\mathbf{R}^n(\sigma_1, \sigma_2, \cdots, \sigma_r)$ is realized on the space of all C^∞-functions $f\colon G_n \to V_\sigma$ satisfying

$$f(pg) = \rho(p)\sigma(p)f(g), \qquad (p, g) \in P \times G_n.$$

Here G_n acts on this space by the right regular action and we topologize this space by the usual Fréchet topology.

When $I_\mathbf{R}^n(\sigma_1, \sigma_2, \cdots, \sigma_r)$ is irreducible, it is known that

$$I_\mathbf{R}^n(\sigma_1, \sigma_2, \cdots, \sigma_r) \simeq I_\mathbf{R}^n(\sigma_{w(1)}, \sigma_{w(2)}, \cdots, \sigma_{w(r)})$$

for any element w of the symmetric group $\mathfrak{S}_r$ of degree r (*cf.* [SV] Corollary 2.8). Hence, in the case of $n = 3$, it suffices to consider Whittaker functions for irreducible generalized principal series representations

$I^3_{\mathbf{R}}(\chi_{(\nu_1,\delta_1)}, \chi_{(\nu_2,\delta_2)}, \chi_{(\nu_3,\delta_3)})$ and $I^3_{\mathbf{R}}(D_{(\nu_1,\kappa_1)}, \chi_{(\nu_2,\delta_2)})$ with $(\nu_i, \delta_i) \in \mathbf{C} \times \{0,1\}$ $(1 \leq i \leq 3)$ and $\kappa_1 \in \mathbf{Z}_{\geq 2}$.

2.7. *The principal series representations of $G_n = GL(n, \mathbf{C})$*

In this subsection, we set $F = \mathbf{C}$. For $\nu \in \mathbf{C}$ and $d \in \mathbf{Z}$, we define the character $\chi_{[\nu,d]} \colon G_1 \to \mathbf{C}^\times$ by

$$\chi_{[\nu,d]}(t) = \left(\frac{t}{|t|}\right)^d |t|^\nu_{\mathbf{C}} \qquad (t \in G_1 = \mathbf{C}^\times).$$

We set

$$M = \{\mathrm{diag}(m_1, m_2, \cdots, m_n) \in G_n\} \simeq (G_1)^n, \qquad P = N_n M.$$

We define the function ρ on P by

$$\rho(p) = |\det(\mathrm{Ad}_{\mathfrak{n}_n}(p))|^{\frac{1}{2}} \qquad (p \in P),$$

where $\mathrm{Ad}_{\mathfrak{n}_n}$ means the adjoint action on $\mathfrak{n}_n$. Then the principal series representation $I^n_{\mathbf{C}}(\chi_{[\nu_1,d_1]}, \chi_{[\nu_2,d_2]}, \cdots, \chi_{[\nu_n,d_n]})$ is realized on the space of all C^∞-functions $f \colon G_n \to \mathbf{C}$ satisfying

$$f(xmg) = \rho(xm) \left\{ \prod_{i=1}^n \chi_{[\nu_i,d_i]}(m_i) \right\} f(g),$$

for $x \in N_n$, $m = \mathrm{diag}(m_1, \cdots, m_n) \in M$ and $g \in G_n$. Here G_n acts on this space by the right regular action R and we topologize this space by the usual Fréchet topology.

When $I^n_{\mathbf{C}}(\chi_{[\nu_1,d_1]}, \chi_{[\nu_2,d_2]}, \cdots, \chi_{[\nu_n,d_n]})$ is irreducible, it is known that

$$I^n_{\mathbf{C}}(\chi_{[\nu_1,d_1]}, \cdots, \chi_{[\nu_n,d_n]}) \simeq I^n_{\mathbf{C}}(\chi_{[\nu_{w(1)},d_{w(1)}]}, \cdots, \chi_{[\nu_{w(n)},d_{w(n)}]})$$

for any element w of the symmetric group $\mathfrak{S}_n$ of degree n (*cf.* [SV] Corollary 2.8). Hence, in the case $n = 3$, it suffices to consider $I^3_{\mathbf{C}}(\chi_{[\nu_1,d_1]}, \chi_{[\nu_2,d_2]}, \chi_{[\nu_3,d_3]})$ for $(\nu_i, d_i) \in \mathbf{C} \times \mathbf{Z}$ $(1 \leq i \leq 3)$ such that $d_1 \geq d_2 \geq d_3$.

3. Whittaker functions on $G_3 = GL(3, \mathbf{R})$

Throughout this section, we set $F = \mathbf{R}$.

3.1. Irreducible representations of $K_3 = O(3)$

For $\lambda \in \mathbf{Z}_{\geq 0}$, let $\tilde{V}_\lambda^{\mathbf{R}}$ be the subspace consisting of degree λ homogeneous polynomials of three variables x_1, x_2, x_3 in the polynomial ring $\mathbf{C}[x_1, x_2, x_3]$. For $k \in K_3$ and $f \in \tilde{V}_\lambda^{\mathbf{R}}$, we set

$$\tilde{\tau}_\lambda^{\mathbf{R}}(k) f(x_1, x_2, x_3) = f((x_1, x_2, x_3) \cdot k).$$

Here $(x_1, x_2, x_3) \cdot k$ is the ordinal product of matrices. We put $r_{\mathbf{R}} = x_1^2 + x_2^2 + x_3^2 \in \tilde{V}_2^{\mathbf{R}}$. Since $r_{\mathbf{R}}$ is K_3-invariant, we can define the quotient representation $\tau_\lambda^{\mathbf{R}}$ of $\tilde{\tau}_\lambda^{\mathbf{R}}$ on $V_\lambda^{\mathbf{R}} = \tilde{V}_\lambda^{\mathbf{R}}/(r_{\mathbf{R}} \cdot \tilde{V}_{\lambda-2}^{\mathbf{R}})$. Here we put $\tilde{V}_\lambda^{\mathbf{R}} = 0$ if $\lambda < 0$.

For $\delta \in \mathbf{Z}/2\mathbf{Z}$, we define the action $\tau_{(\lambda,\delta)}^{\mathbf{R}}$ of K_3 on $V_{(\lambda,\delta)}^{\mathbf{R}} = V_\lambda^{\mathbf{R}}$ by

$$\tau_{(\lambda,\delta)}^{\mathbf{R}}(k) = \det(k)^\delta \tau_\lambda^{\mathbf{R}}(k), \qquad k \in K_3.$$

Then $(\tau_{(\lambda,\delta)}^{\mathbf{R}}, V_{(\lambda,\delta)}^{\mathbf{R}})$ is an irreducible $(2\lambda+1)$-dimensional representation and the set of equivalence classes of irreducible finite dimensional representations of K_3 is exhausted by

$$\{\tau_{(\lambda,\delta)}^{\mathbf{R}} \mid \delta \in \mathbf{Z}/2\mathbf{Z},\ \lambda \in \mathbf{Z}_{\geq 0}\}.$$

We define the basis $\{v_{(\lambda;q)}^{\mathbf{R}}\}_{-\lambda \leq q \leq \lambda}$ of $V_{(\lambda,\delta)}^{\mathbf{R}} = V_\lambda^{\mathbf{R}}$ by

$$v_{(\lambda;q)}^{\mathbf{R}} = \begin{cases} (x_1 + \sqrt{-1}x_2)^q x_3^{\lambda-q} \mod r_{\mathbf{R}} \cdot \tilde{V}_{\lambda-2}^{\mathbf{R}} & \text{if } q \geq 0, \\ (-x_1 + \sqrt{-1}x_2)^{-q} x_3^{\lambda+q} \mod r_{\mathbf{R}} \cdot \tilde{V}_{\lambda-2}^{\mathbf{R}} & \text{if } q \leq 0. \end{cases}$$

If we put $K_{ij} = E_{ij} - E_{ji} \in \mathfrak{k}_3$, then $\{K_{ij} \mid 1 \leq i < j \leq 3\}$ is a basis of $\mathfrak{k}_3$. By direct computation, we see that the action of $\mathfrak{k}_{3\mathbf{C}}$ on $\{v_{(\lambda;q)}^{\mathbf{R}}\}_{-\lambda \leq q \leq \lambda}$ is given by

$$\tau_{(\lambda,\delta)}^{\mathbf{R}}(K_{12} \otimes 1) v_{(\lambda;q)}^{\mathbf{R}} = \sqrt{-1} q v_{(\lambda;q)}^{\mathbf{R}}, \tag{2}$$

$$\tau_{(\lambda,\delta)}^{\mathbf{R}}(K_{13} \otimes 1 \pm K_{23} \otimes \sqrt{-1}) v_{(\lambda;q)}^{\mathbf{R}} = -(q \mp \lambda) v_{(\lambda;q\pm 1)}^{\mathbf{R}}. \tag{3}$$

Here we denote the differential of $\tau_{(\lambda,\delta)}^{\mathbf{R}}$ again by $\tau_{(\lambda,\delta)}^{\mathbf{R}}$ and put $v_{(\lambda;q)}^{\mathbf{R}} = 0$ if $|q| > \lambda$.

We also define a system $\{v_l^{\mathbf{R}}\}_{l \in S_\lambda^{\mathbf{R}}}$ of generators of $V_{(\lambda,\delta)}^{\mathbf{R}} = V_\lambda^{\mathbf{R}}$ by

$$v_{(l_1,l_2,l_3)}^{\mathbf{R}} = x_1^{l_1} x_2^{l_2} x_3^{l_3} \mod r_{\mathbf{R}} \cdot \tilde{V}_{\lambda-2}^3,$$

$$S_\lambda^{\mathbf{R}} = \{l = (l_1, l_2, l_3) \in (\mathbf{Z}_{\geq 0})^3 \mid l_1 + l_2 + l_3 = \lambda\}.$$

By direct computation, we see that the action of K_{ij} on $v_l^{\mathbf{R}}$ is given by

$$\tau_{(\lambda,\delta)}^{\mathbf{R}}(K_{ij}) v_l^{\mathbf{R}} = -l_i v_{l-\mathbf{e}_i+\mathbf{e}_j}^{\mathbf{R}} + l_j v_{l-\mathbf{e}_j+\mathbf{e}_i}^{\mathbf{R}} \tag{4}$$

for $1 \le i \ne j \le 3$ and $l = (l_1, l_2, l_3) \in S_\lambda^{\mathbf{R}}$. Here $\mathbf{e}_1 = (1,0,0)$, $\mathbf{e}_2 = (0,1,0)$, $\mathbf{e}_3 = (0,0,1)$ and we put $v_l^{\mathbf{R}} = 0$ if $l \notin S_\lambda^{\mathbf{R}}$.

Moreover, $\{v_l^{\mathbf{R}}\}_{l \in S_\lambda^{\mathbf{R}}}$ has a symmetry for the action of $\mathfrak{S}_3$. For $w \in \mathfrak{S}_3$, let k_w be the corresponding permutation matrix, that is, the matrix in K_3 whose $(w(i), i)$-th component is 1 for $1 \le i \le 3$ and other components are 0. Then we have

$$\tau_{(\lambda,\delta)}^{\mathbf{R}}(k_w) v_{(l_1,l_2,l_3)}^{\mathbf{R}} = \mathrm{sgn}(w)^\delta v_{(l_{w^{-1}(1)}, l_{w^{-1}(2)}, l_{w^{-1}(3)})}^{\mathbf{R}} \qquad (w \in \mathfrak{S}_3). \tag{5}$$

Lemma 3.1. *(i) For $(\nu_i, \delta_i) \in \mathbf{C} \times \{0,1\}$ $(1 \le i \le 3)$, the set of K_3-types of a principal series representation $\Pi = I_{\mathbf{R}}^3(\chi_{(\nu_1,\delta_1)}, \chi_{(\nu_2,\delta_2)}, \chi_{(\nu_3,\delta_3)})$ is given by*

$$\left\{ \tau_{(\lambda,\lambda+\tilde{\delta})}^{\mathbf{R}} \;\middle|\; \lambda \in \tilde{\delta} + 2\mathbf{Z}_{\ge 0} \text{ or } \lambda \in 3 - \tilde{\delta} + 2\mathbf{Z}_{\ge 0} \right\},$$

where $\tilde{\delta} = \delta_1 + \delta_2 + \delta_3$. Moreover, the set of multiplicity one K_3-types of Π is exhausted by $\tau_{(\tilde{\delta},0)}^{\mathbf{R}}$ and $\tau_{(3-\tilde{\delta},1)}^{\mathbf{R}}$.
(ii) For $\nu_1, \nu_2 \in \mathbf{C}$, $\kappa_1 \in \mathbf{Z}_{\ge 2}$ and $\delta_2 \in \{0,1\}$, the set of K_3-types of a generalized principal series representation $\Pi = I_{\mathbf{R}}^3(D_{(\nu_1,\kappa_1)}, \chi_{(\nu_2,\delta_2)})$ is given by

$$\left\{ \tau_{(\lambda,\lambda+\kappa_1+\delta_2)}^{\mathbf{R}} \;\middle|\; \lambda \in \mathbf{Z}_{\ge \kappa_1} \right\}.$$

Moreover, the set of multiplicity one K_3-types of Π is exhausted by $\tau_{(\kappa_1+\delta_2,0)}^{\mathbf{R}}$ and $\tau_{(\kappa_1+1-\delta_2,1)}^{\mathbf{R}}$.

Proof. Since $G_3 \simeq SL(3,\mathbf{R}) \times \mathbf{R}^\times$, we obtain the assertion from the result of $SL(3,\mathbf{R})$ ([Mi] Corollary 3.6 and Lemma 5.1). □

3.2. *Whittaker functions on $G_3 = GL(3,\mathbf{R})$ at the minimal K_3-types*

Lemma 3.2. *For $s \in \mathbf{C}$, we have*

$$\Gamma_{\mathbf{R}}(s+2) = (2\pi)^{-1} s \Gamma_{\mathbf{R}}(s), \qquad \Gamma_{\mathbf{C}}(s+1) = (2\pi)^{-1} s \Gamma_{\mathbf{C}}(s). \tag{6}$$

Proof. We obtain the assertion from the functional equation $\Gamma(s+1) = s\Gamma(s)$ of the Gamma function. □

Theorem 3.1. *Let Π be an irreducible smooth admissible Fréchet representation of G_3 which has the Whittaker model $\mathcal{W}(\Pi, \psi_{\mathbf{R}})$. We take $(\lambda_\Pi, \delta_\Pi)$*

such that $\tau^{\mathbf{R}}_{(\lambda_\Pi,\delta_\Pi)}$ is the minimal K_3-type of Π. Then there exists the K_3-embedding $v \mapsto W^{(\delta_\Pi)}_\Pi(v;g)$ from $V^{\mathbf{R}}_{(\lambda_\Pi,\delta_\Pi)}$ to $\mathcal{W}(\Pi,\psi_{\mathbf{R}})$, whose radial part is given by

$$W^{(\delta_\Pi)}_\Pi(v^{\mathbf{R}}_l;y) = (\sqrt{-1})^{l_1-l_3}\frac{y_1y_2(y_2y_3)^{\tilde\nu}}{(2\pi\sqrt{-1})^2}\int_{s_2}\int_{s_1} V^{(\delta_\Pi)}_l(s_1,s_2) \times \{\Gamma_{\mathbf{R}}(s_1+s_2+l_1+l_3)\}^{-1}y_1^{-s_1}y_2^{-s_2}ds_1ds_2$$

for $y\in A_3$ and $l=(l_1,l_2,l_3)\in S^{\mathbf{R}}_{\lambda_\Pi}$. Here the paths of the integrations $\int_{s_i}$ are the vertical lines from $\mathrm{Re}(s_i)-\sqrt{-1}\infty$ to $\mathrm{Re}(s_i)+\sqrt{-1}\infty$ with sufficiently large real parts to keep the poles of the integrand on its left. The parameters $\tilde\nu, \lambda_\Pi, \delta_\Pi$ and $V^{(\delta_\Pi)}_l(s_1,s_2)$ are given as follows:
(i) When $\Pi\simeq I^3_{\mathbf{R}}(\chi_{(\nu_1,\delta_1)},\chi_{(\nu_2,\delta_2)},\chi_{(\nu_3,\delta_3)})$, we set $\tilde\nu=\nu_1+\nu_2+\nu_3$ and

$$\lambda_\Pi = \begin{cases} 0 \text{ if } \tilde\delta = 0 \text{ or } 3,\\ 1 \text{ if } \tilde\delta = 1 \text{ or } 2,\end{cases} \qquad \delta_\Pi = \begin{cases} 0 \text{ if } \tilde\delta \le 1,\\ 1 \text{ if } \tilde\delta \ge 2,\end{cases}$$

$$V^{(\delta_\Pi)}_l(s_1,s_2) = \prod_{i=1}^{3}\Gamma_{\mathbf{R}}(s_1+\nu_i+l_1+\gamma_0(l_1)\delta_{\Pi,i}) \times\Gamma_{\mathbf{R}}(s_2-\nu_i+l_3+\gamma_0(l_3)\delta_{\Pi,i}),$$

where $\tilde\delta=\delta_1+\delta_2+\delta_3$ and

$$\gamma_0(m) = \begin{cases} 1 & \text{if } m=0,\\ -1 & \text{otherwise,}\end{cases} \qquad \delta_{\Pi,i} = \begin{cases} \delta_i & \text{if } \delta_\Pi = 0,\\ 1-\delta_i & \text{if } \delta_\Pi = 1.\end{cases}$$

(ii) When $\Pi\simeq I^3_{\mathbf{R}}(D_{(\nu_1,\kappa_1)},\chi_{(\nu_2,\delta_2)})$, we set $\tilde\nu=2\nu_1+\nu_2$, $\lambda_\Pi=\kappa_1$, $\delta_\Pi=\delta_2$ and

$$V^{(\delta_\Pi)}_l(s_1,s_2) = \Gamma_{\mathbf{C}}\left(s_1+\nu_1+\frac{\kappa_1-1}{2}\right)\Gamma_{\mathbf{C}}\left(s_2-\nu_1+\frac{\kappa_1-1}{2}\right) \times\Gamma_{\mathbf{R}}(s_1+\nu_2+l_1)\Gamma_{\mathbf{R}}(s_2-\nu_2+l_3).$$

Proof. In [MIO] Proposition 7.2 and [Mi] Theorem 5.9, we obtain the explicit formulas of Whittaker functions at the minimal $SO(3)$-types of generalized principal series representations of $SL(3,\mathbf{R})$. Since $G_3\simeq SL(3,\mathbf{R})\times\mathbf{R}^\times$, by modifying these results slightly, we obtain the statement in the case of $l_2\le 1$. The statement for general l is obtained by the induction with respect to l_2. Here we use the relation

$$v^{\mathbf{R}}_{l+2\mathbf{e}_2} = -(v^{\mathbf{R}}_{l+2\mathbf{e}_1}+v^{\mathbf{R}}_{l+2\mathbf{e}_3}) \qquad (l\in S^{\mathbf{R}}_{\lambda_\Pi-2})$$

and Lemma 3.2. □

3.3. *Whittaker functions on $G_3 = GL(3,\mathbf{R})$ at the multiplicity one K_3-types*

Let $\mathfrak{p}_3$ be the orthogonal complement of $\mathfrak{k}_3$ in $\mathfrak{g}_3 = \mathfrak{gl}(3,\mathbf{R})$ with respect to the Killing form, that is,

$$\mathfrak{p}_3 = \{X \in \mathfrak{g}_3 \mid {}^tX = X\}.$$

We put $I_3 = E_{11} + E_{22} + E_{33} \in \mathfrak{g}_3$ and decompose $\mathfrak{p}_3 = Z_{\mathfrak{p}_3} \oplus \mathfrak{p}_3^\circ$ with

$$Z_{\mathfrak{p}_3} = \mathbf{R}I_3, \qquad \mathfrak{p}_3^\circ = \{X \in \mathfrak{p}_3 \mid \mathrm{tr}(X) = 0\}.$$

Here $\mathrm{tr}(X)$ means the trace of X.

It is known that $\mathfrak{p}_{3\mathbf{C}}^\circ$ becomes a K_3-module via the adjoint action Ad. Concerning this, we have the following lemma.

Lemma 3.3. *Let $\{X_l\}_{l \in S_2^{\mathbf{R}}}$ be a system of generators of $\mathfrak{p}_{3\mathbf{C}}^\circ$ defined as follows:*

$$X_{2\mathbf{e}_i} = (E_{ii} - \frac{1}{3}I_3) \otimes 1, \quad X_{\mathbf{e}_i+\mathbf{e}_j} = \frac{1}{2}(E_{ij} + E_{ji}) \otimes 1 = (E_{ij} - \frac{1}{2}K_{ij}) \otimes 1$$

for $1 \le i \ne j \le 3$. Then the linear map $\iota_{\mathfrak{p}} \colon V_{(2,0)}^{\mathbf{R}} \to \mathfrak{p}_{3\mathbf{C}}^\circ$ defined by $v_l^{\mathbf{R}} \mapsto X_l$ $(l \in S_2^{\mathbf{R}})$ is a K_3-isomorphism.

Proof. We see that $\iota_{\mathfrak{p}}$ is well-defined because $\sum_{i=1}^3 X_{2\mathbf{e}_i} = 0$. Using the formula (4), we can check

$$\iota_{\mathfrak{p}}(\tau_{(2,0)}^{\mathbf{R}}(K_{ij})v_l^{\mathbf{R}}) = \mathrm{ad}(K_{ij})X_l, \qquad \iota_{\mathfrak{p}}(\tau_{(2,0)}^{\mathbf{R}}(-1_3)v_l^{\mathbf{R}}) = \mathrm{Ad}(-1_3)X_l,$$

for $1 \le i < j \le 3$ and $l \in S_2^{\mathbf{R}}$. Here we denote the differential of Ad by ad. Since the action of K_3 is characterized by the actions of $\mathfrak{k}_3$ and -1_3, we see that $\iota_{\mathfrak{p}}$ is a K_3-isomorphism. □

The tensor product $\mathfrak{p}_{3\mathbf{C}}^\circ \otimes_{\mathbf{C}} V_{(\lambda,\delta)}^{\mathbf{R}} \simeq V_{(2,0)}^{\mathbf{R}} \otimes_{\mathbf{C}} V_{(\lambda,\delta)}^{\mathbf{R}}$ has the irreducible decomposition

$$V_{(2,0)}^{\mathbf{R}} \otimes_{\mathbf{C}} V_{(\lambda,\delta)}^{\mathbf{R}} \simeq \begin{cases} V_{(\lambda+2,\delta)}^{\mathbf{R}} \oplus V_{(\lambda+1,\delta+1)}^{\mathbf{R}} \oplus V_{(\lambda,\delta)}^{\mathbf{R}} \oplus V_{(\lambda-1,\delta+1)}^{\mathbf{R}} \oplus V_{(\lambda-2,\delta)}^{\mathbf{R}}, & \text{if } \lambda \ge 2, \\ V_{(3,\delta)}^{\mathbf{R}} \oplus V_{(2,\delta+1)}^{\mathbf{R}} \oplus V_{(1,\delta)}^{\mathbf{R}}, & \text{if } \lambda = 1, \\ V_{(2,\delta)}^{\mathbf{R}}, & \text{if } \lambda = 0. \end{cases} \tag{7}$$

Lemma 3.4. *For $(\lambda,\delta) \in \mathbf{Z}_{\ge 1} \times (\mathbf{Z}/2\mathbf{Z})$, we define the linear map $I'_{(\lambda,\delta)}$ from $V_{(\lambda+1,\delta+1)}^{\mathbf{R}}$ to $V_{(2,0)}^{\mathbf{R}} \otimes_{\mathbf{C}} V_{(\lambda,\delta)}^{\mathbf{R}}$ by*

$$I'_{(\lambda,\delta)}(v_{(\lambda+1;q)}^{\mathbf{R}}) = (\lambda+1-q)(\lambda-q)(\lambda-1-q)v_{(2;-2)}^{\mathbf{R}} \otimes v_{(\lambda;q+2)}^{\mathbf{R}}$$

$$
\begin{aligned}
&+2(\lambda+1-q)(\lambda-q)(\lambda+2+2q)v^{\mathbf{R}}_{(2;-1)}\otimes v^{\mathbf{R}}_{(\lambda;q+1)}\\
&+6(\lambda+1-q)q(\lambda+1+q)v^{\mathbf{R}}_{(2;0)}\otimes v^{\mathbf{R}}_{(\lambda;q)}\\
&-2(\lambda+1+q)(\lambda+q)(\lambda+2-2q)v^{\mathbf{R}}_{(2;1)}\otimes v^{\mathbf{R}}_{(\lambda;q-1)}\\
&-(\lambda+1+q)(\lambda+q)(\lambda-1+q)v^{\mathbf{R}}_{(2;2)}\otimes v^{\mathbf{R}}_{(\lambda;q-2)}.
\end{aligned}
$$

Then $I'_{(\lambda,\delta)}$ is a K_3-homomorphism.

Proof. Using the formulas (2) and (3), we can check

$$
\begin{aligned}
I'_{(\lambda,\delta)}(\tau^{\mathbf{R}}_{(\lambda+1,\delta+1)}(X)v^{\mathbf{R}}_{(\lambda+1;q)}) &= \tau^{\mathbf{R}}_{(2,0)}\otimes\tau^{\mathbf{R}}_{(\lambda,\delta)}(X)I'_{(\lambda,\delta)}(v^{\mathbf{R}}_{(\lambda+1;q)}),\\
I'_{(\lambda,\delta)}(\tau^{\mathbf{R}}_{(\lambda+1,\delta+1)}(-1_3)v^{\mathbf{R}}_{(\lambda+1;q)}) &= \tau^{\mathbf{R}}_{(2,0)}\otimes\tau^{\mathbf{R}}_{(\lambda,\delta)}(-1_3)I'_{(\lambda,\delta)}(v^{\mathbf{R}}_{(\lambda+1;q)}),
\end{aligned}
$$

for $X\in\{K_{12}\otimes 1,\ K_{13}\otimes 1\pm K_{23}\otimes\sqrt{-1}\}$ and $-\lambda\le q\le\lambda$. Since the action of K_3 is characterized by the actions of $\mathfrak{k}_{3\mathbf{C}}$ and -1_3, we see that $I'_{(\lambda,\delta)}$ is a K_3-homomorphism. □

Lemma 3.5. *Let $\lambda\in\mathbf{Z}_{\ge 1}$ and $\delta,\delta'\in\mathbf{Z}/2\mathbf{Z}$.*
(i) We define the linear map $B^{(\delta')}_{(\lambda,\delta)}$ from $V^{\mathbf{R}}_{(\lambda,\delta+\delta')}$ to $V^{\mathbf{R}}_{(\lambda,\delta)}\otimes_{\mathbf{C}}V^{\mathbf{R}}_{(0,\delta')}$ by $v\mapsto v\otimes 1$. Then $B^{(\delta')}_{(\lambda,\delta)}$ is a K_3-isomorphism.
(ii) There exists a K_3-homomorphism $I^{(1)}_{(\lambda,\delta)}: V^{\mathbf{R}}_{(\lambda+1,\delta+1)}\to V^{\mathbf{R}}_{(2,0)}\otimes_{\mathbf{C}}V^{\mathbf{R}}_{(\lambda,\delta)}$ which satisfies

$$
\begin{aligned}
I^{(1)}_{(\lambda,\delta)}(v^{\mathbf{R}}_{(\lambda+1)\mathbf{e}_i}) = (-1)^i\frac{\lambda-1}{2\lambda+1}\Big\{&v^{\mathbf{R}}_{\mathbf{e}_j+\mathbf{e}_k}\otimes(v^{\mathbf{R}}_{(\lambda-2)\mathbf{e}_i+2\mathbf{e}_j}-v^{\mathbf{R}}_{(\lambda-2)\mathbf{e}_i+2\mathbf{e}_k})\\
&-(v^{\mathbf{R}}_{2\mathbf{e}_j}-v^{\mathbf{R}}_{2\mathbf{e}_k})\otimes v^{\mathbf{R}}_{(\lambda-2)\mathbf{e}_i+\mathbf{e}_j+\mathbf{e}_k}\Big\}\\
+(-1)^i\frac{\lambda+2}{2\lambda+1}\Big\{&v^{\mathbf{R}}_{\mathbf{e}_i+\mathbf{e}_j}\otimes v^{\mathbf{R}}_{(\lambda-1)\mathbf{e}_i+\mathbf{e}_k}-v^{\mathbf{R}}_{\mathbf{e}_i+\mathbf{e}_k}\otimes v^{\mathbf{R}}_{(\lambda-1)\mathbf{e}_i+\mathbf{e}_j}\Big\}
\end{aligned}
$$

for $\{i,j,k\}=\{1,2,3\}$ such that $j<k$.
(iii) We define the linear map $I^{(0)}_{(2,\delta)}$ from $V^{\mathbf{R}}_{(2,\delta)}$ to $V^{\mathbf{R}}_{(2,0)}\otimes_{\mathbf{C}}V^{\mathbf{R}}_{(2,\delta)}$ by

$$
\begin{aligned}
I^{(0)}_{(2,\delta)}(v^{\mathbf{R}}_{2\mathbf{e}_i}) &= v^{\mathbf{R}}_{\mathbf{e}_j+\mathbf{e}_k}\otimes v^{\mathbf{R}}_{\mathbf{e}_j+\mathbf{e}_k}-v^{\mathbf{R}}_{2\mathbf{e}_i}\otimes v^{\mathbf{R}}_{2\mathbf{e}_i}\\
&\quad-\frac{1}{6}\sum_{\{i',j',k'\}=\{1,2,3\}}\{v^{\mathbf{R}}_{\mathbf{e}_{j'}+\mathbf{e}_{k'}}\otimes v^{\mathbf{R}}_{\mathbf{e}_{j'}+\mathbf{e}_{k'}}-v^{\mathbf{R}}_{2\mathbf{e}_{i'}}\otimes v^{\mathbf{R}}_{2\mathbf{e}_{i'}}\},
\end{aligned}
$$

$$
\begin{aligned}
I^{(0)}_{(2,\delta)}(v^{\mathbf{R}}_{\mathbf{e}_j+\mathbf{e}_k}) = \frac{1}{2}\{&v^{\mathbf{R}}_{2\mathbf{e}_i}\otimes v^{\mathbf{R}}_{\mathbf{e}_j+\mathbf{e}_k}+v^{\mathbf{R}}_{\mathbf{e}_j+\mathbf{e}_k}\otimes v^{\mathbf{R}}_{2\mathbf{e}_i}\\
&-v^{\mathbf{R}}_{\mathbf{e}_i+\mathbf{e}_j}\otimes v^{\mathbf{R}}_{\mathbf{e}_i+\mathbf{e}_k}-v^{\mathbf{R}}_{\mathbf{e}_i+\mathbf{e}_k}\otimes v^{\mathbf{R}}_{\mathbf{e}_i+\mathbf{e}_j}\},
\end{aligned}
$$

for $\{i,j,k\}=\{1,2,3\}$. Then $I^{(0)}_{(2,\delta)}$ is a K_3-homomorphism.

Proof. The statement (i) follows immediately from the definition of $\tau^{\mathbf{R}}_{(\lambda,\delta)}$.
From Lemma 3.4, we have

$$I'_{(\lambda,\delta)}(v^{\mathbf{R}}_{(\lambda+1;0)}) = (\lambda+1)\lambda(\lambda-1)\left\{v^{\mathbf{R}}_{(2;-2)}\otimes v^{\mathbf{R}}_{(\lambda;2)} - v^{\mathbf{R}}_{(2;2)}\otimes v^{\mathbf{R}}_{(\lambda;-2)}\right\}$$
$$+ 2(\lambda+1)\lambda(\lambda+2)\left\{v^{\mathbf{R}}_{(2;-1)}\otimes v^{\mathbf{R}}_{(\lambda;1)} - v^{\mathbf{R}}_{(2;1)}\otimes v^{\mathbf{R}}_{(\lambda;-1)}\right\}.$$

Since

$$v^{\mathbf{R}}_{(\lambda+1;0)} = v^{\mathbf{R}}_{(\lambda+1)\mathbf{e}_3},$$

$$\begin{aligned}
&v^{\mathbf{R}}_{(2;-2)}\otimes v^{\mathbf{R}}_{(\lambda;2)} - v^{\mathbf{R}}_{(2;2)}\otimes v^{\mathbf{R}}_{(\lambda;-2)}\\
&= 4\sqrt{-1}\{-v^{\mathbf{R}}_{\mathbf{e}_1+\mathbf{e}_2}\otimes(v^{\mathbf{R}}_{(\lambda-2)\mathbf{e}_3+2\mathbf{e}_1} - v^{\mathbf{R}}_{(\lambda-2)\mathbf{e}_3+2\mathbf{e}_2})\\
&\quad + (v^{\mathbf{R}}_{2\mathbf{e}_1} - v^{\mathbf{R}}_{2\mathbf{e}_2})\otimes v^{\mathbf{R}}_{(\lambda-2)\mathbf{e}_3+\mathbf{e}_1+\mathbf{e}_2}\},
\end{aligned}$$

$$\begin{aligned}
&v^{\mathbf{R}}_{(2;-1)}\otimes v^{\mathbf{R}}_{(\lambda;1)} - v^{\mathbf{R}}_{(2;1)}\otimes v^{\mathbf{R}}_{(\lambda;-1)}\\
&= 2\sqrt{-1}\{v^{\mathbf{R}}_{\mathbf{e}_2+\mathbf{e}_3}\otimes v^{\mathbf{R}}_{(\lambda-1)\mathbf{e}_3+\mathbf{e}_1} - v^{\mathbf{R}}_{\mathbf{e}_1+\mathbf{e}_3}\otimes v^{\mathbf{R}}_{(\lambda-1)\mathbf{e}_3+\mathbf{e}_2}\},
\end{aligned}$$

we see that the K_3-homomorphism $I^{(1)}_{(\lambda,\delta)} = \{4\sqrt{-1}\lambda(\lambda+1)(2\lambda+1)\}^{-1}I'_{(\lambda,\delta)}$ satisfies the condition in the statement (ii) for $(i,j,k) = (3,1,2)$. Using the symmetry (5), we also confirm that $I^{(1)}_{(\lambda,\delta)}$ satisfies the condition for $(i,j,k) = (1,2,3)$ and $(2,1,3)$. Hence we obtain the statement (ii).

The statement (iii) is proved directly as in the proof of Lemma 3.4, by using the formula (4). □

Let Π be an irreducible smooth admissible Fréchet representation of G_3 which has the Whittaker model $\mathcal{W}(\Pi,\psi_{\mathbf{R}})$. We denote by Φ_Π the K_3-embedding $V_{(\lambda_\Pi,\delta_\Pi)} \ni v \mapsto W_\Pi(v;g) \in \mathcal{W}(\Pi,\psi_{\mathbf{R}})$ in Theorem 3.1. We take $\lambda'_\Pi \in \mathbf{Z}_{\geq 0}$ such that $\lambda'_\Pi > \lambda_\Pi$ and $\tau_{(\lambda'_\Pi,\delta_\Pi+1)}$ is a multiplicity one K_3-type of Π, that is,

$$\lambda'_\Pi = \begin{cases} \lambda_\Pi + 3 \text{ if } \Pi \simeq I^3_{\mathbf{R}}(\chi_{(\nu_1,\delta_1)},\chi_{(\nu_2,\delta_2)},\chi_{(\nu_3,\delta_3)}) \text{ and } \delta_1=\delta_2=\delta_3,\\ \lambda_\Pi + 1 \text{ otherwise.}\end{cases}$$

We shall construct a K_3-embedding $V_{(\lambda'_\Pi,\delta_\Pi+1)} \ni v \mapsto W^{(\delta_\Pi+1)}_\Pi(v;g) \in \mathcal{W}(\Pi,\psi_{\mathbf{R}})$. We define the K_3-homomorphism $\iota'_{\mathfrak{p}}$ from $V^{\mathbf{R}}_{(2,0)}\otimes_{\mathbf{C}} V^{\mathbf{R}}_{(2,0)}\otimes_{\mathbf{C}} V^{\mathbf{R}}_{(2,0)}$ to $U(\mathfrak{g}_{3\mathbf{C}})$ by $v_1\otimes v_2\otimes v_3 \mapsto \iota_{\mathfrak{p}}(v_1)\cdot\iota_{\mathfrak{p}}(v_2)\cdot\iota_{\mathfrak{p}}(v_3)$, and set

$$\iota''_{\mathfrak{p}} = \iota'_{\mathfrak{p}}\circ(\mathrm{id}_{V^{\mathbf{R}}_{(2,0)}}\otimes_{\mathbf{C}} I^{(0)}_{(2,0)})\circ I^{(1)}_{(2,0)}\colon V^{\mathbf{R}}_{(3,1)} \to U(\mathfrak{g}_{3\mathbf{C}}).$$

Here K_3 acts on $U(\mathfrak{g}_{3\mathbf{C}})$ by the adjoint action and id_V means the identity map on V. Moreover, we define the natural K_3-homomorphism

$$P\colon U(\mathfrak{g}_{3\mathbf{C}})\otimes_{\mathbf{C}} C^\infty(N_3\backslash G_3;\psi_{\mathbf{R}}) \to C^\infty(N_3\backslash G_3;\psi_{\mathbf{R}})$$

by $X \otimes W \mapsto R(X)W$. For $v \in V^{\mathbf{R}}_{(\lambda'_\Pi,\delta_\Pi+1)}$, we define a Whittaker function $W_\Pi^{(\delta_\Pi+1)}(v;g) \in \mathcal{W}(\Pi,\psi_{\mathbf{R}})$ by

$$W_\Pi^{(\delta_\Pi+1)}(v;g) = (2\pi)^{-3}\left(P \circ (\iota''_{\mathfrak{p}} \otimes \Phi_\Pi) \circ B^{(\delta_\Pi)}_{(3,1)}(v)\right)(g)$$

if $\lambda'_\Pi = \lambda_\Pi + 3 = 3$, and

$$W_\Pi^{(\delta_\Pi+1)}(v;g) = (2\pi)^{-1}\left(P \circ (\iota_{\mathfrak{p}} \otimes \Phi_\Pi) \circ I^{(1)}_{(\lambda_\Pi,\delta_\Pi)}(v)\right)(g)$$

if $\lambda'_\Pi = \lambda_\Pi + 1$. Then we note that $v \mapsto W_\Pi^{(\delta_\Pi+1)}(v;g)$ is a K_3-isomorphism, since it is a composite of K_3-homomorphisms.

Lemma 3.6. *Let (τ, V_τ) be an irreducible representation of K_3 and let $V_\tau \ni v \mapsto W(v;g) \in C^\infty(N_n\backslash G_n;\psi_{\mathbf{R}})$ be a K_3-embedding. For $X \in U(\mathfrak{n}_{3\mathbf{C}})$, $Y \in U(\mathfrak{a}_{3\mathbf{C}})$, $Z \in U(\mathfrak{k}_{3\mathbf{C}})$, $y \in A_3$ and $v \in V_\tau$, we have*

$$R(XYZ)W(v;y) = \psi_{\mathbf{R}}(\mathrm{Ad}(y)X)R(Y)W(\tau(Z)v;y).$$

Here we denote the differentials of $\psi_{\mathbf{R}}$ and τ again by $\psi_{\mathbf{R}}$ and τ, respectively. In particular, for $y \in A_3$ and $v \in V_\tau$, we have

$$\begin{aligned}
&R(X_{(2,0,0)})W(v;y) = \left(\partial_1 - \frac{1}{3}\partial_3\right)W(v;y),\\
&R(X_{(0,2,0)})W(v;y) = \left(-\partial_1 + \partial_2 - \frac{1}{3}\partial_3\right)W(v;y),\\
&R(X_{(0,0,2)})W(v;y) = \left(-\partial_2 + \frac{2}{3}\partial_3\right)W(v;y),\\
&R(E_{i,i+1} \otimes 1)W(v;y) = 2\pi\sqrt{-1}y_iW(v;y) \qquad (i \in \{1,2\}),\\
&R(E_{13} \otimes 1)W(v;y) = 0,
\end{aligned}$$

where $\partial_i = y_i \dfrac{\partial}{\partial y_i}$.

Proof. By the definition of $C^\infty(N_n\backslash G_n;\psi_{\mathbf{R}})$ and easy computation, we obtain the assertion. □

Lemma 3.7. *We use the notation in Theorem 3.1. We set*

$$\tilde{W}_l^{(\delta)}(y_1,y_2) = (y_1y_2)^{-1}(y_2y_3)^{-\tilde{\nu}}W_\Pi^{(\delta)}(v_l^{\mathbf{R}};y) \qquad (y \in A_3,\ \delta \in \mathbf{Z}/2\mathbf{Z}).$$

(i) When $\lambda'_\Pi = \lambda_\Pi + 3 = 3$, we have

$$\begin{aligned}
&\tilde{W}_{3\mathbf{e}_2}^{(\delta_\Pi+1)} = -(2\pi)^{-1}y_1y_2(\partial_1+\partial_2+2)\tilde{W}^{(\delta_\Pi)}_{(0,0,0)},\\
&\tilde{W}_{3\mathbf{e}_1}^{(\delta_\Pi+1)} = -\sqrt{-1}y_1^2y_2\tilde{W}^{(\delta_\Pi)}_{(0,0,0)}, \qquad \tilde{W}_{3\mathbf{e}_3}^{(\delta_\Pi+1)} = \sqrt{-1}y_1y_2^2\tilde{W}^{(\delta_\Pi)}_{(0,0,0)}.
\end{aligned}$$

(ii) When $\lambda'_\Pi = \lambda_\Pi + 1$, we have

$$\begin{aligned}
\tilde{W}^{(\delta_\Pi+1)}_{(\lambda_\Pi+1)\mathbf{e}_2} &= -\frac{\lambda_\Pi - 1}{2\lambda_\Pi+1}(2\pi)^{-1}(\partial_1+\partial_2+\lambda_\Pi+2)\tilde{W}^{(\delta_\Pi)}_{(\lambda_\Pi-2)\mathbf{e}_2+\mathbf{e}_1+\mathbf{e}_3} \\
&\quad + \sqrt{-1}\frac{\lambda_\Pi+2}{2\lambda_\Pi+1}\left\{y_1\tilde{W}^{(\delta_\Pi)}_{(\lambda_\Pi-1)\mathbf{e}_2+\mathbf{e}_3} - y_2\tilde{W}^{(\delta_\Pi)}_{(\lambda_\Pi-1)\mathbf{e}_2+\mathbf{e}_1}\right\}, \\
\tilde{W}^{(\delta_\Pi+1)}_{(\lambda_\Pi+1)\mathbf{e}_1} &= \frac{\lambda_\Pi-1}{2\lambda_\Pi+1}\Big\{(2\pi)^{-1}(-\partial_1+2\partial_2-1+\tilde{\nu})\tilde{W}^{(\delta_\Pi)}_{(\lambda_\Pi-2)\mathbf{e}_1+\mathbf{e}_2+\mathbf{e}_3} \\
&\qquad - \sqrt{-1}y_2(\tilde{W}^{(\delta_\Pi)}_{(\lambda_\Pi-2)\mathbf{e}_1+2\mathbf{e}_2} - \tilde{W}^{(\delta_\Pi)}_{(\lambda_\Pi-2)\mathbf{e}_1+2\mathbf{e}_3})\Big\} \\
&\quad - \sqrt{-1}\frac{\lambda_\Pi+2}{2\lambda_\Pi+1}y_1\tilde{W}^{(\delta_\Pi)}_{(\lambda_\Pi-1)\mathbf{e}_1+\mathbf{e}_3}, \\
\tilde{W}^{(\delta_\Pi+1)}_{(\lambda_\Pi+1)\mathbf{e}_3} &= \frac{\lambda_\Pi-1}{2\lambda_\Pi+1}\Big\{(2\pi)^{-1}(-\partial_2+2\partial_1-1-\tilde{\nu})\tilde{W}^{(\delta_\Pi)}_{(\lambda_\Pi-2)\mathbf{e}_3+\mathbf{e}_1+\mathbf{e}_2} \\
&\qquad + \sqrt{-1}y_1(\tilde{W}^{(\delta_\Pi)}_{(\lambda_\Pi-2)\mathbf{e}_3+2\mathbf{e}_2} - \tilde{W}^{(\delta_\Pi)}_{(\lambda_\Pi-2)\mathbf{e}_3+2\mathbf{e}_1})\Big\} \\
&\quad + \sqrt{-1}\frac{\lambda_\Pi+2}{2\lambda_\Pi+1}y_2\tilde{W}^{(\delta_\Pi)}_{(\lambda_\Pi-1)\mathbf{e}_3+\mathbf{e}_1}.
\end{aligned}$$

Proof. We note that $(y_1y_2)^{-1}(y_2y_3)^{-\tilde{\nu}}W^{(\delta)}_\Pi(v^{\mathbf{R}}_l; y)$ does not depend on y_3, since $\Pi(t1_3) = t^{\tilde{\nu}}$ ($t \in \mathbf{R}_+$). Hence $\tilde{W}^{(\delta)}_l(y_1, y_2)$ is well-defined. Using Lemma 3.6 and the formula (4), we obtain the statement (ii) from Lemma 3.5 (ii).

Next, we prove the statement (i). For $X, Y \in U(\mathfrak{g}_{3\mathbf{C}})$, we denote

$$X \equiv Y \bmod (E_{13} \otimes \mathbf{C})U(\mathfrak{g}_{3\mathbf{C}}) + U(\mathfrak{g}_{3\mathbf{C}})\mathfrak{k}_{3\mathbf{C}}$$

simply by $X \equiv Y$. Using the relation

$$YX_1X_2 \equiv YX_2X_1 \qquad (X_1, X_2 \in \mathfrak{p}^\circ_{3\mathbf{C}},\ Y \in U(\mathfrak{g}_{3\mathbf{C}})),$$

we have

$$\begin{aligned}
(-1)^i 5\iota''_{\mathfrak{p}}(v^{\mathbf{R}}_{3\mathbf{e}_i}) \equiv\ & X_{\mathbf{e}_j+\mathbf{e}_k}(X^2_{\mathbf{e}_i+\mathbf{e}_k} - X^2_{\mathbf{e}_i+\mathbf{e}_j} + X^2_{2\mathbf{e}_k} - X^2_{2\mathbf{e}_j}) \\
& - (X_{2\mathbf{e}_j} - X_{2\mathbf{e}_k})(X_{2\mathbf{e}_i}X_{\mathbf{e}_j+\mathbf{e}_k} - X_{\mathbf{e}_i+\mathbf{e}_j}X_{\mathbf{e}_i+\mathbf{e}_k}) \\
& + 4X_{\mathbf{e}_i+\mathbf{e}_j}(X_{2\mathbf{e}_j}X_{\mathbf{e}_i+\mathbf{e}_k} - X_{\mathbf{e}_i+\mathbf{e}_j}X_{\mathbf{e}_j+\mathbf{e}_k}) \\
& - 4X_{\mathbf{e}_i+\mathbf{e}_k}(X_{2\mathbf{e}_k}X_{\mathbf{e}_i+\mathbf{e}_j} - X_{\mathbf{e}_i+\mathbf{e}_k}X_{\mathbf{e}_j+\mathbf{e}_k})
\end{aligned}$$

for $\{i, j, k\} = \{1, 2, 3\}$ such that $j < k$. Using the relations

$$\begin{aligned}
& ZX_{\mathbf{e}_j+\mathbf{e}_k}Y \equiv Z(E_{jk} \otimes 1)Y - \frac{1}{2}Z(\mathrm{ad}(K_{jk})Y), \\
& X_{2\mathbf{e}_i}(E_{jk} \otimes 1) = (E_{jk} \otimes 1)(X_{2\mathbf{e}_i} + \delta_{ij} - \delta_{jk}), \quad E_1E_2Y \equiv E_2E_1Y
\end{aligned}$$

for $1 \le i \le 3$, $1 \le j < k \le 3$, $E_1, E_2 \in \mathfrak{n}_{3\mathbf{C}}$, $Z \in U(\mathfrak{n}_{3\mathbf{C}} \oplus \mathfrak{a}_{3\mathbf{C}})$ and $Y \in U(\mathfrak{g}_{3\mathbf{C}})$, we have

$$\iota''_{\mathfrak{p}}(v^{\mathbf{R}}_{3\mathbf{e}_2}) \equiv (E_{12} \otimes 1)(E_{23} \otimes 1)(X_{2\mathbf{e}_1} - X_{2\mathbf{e}_3}),$$
$$\iota''_{\mathfrak{p}}(v^{\mathbf{R}}_{3\mathbf{e}_1}) \equiv (E_{12} \otimes 1)^2(E_{23} \otimes 1), \qquad \iota''_{\mathfrak{p}}(v^{\mathbf{R}}_{3\mathbf{e}_3}) \equiv -(E_{12} \otimes 1)(E_{23} \otimes 1)^2.$$

Here we denote by δ_{ij} the Kronecker delta, that is,

$$\delta_{ij} = \begin{cases} 1 \text{ if } i = j, \\ 0 \text{ otherwise.} \end{cases}$$

Since $R(Z)W_{\Pi}^{(\delta_\Pi)}(v^{\mathbf{R}}_{(0,0,0)}; y) = 0$ for $Z \in (E_{13} \otimes \mathbf{C})U(\mathfrak{g}_{3\mathbf{C}}) + U(\mathfrak{g}_{3\mathbf{C}})\mathfrak{k}_{3\mathbf{C}}$, we obtain the statement (i) using Lemma 3.6. □

Theorem 3.2. *Let Π be an irreducible smooth admissible Fréchet representation of G_3 which has the Whittaker model $\mathcal{W}(\Pi, \psi_{\mathbf{R}})$. We take $(\lambda, \delta) \in \mathbf{Z}_{\ge 0} \times (\mathbf{Z}/2\mathbf{Z})$ such that $\tau^{\mathbf{R}}_{(\lambda,\delta)}$ is a multiplicity one K_3-type of Π (in Lemma 3.1). Then there exists a K_3-embedding $v \mapsto W_{\Pi}^{(\delta)}(v; g)$ from $V^{\mathbf{R}}_{(\lambda,\delta)}$ to $\mathcal{W}(\Pi, \psi_{\mathbf{R}})$, which satisfies*

$$W_{\Pi}^{(\delta)}(v^{\mathbf{R}}_l; y) = (\sqrt{-1})^{l_1 - l_3} \frac{y_1 y_2 (y_2 y_3)^{\tilde{\nu}}}{(2\pi\sqrt{-1})^2} \int_{s_2} \int_{s_1} V_l^{(\delta)}(s_1, s_2)$$
$$\times \{\Gamma_{\mathbf{R}}(s_1 + s_2 + l_1 + l_3)\}^{-1} y_1^{-s_1} y_2^{-s_2} ds_1 ds_2$$

for $y \in A_3$ and $l = (l_1, l_2, l_3) \in \{\lambda \mathbf{e}_i \mid 1 \le i \le 3\}$. Here the paths of the integrations $\int_{s_i}$ are the vertical lines from $\mathrm{Re}(s_i) - \sqrt{-1}\infty$ to $\mathrm{Re}(s_i) + \sqrt{-1}\infty$ with sufficiently large real parts to keep the poles of the integrand on its left. The parameter $\tilde{\nu}$ and $V_l^{(\delta)}(s_1, s_2)$ are given as follows:

(i) When $\Pi \simeq I^3_{\mathbf{R}}(\chi_{(\nu_1,\delta_1)}, \chi_{(\nu_2,\delta_2)}, \chi_{(\nu_3,\delta_3)})$, we set $\tilde{\nu} = \nu_1 + \nu_2 + \nu_3$ and

$$V_l^{(\delta)}(s_1, s_2) = \prod_{i=1}^{3} \Gamma_{\mathbf{R}}(s_1 + \nu_i + l_1 + \gamma_0(l_1)\delta'_i)\Gamma_{\mathbf{R}}(s_2 - \nu_i + l_3 + \gamma_0(l_3)\delta'_i).$$

(ii) When $\Pi \simeq I^3_{\mathbf{R}}(D_{(\nu_1,\kappa_1)}, \chi_{(\nu_2,\delta_2)})$, we set $\tilde{\nu} = 2\nu_1 + \nu_2$ and

$$V_l^{(\delta)}(s_1, s_2) = \Gamma_{\mathbf{C}}\left(s_1 + \nu_1 - \frac{1}{2} + l_1 + \gamma_0(l_1)\frac{\kappa_1}{2}\right)$$
$$\times \Gamma_{\mathbf{C}}\left(s_2 - \nu_1 - \frac{1}{2} + l_3 + \gamma_0(l_3)\frac{\kappa_1}{2}\right)$$
$$\times \Gamma_{\mathbf{R}}(s_1 + \nu_2 + l_1 + \gamma_0(l_1)\delta'_2)\Gamma_{\mathbf{R}}(s_2 - \nu_2 + l_3 + \gamma_0(l_3)\delta'_2).$$

Here we use the notation:

$$\gamma_0(m) = \begin{cases} 1 & \text{if } m = 0, \\ -1 & \text{otherwise,} \end{cases} \qquad \delta'_i = \begin{cases} \delta_i & \text{if } \delta \equiv 0 \bmod 2, \\ 1 - \delta_i & \text{if } \delta \equiv 1 \bmod 2. \end{cases}$$

Proof. The formulas for $\delta = \delta_\Pi$ are found in Theorem 3.1. Using Lemmas 3.2 and 3.7, we also obtain the formulas for $\delta = \delta_\Pi + 1$. □

4. Whittaker functions on $G_3 = GL(3, \mathbf{C})$

Throughout this section, we set $F = \mathbf{C}$.

4.1. *Irreducible representations of $K_3 = U(3)$*

We shall recall the construction of irreducible representations of K_3, which is introduced in [GZ] §3.

Let $\Lambda = \{\lambda \in \mathbf{Z}^3 \mid \lambda_1 \geq \lambda_2 \geq \lambda_3\}$. For $\lambda \in \Lambda$, we define the space $\tilde{V}_\lambda^{\mathbf{C}}$ consisting of polynomials of six variables $x_1, x_2, x_3, x_{23}, x_{13}, x_{12}$ which are degree $(\lambda_1 - \lambda_2)$ homogeneous with respect to three variables x_1, x_2, x_3 and are degree $(\lambda_2 - \lambda_3)$ homogeneous with respect to three variables x_{23}, x_{13}, x_{12}.

For $k = (k_{ij}) \in K_3$, we set $\tilde{k} = (\tilde{k}_{ij}) \in G_3$ with

$$\tilde{k}_{ij} = \begin{vmatrix} k_{i_1 j_1} & k_{i_1 j_2} \\ k_{i_2 j_1} & k_{i_2 j_2} \end{vmatrix}$$

where $1 \leq i_1 < i_2 \leq 3,\ 1 \leq j_1 < j_2 \leq 3$ such that $i \notin \{i_1, i_2\},\ j \notin \{j_1, j_2\}$. Then we define the action of K_3 on $\tilde{V}_\lambda^{\mathbf{C}}$ by

$$\begin{aligned} &\tilde{\tau}_\lambda^{\mathbf{C}}(k) f(x_1, x_2, x_3, x_{23}, x_{13}, x_{12}) \\ &= \det(k)^{\lambda_3} f((x_1, x_2, x_3) \cdot k, (x_{23}, x_{13}, x_{12}) \cdot \tilde{k}), \end{aligned}$$

for $k \in K_3$ and $f \in \tilde{V}_\lambda^{\mathbf{C}}$. Here $(x_1, x_2, x_3) \cdot k$ and $(x_{23}, x_{13}, x_{12}) \cdot \tilde{k}$ are the ordinal products of matrices. We put $r_{\mathbf{C}} = x_1 x_{23} - x_2 x_{13} + x_3 x_{12} \in \tilde{V}_{(2,1,0)}^{\mathbf{C}}$. Since $r_{\mathbf{C}}$ is K_3-invariant, we can define the quotient representation $\tau_\lambda^{\mathbf{C}}$ of $\tilde{\tau}_\lambda^{\mathbf{C}}$ on $V_\lambda^{\mathbf{C}} = \tilde{V}_\lambda^{\mathbf{C}} / (r_{\mathbf{C}} \cdot \tilde{V}_{\lambda-(2,1,0)}^{\mathbf{C}})$. Here we put $\tilde{V}_\lambda^{\mathbf{C}} = 0$ if $\lambda \notin \Lambda$.

Then $(\tau_\lambda^{\mathbf{C}}, V_\lambda^{\mathbf{C}})$ is an irreducible representation of K_3 and the set of equivalence classes of irreducible finite dimensional representations of K_3 is exhausted by $\{\tau_\lambda^{\mathbf{C}} \mid \lambda \in \Lambda\}$.

We define a system $\{v_l^{\mathbf{C}}\}_{l \in S_\lambda^{\mathbf{C}}}$ of generators of $V_\lambda^{\mathbf{C}}$ by

$$v_{(l_1, l_2, l_3, l_{23}, l_{13}, l_{12})}^{\mathbf{C}} = x_1^{l_1} x_2^{l_2} x_3^{l_3} x_{23}^{l_{23}} x_{13}^{l_{13}} x_{12}^{l_{12}} \mod r_{\mathbf{C}} \cdot \tilde{V}_{\lambda-(2,1,0)}^{\mathbf{C}},$$

$$S_\lambda^{\mathbf{C}} = \left\{ l = (l_1, l_2, l_3, l_{23}, l_{13}, l_{12}) \in (\mathbf{Z}_{\geq 0})^6 \;\middle|\; \begin{array}{l} l_1 + l_2 + l_3 = \lambda_1 - \lambda_2, \\ l_{23} + l_{13} + l_{12} = \lambda_2 - \lambda_3 \end{array} \right\}.$$

If we put

$$E_{ij}^{\mathfrak{k}} = \frac{1}{2} \left\{ (E_{ij} - E_{ji}) \otimes 1 - (\sqrt{-1}(E_{ij} + E_{ji})) \otimes \sqrt{-1} \right\},$$

then $\{E^{\mathfrak{k}}_{ij} \mid 1 \leq i,j \leq 3\}$ is a basis of $\mathfrak{k}_{3\mathbf{C}}$. By direct computation, we have

$$\tau^{\mathbf{C}}_\lambda(E^{\mathfrak{k}}_{ii})v^{\mathbf{C}}_l = (l_i - l_{jk} + \lambda_2)v^{\mathbf{C}}_l,$$
$$\tau^{\mathbf{C}}_\lambda(E^{\mathfrak{k}}_{ij})v^{\mathbf{C}}_l = l_j v^{\mathbf{C}}_{l-\mathbf{f}_j+\mathbf{f}_i} - (-1)^k l_{jk} v^{\mathbf{C}}_{l-\mathbf{f}_{jk}+\mathbf{f}_{ik}},$$

for $\{i,j,k\} = \{1,2,3\}$ and $l \in S^{\mathbf{C}}_\lambda$. Here we put $v^{\mathbf{C}}_l = 0$ if $l \notin S^{\mathbf{C}}_\lambda$, $l_{ji} = l_{ij}$ for $1 \leq i < j \leq 3$, and

$$\mathbf{f}_1 = (1,0,0,0,0,0), \qquad \mathbf{f}_{23} = \mathbf{f}_{32} = (0,0,0,1,0,0),$$
$$\mathbf{f}_2 = (0,1,0,0,0,0), \qquad \mathbf{f}_{13} = \mathbf{f}_{31} = (0,0,0,0,1,0),$$
$$\mathbf{f}_3 = (0,0,1,0,0,0), \qquad \mathbf{f}_{12} = \mathbf{f}_{21} = (0,0,0,0,0,1).$$

Lemma 4.1 ([HO] Proposition 2.2). *Let Π be a principal series representation $I^3_{\mathbf{C}}(\chi_{[\nu_1,d_1]}, \chi_{[\nu_2,d_2]}, \chi_{[\nu_3,d_3]})$ with $(\nu_1,\nu_2,\nu_3) \in \mathbf{C}^3$ and $d = (d_1,d_2,d_3) \in \Lambda$. A necessary and sufficient condition for a representation $\tau^{\mathbf{C}}_\lambda$ of K_3 to be a constituent of the restriction $\Pi|_{K_3}$ to K_3 is that the convex closure of the subset*

$$\{(\lambda_i,\lambda_j,\lambda_k) \in \mathbf{Z}^3 \mid \{i,j,k\} = \{1,2,3\}\}$$

in $\mathbf{R}^3$ contains the point d. Moreover, the minimal K_3-type $\tau^{\mathbf{C}}_d$ occurs in Π with multiplicity one.

Remark 4.1. (i) For $k \in K_3$, the matrix $\tilde{k}$ is characterized by the equation

$$(x^k_2 \wedge x^k_3, x^k_1 \wedge x^k_3, x^k_1 \wedge x^k_2) = (x_2 \wedge x_3, x_1 \wedge x_3, x_1 \wedge x_2) \cdot \tilde{k}$$

with $(x^k_1, x^k_2, x^k_3) = (x_1,x_2,x_3) \cdot k$. From this, we easily see that the action $\tilde{\tau}^{\mathbf{C}}_\lambda$ is well-defined.
(ii) The system $\{v^{\mathbf{C}}_l\}_{l \in S^{\mathbf{C}}_\lambda}$ of generators has a symmetry for the action of $\mathfrak{S}_3$, which is similar to (5).
(iii) In the paper [HO], we use the Gelfand-Zelevinsky basis $\{f(M)\}_{M \in G(\lambda)}$ instead of $\{v^{\mathbf{C}}_l\}_{l \in S^{\mathbf{C}}_\lambda}$, to describe Whittaker functions on G_3. The correspondence between them is given as follows (*cf.* [GZ] §3):

$$f(M) = v^{\mathbf{C}}_{l(M)}, \qquad M = \begin{pmatrix} \lambda_1 \ \lambda_2 \ \lambda_3 \\ \alpha_1 \ \alpha_2 \\ \beta \end{pmatrix} \in G(\lambda)$$

with

$$l(M) = \begin{cases} (\beta - \alpha_2,\ \delta(M),\ \lambda_1 - \alpha_1,\ \lambda_2 - \alpha_2,\ 0,\ \alpha_2 - \lambda_3) & \text{if } \delta(M) \geq 0, \\ (\alpha_1 - \lambda_2,\ 0,\ \lambda_1 - \alpha_1,\ \alpha_1 - \beta,\ -\delta(M),\ \alpha_2 - \lambda_3) & \text{if } \delta(M) \leq 0, \end{cases}$$

$$\delta(M) = \alpha_1 + \alpha_2 - \beta - \lambda_2.$$

4.2. *Whittaker functions on $G_3 = GL(3,\mathbf{C})$ at the minimal K_3-types*

Theorem 4.1. *Let $\Pi = I^3_{\mathbf{C}}(\chi_{[\nu_1,d_1]},\chi_{[\nu_2,d_2]},\chi_{[\nu_3,d_3]})$ with $(\nu_1,\nu_2,\nu_3)\in\mathbf{C}^3$ and $d=(d_1,d_2,d_3)\in\Lambda$. Then there exists a K_3-embedding $v\mapsto W_\Pi(v;g)$ from $V_d^{\mathbf{C}}$ to $\mathcal{W}(\Pi,\psi_{\mathbf{C}})$, whose radial part is given by*

$$\begin{aligned} W_\Pi\left(v_l^{\mathbf{C}};y\right) &= (\sqrt{-1})^{l_1+l_{12}-l_3-l_{23}}\frac{y_1^2y_2^2(y_2y_3)^{2\tilde{\nu}}}{(2\pi\sqrt{-1})^2}\\ &\quad\times\int_{s_2}\int_{s_1}\prod_{i=1}^{3}\Gamma_{\mathbf{C}}\left(s_1+\nu_i+\frac{\zeta_i(l)}{2}\right)\Gamma_{\mathbf{C}}\left(s_2-\nu_i+\frac{\zeta_i(\hat{l})}{2}\right)\\ &\quad\times\left\{\Gamma_{\mathbf{C}}\left(s_1+s_2+\frac{\zeta_2(l)+\zeta_2(\hat{l})}{2}\right)\right\}^{-1}y_1^{-2s_1}y_2^{-2s_2}ds_1ds_2 \end{aligned}$$

for $y\in A_3$ and $l=(l_1,l_2,l_3,l_{23},l_{13},l_{12})\in S_d^{\mathbf{C}}$. Here we put

$$\zeta_1(l)=l_2+l_3+l_{23},\qquad \zeta_2(l)=l_1+l_{23},\qquad \zeta_3(l)=l_1+l_{12}+l_{13},$$
$$\hat{l}=(l_3,l_2,l_1,l_{12},l_{13},l_{23}),\qquad \tilde{\nu}=\nu_1+\nu_2+\nu_3$$

and the paths of the integrations $\int_{s_i}$ are the vertical lines from $\mathrm{Re}(s_i)-\sqrt{-1}\infty$ to $\mathrm{Re}(s_i)+\sqrt{-1}\infty$ with sufficiently large real parts to keep the poles of the integrand on its left.

Proof. For $l\in S_d^{\mathbf{C}}$ such that $\min\{l_2,l_{13}\}=0$ and a K_3-embedding $V_d^{\mathbf{C}}\ni v\mapsto W'_\Pi(v;g)\in\mathcal{W}(\Pi,\psi_{\mathbf{C}}^{(-\sqrt{-1}/2)})$, we found the explicit formula of $W'_\Pi\left(v_l^{\mathbf{C}};y\right)$ in [HO] Theorem 7.7. Hence, by Remark 2.1, we obtain the explicit formula of $W_\Pi\left(v_l^{\mathbf{C}};y\right)$ for $l\in S_d^{\mathbf{C}}$ such that $\min\{l_2,l_{13}\}=0$. The explicit formulas for general l are obtained by the induction with respect to $\min\{l_2,l_{13}\}$. Here we use the relation

$$v^{\mathbf{C}}_{l+\mathbf{e}_2+\mathbf{e}_{13}}=v^{\mathbf{C}}_{l+\mathbf{e}_1+\mathbf{e}_{23}}+v^{\mathbf{C}}_{l+\mathbf{e}_3+\mathbf{e}_{12}}\qquad (l\in S^{\mathbf{C}}_{d-(2,1,0)})$$

and Lemma 3.2. □

5. The archimedean local theory of the standard L-functions for $GL(n_1)\times GL(n_2)$ $(n_1>n_2)$

Our results on new explicit formulas of Whittaker functions are applicable to the theory of automorphic L-functions. Here we recall the archimedean L- and ϵ-factors via the Langlands parameters, and the archimedean zeta integrals for $GL(n_1)\times GL(n_2)$ $(n_1>n_2)$. See [Ja] and [Kn] for details.

5.1. *The local Langlands correspondence for $GL(n)$ over* $\mathbf{R}$

In this subsection, we set $F = \mathbf{R}$. The Langlands classification of irreducible admissible representations of G_n is given as follows:

Theorem 5.1 ([Kn] Theorem 1). *We take n_i, σ_i, ν_i $(1 \le i \le r)$ as in Subsection 2.6.*
(i) If the parameters ν_i of σ_i satisfy

$$n_1{}^{-1}\mathrm{Re}(\nu_1) \ge n_2{}^{-1}\mathrm{Re}(\nu_2) \ge \cdots \ge n_r{}^{-1}\mathrm{Re}(\nu_r),$$

then a generalized principal series representation $I_{\mathbf{R}}^n(\sigma_1, \sigma_2, \cdots, \sigma_r)$ has a unique irreducible subquotient $J_{\mathbf{R}}^n(\sigma_1, \sigma_2, \cdots, \sigma_r)$.
(ii) The representations $J_{\mathbf{R}}^n(\sigma_1, \sigma_2, \cdots, \sigma_r)$ exhaust the irreducible admissible representations of G_n, up to infinitesimal equivalence.
(iii) The representation $J_{\mathbf{R}}^n(\sigma_1, \sigma_2, \cdots, \sigma_r)$ is infinitesimally equivalent with the representation $J_{\mathbf{R}}^n(\sigma_1', \sigma_2', \cdots, \sigma_{r'}')$ if and only if $r' = r$ and there exists $w \in \mathfrak{S}_r$ such that $\sigma_i' = \sigma_{w(i)}$ for $1 \le i \le r$.

Next, we recall the theory of finite dimensional semisimple representations of the Weil group $W_{\mathbf{R}}$. The Weil group $W_{\mathbf{R}}$ for the field $\mathbf{R}$ is given by $W_{\mathbf{R}} = \mathbf{C}^\times \cup (\mathbf{C}^\times \cdot j) \subset \mathbf{H}^\times$. Here we regard $W_{\mathbf{R}}$ as a subgroup of the multiplicative group $\mathbf{H}^\times$ of the Hamilton quaternion algebra $\mathbf{H} = \mathbf{C} \oplus (\mathbf{C} \cdot j)$ ($j^2 = -1$ and for $z \in \mathbf{C}$, jzj^{-1} equals the complex conjugate of z). We recall the irreducible representations of $W_{\mathbf{R}}$ and the corresponding L- and ϵ-factors.

(1) *Characters.* We define characters ϕ_μ^δ $(\mu \in \mathbf{C}, \delta \in \{0, 1\})$ of $W_{\mathbf{R}}$ by

$$\phi_\mu^\delta(z) = |z|^{2\mu}, \qquad \phi_\mu^\delta(j) = (-1)^\delta.$$

We define the corresponding L- and ϵ-factors by

$$L(s, \phi_\mu^\delta) = \Gamma_{\mathbf{R}}(s + \mu + \delta), \qquad \epsilon(s, \phi_\mu^\delta, \psi_{\mathbf{R}}) = (\sqrt{-1})^\delta.$$

(2) *Two dimensional representations.* We define two dimensional representations $\phi_{\mu,N} \colon W_{\mathbf{R}} \to GL(2, \mathbf{C})$ $(\mu \in \mathbf{C}, N \in \mathbf{Z}_{\ge 0})$ by

$$\phi_{\mu,N}(re^{\sqrt{-1}\theta}) = \begin{pmatrix} r^{2\mu-N}e^{-\sqrt{-1}N\theta} & 0 \\ 0 & r^{2\mu-N}e^{\sqrt{-1}N\theta} \end{pmatrix} \quad (r \in \mathbf{R}_+,\ \theta \in \mathbf{R}),$$

$$\phi_{\mu,N}(j) = \begin{pmatrix} 0 & (-1)^N \\ 1 & 0 \end{pmatrix}.$$

The representation $\phi_{\mu,N}$ is irreducible for $N > 0$, while $\phi_{\mu,0}$ is equivalent to a sum of two characters ϕ_μ^0 and ϕ_μ^1. For $\phi_{\mu,N}$ $(N > 0)$, we define the corresponding L- and ϵ-factors by

$$L(s,\phi_{\mu,N})=\Gamma_{\mathbf{C}}(s+\mu),\qquad \epsilon(s,\phi_{\mu,N},\psi_{\mathbf{R}})=(\sqrt{-1})^{N+1}.$$

The set of equivalence classes of irreducible representations of $W_{\mathbf{R}}$ is exhausted by

$$\Sigma_{\mathbf{R}}=\left\{\phi_\mu^\delta \mid \mu\in\mathbf{C},\ \delta\in\{0,1\}\right\}\cup\left\{\phi_{\mu,N}\mid \mu\in\mathbf{C},\ N\in\mathbf{Z}_{\geq 1}\right\}.$$

For a finite dimensional semisimple representation ϕ of $W_{\mathbf{R}}$, we take an irreducible decomposition $\phi\simeq\bigoplus_{i=1}^m\phi_i$ with $\phi_i\in\Sigma_{\mathbf{R}}$. We define its L- and ϵ-factors by

$$L(s,\phi)=\prod_{i=1}^m L(s,\phi_i),\qquad \epsilon(s,\phi,\psi_{\mathbf{R}})=\prod_{i=1}^m\epsilon(s,\phi_i,\psi_{\mathbf{R}}).$$

By Theorem 5.1 (ii), for an irreducible admissible representation Π of G_n, there exists some $J_{\mathbf{R}}^n(\sigma_1,\sigma_2,\cdots,\sigma_r)$ which is infinitesimally equivalent with Π. Then we define the Langlands parameter $\phi[\Pi]$ of Π by $\phi[\Pi]=\bigoplus_{i=1}^r\phi[\sigma_i]$, where

$$\phi[\chi_{(\nu,\delta)}]\simeq\phi_\nu^\delta,\qquad \phi[D_{(\nu,\kappa)}]\simeq\phi_{\nu+\frac{\kappa-1}{2},\kappa-1}.$$

By Theorem 5.1 (iii), this definition is well-defined and we obtain the following theorem, which is called the local Langlands correspondence over $\mathbf{R}$.

Theorem 5.2 ([Kn] Theorem 2). *The correspondence $\Pi\leftrightarrow\phi[\Pi]$ gives a bijection between the set of infinitesimal equivalence classes of irreducible admissible representations of $G_n=GL(n,\mathbf{R})$ and the set of equivalence classes of n-dimensional semisimple representations of $W_{\mathbf{R}}$.*

5.2. *The local Langlands correspondence for $GL(n)$ over* C

In this subsection, we set $F=\mathbf{C}$. The Langlands classification of irreducible admissible representations of G_n is given as follows:

Theorem 5.3 ([Kn] Theorem 4). *(i) If $\mathrm{Re}(\nu_1)\geq\mathrm{Re}(\nu_2)\geq\cdots\geq\mathrm{Re}(\nu_n)$, then a principal series representation $I_{\mathbf{C}}^n(\chi_{[\nu_1,d_1]},\cdots,\chi_{[\nu_n,d_n]})$ has a unique irreducible subquotient $J_{\mathbf{C}}^n(\chi_{[\nu_1,d_1]},\cdots,\chi_{[\nu_n,d_n]})$.*
(ii) The representations $J_{\mathbf{C}}^n(\chi_{[\nu_1,d_1]},\cdots,\chi_{[\nu_n,d_n]})$ exhaust the irreducible admissible representations of G_n, up to infinitesimal equivalence.
(iii) The representation $J_{\mathbf{C}}^n(\chi_{[\nu_1,d_1]},\cdots,\chi_{[\nu_n,d_n]})$ is infinitesimally equivalent with the representation $J_{\mathbf{C}}^n(\chi_{[\nu_1',d_1']},\cdots,\chi_{[\nu_n',d_n']})$ if and only if there exists $w\in\mathfrak{S}_n$ such that $\chi_{[\nu_i',d_i']}=\chi_{[\nu_{w(i)},d_{w(i)}]}$ for $1\leq i\leq n$.

Next, we recall the theory of finite dimensional semisimple representations of the Weil group $W_{\mathbf{C}}$. The Weil group $W_{\mathbf{C}}$ for the field $\mathbf{C}$ is given by $W_{\mathbf{C}} = \mathbf{C}^{\times} = G_1$. Hence the characters $\chi_{[\nu,d]}$ exhaust the irreducible representations of $W_{\mathbf{C}}$, up to isomorphism. We define the L- and ϵ-factors corresponding to $\chi_{[\nu,d]}$ by

$$L(s,\chi_{[\nu,d]}) = \Gamma_{\mathbf{C}}\left(s+\nu+\frac{|d|}{2}\right), \qquad \epsilon(s,\chi_{[\nu,d]},\psi_{\mathbf{C}}) = (\sqrt{-1})^{|d|}.$$

For a finite dimensional semisimple representation ϕ of $W_{\mathbf{C}}$, we take an irreducible decomposition $\phi \simeq \bigoplus_{i=1}^{m} \chi_{[\nu_i,d_i]}$. We define its L- and ϵ-factors by

$$L(s,\phi) = \prod_{i=1}^{m} L(s,\chi_{[\nu_i,d_i]}), \qquad \epsilon(s,\phi,\psi_{\mathbf{C}}) = \prod_{i=1}^{m} \epsilon(s,\chi_{[\nu_i,d_i]},\psi_{\mathbf{C}}).$$

By Theorem 5.3 (ii), for an irreducible admissible representation Π of G_n, there exists some $J_{\mathbf{C}}^{n}(\chi_{[\nu_1,d_1]},\cdots,\chi_{[\nu_n,d_n]})$ which is infinitesimally equivalent with Π. Then we define the Langlands parameter $\phi[\Pi]$ of Π by $\phi[\Pi] = \bigoplus_{i=1}^{n} \chi_{[\nu_i,d_i]}$. By Theorem 5.3 (iii), this definition is well-defined and we obtain the local Langlands correspondence over $\mathbf{C}$:

Theorem 5.4 ([Kn] Theorem 5). *The correspondence* $\Pi \leftrightarrow \phi[\Pi]$ *gives a bijection between the set of infinitesimal equivalence classes of irreducible admissible representations of* $G_n = GL(n,\mathbf{C})$ *and the set of equivalence classes of* n*-dimensional semisimple representations of* $W_{\mathbf{C}}$.

5.3. *The archimedean zeta integrals for* $GL(n_1) \times GL(n_2)$ $(n_1 > n_2)$

Let n_1 and n_2 be positive integers such that $n_1 > n_2$. For $i = 1,2$, let Π_i be an irreducible smooth admissible Fréchet representation of G_{n_i}. Via the local Langlands correspondence, we define the archimedean L- and ϵ-factors as follows:

$$L(s,\Pi_1\times\Pi_2) = L(s,\phi[\Pi_1]\otimes\phi[\Pi_2]),$$
$$\epsilon(s,\Pi_1\times\Pi_2,\psi_F) = \epsilon(s,\phi[\Pi_1]\otimes\phi[\Pi_2],\psi_F).$$

We assume that Π_1 and Π_2 have the Whittaker models, and take a pair $(W_1,W_2) \in \mathcal{W}(\Pi_1,\psi_F^{(\epsilon)}) \times \mathcal{W}(\Pi_2,\psi_F^{(-\epsilon)})$ with $\epsilon \in \{\pm 1\}$. For $s \in \mathbf{C}$, we define the archimedean zeta integrals $Z(s,W_1,W_2)$ and $Z^{\vee}(s,W_1,W_2)$ by

$$Z(s,W_1,W_2) = \int_{N_{n_2}\backslash G_{n_2}} W_1\begin{pmatrix} h & \\ & 1_{n_1-n_2}\end{pmatrix} W_2(h)|\det(h)|_F^{s-\frac{n_1-n_2}{2}}\,dh,$$

$$Z^{\vee}(s, W_1, W_2) = \int_{N_{n_2}\backslash G_{n_2}} \int_{M_{n_1-n_2-1,n_2}(F)} W_1 \begin{pmatrix} h & & \\ x & 1_{n_1-n_2-1} & \\ & & 1 \end{pmatrix} dx$$

$$\times W_2(h)|\det(h)|_F^{s-\frac{n_1-n_2}{2}} dh,$$

where dh is the right G_{n_2}-invariant measure on $N_{n_2}\backslash G_{n_2}$ and dx is the self-dual Haar measure on $M_{n_1-n_2-1,n_2}(F) \simeq F^{(n_1-n_2-1)n_2}$. Here we note that $Z^{\vee}(s, W_1, W_2) = Z(s, W_1, W_2)$ if $n_1 = n_2 + 1$. There exists some constant $C \in \mathbf{R}$ such that $Z(s, W_1, W_2)$ and $Z^{\vee}(s, W_1, W_2)$ are absolutely convergent for $\mathrm{Re}(s) > C$.

Using the asymptotics of Whittaker functions, Jacquet and Shalika prove the following theorem.

Theorem 5.5 ([Ja] Theorems 2.1, 2.6 and 2.7). *We use the above notation.*
(i) The ratio $Z(s, W_1, W_2)/L(s, \Pi_1 \times \Pi_2)$ extends to an entire function for $s \in \mathbf{C}$ and is bounded in vertical strips.
(ii) For $(W_1, W_2) \in \mathcal{W}(\Pi_1, \psi_F) \times \mathcal{W}(\Pi_2, \psi_F^{(-1)})$, we have the local functional equation

$$\frac{Z^{\vee}(1-s, R(w_{n_1,n_2})W_1^{\vee}, W_2^{\vee})}{L(1-s, \Pi_1^{\vee} \times \Pi_2^{\vee})}$$
$$= \omega_2(-1)^{n_1-1}\epsilon(s, \Pi_1 \times \Pi_2, \psi_F)\frac{Z(s, W_1, W_2)}{L(s, \Pi_1 \times \Pi_2)},$$

where

$$w_{n_1,n_2} = \begin{pmatrix} 1_{n_2} & \\ & w_{n_1-n_2} \end{pmatrix}$$

and ω_2 is the central character of Π_2, that is, $\omega_2(t) = \Pi_2(t1_{n_2})$ $(t \in F^{\times})$.
(iii) For each $s \in \mathbf{C}$, there exists a pair $(W_1, W_2) \in \mathcal{W}(\Pi_1, \psi_F) \times \mathcal{W}(\Pi_2, \psi_F^{(-1)})$ such that $Z(s, W_1, W_2)/L(s, \Pi_1 \times \Pi_2) \neq 0$.
(iv) When $n_1 = n_2 + 1$, there exists a finite subset $\{(W_{1,i}, W_{2,i})\}_{1 \le i \le m}$ of the space $\mathcal{W}(\Pi_1, \psi_F) \times \mathcal{W}(\Pi_2, \psi_F^{(-1)})$ such that each $W_{j,i}$ is K_{n_j}-finite and

$$\sum_{i=1}^{m} Z(s, W_{1,i}, W_{2,i}) = L(s, \Pi_1 \times \Pi_2).$$

6. Calculus of the archimedean zeta integrals

Based on our new explicit formulas of Whittaker functions given in the previous sections, we now evaluate the associated archimedean zeta integrals to get an explicit description of the ratio of them to the L-factors.

6.1. The archimedean zeta integrals for $GL(3) \times GL(1)$

Let Π be an irreducible smooth admissible Fréchet representation of $G_3 = GL(3, F)$ which has the Whittaker model $\mathcal{W}(\Pi, \psi_F^{(\epsilon)})$ ($\epsilon \in \{\pm 1\}$). Let χ be a character of $G_1 = F^\times$ and denote by $\mathbf{1}$ the trivial character of G_1. Then, for $W \in \mathcal{W}(\Pi, \psi_F^{(\epsilon)})$, we have

$$\begin{aligned} &Z(s, W, \chi) = Z(s, W_\chi, \mathbf{1}), && Z^\vee(s, W, \chi) = Z^\vee(s, W_\chi, \mathbf{1}), \\ &L(s, \Pi \times \chi) = L(s, (\Pi \otimes \chi) \times \mathbf{1}), && \epsilon(s, \Pi \times \chi, \psi_F) = \epsilon(s, (\Pi \otimes \chi) \times \mathbf{1}, \psi_F) \end{aligned}$$

where $W_\chi(g) = \chi(\det(g))W(g) \in \mathcal{W}(\Pi \otimes \chi, \psi_F^{(\epsilon)})$. Hence, without loss of generalities, we may assume that $\chi = \mathbf{1}$ and omit the symbol $\mathbf{1}$ from the notations.

In the following, we evaluate the archimedean zeta integrals

$$Z(s, W) = \int_{F^\times} W\begin{pmatrix} h & 0 & 0 \\ 0 & 1 & 0 \\ 0 & 0 & 1 \end{pmatrix} |h|_F^{s-1} d_F^\times h,$$

$$Z^\vee(s, R(w_{3,1})W^\vee) = \int_{F^\times} \left\{ \int_F R(w_{3,1})W^\vee \begin{pmatrix} h & 0 & 0 \\ x & 1 & 0 \\ 0 & 0 & 1 \end{pmatrix} d_F^+ x \right\} |h|_F^{s-1} d_F^\times h$$

for certain $W \in \mathcal{W}(\Pi, \psi_F)$ using the explicit formulas of Whittaker functions in Theorems 3.2 and 4.1. The result is as follows.

Theorem 6.1. *Let Π be an irreducible smooth admissible Fréchet representation of G_3 in Theorems 3.2 and 4.1, which has the Whittaker model $\mathcal{W}(\Pi, \psi_F)$. When $F = \mathbf{C}$, we assume in addition that $d_2 = 0$. We set*

$$W_{[\Pi,t]}(g) = \begin{cases} R(y[t])W_\Pi^{(0)}(v_{\lambda \mathbf{e}_2}^{\mathbf{R}}; g) & \text{if } F = \mathbf{R}, \\ R(y[t])W_\Pi(v_{d_1 \mathbf{f}_2 - d_3 \mathbf{f}_{13}}^{\mathbf{C}}; g) & \text{if } F = \mathbf{C}, \end{cases}$$

with the element $y[t] = \mathrm{diag}(t, 1, t)$ ($t \in \mathbf{R}_+$) of A_3. Then both of the ratios $\dfrac{\epsilon(s, \Pi, \psi_F)Z(s, W_{[\Pi,t]})}{L(s, \Pi)}$ and $\dfrac{Z^\vee(1-s, R(w_{3,1})W_{[\Pi,t]}^\vee)}{L(1-s, \Pi^\vee)}$ equal to the integral

$$\frac{\epsilon(s, \Pi, \psi_F)}{2\pi\sqrt{-1}} \int_u \frac{L(u+s, \Pi^\vee)}{\Gamma_F(u+2s)} |t|_F^u du,$$

where the path of the integration $\int_u$ is the vertical line from $\mathrm{Re}(u) - \sqrt{-1}\infty$ to $\mathrm{Re}(u) + \sqrt{-1}\infty$ with sufficiently large real parts to keep the poles of the integrand on its left. Therefore, the local functional equation

$$\frac{Z^\vee(1-s, R(w_{3,1})W_{[\Pi,t]}^\vee)}{L(1-s, \Pi^\vee)} = \epsilon(s, \Pi, \psi_F) \frac{Z(s, W_{[\Pi,t]})}{L(s, \Pi)}$$

holds and for each $s \in \mathbf{C}$ there is some $t \in \mathbf{R}_+$ such that $\dfrac{Z(s, W_{[\Pi,t]})}{L(s,\Pi)} \neq 0.$

Remark 6.1. (i) For simplicity, we need the assumption for Π in the case of $F = \mathbf{C}$. It seems to be possible to generalize the result for general Π. However, the computation becomes much harder.
(ii) For the spherical principal series representation Π, this result is obtained by Hoffstein and Murty [HM].

Here, for a representation Π in Theorem 6.1, we write down $\Pi^\vee$, $\epsilon(s,\Pi,\psi_F)$, $L(s,\Pi)$ and $L(s,\Pi^\vee)$ as follows:
(1) When $\Pi \simeq I^3_{\mathbf{R}}(\chi_{(\nu_1,\delta_1)}, \chi_{(\nu_2,\delta_2)}, \chi_{(\nu_3,\delta_3)})$, we have

$$\Pi^\vee \simeq I^3_{\mathbf{R}}(\chi_{(-\nu_1,\delta_1)}, \chi_{(-\nu_2,\delta_2)}, \chi_{(-\nu_3,\delta_3)}), \quad \epsilon(s,\Pi,\psi_{\mathbf{R}}) = (\sqrt{-1})^{\delta_1+\delta_2+\delta_3},$$
$$L(s,\Pi) = \prod_{1\le i\le 3} \Gamma_{\mathbf{R}}(s+\nu_i+\delta_i), \quad L(s,\Pi^\vee) = \prod_{1\le i\le 3} \Gamma_{\mathbf{R}}(s-\nu_i+\delta_i).$$

(2) When $\Pi \simeq I^3_{\mathbf{R}}(D_{(\nu_1,\kappa_1)}, \chi_{(\nu_2,\delta_2)})$, we have

$$\Pi^\vee \simeq I^3_{\mathbf{R}}(D_{(-\nu_1,\kappa_1)}, \chi_{(-\nu_2,\delta_2)}), \quad \epsilon(s,\Pi,\psi_{\mathbf{R}}) = (\sqrt{-1})^{\kappa_1+\delta_2},$$
$$L(s,\Pi) = \Gamma_{\mathbf{C}}\left(s+\nu_1+\frac{\kappa_1-1}{2}\right)\Gamma_{\mathbf{R}}(s+\nu_2+\delta_2),$$
$$L(s,\Pi^\vee) = \Gamma_{\mathbf{C}}\left(s-\nu_1+\frac{\kappa_1-1}{2}\right)\Gamma_{\mathbf{R}}(s-\nu_2+\delta_2).$$

(3) When $\Pi \simeq I^3_{\mathbf{C}}(\chi_{[\nu_1,d_1]}, \chi_{[\nu_2,0]}, \chi_{[\nu_3,d_3]})$ with $d_1 \ge 0 \ge d_3$, we have

$$\Pi^\vee \simeq I^3_{\mathbf{C}}(\chi_{[-\nu_1,-d_1]}, \chi_{[-\nu_2,0]}, \chi_{[-\nu_3,-d_3]}), \quad \epsilon(s,\Pi,\psi_{\mathbf{C}}) = (\sqrt{-1})^{d_1-d_3},$$
$$L(s,\Pi) = \Gamma_{\mathbf{C}}\left(s+\nu_1+\frac{d_1}{2}\right)\Gamma_{\mathbf{C}}(s+\nu_2)\Gamma_{\mathbf{C}}\left(s+\nu_3-\frac{d_3}{2}\right),$$
$$L(s,\Pi^\vee) = \Gamma_{\mathbf{C}}\left(s-\nu_1+\frac{d_1}{2}\right)\Gamma_{\mathbf{C}}(s-\nu_2)\Gamma_{\mathbf{C}}\left(s-\nu_3-\frac{d_3}{2}\right).$$

6.2. *The proof of Theorem 6.1*

In order to prove Theorem 6.1, we prepare some computational lemmas.

Lemma 6.1. *(i) For $y \in F^\times$ and $\mathrm{Re}(\nu) > 1/2$, we have*

$$\int_F |x\bar{x}+1|_F^{-\nu}\psi_F(-yx)d_F^+x$$
$$= \frac{1}{\Gamma_F(2\nu)}\frac{1}{4\pi\sqrt{-1}}\int_s \Gamma_F(s)\Gamma_F(s+2\nu-1)|y|_F^{-s}ds.$$

Here the path of the integration $\int_s$ is the vertical line from $\mathrm{Re}(s)-\sqrt{-1}\infty$ *to* $\mathrm{Re}(s)+\sqrt{-1}\infty$ *with sufficiently large real part to keep the poles of the integrand on its left.*
(ii) For $z \in \mathbf{C}$,

$$\Gamma_{\mathbf{R}}(z)\Gamma_{\mathbf{R}}(z+1) = \Gamma_{\mathbf{C}}(z).$$

(iii) For $a, b, c, p, q \in \mathbf{C}$, *we have*

$$\frac{1}{4\pi\sqrt{-1}} \int_z \frac{\Gamma_F(-z+a)\Gamma_F(-z+b)\Gamma_F(-z+c)\Gamma_F(z+p)\Gamma_F(z+q)}{\Gamma_F(-z+a+b+c+p+q)} dz$$
$$= \frac{\Gamma_F(p+a)\Gamma_F(p+b)\Gamma_F(p+c)\Gamma_F(q+a)\Gamma_F(q+b)\Gamma_F(q+c)}{\Gamma_F(p+q+b+c)\Gamma_F(p+q+c+a)\Gamma_F(p+q+a+b)}.$$

Here the path of the integration $\int_z$ is the vertical line from $\mathrm{Re}(z)-\sqrt{-1}\infty$ *to* $\mathrm{Re}(z)+\sqrt{-1}\infty$ *with the real part to keep the poles of* $\Gamma_F(z+p)\Gamma_F(z+q)$ *on its left and the poles of* $\Gamma_F(-z+a)\Gamma_F(-z+b)\Gamma_F(-z+c)$ *on its right.*
(iv) For $a, b, p, q \in \mathbf{C}$, *we have*

$$\frac{1}{4\pi\sqrt{-1}} \int_z \frac{\Gamma_{\mathbf{C}}(-z+a)\Gamma_{\mathbf{R}}(-z+b)\Gamma_{\mathbf{R}}(z+p)\Gamma_{\mathbf{R}}(z+q)}{\Gamma_{\mathbf{R}}(-z+2a+b+p+q+1)} dz$$
$$= \frac{\Gamma_{\mathbf{C}}(p+a)\Gamma_{\mathbf{R}}(p+b)\Gamma_{\mathbf{C}}(q+a)\Gamma_{\mathbf{R}}(q+b)}{\Gamma_{\mathbf{C}}(p+q+a+b)\Gamma_{\mathbf{R}}(p+q+2a+1)}.$$

Here the path of the integration $\int_z$ is the vertical line from $\mathrm{Re}(z)-\sqrt{-1}\infty$ *to* $\mathrm{Re}(z)+\sqrt{-1}\infty$ *with the real part to keep the poles of* $\Gamma_{\mathbf{R}}(z+p)\Gamma_{\mathbf{R}}(z+q)$ *on its left and the poles of* $\Gamma_{\mathbf{C}}(-z+a)\Gamma_{\mathbf{R}}(-z+b)$ *on its right.*
(v) Let f be a measurable function in $L^1(G_1) = L^1(F^\times)$ *which is K_1-invariant, that is,* $f(gk) = f(g)$ $(k \in K_1)$. *Then we have*

$$\int_{F^\times} f(\xi) d_F^\times \xi = \int_{\mathbf{R}_+} f(\xi) \frac{d_F \xi}{\xi}.$$

Proof. The statement (i) is obtained from the formulas [EMOT] Chap.I 1.3 (7), [Pr] Part 0 (0.1.33) and the Mellin-Barnes type integral expression of the K-Bessel function $K_\nu(z)$ (See, for example, [Bu2] p.66 (6.5) and Lemma 1.9.1):

$$K_\nu(z) = \frac{1}{4 \cdot 2\pi\sqrt{-1}} \int_s \Gamma\left(\frac{s+\nu}{2}\right)\Gamma\left(\frac{s-\nu}{2}\right)\left(\frac{z}{2}\right)^{-s} ds.$$

The statements (ii) and (iii) are obtained immediately from the duplication formula ([WW] §12.15) and Barnes' second lemma ([Ba] §6.2) of the gamma function, respectively. The statement (iv) is obtained from (ii) and (iii). The statement (v) follows immediately from the definition of $d_F^\times \xi$ and $d_F \xi$. □

Lemma 6.2. *Let $C \in \mathbf{R}$.*
(i) Let $\varphi(\xi)$ is a continuous function on $\mathbf{R}_+$ such that $\varphi_\sigma(t) = \varphi(e^t)e^{\sigma t}$ is contained in $L^1(\mathbf{R}) \cap L^2(\mathbf{R})$ for any $\sigma > C$. Then we have

$$\frac{1}{2\pi\sqrt{-1}} \int_s \left\{ \int_{\mathbf{R}_+} \varphi(\xi)|\xi|_F^s \frac{d_F\xi}{\xi} \right\} |\xi_0|_F^s ds = \varphi(\xi_0)$$

for $\xi_0 \in \mathbf{R}_+$. Here the path of the integration $\int_s$ is the vertical line from $\sigma - \sqrt{-1}\infty$ to $\sigma + \sqrt{-1}\infty$ with $\sigma > C$.
(ii) Let $\Phi(s)$ be a holomorphic function on $\{s \in \mathbf{C} \mid \mathrm{Re}(s) > C\}$ such that $\Phi_\sigma(t) = \Phi(\sigma + \sqrt{-1}t)$ is contained in $L^1(\mathbf{R}) \cap L^2(\mathbf{R})$ for any $\sigma > C$. Then we have

$$\int_{\mathbf{R}_+} \left\{ \frac{1}{2\pi\sqrt{-1}} \int_s \Phi(s)|\xi|_F^{-s} ds \right\} |\xi|_F^{s_0} \frac{d_F\xi}{\xi} = \Phi(s_0)$$

for $s_0 \in \mathbf{C}$ such that $\mathrm{Re}(s_0) > C$. Here the path of the integration $\int_s$ is the vertical line from $\sigma - \sqrt{-1}\infty$ to $\sigma + \sqrt{-1}\infty$ with $\sigma > C$.

Proof. The statement (i) follows immediately from the Mellin inversion formula. We deduce the statement (ii) from the Fourier inversion formula, similar to the proof of the Mellin inversion formula (See, for example, [Bu2] §1.5). □

For a representation Π in Theorem 6.1, we define $\lambda[\Pi] \in \mathbf{Z}$, $W^*_{[\Pi,1]} \in \mathcal{W}(\Pi, \psi_F)$ and $M(s,\Pi)$ as follows:
(1) When $\Pi \simeq I^3_{\mathbf{R}}(\chi_{(\nu_1,\delta_1)}, \chi_{(\nu_2,\delta_2)}, \chi_{(\nu_3,\delta_3)})$, we set

$$\lambda[\Pi] = \delta_1 + \delta_2 + \delta_3, \quad W^*_{[\Pi,1]}(g) = W^{(0)}_{\Pi}(v^{\mathbf{R}}_{\lambda[\Pi]\mathbf{e}_1}; g),$$
$$M(s,\Pi) = \prod_{1\le i\le 3} \Gamma_{\mathbf{R}}(s + \nu_i + \lambda[\Pi] - \delta_i).$$

(2) When $\Pi \simeq I^3_{\mathbf{R}}(D_{(\nu_1,\kappa_1)}, \chi_{(\nu_2,\delta_2)})$, we set

$$\lambda[\Pi] = \kappa_1 + \delta_2, \quad W^*_{[\Pi,1]}(g) = W^{(0)}_{\Pi}(v^{\mathbf{R}}_{\lambda[\Pi]\mathbf{e}_1}; g),$$
$$M(s,\Pi) = \Gamma_{\mathbf{C}}\left(s + \nu_1 + \frac{\kappa_1 - 1}{2} + \delta_2\right)\Gamma_{\mathbf{R}}(s + \nu_2 + \kappa_1).$$

(3) When $\Pi \simeq I^3_{\mathbf{C}}(\chi_{[\nu_1,d_1]}, \chi_{[\nu_2,0]}, \chi_{[\nu_3,d_3]})$ with $d_1 \ge 0 \ge d_3$, we set

$$\lambda[\Pi] = (d_1 - d_3)/2, \quad W^*_{[\Pi,1]}(g) = (-1)^{-d_3} W_{\Pi}(v^{\mathbf{C}}_{d_1\mathbf{f}_1 - d_3\mathbf{f}_{23}}; g),$$
$$M(s,\Pi) = \Gamma_{\mathbf{C}}\left(s + \nu_1 - \frac{d_3}{2}\right)\Gamma_{\mathbf{C}}\left(s + \nu_2 + \frac{d_1 - d_3}{2}\right)\Gamma_{\mathbf{C}}\left(s + \nu_3 + \frac{d_1}{2}\right).$$

Lemma 6.3. *(i) For $y \in A_3$, we have*

$$\begin{aligned} W_{[\Pi,1]}(y) =& |y_1y_2|_F \\ & \times \frac{|y_2y_3|_F^{\tilde{\nu}}}{(2\pi\sqrt{-1})^2}\int_{s_2}\int_{s_1}\frac{L(s_1,\Pi)L(s_2,\Pi^\vee)}{\Gamma_F(s_1+s_2)}|y_1|_F^{-s_1}|y_2|_F^{-s_2}ds_1ds_2, \\ W^*_{[\Pi,1]}(y) =& \epsilon(s,\Pi,\psi_F)|y_1y_2|_F \\ & \times \frac{|y_2y_3|_F^{\tilde{\nu}}}{(2\pi\sqrt{-1})^2}\int_{s_2}\int_{s_1}\frac{M(s_1,\Pi)L(s_2,\Pi^\vee)}{\Gamma_F(s_1+s_2+\lambda[\Pi])}|y_1|_F^{-s_1}|y_2|_F^{-s_2}ds_1ds_2. \end{aligned}$$

Here the paths of the integrations $\int_{s_i}$ are the vertical lines from $\mathrm{Re}(s_i)-\sqrt{-1}\infty$ to $\mathrm{Re}(s_i)+\sqrt{-1}\infty$ with sufficiently large real parts to keep the poles of the integrand on its left.
(ii) For $s_1, s_2 \in \mathbf{C}$, we have

$$\begin{aligned} &\frac{1}{4\pi\sqrt{-1}}\int_u \frac{\Gamma_F(u)\Gamma_F(u-2s_1+s_2+1)L(-u+s_1,\Pi^\vee)}{\Gamma_F(-u+s_1+s_2-\tilde{\nu}+1+\lambda[\Pi])}du \\ &= \frac{L(s_1,\Pi^\vee)L(1-s_1+s_2,\Pi^\vee)}{M(s_2-\tilde{\nu}+1,\Pi)}. \end{aligned}$$

Here the path of the integration $\int_u$ is the vertical line from $\mathrm{Re}(u)-\sqrt{-1}\infty$ to $\mathrm{Re}(u)+\sqrt{-1}\infty$ with the real part to keep the poles of $\Gamma_F(u)\Gamma_F(u-2s_1+s_2+1)$ on its left and the poles of $L(-u+s_1,\Pi^\vee)$ on its right.

Proof. The statement (i) follows immediately from Theorems 3.2 and 4.1. We obtain the statement (ii) from Lemma 6.1 (iii), (iv). □

Proof of Theorem 6.1. By the definition of $W_{[\Pi,t]}$, we note that the function $G_1 \ni h \mapsto W_{[\Pi,t]}(\mathrm{diag}(h,1,1)) \in \mathbf{C}$ is K_1-invariant. Moreover, by Lemma 6.3 (i), for $h, t \in \mathbf{R}_+$, we have

$$\begin{aligned} W_{[\Pi,t]}\begin{pmatrix} h&0&0\\0&1&0\\0&0&1\end{pmatrix} &= W_{[\Pi,1]}\begin{pmatrix} ht&0&0\\0&1&0\\0&0&t\end{pmatrix} \\ &= \frac{|h|_F}{2\pi\sqrt{-1}}\int_{s_2}\int_{s_1}\frac{L(s_1,\Pi)L(s_1+s_2,\Pi^\vee)}{\Gamma_F(2s_1+s_2)}|h|_F^{-s_1}|t|_F^{s_2}ds_1ds_2. \end{aligned}$$

Here we modify the integral expression by the substitution $s_2 \to s_2 + s_1$. Therefore, by Lemmas 6.1 (v) and 6.2 (ii), we have

$$Z(s,W_{[\Pi,t]}) = \int_{\mathbf{R}_+} W_{[\Pi,t]}\begin{pmatrix} h&0&0\\0&1&0\\0&0&1\end{pmatrix}|h|_F^{s-1}\frac{d_Fh}{h}$$

$$= \frac{L(s,\Pi)}{2\pi\sqrt{-1}} \int_{s_2} \frac{L(s_2+s,\Pi^\vee)}{\Gamma_F(s_2+2s)} |t|_F^{s_2} ds_2.$$

Next, we evaluate $Z^\vee(s, R(w_{3,1})W^\vee_{[\Pi,t]})$. We set

$$\tilde{Z}^\vee(s_1,s_2) = \int_{\mathbf{R}_+} Z^\vee(s_1, R(w_{3,1})W^\vee_{[\Pi,t^{-1}]})|t|_F^{s_2} \frac{d_F t}{t}$$

$$= \int_{\mathbf{R}_+}\int_{F^\times}\int_F R(w_{3,1})W^\vee_{[\Pi,t^{-1}]}\begin{pmatrix} h & 0 & 0\\ x & 1 & 0\\ 0 & 0 & 1\end{pmatrix} |h|_F^{s_1-1}|t|_F^{s_2} d_F^+x\ d_F^\times h \frac{d_F t}{t}.$$

By direct computation, we have

$$w_3\ {}^t\left\{\begin{pmatrix} h & 0 & 0\\ x & 1 & 0\\ 0 & 0 & 1\end{pmatrix} w_{3,1}\right\}^{-1} y[t^{-1}]$$

$$= \begin{pmatrix} 1 & 0 & 0\\ 0 & 1 & -h\bar{x}/\Delta^2\\ 0 & 0 & 1\end{pmatrix}\begin{pmatrix} 1 & 0 & 0\\ 0 & 1/(t\Delta) & 0\\ 0 & 0 & \Delta/(ht)\end{pmatrix}\begin{pmatrix} 0 & 1 & 0\\ \bar{x}/\Delta & 0 & 1/\Delta\\ 1/\Delta & 0 & -x/\Delta\end{pmatrix},$$

where $\Delta = \sqrt{1+x\bar{x}}$. Since

$$\tau^{\mathbf{R}}_{(\lambda,0)}\begin{pmatrix} 0 & 1 & 0\\ x/\Delta & 0 & 1/\Delta\\ 1/\Delta & 0 & -x/\Delta\end{pmatrix} v^{\mathbf{R}}_{\lambda\mathbf{e}_2} = v^{\mathbf{R}}_{\lambda\mathbf{e}_1} \quad (x\in\mathbf{R}),$$

$$\tau^{\mathbf{C}}_{(d_1,0,d_3)}\begin{pmatrix} 0 & 1 & 0\\ \bar{x}/\Delta & 0 & 1/\Delta\\ 1/\Delta & 0 & -x/\Delta\end{pmatrix} v^{\mathbf{C}}_{d_1\mathbf{f}_2-d_3\mathbf{f}_{13}} = (-1)^{-d_3} v^{\mathbf{C}}_{d_1\mathbf{f}_1-d_3\mathbf{f}_{23}} \quad (x\in\mathbf{C}),$$

we have

$$R(w_{3,1})W^\vee_{[\Pi,t^{-1}]}\begin{pmatrix} h & 0 & 0\\ x & 1 & 0\\ 0 & 0 & 1\end{pmatrix} = \psi_F\left(-h\bar{x}/\Delta^2\right) W^*_{[\Pi,1]}\begin{pmatrix} 1 & 0 & 0\\ 0 & 1/(t\Delta) & 0\\ 0 & 0 & \Delta/(ht)\end{pmatrix}.$$

Therefore we have

$$\tilde{Z}^\vee(s_1,s_2) = \int_{\mathbf{R}_+}\int_{F^\times}\int_F \psi_F\left(-h\bar{x}/\Delta^2\right) W^*_{[\Pi,1]}\begin{pmatrix} 1 & 0 & 0\\ 0 & 1/(t\Delta) & 0\\ 0 & 0 & \Delta/(ht)\end{pmatrix}$$
$$\times |h|_F^{s_1-1}|t|_F^{s_2} d_F^+x\ d_F^\times h \frac{d_F t}{t}$$

$$= \int_{(\mathbf{R}_+)^2}\int_F \psi_F\left(-h\bar{x}/\Delta^2\right) W^*_{[\Pi,1]}\begin{pmatrix} 1 & 0 & 0\\ 0 & 1/(t\Delta) & 0\\ 0 & 0 & \Delta/(ht)\end{pmatrix}$$

$$\times |h|_F^{s_1-1}|t|_F^{s_2} d_F^+ x \frac{d_F h}{h}\frac{d_F t}{t}.$$

Here the second equality is obtained by Lemma 6.1 (v). Substituting $(x, h, t) \to (\bar{x}, \Delta^2 h, \Delta^{-1} t)$, we have

$$\tilde{Z}^\vee(s_1, s_2) = \int_{(\mathbf{R}_+)^2} \left\{ \int_F |x\bar{x}+1|_F^{s_1-\frac{s_2}{2}-1} \psi_F(-hx) d_F^+ x \right\}$$
$$\times W^*_{[\Pi,1]} \begin{pmatrix} 1 & 0 & 0 \\ 0 & 1/t & 0 \\ 0 & 0 & 1/(ht) \end{pmatrix} |h|_F^{s_1-1}|t|_F^{s_2} \frac{d_F h}{h}\frac{d_F t}{t}.$$

We apply Lemma 6.1 (i) to this equation with $(y, \nu) = (h, -s_1 + \frac{s_2}{2} + 1)$. Then we have

$$\tilde{Z}^\vee(s_1, s_2) = \frac{1}{\Gamma_F(-2s_1+s_2+2)} \frac{1}{4\pi\sqrt{-1}} \int_u \Gamma_F(u)\Gamma_F(u-2s_1+s_2+1)$$
$$\times \int_{(\mathbf{R}_+)^2} W^*_{[\Pi,1]} \begin{pmatrix} 1 & 0 & 0 \\ 0 & 1/t & 0 \\ 0 & 0 & 1/(ht) \end{pmatrix} |h|_F^{-u+s_1-1}|t|_F^{s_2} \frac{d_F h}{h}\frac{d_F t}{t} du.$$

By Lemmas 6.2 (ii) and 6.3 (i), we know that

$$\int_{(\mathbf{R}_+)^2} W^*_{[\Pi,1]} \begin{pmatrix} 1 & 0 & 0 \\ 0 & 1/t & 0 \\ 0 & 0 & 1/(ht) \end{pmatrix} |h|_F^{-u+s_1-1}|t|_F^{s_2} \frac{d_F h}{h}\frac{d_F t}{t}$$
$$= \epsilon(s, \Pi, \psi_F) \frac{M(s_2-\tilde{\nu}+1, \Pi) L(-u+s_1, \Pi^\vee)}{\Gamma_F(-u+s_1+s_2-\tilde{\nu}+1+\lambda[\Pi])}.$$

Therefore, we have

$$\tilde{Z}^\vee(s_1, s_2) = \epsilon(s, \Pi, \psi_F) \frac{M(s_2-\tilde{\nu}+1, \Pi)}{\Gamma_F(-2s_1+s_2+2)}$$
$$\times \frac{1}{4\pi\sqrt{-1}} \int_u \frac{\Gamma_F(u)\Gamma_F(u-2s_1+s_2+1) L(-u+s_1, \Pi^\vee)}{\Gamma_F(-u+s_1+s_2-\tilde{\nu}+1+\lambda[\Pi])} du.$$

By Lemma 6.3 (ii), we have

$$\tilde{Z}^\vee(s_1, s_2) = \epsilon(s, \Pi, \psi_F) \frac{L(s_1, \Pi^\vee) L(1-s_1+s_2, \Pi^\vee)}{\Gamma_F(-2s_1+s_2+2)}.$$

By Lemma 6.2 (i), we have

$$Z^\vee(1-s, R(w_{3,1}) W^\vee_{[\Pi,t^{-1}]}) = \frac{1}{2\pi\sqrt{-1}} \int_{s_2} \tilde{Z}^\vee(1-s, s_2) |t|_F^{-s_2} ds_2$$
$$= \epsilon(s, \Pi, \psi_F) \frac{L(1-s, \Pi^\vee)}{2\pi\sqrt{-1}} \int_{s_2} \frac{L(s_2+s, \Pi^\vee)}{\Gamma_F(s_2+2s)} |t|_F^{-s_2} ds_2. \qquad \square$$

6.3. *The archimedean zeta integrals for* $GL(3) \times GL(2)$

Let us consider the archimedean zeta integrals for $G_3 \times G_2$. In the case of $n_1 = n_2 + 1$, we expect that the archimedean zeta integrals for appropriate Whittaker functions equal to the associated L-factors. Actually, there are the following results.

Theorem 6.2 ([St1]). *We set $F = \mathbf{R}$ and assume that $n_1 = n_2 + 1$. Let Π_i be an irreducible spherical principal series representation $I_{\mathbf{R}}^{n_i}(\chi_{[\nu_1,0]}, \chi_{[\nu_2,0]}, \cdots, \chi_{[\nu_{n_i},0]})$ of G_{n_i} for $i = 1, 2$. If we take spherical Whittaker functions $W_1 \in \mathcal{W}(\Pi_1, \psi_{\mathbf{R}})$ and $W_2 \in \mathcal{W}(\Pi_2, \psi_{\mathbf{R}}^{(-1)})$ with some normalization, then we have*

$$Z(s, W_1, W_2) = L(s, \Pi_1 \times \Pi_2).$$

Theorem 6.3 ([Po]). *Let Π_1 be an irreducible smooth admissible Fréchet representation of G_2 which has Whittaker model. Let χ_2 be a character of G_1. Then there exists a K_2-finite function $W_1 \in \mathcal{W}(\Pi_1, \psi_F)$ such that*

$$Z(s, W_1, \chi_2) = L(s, \Pi_1 \times \chi_2).$$

In addition to the above, we can announce the following theorem which supports our expectation.

Theorem 6.4 ([HIM]). *We set $F = \mathbf{R}$. Let Π_1 and Π_2 be irreducible smooth admissible Fréchet representations of G_3 and G_2, respectively. We assume that Π_1 and Π_2 have the Whittaker models. Then there exists a pair $(W_1, W_2) \in \mathcal{W}(\Pi_1, \psi_{\mathbf{R}}) \times \mathcal{W}(\Pi_2, \psi_{\mathbf{R}}^{(-1)})$ such that W_1 is K_3-finite, W_2 is K_2-finite and*

$$Z(s, W_1, W_2) = L(s, \Pi_1 \times \Pi_2),$$
$$Z(s, W_1^\vee, W_2^\vee) = \epsilon(s, \Pi_1 \times \Pi_2, \psi_{\mathbf{R}}) L(s, \Pi_1^\vee \times \Pi_2^\vee).$$

We will publish the above result elsewhere together with the study in the case of $F = \mathbf{C}$.

References

[Ba]. W. N. Bailey. *Generalized hypergeometric series.* Cambridge Tracts in Mathematics and Mathematical Physics, No. 32. Stechert-Hafner, Inc., New York, 1964.

[Bu1]. Daniel Bump. Barnes' second lemma and its application to Rankin-Selberg convolutions. *Amer. J. Math.*, Vol. 110, No. 1, pp. 179–185, 1988.

[Bu2]. Daniel Bump. *Automorphic forms and representations*, Vol. 55 of *Cambridge Studies in Advanced Mathematics.* Cambridge University Press, Cambridge, 1997.

[Co]. James W. Cogdell. Lectures on L-functions, converse theorems, and functoriality for GL_n. In *Lectures on automorphic L-functions*, Vol. 20 of *Fields Inst. Monogr.*, pp. 1–96. Amer. Math. Soc., Providence, RI, 2004.

[EMOT]. A. Erdélyi, W. Magnus, F. Oberhettinger, and F. G. Tricomi. *Tables of integral transforms. Vol. I.* McGraw-Hill Book Company, Inc., New York-Toronto-London, 1954. Based, in part, on notes left by Harry Bateman.

[GZ]. I. M. Gel′fand and A. V. Zelevinskiĭ. A canonical basis in irreducible representations of gl_3 and its applications. In *Group-theoretic methods in physics, Vol. 2 (Russian) (Jūrmala, 1985)*, pp. 31–45. "Nauka", Moscow, 1986.

[HIM]. Miki Hirano, Taku Ishii, and Tadashi Miyazaki. Calculus of archimedean zeta integrals for $\mathrm{GL}(3) \times \mathrm{GL}(2)$. *in preparation.*

[HO]. Miki Hirano and Takayuki Oda. Calculus of principal series Whittaker functions on $\mathrm{GL}(3, \mathbf{C})$. *J. Funct. Anal.*, Vol. 256, No. 7, pp. 2222–2267, 2009.

[HM]. Jeff Hoffstein and M. Ram Murty. L-series of automorphic forms on $\mathrm{GL}(3, \mathbf{R})$. In *Théorie des nombres (Quebec, PQ, 1987)*, pp. 398–408. de Gruyter, Berlin, 1989.

[Ja]. Hervé Jacquet. Archimedean Rankin-Selberg integrals. In *Automorphic forms and L-functions II. Local aspects*, Vol. 489 of *Contemp. Math.*, pp. 57–172. Amer. Math. Soc., Providence, RI, 2009.

[Kn]. A. W. Knapp. Local Langlands correspondence: the Archimedean case. In *Motives (Seattle, WA, 1991)*, Vol. 55 of *Proc. Sympos. Pure Math.*, pp. 393–410. Amer. Math. Soc., Providence, RI, 1994.

[MIO]. Hiroyuki Manabe, Taku Ishii, and Takayuki Oda. Principal series Whittaker functions on $\mathrm{SL}(3, R)$. *Japan. J. Math. (N.S.)*, Vol. 30, No. 1, pp. 183–226, 2004.

[Mi]. Tadashi Miyazaki. Whittaker functions for generalized principal series representations of $\mathrm{SL}(3, \mathbf{R})$. *Manuscripta Math.*, Vol. 128, No. 1, pp. 107–135, 2009.

[Po]. Alexandru A. Popa. Whittaker newforms for Archimedean representations. *J. Number Theory*, Vol. 128, No. 6, pp. 1637–1645, 2008.

[Pr]. Nikolai Proskurin. *Cubic metaplectic forms and theta functions*, Vol. 1677 of *Lecture Notes in Mathematics.* Springer-Verlag, Berlin, 1998.

[Sha]. J. A. Shalika. The multiplicity one theorem for GL_n. *Ann. of Math. (2)*, Vol. 100, pp. 171–193, 1974.

[SV]. Birgit Speh and David A. Vogan, Jr. Reducibility of generalized principal series representations. *Acta Math.*, Vol. 145, No. 3-4, pp. 227–299, 1980.

[St1]. Eric Stade. Mellin transforms of $\mathrm{GL}(n, \mathbb{R})$ Whittaker functions. *Amer. J. Math.*, Vol. 123, No. 1, pp. 121–161, 2001.

[St2]. Eric Stade. Archimedean L-factors on $\mathrm{GL}(n) \times \mathrm{GL}(n)$ and generalized Barnes integrals. *Israel J. Math.*, Vol. 127, pp. 201–219, 2002.

[WW]. E. T. Whittaker and G. N. Watson. *A course of modern analysis.* Cambridge Mathematical Library. Cambridge University Press, Cambridge, 1996. An introduction to the general theory of infinite processes and of analytic functions; with an account of the principal transcendental functions, Reprint of the fourth (1927) edition.

Arithmetic properties of p-adic elliptic logarithmic functions

Noriko Hirata-Kohno

Department of Mathematics, College of Science and Technology, Nihon University, Suruga-Dai, Kanda, Chiyoda, Tokyo 101-8308, Japan
E-mail: hirata@math.cst.nihon-u.ac.jp

Dedicated to Professor Takayuki Oda for his 60th birthday

In this article, we present arithmetic properties of p-adic elliptic logarithmic functions and p-adic elliptic polylogarithmic functions defined by using a formal group on elliptic curves.

Keywords: elliptic curve; p-adic elliptic logarithmic function; p-adic elliptic polylogarithmic function; formal group

1. Introduction

Let K be a number field of finite degree over $\mathbb{Q}$. Let $\mathcal{E}$ be an elliptic curve defined by the Weierstraß equation: $Y^2 = X^3 - AX - B$ $(A, B \in K)$. Let S be a finite set of places in $\mathbb{Q}$ including the infinite place. Consider the set of rational S-integers: $\mathfrak{O}_S = \{x \in \mathbb{Q} : |x|_v \leq 1 \, for \, \forall v \notin S\}$. Denote the set of S-integer points of $\mathcal{E}$ by $\mathcal{E}(\mathfrak{O}_S) = \{(X, Y, 1) \in \mathbb{P}^2(\mathbb{Q}) : |X|_v \leq 1, \, |Y|_v \leq 1 \, for \, \forall v \notin S\} \cup \{(0, 1, 0)\}$.
The finiteness of the set of the integer points $\mathcal{E}(\mathbb{Z})$ was proven by C. L. Siegel in 1929 and generalized by K. Mahler to the finiteness of $\mathcal{E}(\mathfrak{O}_S)$ with arbitrary finite S. By using Baker's theory on linear forms in logarithms, the problem to effectively determine the points of $\mathcal{E}(\mathbb{Z})$ is reduced to the question of finding explicit lower bounds for linear forms in ordinary logarithms. Similarly, to effectively determine the points of $\mathcal{E}(\mathfrak{O}_S)$, those for linear forms in p-adic logarithms are required.
Alternatively, using the theory of linear forms in *elliptic* logarithms, we are able to supply an improved bound for the height of the integer points in $\mathcal{E}(\mathbb{Z})$, provided that a basis of the Mordell-Weil group $\mathcal{E}(\mathbb{Q})$ be known. Such a direct approach (not *via* unit equations) was proposed by S. Lang [16] [17] and developed by D. W. Masser [20], D. Zagier [35]. Instead of the

exponential function, we consider the Weierstraß elliptic function $\wp$ viewed as the exponential map of the elliptic curve. An effective lower bound for linear forms in arbitrary n elliptic logarithms without specific condition on the elliptic curve was first obtained by P. Philippon and M. Waldschmidt [21], and the best possible estimate concerning with the height of the coefficients of the linear forms, was obtained by S. David and the author [9] in 2009. The latter [9] solves a conjecture of S. Lang dating back to 1964. For a historical survey of the subject, the readers are invited to see [1] [7] [9] [14] [30] [33].

It is thus natural to try to get an explicit lower bound for linear forms in p-adic elliptic logarithms, for the purpose of determining the element of $\mathcal{E}(\mathfrak{O}_S)$. In 1996, G. Rémond and F. Urfels [22] obtained an explicit estimate for a linear combination of two p-adic elliptic logarithms. This estimate is refined by R. Takada and the author [14] by combining the argument of [9] and that of [22]. We note that explicit lower bounds for linear forms in p-adic logarithms, not elliptic, are mainly obtained by Kunrui Yu [31] [33] [34].

The main idea to allow us to arrive at the best possible estimate in [14] is the use of p-adic *elliptic logarithmic function*, as is employed in [5] [7] [9], instead of p-adic *elliptic function* in the traditional proof of Diophantine approximations. Indeed, the algebraic Taylor coefficients at the origin of p-adic elliptic logarithmic function have better estimate of the height than those of p-adic elliptic function. Thus p-adic elliptic logarithmic function plays a crucial role from an arithmetic viewpoint.

Our aim here is then to collect arithmetic properties of p-adic elliptic logarithmic function, moreover, to introduce p-adic elliptic polylogarithmic function in a natural way with an estimate of height of the algebraic Taylor coefficients of the function.

2. Definition of p-adic elliptic logarithmic and polylogarithmic functions

We introduce the following notation. Let $\overline{\mathbb{Q}}$ be the algebraic closure of $\mathbb{Q}$ in $\mathbb{C}$ and K be a number field of finite degree D over $\mathbb{Q}$ with $\mathfrak{O} = \mathfrak{O}_K$ the ring of integers. Let p be a rational prime $\in \mathbb{Q}$ and $|\cdot|_\infty$ be an Archimedean valuation on K. For a place v of K over p, we write the valuation $|\cdot|_v$ normalized such that $|x|_v = p^{-ord_p(x)}$ for $x \in \mathbb{Q}$. Denote K_v the completion of K by v, and write $\mathbb{Q}_p$ the completion of $\mathbb{Q}$ by p. The field K_v is a finite extension of $\mathbb{Q}_p$ of local degree $n_v = [K_v : \mathbb{Q}_p]$ with $\sum_{v|p} n_v = D$. Write

$\mathbb{C}_p$ the completion of the algebraic closure of K_v. It is well-known that the algebraic closure of K_v is not complete itself but $\mathbb{C}_p$ is an algebraically closed complete field of characteristic 0, in which the algebraic closure of K_v is dense and that there are D distinct embeddings of K into $\mathbb{C}_p$. Denote again by $|\cdot|_v$ the extension of $|\cdot|_v$ on $\mathbb{C}_p$.

For $\underline{x} \in \mathbb{P}^N(\overline{\mathbb{Q}})$ having coordinates $\underline{x} = (x_0, \ldots, x_N) \in \mathbb{P}^N(K)$, define *the absolute logarithmic height* of $\underline{x}$ by

$$h(\underline{x}) = \frac{1}{[K:\mathbb{Q}]} \sum_v n_v \log(\max\{|x_0|_v, \ldots, |x_N|_v\})$$

where the sum runs over all the normalized places of K. This height is independent of the choice of the projective coordinates and of the choice of the field containing $x_0, \ldots, x_N$. Set $h(a) = h(1:a)$ with $(1:a) \in \mathbb{P}^1(\overline{\mathbb{Q}})$, the absolute logarithmic height of the algebraic number $a \in \overline{\mathbb{Q}}$.

Let $\mathcal{E}$ be an elliptic curve defined over K, equipped with a Weierstraß model:

$$Y^2 = X^3 - AX - B \ \ (A, B \in K, \ \Delta = 4A^3 - 27B^2 \neq 0).$$

We may assume $A, B \in \mathfrak{O}$; since, if either A or B does not lie in $\mathfrak{O}$, then there exists a suitable $c \in \mathfrak{O}$ such that the elliptic curve $\mathcal{E}' : Y^2 = X^3 - A'X - B'$ with $A' = c^4 A \in \mathfrak{O}$, $B' = c^6 B \in \mathfrak{O}$. The discriminant of $\mathcal{E}'$ is $\Delta' = c^{12}\Delta$, thus $\mathcal{E}'$ is isomorphic to $\mathcal{E}$ since the j-invariant remains equal under these multiplications. We denote further by

$$h = h(\mathcal{E}) := \max\{1, h(1, A, B)\}$$

the height of the elliptic curve $\mathcal{E}$.

We recall here the Lutz-Weil p-adic elliptic function which corresponds to the p-adic version of the Weierstraß elliptic function $\wp$. Put $\lambda_p = \dfrac{1}{p-1}$ if $p \neq 2$, and $\lambda_2 = 3$. We set $\mathcal{C}_p := \{z \in \mathbb{C}_p : |z|_v < p^{-\lambda_p}\}$ and $\mathcal{C}_v := \mathcal{C}_p \cap K_v$. E. Lutz [18] and A. Weil [29] showed (see also [4]) that there exist two solutions φ and $-\varphi$ to the differential equation $(\varphi')^2 = 1 - A\varphi^4 - B\varphi^6$ with $\varphi(0) = 0$, defined over $\mathcal{C}_v \to K_v$, analytic in $\mathcal{C}_v$ (p. 151 of [2]). In fact putting $\varphi^2 = \dfrac{1}{\wp_\varphi}$ we have $\left(\dfrac{\wp'_\varphi}{2}\right)^2 = \wp_\varphi^3 - A\wp_\varphi - B$ and $\varphi'(0) = 1$. The function $\varphi(z)$ is called *the Lutz-Weil p-adic elliptic function*. The structure of the p-adic Lie-group $\mathcal{E}(\mathbb{C}_p) \subset \mathbb{P}^2(\mathbb{C}_p)$ is given, and the exponential map is expressed by the Lutz-Weil p-adic elliptic function as follows. The domain of the definition of the function φ is enlarged to $\mathcal{C}_p$.

Definition 1. For the p-adic Lie-group $\mathcal{E}(\mathbb{C}_p) \subset \mathbb{P}^2(\mathbb{C}_p)$ we have the exponential map:

$$\exp = \exp_{\mathcal{E}} : \mathcal{C}_p \to \mathcal{E}(\mathbb{C}_p) \subset \mathbb{P}^2(\mathbb{C}_p)$$

$$z \mapsto (\varphi(z), -\varphi'(z), \varphi^3(z)) = (\frac{\varphi(z)}{\varphi'(z)}, -1, \frac{\varphi^3(z)}{\varphi'(z)}).$$

Our elliptic curve is parameterized by $(X, Y, Z) = (\varphi, -\varphi', \varphi^3)$. This p-adic exponential map is locally analytic only. The function φ is odd and injective; indeed, $|\varphi(z)|_v = |z|_v, |\varphi'(z)|_v = 1$ for any $z \in \mathcal{C}_p$, hence $\exp_{\mathcal{E}}$ has no period. There are corresponding addition formula and derivation formula, similar to those of $\wp$. We now define the p-adic logarithmic function in elliptic case as a reversed function of the $\exp_{\mathcal{E}}$ with an expression of formal group over $\mathfrak{O}$, following [17] [26].

For a point $P = (X, Y, 1) \in \mathcal{E}(K)$, let $t = t(P) = -\dfrac{X}{Y}$, $\omega(t) = -\dfrac{1}{Y}$. We have $P = (X, Y, 1) = (t, -1, \omega(t)) = \left(\dfrac{\varphi(z)}{\varphi'(z)}, -1, \dfrac{\varphi^3(z)}{\varphi'(z)}\right)$. Putting $\psi(z) = \varphi^2(z)$, we have $\psi(z) = \dfrac{\omega(t)}{t}$ and $\dfrac{\psi'(z)}{2} = \dfrac{\omega(t)}{t^2}$.

Proposition 1. *We have* $\omega(t) = \sum_{n \geq 3} A_n t^n$ *where* $A_n \in \mathbb{Z}[A, B]$ *is homogeneous of degree* $n - 3$ *with* $A_3 = 1$ *and for* $n \geq 3$*;*

$$A_n = \sum_{\substack{4\lambda+6\mu=n-3, \\ \lambda,\mu \geq 0}} a^{(n)}_{\lambda,\mu} A^\lambda B^\mu,$$

$$a^{(n)}_{\lambda,\mu} \in \mathbb{Z}, \; |a^{(n)}_{\lambda,\mu}|_\infty \leq \frac{3^3 \cdot 8^{n-3}}{n^3(\lambda+1)^3(\mu+1)^3} \quad (n \geq 3, \lambda \geq 0, \mu \geq 0).$$

As an arithmetic estimate, we have

$$h(A_n) \leq 2.25n + (n-3)h.$$

Proof. It is shown as Lemma 3-1 in [14] (replacing the height of the j-invariant by the height of the elliptic curve). □

The function $z = z(t)$ corresponds to the logarithmic function introduced in [17] (*see* [8] [26]) because it is a reversed function of $\exp_{\mathcal{E}}(z) =$

$\left(\frac{1}{\varphi^2(z)}, \frac{-\varphi'(z)}{\varphi^3(z)}, 1\right) = (t, -1, \omega(t))$. Proposition [1] then implies the estimate of the height of Taylor coefficients for the functions $\psi(z) = \varphi^2(z) = \frac{\omega(t)}{t} = \sum_{n\geq 3} A_n t^{n-1}$ and $\frac{\psi'(z)}{2} = \frac{\omega(t)}{t^2} = \sum_{n\geq 3} A_n t^{n-2}$.

By writing X, Y in terms of t and $\omega(t)$, the differential form $\Omega(t) = \frac{dX}{2Y}$ is viewed as a formal power series in t, and we define as in [17] [26] the formal integral $\log_{\mathcal{E}}(t) = \int \Omega(t)$. With this formal integral we have;

$$\int \Omega(t) = \int \frac{dX}{2Y} = \int \frac{d\left(\frac{1}{\varphi^2}\right) z(t)}{\left(\frac{-2\varphi'}{\varphi^3}\right) z(t)} dt = \int \frac{\left(\frac{-2\varphi'}{\varphi^3}\right) z(t)}{\left(\frac{-2\varphi'}{\varphi^3}\right) z(t)} z'(t)dt = z(t),$$

the local reversed function around the origin, of the function $t = t(P) = -\frac{X}{Y} = \frac{\varphi(z)}{\varphi'(z)}$.

Definition 2. Put $\log_{\mathcal{E}}(t) = z(t)$. We call the function an elliptic p-adic logarithmic function associated to $\mathcal{E}$.

Now we define p-adic elliptic polylogarithmic function in a formal way as follows. Recall that an usual (not elliptic) polylogarithmic function has the property: $Li_1(t) = -\log(1-t)$ and $Li_k(t) = \int_0^t \frac{Li_{k-1}(t)}{t} dt$ $(k = 2, 3, \cdots)$. Here we consider a formal p-adic elliptic analogue as follows.

Definition 3. Let $t \in \mathbb{C}_p$ with $|t|_p < 1$. Define a p-adic elliptic k-th polylogarithmic function by $Li_{\mathcal{E},1}(t) = \log_{\mathcal{E}}(t)$ and $Li_{\mathcal{E},k}(t) = \int_0^t \frac{Li_{\mathcal{E},k-1}(t)}{t} dt$ $(k = 2, 3, \cdots)$.

3. Arithmetic properties of p-adic elliptic logarithmic functions and p-adic elliptic polylogarithmic functions

We give bounds for the height of the Taylor coefficients of our p-adic elliptic logarithmic function and p-adic elliptic polylogarithmic function, in a formal manner. We follow the proof in [8] [14] to estimate the height of the Taylor coefficients of p-adic elliptic logarithmic function.

Proposition 2. *The Taylor expansion of $Li_{\mathcal{E},k}(t)$ $(k = 1, 2, 3, \cdots)$ is given by*

$$Li_{\mathcal{E},k}(t) = \sum_{n\geq 1} \frac{B_n}{n^k} t^n$$

where $B_1 = 1, B_n = \dfrac{C_n}{2}, C_n = \displaystyle\sum_{\substack{4\lambda+6\mu=n-1,\\ \lambda,\mu\geq 0}} b^{(n)}_{\lambda,\mu} A^\lambda B^\mu$ $(n \geq 1)$ *with* $b^{(n)}_{\lambda,\mu} \in \mathbb{Z}$ *and*

$$|b^{(n)}_{\lambda,\mu}|_\infty \leq \frac{(2^5\cdot 3\cdot 5^2)^n}{(n+2)^3(\lambda+1)^3(\mu+1)^3} \quad (n \geq 1, \lambda \geq 0, \mu \geq 0).$$

The height is bounded as follows:

$$h(C_n) \leq 8.8n + (n-1)h.$$

The domain of convergence of $Li_{\mathcal{E},k}(t)$ $(k = 1, 2, 3, \cdots)$ is given by $\{t \in \mathbb{C}_p : |t|_p < 1\}$.

Proof. We trace the proof in [8] [14] to make an analogue in the case of p-adic elliptic polylogarithmic functions. We see that the upper bound in [8];

$$|b^{(n)}_{\lambda,\mu}|_\infty \leq \frac{(2^5\cdot 3\cdot 5^2)^n}{(n+2)^2(\lambda+1)^3(\mu+1)^3} \quad (n \geq 1, \lambda \geq 0, \mu \geq 0)$$

should be replaced by a better estimate (proven below)

$$|b^{(n)}_{\lambda,\mu}|_\infty \leq \frac{(2^5\cdot 3\cdot 5^2)^n}{(n+2)^3(\lambda+1)^3(\mu+1)^3} \quad (n \geq 1, \lambda \geq 0, \mu \geq 0).$$

Put

$$\sum_{n\geq -3} D_n t^n = \frac{1}{\omega(t)} = \frac{1}{\sum_{n\geq 3} A_n t^n}.$$

We have $D_{-3} = 1$ and the induction formula

$$\sum_{i+j=n} A_i D_j = 0 \quad (n \geq 1).$$

By induction, for all $n \geq -3$ we have:

$$D_n = \sum_{\substack{4\lambda+6\mu=n+3\\ \lambda\geq 0,\mu\geq 0}} d^{(n)}_{\lambda,\mu} A^\lambda B^\mu, \; d^{(n)}_{\lambda,\mu} \in \mathbb{Z}, \; \left|d^{(n)}_{\lambda,\mu}\right|_\infty \leq \frac{(2^5\cdot 3\cdot 5^2)^{n+3}}{(n+4)^3(\lambda+1)^3(\mu+1)^3}.$$

Recall $z(t) = -\frac{t}{2} + \frac{1}{2} \int \frac{t\omega'(t)}{\omega(t)} dt$ and replace $\omega'(t)$ by $\sum_i iA_i t^{i-1}$ and $1/\omega(t)$ by $\sum_j D_j t^j$. Then by

$$C_n = \sum_{i=3}^{n+2} iA_i D_{n-1-i}$$

with $D_{-3} = A_3 = 1$, we have

$$C_n = \sum_{4\lambda+6\mu=n-1} b_{\lambda,\mu}^{(n)} A^\lambda B^\mu$$

with $b_{\lambda,\mu}^{(n)} \in \mathbb{Z}$.

We show an upper bound for $\left| b_{\lambda,\mu}^{(n)} \right|_\infty$. By definition, we have

$$b_{\lambda,\mu}^{(n)} = \sum_{i=3}^{n+2} \sum_{\lambda_1+\lambda_2=\lambda} \sum_{\mu_1+\mu_2=\mu} ia_{\lambda_1,\mu_1}^{(i)} d_{\lambda_2,\mu_2}^{(n-1-i)}.$$

We combine this expression with the statements above to get:

$$\left| b_{\lambda,\mu}^{(n)} \right|_\infty \leq \sum_{i,\lambda_1,\lambda_2,\mu_1,\mu_2} \frac{3^3 \cdot 8^{i-3} \cdot (2^5 \cdot 3 \cdot 5^2)^{n+2-i}}{i^2(n+3-i)^3(\lambda_1+1)^3(\mu_1+1)^3(\lambda_2+1)^3(\mu_2+1)^3}.$$

Using Lemma 2.1 in [8], we have:

$$\sum_{\lambda_1+\lambda_2=\lambda} \sum_{\mu_1+\mu_2=\mu} \frac{1}{(\lambda_1+1)^3(\mu_1+1)^3(\lambda_2+1)^3(\mu_2+1)^3} \leq \frac{13^2}{2^4(\lambda+1)^3(\mu+1)^3}.$$

Then we get

$$\left| b_{\lambda,\mu}^{(n)} \right|_\infty \leq \frac{3^3 \cdot 13^2 \cdot (2^5 \cdot 3 \cdot 5^2)^{n+2}}{2^{13}(\lambda+1)^3(\mu+1)^3} \sum_{i=3}^{n+2} \frac{1}{300^i i^2 (n+3-i)^3}.$$

Since

$$\sum_{i=3}^{n+2} \frac{1}{300^i i^2(n+3-i)^3} \leq \frac{13}{4} \cdot \frac{1}{(n+2)^3} \sum_{i=3}^{n+2} i \cdot 300^{-i} \leq \frac{13 \cdot 898}{4 \cdot 299^2 \cdot 300^2 (n+2)^3},$$

we have

$$\left| b_{\lambda,\mu}^{(n)} \right|_\infty \leq \frac{3^3 \cdot 13^3 \cdot 898 \cdot (2^5 \cdot 3 \cdot 5^2)^{n+2}}{2^{15} \cdot 299^2 \cdot 300^2 (n+2)^3 (\lambda+1)^3(\mu+1)^3} \leq \frac{(2^5 \cdot 3 \cdot 5^2)^n}{(n+2)^3(\lambda+1)^3(\mu+1)^3}$$

that is the required upper bound reproven (with corrections of [8]).

Consider any place v satisfying $|C_n|_v > 1$. If v is an infinite place, we have

$$\begin{aligned}
|C_n|_v &= \left| \sum_{4\lambda+6\mu=n-1,\lambda,\mu\in\mathbb{Z},\lambda,\mu\geq 0} b_{\lambda,\mu}^{(n)} A^\lambda B^\mu \right|_v \\
&\leq \sum_{4\lambda+6\mu=n-1,\lambda,\mu\in\mathbb{Z},\lambda,\mu\geq 0} |b_{\lambda,\mu}^{(n)}|_v |A|_v^\lambda |B|_v^\mu \\
&\leq (2^5 \cdot 3 \cdot 5^2)^n \times \sum_{4\lambda+6\mu=n-1,\lambda,\mu\in\mathbb{Z},\lambda,\mu\geq 0} |A|_v^\lambda |B|_v^\mu \\
&\leq (2^5 \cdot 3 \cdot 5^2)^n n \max\{1, |A|_v, |B|_v\}^{n-1}.
\end{aligned}$$

For a finite place v, since $b_{\lambda,\mu}^{(n)} \in \mathbb{Z}$, we have

$$\begin{aligned}
|C_n|_v &\leq \max_{4\lambda+6\mu=n-1} |b_{\lambda,\mu}^{(n)} A^\lambda B^\mu|_v \leq \max_{4\lambda+6\mu=n-1} |A^\lambda B^\mu|_v \\
&\leq \max\{1, |A|_v, |B|_v\}^{n-1}.
\end{aligned}$$

Then we obtain the estimate of $h(C_n)$ by definition of the height.

Since our p-adic elliptic polylogarithmic function is defined by $Li_{\mathcal{E},k}(t) = \int_0^t \frac{Li_{\mathcal{E},k-1}(t)}{t} dt \quad (k = 1, 2, 3, \cdots)$, a formal integration gives the estimates of their Taylor coefficients at the origin.

The radius of convergence is obtained by Hadamard's formula. □

References

1. A. Baker, *Transcendental Number Theory*, Cambridge University. Press, 1975.
2. D. Bertrand, *Algebraic Values of p-adic Elliptic Functions*, in: Transcendence theory: advances and applications, (eds. A. Baker and D.W. Masser), Academic Press, 1977, 149–159.
3. D. Bertrand, *Problèmes locaux*, Appendix to *Nombres transcendants et groupes algébriques by M. Waldschmidt*, Astérisque, **69/70**, Soc. Math. de France, 1987.
4. D. Bertrand, *Approximations diophantiennes p-adiques sur les courbes elliptiques admettant une multiplication complexe*, Compositio Math. **37**, 1978, 21–50.
5. G. V. Chudnovsky, *Contributions to the theory of transcendental numbers*, Math. Soc. Math. Surveys Monographs, **19**, 1984.
6. S. David, *Minorations de formes linéaires de logarithmes elliptiques*, Mémoires, Nouvelle série 62, Supplément au Bulletin de la Soc. Math. de France, Tome **123**, Fascicule **3**, 1995.

7. S. David and N. Hirata-Kohno, *Recent progress on linear forms in elliptic logarithms*, in: A Panorama of Number Theory, (ed. G. Wüstholz), Cambridge University Press, 2002, 26–37.
8. S. David and N. Hirata-Kohno, *Logarithmic Functions and Formal Groups of Elliptic Curves*, In : Diophantine Equations, Tata Institute of Fundamental Research, Studies in Mathematics, Narosa Publishing House, 2008, 243–256.
9. S. David and N. Hirata-Kohno, *Linear Forms in Elliptic Logarithms*, J. für die reine angew. Math., **628**, 2009, 37–89.
10. J. Gebel, A. Pethő and H. G. Zimmer, *Computing integer points on elliptic curves*, Acta Arith., **68**, 1994, 171–192.
11. N. Hirata-Kohno, *Formes linéaires de logarithmes de points algébriques sur les groupes algébriques*, Inventiones Math., **104**, 1991, 401–433.
12. N. Hirata-Kohno, *Approximations simultanées sur les groupes algébriques commutatifs*, Compositio Math., **86**, 1993, 69–96.
13. N. Hirata-Kohno, *Linear forms in p-adic elliptic logarithms*, preprint.
14. N. Hirata-Kohno and R. Takada, *Linear forms in two elliptic logarithms in the p-adic case*, Kyushu Journal of Mathematics, **64**, No. 2, 2010, 239–260.
15. N. Koblitz, *p-adic Numbers, p-adic Analysis and Zeta-Functions*, Grad. Text in Math., **58**, Springer, Second edition, 1984.
16. S. Lang, *Diophantine approximation on toruses*, Amer. J. Math., **86**, 1964, 521–533.
17. S. Lang, *Elliptic curves: Diophantine analysis*, Grundlehren der Math. Wissenschaften, **231**, Springer, 1978.
18. E. Lutz, *Sur l'équation $y^2 = x^3 - Ax - B$ dans les corps p-adiques*, J. für die reine angew. Math., **177**, 1937, 237–247.
19. K. Mahler, *p-adic numbers and their functions*, Cambridge University Press, Second edition, 1965.
20. D. W. Masser, *Linear forms in algebraic points of Abelian functions III*, Proc. London Math. Soc., **33**, 1976, 549–564.
21. P. Philippon and M. Waldschmidt, *Formes linéaires de logarithmes sur les groupes algébriques commutatifs*, Illinois J. Math. **32**, 1988, 281–314.
22. G. Rémond and F. Urfels, *Approximation diophantienne de logarithmes elliptiques p-adiques*, Journal of Number Theory, **57**, 1996, 133–169.
23. W. H. Schikhof, *Ultrametric calculus*, Cambridge Studies in Advanced Mathematics, **4**, Cambridge University Press, 1984.
24. L. Schnirelmann, *Sur les fonctions dans les corps normés et algébriquement fermés* [in Russian], Bull. Acad. Sc. URSS, Série Math., 1938, 487– 498.
25. J. -P. Serre, *Quelques propriétés des groupes algébriques commutatifs*, Appendix to *Nombres transcendants et groupes algébriques by M. Waldschmidt*, Astérisque, **69/70**, Soc. Math. de France, 1987.
26. J. H. Silverman, *The arithmetic of elliptic curves*, Grad. Text in Math., **106**, Springer, 1986.
27. R. J. Stroeker and N. Tzanakis, *Solving elliptic diophantine equations by estimating linear forms in elliptic logarithms*, Acta Arith., **67**, 1994, 177–196.
28. M. Waldschmidt, *Nombres transcendants et groupes algébriques*, Astérisque,

69/70, Soc. Math. de France, édition originale 1979, seconde édition 1987.
29. A. Weil, *Sur les fonctions elliptiques p-adiques*, C. R. Acad. Sc. Paris, t. **203**, 1936, 22–24.
30. G. Wüstholz, *A Panorama of Number Theory*, Cambridge University Press, 2002.
31. Kunrui Yu, *P-adic logarithmic forms and group varieties I*, J. für die reine angew. Math., **502**, 1998, 29–92.
32. Kunrui Yu, *p-adic logarithmic forms and group varieties II*, Acta Arith., **89**, 1999, 337–378.
33. Kunrui Yu, *Report on p-adic logarithmic forms*, in: A Panorama of Number Theory, (ed. G. Wüstholz), Cambridge University Press, 2002, 11–25.
34. Kunrui Yu, *P-adic logarithmic forms and group varieties III*, Forum Math., **19**, 2007, 187–280.
35. D. Zagier, *Large integral points on elliptic curves*, Math. Comp., **48**, 1987, 425–436.

Spherical functions on $U(2n)/\left(U(n) \times U(n)\right)$ and hermitian Siegel series

Yumiko Hironaka

Department of Mathematics,
Faculty of Education and Integrated Sciences, Waseda University,
Nishi-Waseda, Tokyo, 169-8050, Japan
E-mail: hironaka@waseda.jp

Dedicated to Professor Takayuki Oda on his sixtieth birthday

Keywords: spherical functions; unitary groups; hermitian Siegel series

0. Introduction

Let $\mathbb{G}$ be a reductive algebraic group and $\mathbb{X}$ an affine algebraic variety which is $\mathbb{G}$-homogeneous, where everything is assumed to be defined over a non-archimedian local field k of characteristic 0. We denote by G and X the sets of k-rational points of $\mathbb{G}$ and $\mathbb{X}$, respectively, take a maximal compact subgroup K of G, and consider the Hecke algebra $\mathcal{H}(G, K)$. Then, a nonzero K-invariant function on X is called *a spherical function on* X if it is an $\mathcal{H}(G, K)$-common eigenfunction.

Spherical functions on homogeneous spaces comprise an interesting topic to investigate and a basic tool to study harmonic analysis on G-space X. They have been studied as spherical vectors of distinguished models, Shalika functions and Whitakker-Shintani functions, there is a close relation to the theory of automorphic forms, and spherical functions may appear in local factor of global object like Rankin-Selberg convolution and Eisenstein series. The theory of spherical functions has also an application of classical number theory, e.g. local densities of representations of quadratic forms or hermitian forms.

2010 Mathematics Subject Classification: Primary 11F85; Secondary 11E95, 11F70, 22E50.

This research is partially supported by Grant-in-Aid for Scientific Research (C):20540029.

To obtain explicit formulas of spherical functions is one of basic problems, and it has been done for the group cases by I. G. Macdonald and afterwards by W. Casselman by a representation theoretical method ([13], [2]). For homogeneous spaces, there are results mainly for the case that the space of spherical functions attached to each Satake parameter is of dimension one (e.g., [3], [11], [16]). The author gave general expressions of spherical functions on the basis of data of the group G and functional equations of spherical functions when the dimension is not necessarily one, and a sufficient condition to have functional equations with respect to the Weyl group of G (cf. [7]).

In the present paper, we investigate spherical functions on spaces X_T for hermitian form T, where X_T is a homogeneous space of the unitary group $\mathbb{G} = U(2n)$ and isomorphic to $U(2n)/U(n) \times U(n)$ over the algebraic closure of k (n is the size of T). Here and henceforth we fix an unramified quadratic extension k' of k and consider hermitian forms and unitary groups with respect to the extension k'/k.

In §1, we introduce the space $X_T = \mathfrak{X}_T/U(T)$ for each hermitian matrix T of size n, and construct spherical functions $\omega_T(x;s) = \omega_T(x;z)$ on X_T, where $x \in X_T$ and $s, z \in \mathbb{C}^n$ are related by (1.11).

We give the functional equations of $\omega_T(x;s)$ with respect to the Weyl group W and determine the location of their possible poles and zeros (Theorem 2.6, Theorem 2.9). The Weyl group W is isomorphic to $S_n \ltimes (\pm 1)^n$, and S_n acts on the variable $z = (z_1, \ldots, z_n)$ by permutation of indices, and we may apply previous results on the spherical functions of hermitian forms to obtain the functional equations with respect to S_n. As for $\tau \in W$ corresponding to the remaining simple root, we need to consider the standard parabolic subgroup P associated to τ and enlarge the space $\mathfrak{X}_T$ into $\widetilde{\mathfrak{X}_T}$ on which $P \times GL_1(k')$ acts. Different from the cases of the other simple roots, i.e., transpositions $(i\ i+1)$, $1 \leq i \leq n-1$, the functional equation with respect to τ does not come from that of a prehomogeneous vector space (cf. Remark 1.7, Theorem 2.3).

Next we apply the general expression given in [7] to the present case, and obtain the explicit formula for $\omega_T(x_T;z)$ for some diagonal T and a particular point x_T (Theorem 3.1). Then, by sliding, we have the explicit formulas for general T at many points (Theorem 3.3).

In §4, we consider the spherical Fourier transform on the Schwartz space $\mathcal{S}(K\backslash X_T)$, which is an integral transform employing the spherical function as kernel function, and show that the image is a free $\mathcal{H}(G,K)$-module of rank 2^{n-1}.

In §5, as an application, we consider hermitian Siegel series $b_\pi(T;t)$, relate them to our spherical functions $\omega_T(x;s)$. Then we give the 'common factor' of $b_\pi(T;t)$ and the functional equations of $b_\pi(T;t)$ by using the results in §2. A similar study for (symmetric) Siegel series has been done by F. Sato and the author, but in that case we could not obtain the explicit Gamma factor from spherical functions. In the present case, we give the explicit functional by a specialization of functional equations of spherical functions $\omega_T(x;z)$. The existence of the functional equations was known in an abstract form as functional equations of Whitakker functions of p-adic groups by M. L. Karel, and explicit formulas have been given recently by T. Ikeda (more precisely, see remarks in §5).

1. Spherical function $\omega_T(x;s)$ on $\mathfrak{X}_T$ and X_T

Let k' be an unramified quadratic extension of a p-adic field k with involution $*$, and for each $A = (a_{ij}) \in M_{mn}(k')$, we denote by A^* the matrix $(a_{ji}{}^*) \in M_{nm}(k')$. We fix a unit $\epsilon \in \mathcal{O}_k^\times$ such that $k' = k(\sqrt{\epsilon})$ and $\varepsilon - 1 \in 4\mathcal{O}_k^\times$ (cf. [15], 63.3 and 63.4), and set

$$\xi = \frac{1+\sqrt{\epsilon}}{2}. \tag{1.1}$$

Then $\{1,\ \xi\}$ forms an $\mathcal{O}_k$-basis for $\mathcal{O}_{k'}$, and $\{\alpha \in \mathcal{O}_{k'} \mid \alpha^* = -\alpha\} = \sqrt{\epsilon}\mathcal{O}_k$. We fix a prime element π of k, and denote by $v_\pi(\ )$ the additive value on k, by $|\,|$ the normalized absolute value on $k^\times$ with $|\pi|^{-1} = q$ being the cardinality of the residue class field of k.

We set

$$\mathcal{H}_m = \{\, A \in M_m(k') \mid A^* = A\}, \quad \mathcal{H}_m^{nd} = \mathcal{H}_m \cap GL_m(k').$$

For $A \in \mathcal{H}_m$ and $X \in M_{mn}(k')$, we write

$$A[X] = X^* \cdot A = X^*AX \in \mathcal{H}_n,$$

and define the unitary group of A by

$$U(A) = \{g \in GL_m(k') \mid A[g] = A\}.$$

In particular we set

$$G = U(H_n) \quad \text{with } H_n = \begin{pmatrix} 0 & 1_n \\ 1_n & 0 \end{pmatrix}, \qquad U(m) = U(1_m).$$

For $T\in\mathcal{H}_n^{nd}$, we set

$$\begin{aligned}\mathfrak{X}_T &= \{x\in M_{2n,n}(k') \mid H_n[x]=T\} \ni x_T = \begin{pmatrix}\xi T\\ 1_n\end{pmatrix},\\ X_T &= \mathfrak{X}_T/U(T).\end{aligned} \tag{1.2}$$

The group G acts on $\mathfrak{X}_T$, as well as on X_T, through left multiplication, which is transitive by Witt's theorem for hermitian matrices (cf. [17], Ch.7, §9).

Lemma 1.1. *The stabilizer G_0 of G at $x_TU(T)\in X_T$ is isomorphic to $U(T)\times U(T)$:*

$$U(T)\times U(T) \xrightarrow{\sim} G_0,\ (h_1,h_2)\longmapsto \widetilde{T}^{-1}\begin{pmatrix}h_1^{*-1} & 0\\ 0 & h_2^{*-1}\end{pmatrix}\widetilde{T},$$

where

$$\widetilde{T} = \begin{pmatrix}1_n & \xi^*T\\ 1_n & -\xi T\end{pmatrix}\in GL_{2n}(k').$$

Proof. Since $\widetilde{T}x_T = \begin{pmatrix}T\\ 0\end{pmatrix}$, we have, for any $h\in U(T)$,

$$x_Th = \widetilde{T}^{-1}\widetilde{T}x_Th = \widetilde{T}^{-1}\begin{pmatrix}Th\\ 0\end{pmatrix} = \widetilde{T}^{-1}\begin{pmatrix}h^{*-1} & 0\\ 0 & 1\end{pmatrix}\widetilde{T}x_T. \tag{1.3}$$

Take any $g\in G$ such that $gx_T = x_T$. Then $\widetilde{T}g\widetilde{T}^{-1}$ stabilizes $\widetilde{T}x_T$ and belongs to $U(H_n[\widetilde{T}^{-1}])$, where

$$H_n[\widetilde{T}^{-1}] = H_n\left[\begin{pmatrix}\xi 1_n & \xi^* 1_n\\ T^{-1} & -T^{-1}\end{pmatrix}\right] = \begin{pmatrix}T^{-1} & 0\\ 0 & -T^{-1}\end{pmatrix},$$

and we get

$$\widetilde{T}g\widetilde{T}^{-1} = \begin{pmatrix}1_n & 0\\ 0 & d\end{pmatrix},\quad \text{for some } d\in U(-T^{-1}).$$

Hence, together with (1.3), we obtain the isomorphism stated as above. ∎

We fix the Borel subgroup B of G as

$$B = \left\{\begin{pmatrix}b & 0\\ 0 & b^{*-1}\end{pmatrix}\begin{pmatrix}1_n & a\\ 0 & 1_n\end{pmatrix} \;\middle|\; \begin{array}{l}b \text{ is upper triangular of size } n,\\ a+a^*=0\end{array}\right\}, \tag{1.4}$$

ant the maximal compact sugbroup $K = G\cap GL_{2n}(\mathcal{O}_{k'})$ of G, which satisfy $G=KB=BK$. We fix the Haar measure dk on K and the left invariant measure dp on B normalized by $\int_K dk = \int_{K\cap B} dp = 1$. For each element $x \in \mathfrak{X}_T$, we denote by x_2 the lower half n by n block of x. We define relative B-invariants on $\mathfrak{X}_T$ by

$$f_{T,i}(x) = d_i(x_2 \cdot T^{-1}) = d_i(x_2 T^{-1} x_2^*), \quad 1 \leq i \leq n, \tag{1.5}$$

where $d_i(y)$ is the determinant of the upper left i by i block of a matrix y. It is easy to see, for $b \in B$,

$$f_{T,i}(bx) = \psi_i(b) f_{T,i}(x), \quad \psi_i(b) = N(d_i(b))^{-1}, \tag{1.6}$$

where $N = N_{k'/k}$. Hence $f_{T,i}(x)$, $1 \leq i \leq n$ are relative B-invariants on $\mathfrak{X}_T$ associated with rational characters ψ_i of B, and we may regard them as relative B-invariants on X_T, since $f_{T,i}(xh) = f_{T,i}(x)$ for any $h \in U(T)$. We set

$$\mathfrak{X}_T^{op} = \{x \in X_T \mid f_{T,i}(x) \neq 0,\ 1 \leq i \leq n\}, \quad X_T^{op} = \mathfrak{X}_T^{op}/U(T). \tag{1.7}$$

Remark 1.2. Though we may realize above objects as the sets of k-rational points of algebraic sets defined over k and develop the arguments, we take down to earth way for simplicity of notations. We only note here that X_T^{op} (resp. $\mathfrak{X}_T^{op}$) becomes a Zariski open B-orbit in X_T (resp. $B \times U(T)$-orbit in $\mathfrak{X}_T^{op}$) over the algebraic closure of k.

We introduce a spherical function $\omega_T(x; s)$ on $\mathfrak{X}_T$ as well as on $X_T = \mathfrak{X}_T/U(T)$. For $x \in \mathfrak{X}_T$ and $s \in \mathbb{C}^n$, set

$$\omega_T(x; s) = \omega_T^{(n)}(x; s) = \int_K |f_T(kx)|^{s+\varepsilon} dk, \tag{1.8}$$

where k runs over the set $\{k \in K \mid kx \in \mathfrak{X}_T^{op}\}$,

$$\varepsilon = \varepsilon_0 + (\frac{\pi\sqrt{-1}}{\log q}, \ldots, \frac{\pi\sqrt{-1}}{\log q}), \quad \varepsilon_0 = (-1, \ldots, -1, -\frac{1}{2}) \in \mathbb{C}^n,$$

$$f_T(x) = \prod_{i=1}^n f_{T,i}(x), \qquad |f_T(x)|^s = \prod_{i=1}^n |f_{T,i}(x)|^{s_i}.$$

The right hand side of (1.8) is absolutely convergent if $\mathrm{Re}(s_i) \geq -\mathrm{Re}(\varepsilon_i) = -\varepsilon_{0,i}$, $1 \leq i \leq n$, and continued to a rational function of $q^{s_1}, \ldots, q^{s_n}$. We note here that

$$|\psi(p)|^\varepsilon \left(= \prod_{i=1}^n |\psi_i(p)|^{\varepsilon_i}\right) = |\psi(p)|^{\varepsilon_0} = \delta^{\frac{1}{2}}(p),$$

where δ is the modulus character on B (i.e., $d(pp') = \delta(p')^{-1} dp$).

We denote by $\mathcal{C}^\infty(K\backslash X_T)$ the space of left K-invariant functions on X_T, which can be identified with the space $\mathcal{C}^\infty(K\backslash \mathfrak{X}_T/U(T))$ of left K-invariant right $U(T)$-invariant functions on $\mathfrak{X}_T$. The function $\omega_T(x;z)$ can be regarded as a function in $\mathcal{C}^\infty(K\backslash X_T)$ and becomes a common eigenfunction by the action of the Hecke algebra $\mathcal{H}(G,K)$ (cf. [5] §1, or [7] §1). In detail, the Hecke algebra $\mathcal{H}(G,K)$ is the commutative $\mathbb{C}$-algebra consisting of compactly supported two-sided K-invariant functions on G, acting on $\mathcal{C}^\infty(K\backslash X_T)$ by the convolution product

$$(\phi * \Psi)(x) = \int_G \phi(g)\Psi(g^{-1}x)dg, \qquad (\phi \in \mathcal{H}(G,K),\ \Psi \in \mathcal{C}^\infty(K\backslash X_T)), \tag{1.9}$$

where dg is the Haar measure on G normalized by $\int_K dg = 1$, and we see

$$(\phi * \omega_T(\ ;s))(x) = \lambda_s(\phi)\omega_T(x;s), \quad (\phi \in \mathcal{H}(G,K)) \tag{1.10}$$

where λ_s is the $\mathbb{C}$-algebra homomorphism defined by

$$\begin{aligned} &\lambda_s : \mathcal{H}(G,K) \longrightarrow \mathbb{C}(q^{s_1},\dots,q^{s_n}),\\ &\phi \longmapsto \int_B \phi(p)\,|\psi(p)|^{-s+\varepsilon}\,dp, \quad \left(|\psi(p)|^{-s+\varepsilon} = \prod_{i=1}^n |\psi_i(p)|^{-s_i+\varepsilon_i}\right). \end{aligned}$$

We introduce a new variable z which is related to s by

$$s_i = -z_i + z_{i+1} \quad (1 \le i \le n-1), \qquad s_n = -z_n \tag{1.11}$$

and write $\omega_T(x;z) = \omega_T(x;s)$. The Weyl group W of G relative to the maximal k-split torus in B acts on rational characters of B as usual (i.e., $\sigma(\psi)(b) = \psi(n_\sigma^{-1}bn_\sigma)$ by taking a representative n_σ of σ), so W acts on $z\in\mathbb{C}^n$ and on $s\in\mathbb{C}^n$ as well. We will determine the functional equations of $\omega_T(x;s)$ with respect to this Weyl group action. The group W is isomorphic to $S_n \ltimes C_2^n$, S_n acts on z by permutation of indices, and W is generated by S_n and $\tau : (z_1,\dots,z_n) \longmapsto (z_1,\dots,z_{n-1},-z_n)$. Keeping the relation (1.11), we also write $\lambda_z(\phi) = \lambda_s(\phi)$; then λ_z gives a $\mathbb{C}$-algebra isomorphism (the Satake isomorphism)

$$\begin{aligned} &\lambda_z : \mathcal{H}(G,K) \xrightarrow{\sim} \mathbb{C}[q^{\pm 2z_1},\dots,q^{\pm 2z_n}]^W,\\ &\phi \longmapsto \int_B \phi(p)\prod_{i=1}^n |N(p_i)|^{-z_i}\,\delta^{\frac12}(p)dp, \end{aligned} \tag{1.12}$$

where p_i is the i-th diagonal component of $p\in B$, and the right hand side is the invariant subring of the Laurent polynomial ring by W.

Proposition 1.3. *Set* $\mathcal{U} = (\mathbb{Z}/2\mathbb{Z})^{n-1}$ *and*

$$\widetilde{u} = (u_1 \frac{\pi\sqrt{-1}}{\log q}, \dots, u_{n-1}\frac{\pi\sqrt{-1}}{\log q}, 0) \in \mathbb{C}^n, \qquad u = (u_1, \dots . u_{n-1}) \in \mathcal{U}.$$

Then $\omega_T(x; z + \widetilde{u})$, $u \in \mathcal{U}$, *are linearly independent for generic* $z \in \mathbb{C}^n$ *and correspond to the same eigenvalue through* $\lambda_z : \mathcal{H}(G, K) \longrightarrow \mathbb{C}$.

Proof. The set $\mathfrak{X}_T^{op}$ is decomposed into the disjoint union of B-orbits as follows:

$$\mathfrak{X}_T^{op} = \bigsqcup_{u \in \mathcal{U}} \mathfrak{X}_{T,u},$$

$$\mathfrak{X}_{T,u} = \{ x \in \mathfrak{X}_T^{op} \mid v_\pi(f_{T,i}(x)) \equiv u_1 + \cdots + u_i \pmod 2, \ 1 \le i \le n-1 \}.$$

We consider finer spherical functions for $u \in \mathcal{U}$

$$\omega_{T,u}(x; s) = \int_K |f_T(kx)|_u^{s+\varepsilon}\, dk, \quad |f_T(y)|_u^{s+\varepsilon} = \begin{cases} |f_T(y)|^{s+\varepsilon} & \text{if } y \in \mathfrak{X}_{T,u}, \\ 0 & \text{otherwise} . \end{cases}$$

Then $\{\omega_{T,u}(x, z) \mid u \in \mathcal{U}\}$ are linearly independent for generic z associated with the same λ_z, where we keep the relation (1.11) between s and z. For each character χ of $\mathcal{U}$, we may represent as follows

$$\sum_{u \in \mathcal{U}} \chi(u)\omega_{T,u}(x; s) = \omega_T(x; z_\chi), \tag{1.13}$$

where z_χ is obtained by adding $\frac{\pi\sqrt{-1}}{\log q}$ to z_i for suitable i according to χ, and they are linearly independent (for generic z) as varying characters χ. The result follows from this, since $\{\omega_T(x; z_\chi) \mid \chi \text{ is a character of } \mathcal{U}\} = \{\omega_T(x; z + \widetilde{u}) \mid u \in \mathcal{U}\}$. ∎

We note here the relation between $\omega_T(x; s)$ and $\omega_{T'}(y; s)$ when T and T' are equivalent under the action of $GL_n(k')$, which is easy to see.

Proposition 1.4. *For* $T \in \mathcal{H}_n^{nd}$ *and* $h \in GL_n(k')$, *we set* $T' = T[h] \, (= h^*Th)$. *Then*

$$\mathfrak{X}_{T'} = (\mathfrak{X}_T)\, h, \quad X_{T'} = \mathfrak{X}_T h / U(T') \quad \textit{and} \quad f_{T',i}(xh) = f_{T,i}(x) \quad (x \in \mathfrak{X}_T),$$

and

$$\omega_{T'}(xh; s) = \omega_T(x; s), \qquad (x \in \mathfrak{X}_T).$$

By using a result on spherical functions on the space of hermitian forms, we obtain the following theorem.

Theorem 1.5. *Set*

$$G_1(z)=\prod_{1\le i<j\le n}\frac{q^{z_j}+q^{z_i}}{q^{z_j}-q^{z_i-1}}. \tag{1.14}$$

Then, for any $T\in\mathcal{H}_n^{nd}$, the function $G_1(z)\cdot\omega_T(x;z)$ is invariant under the action of S_n on z.

Proof. By the embedding

$$K_0=GL_n(\mathcal{O}_{k'})\longrightarrow K,\quad h\longmapsto\widetilde{h}=\begin{pmatrix}h^{*-1}&0\\0&h\end{pmatrix},$$

and the normalized Haar measure dh on K_0, we obtain, for $s\in\mathbb{C}^n$ satisfying $\mathrm{Re}(s_i)\ge-\mathrm{Re}(\varepsilon_i)$, $1\le i\le n$,

$$\begin{aligned}\omega_T(x;z)=\omega_T(x;s)&=\int_{K_0}dh\int_K|f_T(kx)|^{s+\varepsilon}dk\\&=\int_{K_0}dh\int_K\left|f_T(\widetilde{h}kx)\right|^{s+\varepsilon}dk=\int_K\int_{K_0}\left|f_T(\widetilde{h}kx)\right|^{s+\varepsilon}dhdk\\&=\int_K\zeta^{(n)}(D(kx);s)dk.\end{aligned}$$

Here $D(kx)=(kx)_2\cdot T^{-1}\in\mathcal{H}_n^{nd}$ for $\{k\in K\mid kx\in\mathfrak{X}_T^{op}\}$, and $\zeta^{(n)}(y;s)$ is a spherical function on $\mathcal{H}_n^{nd}$ defined by

$$\zeta^{(n)}(y;s)=\int_{K_0}\prod_{i=1}^n|d_i(h\cdot y)|^{s_i+\varepsilon_i}dh,\quad(h\cdot y=hyh^*),$$

where h runs over the set $\{h\in K_0\mid d_i(h\cdot y)\ne0,\ 1\le i\le n\}$. Keeping the relation between s and z as before, the assertion of Theorem 1.5 follows from the next proposition. ∎

Proposition 1.6. *(cf. [4] or [6]) For any $y\in\mathcal{H}_n^{nd}$, the function $G_1(z)\cdot\zeta^{(n)}(y;s)$ is holomorphic for $z\in\mathbb{C}^n$ and invariant under the action of S_n, where $G_1(z)$ is defined as in (1.14).*

In [6] §4.2, we considered a modified function

$$\omega^{(H)}(y;s)=\int_{K_0}\prod_{i=1}^n\chi_\pi(d_i(h\cdot y))\,|d_i(h\cdot y)|^{s_i+\varepsilon_i'}dh,$$

where $\chi_\pi(a) = (-1)^{v_\pi(a)}$ for $a \in k^\times$ and $\varepsilon' = (-1,\ldots,-1,\frac{n-1}{2})$. The function $\zeta^{(n)}(x;s)$ satisfies the same functional properties as $\omega^{(H)}(y;s)$, since $\omega^{(H)}(y;s) = |\det(y)|^{\frac{n}{2}}\,\zeta^{(n)}(y;s)$.

Remark 1.7. For the transposition $\tau_i = (i\ i+1) \in W,\quad 1 \le i \le n-1$, the following functional equations hold by Theorem 1.5

$$\omega_T(x;z) = \frac{1-q^{z_i - z_{i+1}-1}}{q^{z_i - z_{i+1}} - q^{-1}} \times \omega_T(x;\tau_i(z)), \quad 1 \le i \le n-1. \tag{1.15}$$

On the other hand, one may obtain (1.15) directly in the similar way to the case of τ in §2, where the sufficient condition in [7]-§3 for having a functional equation with respect to τ_i is satisfied and the Gamma factor in (1.15) is essentially the same to that of the zeta function of prehomogeneous vector space $(U \times GL_1(k'), (k')^2)$, where $U \cong U(2)$ or $U(Diag(1,\pi))$. Then Theorem 1.5 follows from (1.15), through the similar line to the proof of Proposition 1.6. In fact, Proposition 1.6 was proved by using functional equations of type (1.15).

2. Functional equations, possible zeros and poles

We calculate the functional equation of $\omega_T(x;z)$ for $\tau \in W$, and give the functional equations with respect to the whole W.

2.1. First we calculate the spherical function for $n = 1$. We note the data for $n = 1$, which will be used also in §2.2.

$$G = U(1,1) = \left\{ \begin{pmatrix} \alpha & 0 \\ 0 & \alpha^{*-1} \end{pmatrix} \begin{pmatrix} 1 & w\sqrt{\epsilon} \\ v\sqrt{\epsilon} & 1+vw\epsilon \end{pmatrix} \;\middle|\; \alpha \in k'^\times,\ v,w \in k \right\}$$
$$\cup \left\{ \begin{pmatrix} \alpha & 0 \\ 0 & \alpha^{*-1} \end{pmatrix} \begin{pmatrix} 0 & 1 \\ 1 & v\sqrt{\epsilon} \end{pmatrix} \;\middle|\; \alpha \in k'^\times,\ v \in k \right\},$$

$K = K_1 = K_{1,1} \cup K_{1,2}$, where

$$K_{1,1} = \left\{ \begin{pmatrix} \alpha & 0 \\ 0 & \alpha^{*-1} \end{pmatrix} \begin{pmatrix} 1 & v/\sqrt{\epsilon} \\ u\sqrt{\epsilon} & 1+uv \end{pmatrix} \;\middle|\; \alpha \in \mathcal{O}_{k'}^\times,\ u,v \in \mathcal{O}_k \right\},$$

$$K_{1,2} = \left\{ \begin{pmatrix} \alpha & 0 \\ 0 & \alpha^{*-1} \end{pmatrix} \begin{pmatrix} \pi u\sqrt{\epsilon} & 1+\pi uv \\ 1 & v/\sqrt{\epsilon} \end{pmatrix} \;\middle|\; \alpha \in \mathcal{O}_{k'}^\times,\ u,v \in \mathcal{O}_k \right\}, \tag{2.1}$$

and

$$\omega_T^{(1)}(x;s) = \int_{K_1} \chi_\pi(f_1(hx))\,|f_1(hx)|^{s-\frac{1}{2}}\,dh,$$

where $f_1(x) = \det(T)^{-1}N(x_2)$ for $x \in \mathfrak{X}_T$, $\chi_\pi(a) = (-1)^{v_\pi(a)}$ for $a \in k^\times$ and dh is the Haar measure on K_1.

Proposition 2.1.
(i) *The set*

$$\left\{ x_e = \begin{pmatrix} \pi^e \\ \xi\pi^{t-e} \end{pmatrix} \;\middle|\; e \in \mathbb{Z},\ 2e \le t \right\}, \quad \left(\xi = \frac{1+\sqrt{\epsilon}}{2}\right)$$

forms a complete set of representatives of $K_1\backslash\mathfrak{X}_T$ for $T = \pi^t$.
(ii) *For $x_e \in \mathfrak{X}_T$ with $T = \pi^t$ as above, one has*

$$\omega_T^{(1)}(x_e; s) = \frac{(-1)^t q^{e-\frac{1}{2}t}}{1+q^{-1}} \times \frac{q^{(t-2e+1)s}(1-q^{-2s-1}) - q^{-(t-2e+1)s}(1-q^{2s-1})}{q^s - q^{-s}}.$$

(iii) *For any $T \in \mathcal{H}_1^{nd}$, $\omega_T^{(1)}(x; s)$ is holomorphic for all $s \in \mathbb{C}$ and satisfies the functional equation*

$$\omega_T^{(1)}(x; s) = \omega_T^{(1)}(x; -s).$$

Proof. We recall that $\{1, \xi\}$ forms an $\mathcal{O}_k$-basis of $\mathcal{O}_{k'}$ and $Tr_{k'/k}(\xi) = 1$. Multiplying a suitable element in K_1 of type

$$\begin{pmatrix} 1 & 0 \\ u\sqrt{\varepsilon} & 1 \end{pmatrix}\begin{pmatrix} \alpha & 0 \\ 0 & \alpha^{*-1} \end{pmatrix} \quad \text{or} \quad \begin{pmatrix} 1 & 0 \\ u\sqrt{\varepsilon} & 1 \end{pmatrix}\begin{pmatrix} 0 & \alpha \\ \alpha^{*-1} & 0 \end{pmatrix} \quad (u \in \mathcal{O}_k,\ \alpha \in \mathcal{O}_{k'}^{\times})$$

one may make any $x \in \mathfrak{X}_T$ into some x_e in the given set, and the explicit formula in (ii) shows there is no redundancy within it.

Take x_e as above. For $h \in K_{1,1}$ written as in (2.1), since we have

$$\begin{aligned}(hx_e)_2 &= \alpha^{*-1}(u\sqrt{\epsilon}\pi^e + (1+uv)\xi\pi^{t-2e})) \\ &= \alpha^{*-1}\pi^e\left(-u + ((1+uv)\pi^{t-2e} + 2u)\xi\right),\end{aligned}$$

we see

$$v_\pi(f_1(kx_e)) = -t + 2e + 2\min\{v_\pi(u),\, t-2e\}.$$

Since $vol(K_{1,1}) = (1+q^{-1})^{-1}$, we obtain

$$\begin{aligned}&\int_{K_{1,1}} \chi_\pi(f_1(hx_e))\,|f_1(hx_e)|^{s-\frac{1}{2}}\,dh \\ &= \frac{(-1)^t q^{(t-2e)(s-\frac{1}{2})}}{1+q^{-1}} \cdot \sum_{r\ge 0} q^{-r}(1-q^{-1})q^{-2\min\{r,\,t-2e\}(s-\frac{1}{2})} \\ &= \frac{(-1)^t q^{(t-2e)(s-\frac{1}{2})}}{1+q^{-1}} \cdot \left(\frac{(1-q^{-1})(1-q^{-2(t-2e)s})}{1-q^{-2s}} + q^{-2(t-2e)s}\right).\end{aligned}$$

For $h \in K_{1,2}$ written as in (2.1), since we have

$$(hx_e)_2 = \alpha^{*-1}(\pi^e + v/\sqrt{\varepsilon}\pi^{t-e}\xi) = (\alpha^*\sqrt{\varepsilon})^{-1}\pi^e\left(-1 + (2+v\pi^{t-2e})\xi\right),$$

we see $v_\pi(f_1(hx_e)) = -t+2e$ and

$$\int_{K_{1,2}} \chi_\pi(f_1(hx_e))\,|f_1(hx_e)|^{s-\frac{1}{2}}\,dh = \frac{q^{-1}}{1+q^{-1}}\cdot(-1)^t q^{(t-2e)(s-\frac{1}{2})}.$$

Thus we obtain

$$\begin{aligned}
&\omega_T^{(1)}(x_e;s)\\
&= \frac{(-1)^t q^{(t-2e)(s-\frac{1}{2})}}{1+q^{-1}}\frac{1}{1-q^{-2s}}\cdot\left(1-q^{-2s-1}+q^{-2(t-2e)s-1}-q^{-2(t-2e+1)s)}\right)\\
&= \frac{(-1)^t q^{e-\frac{1}{2}t}}{1+q^{-1}}\cdot\frac{1}{q^s-q^{-s}}\cdot\left(q^{(t-2e+1)s}(1-q^{-2s-1})-q^{-(t-2e+1)s}(1-q^{2s-1})\right),
\end{aligned}$$

which proves (ii) and (iii) for $T=\pi^t$. Then, by Proposition 1.4 we obtain the assertion (iii) for general $T \in \mathcal{H}_1^{nd}$, since $N(\mathcal{O}_{k'}^\times) = \mathcal{O}_k^\times$. ∎

Remark 2.2. In z-variable, the assertion in Proposition 2.1 becomes as follows, where $z=-s$ and $W=\{1,\tau\}$. For $T=\pi^t$,

$$\begin{aligned}
\omega_T^{(1)}(x_e;z) &= \frac{(-1)^t q^{e-\frac{1}{2}t}}{1+q^{-1}}\\
&\quad\times\left\{\frac{q^{-(t-2e)z}(1-q^{2z-1})}{1-q^{2z}}+\frac{q^{(t-2e)z}(1-q^{-2z-1})}{1-q^{-2z}}\right\},\quad (2e\le t);
\end{aligned}$$

and for any $T \in \mathcal{H}_1^{nd}$ and $x \in \mathfrak{X}_T$,

$$\omega_T^{(1)}(x;z) = \omega_T^{(1)}(x;\tau(z)).$$

2.2. Assume that $n \ge 2$ and set

$$w_\tau = \left(\begin{array}{cc|cc} 1_{n-1} & & & \\ & 0 & & 1 \\ \hline & & 1_{n-1} & \\ & 1 & & 0 \end{array}\right) \in G,$$

then w_τ gives the element $\tau \in W$ such that $\tau(z) = (z_1,\dots,z_{n-1},-z_n)$. The main purpose of this subsection is to prove the following.

Theorem 2.3. *For any $T \in \mathcal{H}_n^{nd}$, the spherical function satisfies*

$$\omega_T(x;z) = \omega_T(x;\tau(z)).$$

The standard parabolic subgroup P attached to τ, in the sense of [1] §21.11, is given as follows:

$$P = B \cup Bw_\tau B$$
$$= \left\{ \left(\begin{array}{cc|cc} q & & & \\ & a & & b \\ \hline & & q^{*-1} & \\ & c & & d \end{array}\right) \left(\begin{array}{cc|cc} 1_{n-1} & \alpha & & \\ & 1 & & \\ \hline & & 1_{n-1} & \\ & & -\alpha^* & 1 \end{array}\right) \left(\begin{array}{c|cc} 1_n & \gamma & \beta \\ & -\beta^* & 0 \\ \hline & \multicolumn{2}{c}{1_n} \end{array}\right) \in G \;\middle|\; \begin{array}{l} q \text{ is upper triangular in } GL_{n-1}(k'), \\ \begin{pmatrix} a & b \\ c & d \end{pmatrix} \in U(H_1),\ \alpha, \beta \in M_{n-1,1}(k'), \\ \gamma \in M_{n-1}(k'),\ \gamma + \gamma^* = 0 \end{array} \right\}, \tag{2.2}$$

where each empty place in the above expression means zero-entry.

Since it suffices to show Theorem 2.3 for diagonal T's (cf. Proposition 1.4), we fix a diagonal $T \in \mathcal{H}_n^{nd}$ and write $f_i(x) = f_{T,i}(x)$ for simplicity of notations. We consider the following action of $\widetilde{P} = P \times GL_1$ on $\widetilde{\mathfrak{X}}_T = \mathfrak{X}_T \times V$ with $V = M_{21}(k')$:

$$(p, r) \star (x, v) = (px, \rho(p)vr^{-1}), \qquad (p, r) \in \widetilde{P},\ (x, v) \in \widetilde{\mathfrak{X}}_T,$$

where $\rho(p) = \begin{pmatrix} a & b \\ c & d \end{pmatrix}$ for the decomposition of $p \in P$ as in (2.2). We define

$$g(x, v) = \det \left[\left(\begin{array}{c|c} 1_{n-1} & 0 \\ \hline 0 & {}^t v \end{array}\right) \begin{pmatrix} x_2 \\ -y \end{pmatrix} \cdot T^{-1} \right], \qquad (x, v) \in \widetilde{\mathfrak{X}}_T, \tag{2.3}$$

where the first matrix in the right hand side is of size $(n, n+1)$, x_2 is the lower half n by n block of x (the same as before) and y is the n-th row of x.

Lemma 2.4. *Let $g(x, v)$ be the function on $\widetilde{\mathfrak{X}}_T = \mathfrak{X}_T \times V$ defined by (2.3).*
(i) *$g(x, v)$ is a relative $\widetilde{P}$-invariant on $\widetilde{\mathfrak{X}}_T$ associated with character $\widetilde{\psi}$:*

$$\widetilde{\psi}(p, r) = \psi_{n-1}(p) N(r)^{-1}, \quad (p, r) \in \widetilde{P} = P \times GL_1,$$

where ψ_{n-1} is well-defined on P, and satisfies

$$g(x, v_0) = f_n(x), \qquad v_0 = \begin{pmatrix} 1 \\ 0 \end{pmatrix} \in V$$

(ii) *$g(x, v)$ is expressed as*

$$g(x, v) = D(x)[v], \tag{2.4}$$

with some hermitian matrix

$$D(x) = \begin{pmatrix} a(x) & \beta(x) \\ \beta(x)^* & d(x) \end{pmatrix} \qquad (a(x), d(x) \in k,\ \beta(x) \in k'), \tag{2.5}$$

such that $\det(D(x)) = 0$ *and* $\mathrm{Tr}(\beta(x)) = -f_{n-1}(x)$ *for* $x \in \mathfrak{X}_T$*, where* Tr *is the trace* $\mathrm{Tr}_{k'/k}$.

Proof. (i) It is easy to see that $g((1,r) \star (x,v)) = N(r)^{-1} g(x,v)$. In order to examine the action of P, we write an element $p \in P$ and $x \in \mathfrak{X}_T^{op}$ as follows

$$p = \left(\begin{array}{cc|cc} q & \cdot & \cdot & \cdot \\ 0 & a & t & b \\ \hline 0 & 0 & q^{*-1} & 0 \\ 0 & c & \mu & d \end{array}\right), \qquad x = \begin{pmatrix} \cdot \\ y \\ x' \\ z \end{pmatrix},$$

where $q \in GL_{n-1},\ t, \mu \in M_{1,n-1},\ \rho(p) = \begin{pmatrix} a & b \\ c & d \end{pmatrix} \in U(H_1),\ x' \in M_{n-1,n}$, and $y, z \in M_{1,n}$. Then we obtain

$$\begin{aligned} g((p,1) \star (x,v)) &= \det\left[\left(\begin{array}{c|c} 1_{n-1} & 0 \\ \hline 0 & {}^t v \begin{pmatrix} a & c \\ b & d \end{pmatrix} \end{array}\right) \begin{pmatrix} q^{*-1}x' \\ cy + \mu x' + dz \\ -(ay + tx' + bz) \end{pmatrix} \cdot T^{-1}\right] \\ &= \det\left[\left(\begin{array}{c|c} 1_{n-1} & 0 \\ \hline 0 & {}^t v \begin{pmatrix} a & c \\ b & d \end{pmatrix} \end{array}\right) \left(\begin{array}{c|cc} q^{*-1} & 0 & \\ \hline \mu & d & -c \\ -t & -b & a \end{array}\right) \begin{pmatrix} x' \\ z \\ -y \end{pmatrix} \cdot T^{-1}\right] \\ &= \det\left[\left(\begin{array}{c|c} q^{*-1} & 0 \\ \hline {}^t v \begin{pmatrix} a & c \\ b & d \end{pmatrix} \begin{pmatrix} \mu \\ -t \end{pmatrix} & \varepsilon\, {}^t v \end{array}\right) \begin{pmatrix} x' \\ z \\ -y \end{pmatrix} \cdot T^{-1}\right] \\ &= \det\left[\left(\begin{array}{c|c} q^{*-1} & 0 \\ \hline {}^t v \begin{pmatrix} a & c \\ b & d \end{pmatrix} \begin{pmatrix} \mu \\ -t \end{pmatrix} & \varepsilon \end{array}\right) \left(\begin{array}{c|c} 1_{n-1} & 0 \\ \hline 0 & {}^t v \end{array}\right) \begin{pmatrix} x' \\ z \\ -y \end{pmatrix} \cdot T^{-1}\right] \\ &= N(\det(q))^{-1} g(x,v) = \psi_{n-1}(p) g(x,v), \end{aligned}$$

where $\varepsilon = ad - bc \in \mathcal{O}_{k'}^1$. Hence $g(x,v)$ is a relative $\widetilde{P}$-invariant on $\widetilde{\mathfrak{X}}_T$ associated with character $\widetilde{\psi}$.

(ii) Since $g(x,v)$ is a linear form with respect to both v_1, v_2 and v_1^*, v_2^*, and $g(x,v)^* = g(x,v)$, we have an expression (2.4) with some $D(x) \in \mathcal{H}_2$.

Writing $T = Diag(t_1, \dots, t_n)$, we have

$$\begin{aligned} g(x_T, v) &= (t_1\cdots t_n)^{-1}(v_1 - \xi t_n v_2)(v_1^* - \xi^* t_n v_2^*) \\ &= \begin{pmatrix} (t_1\cdots t_n)^{-1} & -\xi(t_1\cdots t_{n-1})^{-1} \\ -\xi^*(t_1\cdots t_{n-1})^{-1} & N(\xi)(t_1\cdots t_{n-1})^{-1}t_n \end{pmatrix}[v], \\ &= \begin{pmatrix} (t_1\cdots t_n)^{-1} & -\xi f_{n-1}(x_T) \\ -\xi^* f_{n-1}(x_T) & N(\xi)(t_1\cdots t_{n-1})^{-1}t_n \end{pmatrix}[v], \end{aligned} \tag{2.6}$$

in particular $\det(D(x_T)) = 0$. Since $g(x, v)$ is a relative $\widetilde{P}$-invariant associated with $\widetilde{\psi}$ by (i), we see

$$D(px) = \psi(p)D(x)[\rho(p)^{-1}], \quad (p \in P) \tag{2.7}$$

and we have

$$\det(D(px_T)) = 0.$$

Since $\mathfrak{X}_T^{op}$ is a B-orbit over the algebraic closure of k(cf. Remark 1.2), we have

$$\det(D(x)) = 0, \qquad \text{for any } x \in \mathfrak{X}_T^{op}.$$

For an element b of B, $\rho(b)$ can be written as follows (cf. (1.4))

$$\rho(b) = \begin{pmatrix} \gamma & \gamma u\sqrt{\epsilon} \\ 0 & \gamma^{*-1} \end{pmatrix}, \qquad \gamma \in k'^{\times},\ u \in k,$$

and when we express $D(x)$ and $D(bx)$ as in (2.5), we have by (2.7)

$$\beta(bx) = \psi_{n-1}(b)(-a(x)u\sqrt{\varepsilon} + \beta(x)),$$

hence $\mathrm{Tr}(\beta(bx)) = \psi_{n-1}(b)\mathrm{Tr}(\beta(x))$ and $\mathrm{Tr}(\beta(x_T)) = -f_{n-1}(x_T)$ by (2.6). Thus $\mathrm{Tr}(\beta(x)) = -f_{n-1}(x)$ for every $x \in \mathfrak{X}_T^{op}$. ∎

For $A \in \mathcal{H}_2$ and $s \in \mathbb{C}$, we define

$$\zeta_{K_1}(A; s) = \int_{K_1} |d_1(h\cdot A)|^{s-\frac{1}{2}}\, dh,$$

where $h \cdot A = hAh^*$ and dh is the normalized Haar measure on $K_1 = U(H_1) \cap GL_2(\mathcal{O}_{k'})$, which is absolutely convergent if $\mathrm{Re}(s) \geq \frac{1}{2}$.

Lemma 2.5. *Assume $x \in \mathfrak{X}_T^{op}$ and $D(x)$ is given by (2.4). Set $m = \min\{v_\pi(a(x)),\ v_\pi(d(x))\}$ and $t = v_\pi(\beta(x)) - m$ for the expression of $D(x)$ as in (2.5). Then $t \geq 0$ and*

$$\zeta_{K_1}(D(x); s) = \frac{q^{\frac{m}{2}}}{1+q^{-1}} \cdot |f_{n-1}(x)|^s \cdot \frac{q^{(t+1)s}(1-q^{-2s-1}) - q^{-(t+1)s}(1-q^{2s-1})}{q^s - q^{-s}}.$$

In particular, one has the functional equation

$$\zeta_{K_1}(D(x);s) = |f_{n-1}(x)|^{2s} \cdot \zeta_{K_1}(D(x);-s). \tag{2.8}$$

Proof. Take an $x \in \mathfrak{X}_T^{op}$, write $D(x)$ as in (2.5), and set m as above. Then $\beta(x)$ can be written as

$$\begin{aligned} &\beta(x) = b_1 + \xi b_2, \quad b_1, b_2 \in k,\ m \le \min\{b_1, b_2\}, \\ &\mathrm{Tr}(\beta(x)) = 2b_1 + b_2 = -f_{n-1}(x). \end{aligned}$$

Then, by the action of K_1 on $\mathcal{H}_2$, we see $D(x)$ is K_1-equivalent to

$$\pi^m \begin{pmatrix} 1 & \xi b \\ \xi^* b & N(\xi)b^2 \end{pmatrix}, \quad b = \pi^{-m}\mathrm{Tr}(\beta(x)) \in \mathcal{O}_k, \tag{2.9}$$

and if k is nondyadic, it is K_1-equivalent to

$$\pi^m \begin{pmatrix} 1 & \frac{1}{2}b \\ \frac{1}{2}b & \frac{1}{4}b^2 \end{pmatrix}, \quad b = \pi^{-m}\mathrm{Tr}(\beta(x)) \in \mathcal{O}_k. \tag{2.10}$$

We denote by A the matrix given in (2.9) (resp. in (2.10)) if k is dyadic (resp. nondyadic), then $\zeta_{K_1}(D(x);s) = \zeta_{K_1}(A;s)$. We recall the data for $K_1 = K_{1,1} \cup K_{1,2}$ in (2.1).

For $h = \begin{pmatrix} \alpha & 1 \\ \alpha^{*-1} & 1 \end{pmatrix} \begin{pmatrix} 1 & v/\sqrt{\epsilon} \\ u\sqrt{\epsilon} & (1+uv) \end{pmatrix} \in K_{1,1}$, we have

$$d_1(h \cdot A) = \begin{cases} \pi^m N(\alpha)(1 - \dfrac{b^2v^2}{4\epsilon}) & \text{if } k \text{ is nondyadic,} \\ \dfrac{\pi^m N(\alpha)}{\epsilon}(\epsilon - \epsilon b v - N(\xi)b^2v^2) & \text{if } k \text{ is dyadic,} \end{cases}$$

and $v_\pi(d_1(h \cdot A)) = m$ for any $h \in K_{1,1}$, where we recall that $\epsilon \in 1 + 4\mathcal{O}_k^\times$.

For $h = \begin{pmatrix} \alpha & 1 \\ \alpha^{*-1} & 1 \end{pmatrix} \begin{pmatrix} \pi u\sqrt{\epsilon} & 1+\pi uv \\ 1 & v/\sqrt{\epsilon} \end{pmatrix} \in K_{1,2}$ (cf. (2.1)), we have

$$d_1(h \cdot A) =$$
$$\begin{cases} \pi^m N(\alpha)(-\epsilon\pi^2u^2 + (1+\pi uv)^2 b^2/4) & \text{if } k \text{ is non dyadic,} \\ \pi^m N(\alpha)(-\epsilon\pi^2u^2 + (1+\pi uv)\epsilon\pi u b + (1+\pi uv)^2 N(\xi)b^2) & \text{if } k \text{ is dyadic,} \end{cases}$$

and $v_\pi(d_1(h \cdot A)) = m + 2\min\{v_\pi(b), v_\pi(u)+1\}$.

Set $t = v_\pi(b)$. If $t = 0$, it is clear that $\zeta_{K_1}(A; s) = q^{-m(s-\frac{1}{2})}$. If $t > 0$, then we obtain

$$\begin{aligned}
\zeta_{K_1}(A; s) &= \frac{1}{1+q^{-1}} q^{-m(s-\frac{1}{2})} \\
&\quad + \frac{q^{-1}}{1+q^{-1}} \left(\sum_{\ell=0}^{t-1} q^{-\ell}(1-q^{-1}) q^{-(m+2+2\ell)(s-\frac{1}{2})} + q^{-t} q^{-(m+2t)(s-\frac{1}{2})} \right) \\
&= \frac{q^{-m(s-\frac{1}{2})}}{(1+q^{-1})} \times \left(1 + \frac{q^{-2s} - q^{-2s-1} + q^{-2ts-1} - q^{-2(t+1)s}}{1 - q^{-2s}} \right) \\
&= \frac{q^{-(m+t)s+\frac{m}{2}}}{(1+q^{-1})} \times \frac{q^{(t+1)s}(1-q^{-2s-1}) - q^{-(t+1)s}(1-q^{2s-1})}{q^s - q^{-s}},
\end{aligned}$$

and the latter two expressions are valid also for $t = 0$. Since $\pi^m b = \mathrm{Tr}\beta(x) = -f_{n-1}(x)$, we have

$$\begin{aligned}
&\zeta_{K_1}(D(x); s) = \zeta_{K_1}(A; s) \\
&= \frac{q^{\frac{m}{2}}}{1+q^{-1}} |f_{n-1}(x)|^s \times \frac{q^{(t+1)s}(1-q^{-2s-1}) - q^{-(t+1)s}(1-q^{2s-1})}{q^s - q^{-s}}.
\end{aligned}$$

The identity (2.8) follows from the above explicit formula. ∎

Now we will prove Theorem 2.3. We consider the embedding

$$K_1 \longrightarrow K = K_n, \quad h = \begin{pmatrix} a & b \\ c & d \end{pmatrix} \longmapsto \widetilde{h} = \left(\begin{array}{cc|cc} 1_{n-1} & & & \\ & a & & b \\ \hline & & 1_{n-1} & \\ & c & & d \end{array} \right).$$

Then we have

$$\begin{aligned}
\omega_T(x; s) &= \int_{K_1} dh \int_K |f(kx)|^{s+\varepsilon} \, dk = \int_{K_1} dh \int_K \left| f(\widetilde{h}kx) \right|^{s+\varepsilon} dk \\
&= \int_K \chi_\pi(\prod_{i<n} f_i(kx)) \prod_{i<n} |f_i(kx)|^{s_i - 1} \left(\int_{K_1} \chi_\pi(f_n(\widetilde{h}kx)) \left| f_n(\widetilde{h}kx) \right|^{s_n - \frac{1}{2}} dh \right) dk.
\end{aligned}$$

By definition of $f_n(x)$ and $g(x, v)$ and Lemma 2.4, we have for $h = \begin{pmatrix} a & b \\ c & d \end{pmatrix} \in K_1$

$$\begin{aligned}
f_n(\widetilde{h}x) &= \det \left[\begin{pmatrix} x' \\ cy + dz \end{pmatrix} \cdot T^{-1} \right] = g(x, \begin{pmatrix} d \\ -c \end{pmatrix}) \\
&= (d^* \ \ -c^*) D(x) \begin{pmatrix} d \\ -c \end{pmatrix} = d_1(h^{*-1} \cdot D(x)).
\end{aligned}$$

Since $\{h^{*-1} \mid h \in K_1\} = K_1$, we have

$$\omega_T(x;s) = \int_K \chi_\pi(\prod_{i<n} f_i(kx)) \prod_{i<n} |f_i(kx)|^{s_i-1} \zeta_{K_1}(D(kx); s_n + \tfrac{\pi\sqrt{-1}}{\log q})dk,$$

and by Lemma 2.5, we obtain

$$\begin{aligned}
&\omega_T(x;s) \\
&= \int_K \chi_\pi(\prod_{i<n} f_i(kx)) \prod_{i\leq n-2} |f_i(kx)|^{s_i-1} \cdot |f_{n-1}(kx)|^{s_{n-1}+2s_n-1} \\
&\quad \times \zeta_{K_1}(D(kx); -s_n + \frac{\pi\sqrt{-1}}{\log q})dk \\
&= \omega_T(x; s_1, \ldots, s_{n-2}, s_{n-1} + 2s_n, -s_n).
\end{aligned}$$

In variable z, we have

$$\omega_T(x;z) = \omega_T(x;\tau(z)), \qquad \tau(z) = (z_1, \ldots, z_{n-1}, -z_n),$$

which completes the proof. ∎

2.3. In order to describe functional equations of $\omega_T(x;z)$ with respect to W, we prepare some notations. We denote by Σ the set of roots of G with respect to the k-split torus of G contained in B and by Σ^+ the set of positive roots with respect to B. We may understand Σ as a subset in $\mathbb{Z}^n$, and set

$$\begin{aligned}
&\Sigma^+ = \Sigma_s^+ \cup \Sigma_\ell^+, \\
&\Sigma_s^+ = \{e_i - e_j,\ e_i + e_j \mid 1 \leq i < j \leq n\}, \quad \Sigma_\ell^+ = \{2e_i \mid 1 \leq i \leq n\},
\end{aligned}$$

where e_i is the i-th unit vector in $\mathbb{Z}^n$, $1 \leq i \leq n$. The set

$$\Sigma_0 = \{e_i - e_{i+1} \mid 1 \leq i \leq n-1\} \cup \{2e_n\}$$

forms the set of simple roots, and we denote by Δ the set of reflections associated with elements in Σ_0. Then

$$\Delta = \{\tau_i = (i\ i+1) \in S_n \mid 1 \leq i \leq n-1\} \cup \{\tau\},$$

which generates W. For each $\sigma \in W$, we set

$$\Sigma_s^+(\sigma) = \{\alpha \in \Sigma_s^+ \mid -\sigma(\alpha) \in \Sigma^+\}. \tag{2.11}$$

We consider a pairing on $\mathbb{Z}^n \times \mathbb{C}^n$ given by

$$\langle t, z \rangle = \sum_{i=1}^n t_i z_i, \qquad (t \in \mathbb{Z}^n,\ z \in \mathbb{C}^n),$$

which gives a W-invariant pairing on $\Sigma \times \mathbb{C}^n$, i.e.,

$$\langle \alpha,\, z\rangle = \langle \sigma(\alpha),\, \sigma(z)\rangle\,, \qquad (\alpha \in \Sigma,\; z \in \mathbb{C}^n,\; \sigma \in W).$$

Theorem 2.6. *For $T \in \mathcal{H}_n^{nd}$ and $\sigma \in W$, the spherical function $\omega_T(x;z)$ satisfies the following functional equation*

$$\omega_T(x;z) = \Gamma_\sigma(z)\cdot \omega_T(x;\sigma(z)), \tag{2.12}$$

where

$$\Gamma_\sigma(z) = \prod_{\alpha\in\Sigma_s^+(\sigma)} \frac{1-q^{\langle\alpha,\,z\rangle-1}}{q^{\langle\alpha,\,z\rangle}-q^{-1}},$$

and we understand $\Gamma_\sigma(z) = 1$ if $\Sigma_s^+(\sigma) = \emptyset$. In particular, the Gamma factor $\Gamma_\sigma(z)$ does not depend on x nor T.

We note here that the factor $\langle\alpha,\, z\rangle$ for $\alpha = e_i \pm e_j$ $(i<j)$ in s-variable:

$$\langle\alpha,\, z\rangle = \begin{cases} -(s_i+\cdots+s_{j-1}) & \text{if } \alpha = e_i - e_j \\ \\ -(s_i+\cdots+s_{j-1}+2(s_j+\cdots+s_n)) & \text{if } \alpha = e_i + e_j \end{cases}. \tag{2.13}$$

Proof of Theorem 2.6. We define the Gamma factor $\Gamma_\sigma(z)$ by the equation (2.12). Then it is a rational function of $q^{z_1},\ldots,q^{z_n}$ since $\omega_T(x;z)$ and $\omega_T(x;\sigma(z))$ are those functions, and Gamma factors satisfy the cocycle relations

$$\Gamma_{\sigma_2\sigma_1}(z) = \Gamma_{\sigma_2}(\sigma_1(z))\cdot\Gamma_{\sigma_1}(z), \qquad (\sigma_1,\sigma_2\in W). \tag{2.14}$$

For convenience we set for $\alpha \in \Sigma$

$$f_\alpha(\langle\alpha,\, z\rangle) = \begin{cases} 1 & \text{if } \alpha = \pm 2e_i,\ (1\le i\le n) \\ \dfrac{1-q^{\langle\alpha,\,z\rangle-1}}{q^{\langle\alpha,\,z\rangle}-q^{-1}} & \text{otherwise}\,. \end{cases} \tag{2.15}$$

For an element $\sigma\in\Delta$ associated with some $\alpha_0\in\Sigma_0$,

$$\Sigma_s^+(\sigma) = \begin{cases} \{\alpha_0\} & \text{if } \alpha_0\in\Sigma_s^+ \\ \emptyset & \text{if } \alpha_0\in\Sigma_\ell^+ \text{ (i.e., } \alpha_0 = 2e_n), \end{cases}$$

and, by Remark 1.7, Remark 2.2 and Theorem 2.3,

$$\Gamma_\sigma(z) = f_{\alpha_0}(\langle\alpha_0,\, z\rangle),$$

which is independent of x nor T. In general, assume that $\sigma \in W$ has the following shortest expression

$$\sigma = \sigma_\ell \cdots \sigma_1,$$

where $\sigma_i \in \Delta$ is the reflection associated with $\alpha_i \in \Sigma_0$. Then we see

$$\left\{\alpha \in \Sigma^+ \mid \sigma(\alpha) < 0\right\} = \{\alpha_1\} \cup \{\sigma_1 \cdots \sigma_{k-1}(\alpha_k) \mid 2 \leq k \leq \ell\}.$$

By using (2.14), (2.15) and the W-invariancy of the pairing $\langle\ ,\ \rangle$, we obtain

$$\begin{aligned}\Gamma_\sigma(z) &= \Gamma_{\sigma_\ell}(\sigma_{\ell-1} \cdots \sigma_1(z)) \cdots \Gamma_{\sigma_2}(\sigma_1(z)) \cdot \Gamma_{\sigma_1}(z) \\ &= f_{\alpha_\ell}(\langle \alpha_\ell,\, \sigma_{\ell-1} \cdots \sigma_1(z)\rangle) \cdots f_{\alpha_2}(\langle \alpha_2,\, \sigma_1(z)\rangle) \cdot f_{\alpha_1}(\langle \alpha_1,\, z\rangle) \\ &= f_{\alpha_\ell}(\langle \sigma_1 \cdots \sigma_{\ell-1}(\alpha_\ell),\, z\rangle) \cdots f_{\alpha_2}(\langle \sigma_1(\alpha_2),\, z\rangle) \cdot f_{\alpha_1}(\langle \alpha_1,\, z\rangle) \\ &= \prod_{\alpha \in \Sigma_s^+(\sigma)} f_\alpha(\langle \alpha,\, z\rangle),\end{aligned}$$

which completes the proof. ∎

We will use the following explicit value $\Gamma_\rho(z)$ for a particular $\rho \in W$ in §5.

Corollary 2.7. *Set $\rho \in W$ by*

$$\rho(z_1, \ldots, z_n) = (-z_n, -z_{n-1}, \ldots, -z_1).$$

Then

$$\Gamma_\rho(z) = \prod_{1 \leq i < j \leq n} \frac{1 - q^{z_i + z_j - 1}}{q^{z_i + z_j} - q^{-1}}.$$

Proof. Since

$$\Sigma_s^+(\rho) = \left\{e_i + e_j \mid 1 \leq i < j \leq n\right\},$$

the assertion follows from Theorem 2.6. ∎

Remark 2.8. The above ρ gives the functional equation of the hermitian Siegel series (cf. §5), and it is interesting that such ρ corresponds to the unique automorphism of the extended Dynkin diagram of the root system of type (C_n), which was pointed out by Y. Komori.

2.4. By using the functional equations (Theorem 2.6) and the previous results on hermitian forms (Proposition 1.6), we obtain the following theorem, which gives us the information of the location of possible poles and zeros.

Theorem 2.9. *Set*

$$G(z) = \prod_{\alpha\in\Sigma_s^+} \frac{1+q^{\langle\alpha,\, z\rangle}}{1-q^{\langle\alpha,\, z\rangle-1}}.$$

Then, for any $T\in\mathcal{H}_n^{nd}$, the function $G(z)\cdot\omega_T(x;z)$ is holomorphic for all z in $\mathbb{C}^n$ and W-invariant. In particular it is an element in $\mathbb{C}[q^{\pm z_1},\dots,q^{\pm z_n}]^W$.

We denote by $\mathcal{S}(K\backslash X_T)$ the subspace of $\mathcal{C}^\infty(K\backslash X_T)$ consisting of compactly supported functions, which can be regarded as functions on $\mathfrak{X}_T$ of compactly supported functions modulo $U(T)$ on $\mathfrak{X}_T$ modulo. Keeping the relation (1.11) for s and z, we consider the following integral

$$\Phi_T(z;\xi) = \int_{X_T^{op}} \xi(x)\,|f_T(x)|^{s+\varepsilon}\,dx, \qquad (\xi\in\mathcal{S}(K\backslash X_T)) \tag{2.16}$$

where dx is the G-invariant measure on X_T, and the right hand side is absolutely convergent for

$$\begin{aligned} s\in\mathcal{D}_0 &= \{\, s\in\mathbb{C}^n \mid \mathrm{Re}(s_i)\ge -\mathrm{Re}(\varepsilon_i),\ 1\le i\le n\} \\ &= \left\{ z\in\mathbb{C}^n \,\middle|\, -\frac{1}{2}\ge \mathrm{Re}(z_n),\ \mathrm{Re}(z_{i+1})\ge\mathrm{Re}(z_i)+1,\ (1\le i\le n-1)\right\}. \end{aligned}$$

When ξ is the characteristic function of Kx, $\Phi_T(z;\xi)$ is a constant multiple of $\omega_T(x;z)$, and any ξ in $\mathcal{S}(K\backslash X_T)$ is a finite linear sum of those characteristic functions. Thus we see that $\Phi_T(z;\xi)$ is a rational function of $q^{z_1},\dots,q^{z_n}$ and satisfy the same functional equations for $\omega_T(x;z)$, i.e.,

$$\Phi_T(z;\xi) = \Gamma_\sigma(z)\cdot\Phi_T(\sigma(z);\xi), \qquad (\sigma\in W,\ \xi\in\mathcal{S}(K\backslash X_T)). \tag{2.17}$$

Since $G(\sigma(z)) = G(z)\cdot\Gamma_\sigma(z)$ for $\sigma\in\Delta$, we see $G(z)\cdot\Phi_T(z;\xi)$ is invariant under the action of Δ, hence it is W-invariant by cocycle relations. Since $G(z)$ is holomorphic for $z\in\mathcal{D}_0$, we see $G(z)\cdot\Phi_T(z;\xi)$ is holomorphic for

$$z\in\bigcup_{\sigma\in W}\sigma(\mathcal{D}_0).$$

On the other hand, in a similar manner to the proof of Theorem 1.5, we see

$$\Phi_T(z;\xi) = \int_{X_T^{op}} \xi(x)\zeta^{(n)}(D(x);s)dx,$$

where $D(x) = x_2\cdot T^{-1}$ and $\zeta^{(n)}(y;s)$ is the spherical function on $\mathcal{H}_n^{nd}$ (cf. the proof of Theorem 1.5), and recall that $G_1(z)\cdot\zeta^{(n)}(y;z)$ is holomorphic for $z \in \mathbb{C}^n$. Setting

$$G(z) = G_1(z) \cdot G_2(z), \qquad G_2(z) = \prod_{1 \le i < j \le n} \frac{1 + q^{z_i + z_j}}{1 - q^{z_i + z_j - 1}},$$

we see $G(z) \cdot \Phi_T(z; \xi)$ is holomorphic for

$$z \in \mathcal{D}_1 = \{ z \in \mathbb{C}^n \mid \mathrm{Re}(z_i + z_j) \neq 1,\ 1 \le i < j \le n \},$$

since $G_2(z)$ is holomorphic for $z \in \mathcal{D}_1$ and ξ is compactly supported. Since $G(z) \cdot \Phi_T(z; \xi)$ is W-invariant, it is holomorphic for

$$z \in \widetilde{\mathcal{D}} = \bigcup_{\sigma \in W} \sigma(\mathcal{D}_0 \cup \mathcal{D}_1).$$

Since $\widetilde{\mathcal{D}}$ is connected, $G(z) \cdot \Phi_T(z; \xi)$ is holomorphic in the convex hull $\mathbb{C}^n$ of $\widetilde{\mathcal{D}}$.

Taking the characteristic function of Kx for ξ, we obtain the theorem. ∎

3. Explicit formulas

3.1. Set

$$\Lambda_n^+ = \{ \lambda \in \mathbb{Z}^n \mid \lambda_1 \ge \lambda_2 \ge \cdots \ge \lambda_n \ge 0 \}, \tag{3.1}$$

and, for each $\lambda \in \Lambda_n^+$,

$$\begin{aligned} \pi^\lambda &= Diag(\pi^{\lambda_1}, \ldots, \pi^{\lambda_n}) \in \mathcal{H}_n^{nd}, \\ x_\lambda &= \begin{pmatrix} \xi \pi^\lambda \\ 1_n \end{pmatrix} \in \mathfrak{X}_{\pi^\lambda}, \\ \omega_\lambda(x; z) &= \omega_T(x; z) \quad \text{for } T = \pi^\lambda. \end{aligned} \tag{3.2}$$

Then we obtain

Theorem 3.1. *For $\lambda \in \Lambda_n^+$, one has the following explicit expression:*

$$\begin{aligned} \omega_\lambda(x_\lambda; z) = &\frac{(-1)^{\sum_i \lambda_i(n-i+1)} q^{-\sum_i \lambda_i(n-i+\frac{1}{2})} (1 - q^{-2})^n}{\prod_{i=1}^{2n} (1 - (-q^{-1})^i)} \\ &\times \frac{1}{G(z)} \times \sum_{\sigma \in W} q^{-\langle \lambda, \sigma(z) \rangle} H(\sigma(z)), \end{aligned}$$

where $G(z)$ is the same as in Theorem 2.9 and

$$H(z) = \prod_{\alpha \in \Sigma_s^+} \frac{1 + q^{\langle \alpha, z \rangle - 1}}{1 - q^{\langle \alpha, z \rangle}} \prod_{\alpha \in \Sigma_\ell^+} \frac{1 - q^{\langle \alpha, z \rangle - 1}}{1 - q^{\langle \alpha, z \rangle}}.$$

Remark 3.2. By Theorem 2.9, the main part

$$H_\lambda(z) = \sum_{\sigma \in W} \sigma\left(q^{-\langle \lambda,\, z\rangle} H(z)\right) = \sum_{\sigma \in W} q^{-\langle \lambda,\, \sigma(z)\rangle} H(\sigma(z))$$

of $\omega_\lambda(x_\lambda; z)$ belongs to $\mathbb{C}[q^{\pm z_1}, \ldots, q^{\pm z_n}]^W$. Further we see directly in a standard way that the set $\{ H_\lambda(z) \mid \lambda \in \Lambda_n^+\}$ forms its $\mathbb{C}$-basis. On the other hand, $H_\lambda(z)$ is a special case of P_λ (up to a scalar factor) introduced by I. G. Macdonald ([14] §10) in a generous context of orthogonal polynomials associated with root systems.

We will prove the above theorem by using a general expression formula given in [7] (or in [5]) of spherical functions on homogeneous spaces, which is based on functional equations of finer spherical functions and some data depending only on the group G. We need to check the assumptions there. Let $\mathbb{G}$ be a connected reductive linear algebraic group and $\mathbb{X}$ be an affine algebraic variety which is $\mathbb{G}$-homogeneous, where everything is assumed to be defined over a p-adic field k. For an algebraic set, we use the same ordinary letter to indicate the set of k-rational points. Let K be a special good maximal compact open subgroup of G, and $\mathbb{B}$ a minimal parabolic subgroup of $\mathbb{G}$ defined over k satisfying $G = KB = BK$. We denote by $\mathfrak{X}(\mathbb{B})$ the group of rational character of $\mathbb{B}$ defined over k and by $\mathfrak{X}_0(\mathbb{B})$ the subgroup consisting of those characters associated with some relative $\mathbb{B}$-invariant on $\mathbb{X}$ defined over k. In these situation, the assumptions are the following:

($A1$) $\mathbb{X}$ has only a finite number of $\mathbb{B}$-orbits (hence, there is only one open orbit).

($A2$) A basic set of relative $\mathbb{B}$-invariants on $\mathbb{X}$ defined over k can be taken by regular functions on $\mathbb{X}$.

($A3$) For $y \in \mathbb{X}$ not contained in the open orbit, there exists some ψ in $\mathfrak{X}_0(\mathbb{B})$ whose restriction to the identity component of the stabilizer $\mathbb{H}_y$ of $\mathbb{G}$ at y is not trivial.

($A4$) The rank of $\mathfrak{X}_0(\mathbb{B})$ coincides with that of $\mathfrak{X}(\mathbb{B})$.

In the present situation, as is noted in Remark 1.2, we may understand $\mathbb{G} = U(H_n)$ as an algebraic group defined over k, $G = \mathbb{G}(k)$, $B = \mathbb{B}(k)$ for the Borel subgroup defined over k, $K = \mathbb{G}(\mathcal{O}_k)$, and $X = X_T$ as the set of k-rational points of the affine algebraic variety $\mathbb{X} = \mathfrak{X}_T/U(T)$, and we recall that relative invariants $f_{T,i}(x)$ and the spherical function $\omega_T(x;s)$ can be regarded as functions on X_T.

It is easy to see the present $(\mathbb{X},\mathbb{B})$ satisfies the conditions $(A1)$, $(A2)$ and $(A4)$ (cf. Lemma 1.1, (1.5) and (1.6)), in particular, the unique Zariski open $\mathbb{B}$-orbit is given by $\mathbb{X}^{op} = \{x \in \mathbb{X} \mid f_{T,i}(x) \neq 0,\ 1 \leq i \leq n\}$ (cf. (1.7)). We admit the condition $(A3)$, which is proved in §3.2, and give a proof of Theorem 3.1.

We recall the notation in the proof of of Proposition 1.3. By the functional equation of $\omega_T(x;z)$ (Theorem 2.6), we have for each $\sigma \in W$

$$\begin{aligned}\omega_T(x;z_\chi) &= \Gamma_\sigma(z_\chi)\omega_T(x;\sigma(z_\chi))\\ &= \Gamma_\sigma(z_\chi)\omega_T(x;\sigma(z)_{\sigma(\chi)}),\end{aligned} \tag{3.3}$$

by taking a suitable character $\sigma(\chi)$ of $\mathcal{U}$. When χ is the trivial character $\mathbf{1}$, the equation (3.3) coincides with the original functional equation of $\omega_T(x;z)$ and $\Gamma_\sigma(z_\mathbf{1}) = \Gamma_\sigma(z)$. By (1.13) and (3.3), we obtain vector-wise functional equations for finer spherical functions $\omega_{T,u}(x;z)$ for $\sigma \in W$

$$(\omega_{T,u}(x;z))_{u\in\mathcal{U}} = A^{-1}\cdot G(\sigma,z)\cdot \sigma A \cdot (\omega_{T,u}(x;\sigma(z)))_{u\in\mathcal{U}}, \tag{3.4}$$

where

$$A = (\chi(u))_{\chi,u}, \quad \sigma A = (\sigma(\chi)(u))_{\chi,u} \in GL_{2^{n-1}}(\mathbb{Z}),$$

χ runs over characters of $\mathcal{U}$, $u \in \mathcal{U}$, and $G(\sigma,z)$ is the diagonal matrix of size 2^{n-1} whose (χ,χ)-component is $\Gamma_\sigma(z_\chi)$. We denote by U the Iwahori subgroup of K compatible with B and take the normalized Haar measure du on U. It is easy to see

$$Ux_\lambda \subset Bx_\lambda \quad \text{and} \quad |f_T(ux_\lambda)|^s = |f_T(x_\lambda)|^s,$$

which means $x_\lambda \in \mathcal{R}^+$ in the sense of (2.8) in [7]. We set

$$\delta_u(x_\lambda,z) = \int_U |f_T(ux_\lambda)|_u^{s+\varepsilon}\, du = \left\{\begin{matrix} |f_T(x_\lambda)|^{s+\varepsilon} \text{ if } x_\lambda \in \mathfrak{X}_{T,u} \\ \\ 0 \qquad\qquad \text{otherwise.}\end{matrix}\right\}$$

$$= \begin{cases} (-1)^{\sum_i \lambda_i(n-i+1)} q^{-\sum_i \lambda_i(n-i+\frac{1}{2})} q^{-<\lambda,z>} & \text{if } x_\lambda \in \mathfrak{X}_{T,u} \\ \\ 0 & \text{otherwise.}\end{cases} \tag{3.5}$$

Applying Theorem 2.6 in [7] to our present case, we obtain

$$\begin{aligned}&(\omega_{T,u}(x_\lambda;z))_{u\in\mathcal{U}}\\ &= \frac{1}{Q}\sum_{\sigma\in W} \gamma(\sigma(z))\left(A^{-1}\cdot G(\sigma,z)\cdot \sigma A\right)(\delta_u(x_\lambda,\sigma(z)))_{u\in\mathcal{U}},\end{aligned} \tag{3.6}$$

where

$$Q=\sum_{\sigma\in W}[U\sigma U:U]^{-1}=\prod_{i=1}^{2n}\left(1-(-1)^i q^{-i}\right)\Big/(1-q^{-2})^n,$$

$$\gamma(z)=\prod_{\alpha\in\Sigma_s^+}\frac{1-q^{2\langle\alpha,\,z\rangle-2}}{1-q^{2\langle\alpha,\,z\rangle}}\cdot\prod_{\alpha\in\Sigma_\ell^+}\frac{1-q^{\langle\alpha,\,z\rangle-1}}{1-q^{\langle\alpha,\,z\rangle}}.$$

By (3.4), (3.5), (3.6), and the orthogonal relation of characters, we obtain

$$\begin{aligned}\omega_T(x_\lambda;z)&=\sum_{u\in\mathcal{U}}\mathbf{1}(u)\omega_{T,u}(x_\lambda;z)\\&=\frac{(-1)^{-\sum_i\lambda_i(n-i+1)}q^{-\sum_i\lambda_i(n-i+\frac12)}}{Q}\times\sum_{\sigma\in W}\gamma(\sigma(z))\Gamma_\sigma(z)q^{-<\lambda,\sigma(z)>}.\end{aligned}$$

Since we have

$$\Gamma_\sigma(z)=\frac{G(\sigma(z))}{G(z)}\quad\text{(by Theorem 2.9)},$$

$$\gamma(z)\cdot G(z)=\prod_{\alpha\in\Sigma_s^+}\frac{1+q^{\langle\alpha,\,z\rangle-1}}{1-q^{\langle\alpha,\,z\rangle}}\times\prod_{\alpha\in\Sigma_\ell^+}\frac{1-q^{\langle\alpha,\,z\rangle-1}}{1-q^{\langle\alpha,\,z\rangle}}=H(z),$$

we obtain

$$\begin{aligned}\omega_T(x_\lambda;z)=&\frac{(-1)^{\sum_i\lambda_i(n-i+1)}q^{-\sum_i\lambda_i(n-i+\frac12)})(1-q^{-2})^n}{\prod_{i=1}^{2n}(1-(-q^{-1})^i)}\\&\times\frac{1}{G(z)}\times\sum_{\sigma\in W}\sigma\left(q^{-\langle\lambda,\,z\rangle}H(z)\right),\end{aligned}$$

which proves the theorem. ∎

By Theorem 3.1 and Proposition 1.4, we get the explicit formula of $\omega_T(x;s)$ at many points. For $\lambda\in\Lambda_n^+$ and $T\in\mathcal{H}_n^{nd}$, it is known that T and π^λ belong to the same $GL_n(k')$-orbit in $\mathcal{H}_n^{nd}$ if and only if

$$v_\pi(\det T)\equiv|\lambda|\pmod 2,$$

where $|\lambda|=\sum_{i=1}^n\lambda_i$. And then, there exists some $h_\lambda\in GL_n(k')$ for which $\pi^\lambda[h_\lambda]=T$ and $x_\lambda h_\lambda\in\mathfrak{X}_T$. Hence we have the following.

Theorem 3.3. *Let $T\in\mathcal{H}_n^{nd}$ and $\lambda\in\Lambda_n^+$ and assume that $v_\pi(\det T)\equiv|\lambda|$ (mod 2). Taking $h_\lambda\in GL_n(k')$ for which $\pi^\lambda[h_\lambda]=T$, one has $x_\lambda h_\lambda\in\mathfrak{X}_T$*

and

$$\begin{aligned}\omega_T(x_\lambda h_\lambda; z) &= \omega_\lambda(x_\lambda; z)\\ &= \frac{(-1)^{\sum_i \lambda_i(n-i+1)} q^{-\sum_i \lambda_i(n-i+\frac{1}{2})}(1-q^{-2})^n}{\prod_{i=1}^{2n}(1-(-q^{-1})^i)}\\ &\quad \times \frac{1}{G(z)} \cdot \sum_{\sigma\in W} \sigma\left(q^{-\langle\lambda,\, z\rangle} H(z)\right).\end{aligned}$$

Further, each of such λ's gives a different K-orbit

$$Kx_\lambda h_\lambda U(T) \quad in \;\; K\backslash X_T \left(= K\backslash \mathfrak{X}_T/U(T)\right).$$

The latter statement follows from the explicit formula, since different λ gives the different value $\omega_T(x_\lambda h_\lambda; z)$ as a rational function of $q^{z_1}, \dots, q^{z_n}$.

3.2. In this subsection we prove the present $(\mathbb{X}, \mathbb{B})$ satisfies the condition $(A3)$. We consider the action of $G \times U(T)$ on $\mathfrak{X}_T$ defined by $(g,h)\circ x = gxh^{-1}$. Then, the stabilizer B_y of B at $yU(T) \in X_T$ coincides with the image $B_{(y)}$ of the projection to B of the stabilizer $(B\times U(T))_y$ at $y \in \mathfrak{X}_T$ to B. Hence, in our case, the condition $(A3)$ is equivalent to the following:

(C) : For each $y \in \mathfrak{X}_T$ not contained in $\mathfrak{X}_T^{op}$, there exists $\psi \in \mathfrak{X}(\mathbb{B})$ whose restriction to the identity component of $B_{(y)}$ is not trivial.

It suffices to prove the condition $(A3)$ (or (C)) over the algebraic closure $\overline{k}$ of k, since, for a connected linear algebraic group $\mathbb{H}$, $\mathbb{H}(k)$ is dense in $\mathbb{H}(\overline{k})$. Then, we need to consider only for the case $T = 1_n$, since $\mathfrak{X}_T$ is isomorphic to $\mathfrak{X}_{T[g]}$ by $x \longmapsto xg$ and $B_{(x)} = B_{(xg)}$ for $g \in GL_n$; and for simplicity of notation, we write $f_i(x)$ instead of $f_{T,i}(x)$. Until the end of this subsection, we consider algebraic sets over $\overline{k}$, extend the involution $*$ on k' to $\overline{k}$, indicate it by $^{-}$, and write $\overline{x} = (\overline{x_{ij}}) \in M_{\ell m}(\overline{k})$ for $x = (x_{ij}) \in M_{\ell m}(\overline{k})$.

Then, our situation is the following:

$$\begin{aligned}&\mathfrak{X} = \mathfrak{X}_{1_n} = \{x \in M_{2n,n} \mid H_n[x] = 1_n\},\\ &(U(H_n)\times U(1_n))\times \mathfrak{X} \longrightarrow \mathfrak{X}, \quad ((g,h),x) \longmapsto (g,h)\circ x = gxh^{-1},\end{aligned}$$

and B is the Borel subgroup of $U(H_n)$ (as in (1.3)). We introduce a $(GL_{2n}\times GL_n)$-set $\widetilde{\mathfrak{X}}$ as follows:

$$\widetilde{\mathfrak{X}} = \left\{ (x,y) \in M_{2n,n}\oplus M_{2n,n} \;\middle|\; {}^t y H_n x = 1_n\right\} \tag{3.7}$$

$$(g,h)\star(x,y) = (gxh^{-1}, \dot{g}y{}^t h), \qquad ((g,h)\in GL_{2n}\times G_n,\; \dot{g} = H_n{}^t g^{-1} H_n).$$

We write an element of $\widetilde{\mathfrak{X}}$ as $(x,y) = (\begin{pmatrix} x_1 \\ x_2 \end{pmatrix}, \begin{pmatrix} y_1 \\ y_2 \end{pmatrix})$ with $x_i, y_i \in M_n$, then the above condition is the same with

$$ {}^t x_1 y_2 + {}^t x_2 y_1 = 1_n. $$

We fix a Borel subgroup P of GL_{2n} by

$$ P = \left\{ \begin{pmatrix} p & r \\ 0 & q \end{pmatrix} \in GL_{2n} \;\middle|\; {}^t p,\ q \in B_n,\ \ r \in M_n \right\}, $$

where B_n is the Borel subgroup of GL_n consisting of the lower triangular matrices. The involution $g \longmapsto \dot{g} = H_n {}^t g^{-1} H_n$ on GL_{2n} induces an involution on P :

$$ \begin{pmatrix} p & r \\ 0 & q \end{pmatrix} \longmapsto \begin{pmatrix} {}^t q^{-1} & -{}^t q^{-1}\, {}^t r\, {}^t p^{-1} \\ 0 & {}^t p^{-1} \end{pmatrix}. \tag{3.8} $$

Since $\dot{g} = \overline{g}$ for $g \in U(H_n)$ and ${}^t h = \overline{h}^{-1}$ for $h \in U(1_n)$, the embedding $\iota : \mathfrak{X} \longmapsto \widetilde{\mathfrak{X}},\ x \longmapsto (x, \overline{x})$ is compatible with the actions, i.e., we have the commutative diagram

$$ \begin{array}{ccc} (U(H_n) \times U(1_n)) \times \mathfrak{X} & \xrightarrow{\circ} & \mathfrak{X} \\ \Big\downarrow \text{incl.} \quad \Big\downarrow \iota & & \Big\downarrow \iota \\ (GL_{2n} \times GL_n) \times \widetilde{\mathfrak{X}} & \xrightarrow{\star} & \widetilde{\mathfrak{X}}. \end{array} $$

For $(x,y) \in \widetilde{\mathfrak{X}}$ and $p \in P$, set

$$ \widetilde{f}_i(x,y) = d_i(x_2 {}^t y_2), \quad \widetilde{\psi}_i(p) = \prod_{1 \le j \le i} p_j^{-1} p_{n+j}, \quad (1 \le i \le n), \tag{3.9} $$

where p_j is the j-th diagonal component of p. Then $\widetilde{f}_i(x,y)$'s are relative P-invariants on $\widetilde{\mathfrak{X}}$ associated with characters $\widetilde{\psi}_i$, $\widetilde{f}_i(x, \overline{x}) = f_i(x)$ for $x \in \mathfrak{X}$, and $\widetilde{\psi}_i|_B = \psi_i$. We set

$$ \mathcal{S} = \left\{ (x,y) \in \widetilde{\mathfrak{X}} \cap (P \times GL_n) \star \iota(\mathfrak{X}) \;\middle|\; \prod_{i=1}^{n} \widetilde{f}_i(x,y) = 0 \right\}. $$

For $\alpha = (x,y) \in \widetilde{\mathfrak{X}}$, we denote by H_α the stabilizer of $P \times GL_n$ at α, and by P_α the identity component of the image of H_α by the projection to P. In order to prove the condition (C), it suffices to show the following:

$(\widetilde{C})$: For each $\alpha \in \mathcal{S}$, there exists some $\psi \in \langle \widetilde{\psi}_i \mid 1 \le i \le n \rangle$ whose restriction to P_α is not trivial.

We have only to consider $(\widetilde{C})$ for representatives under the action of $P \times GL_n$. In the following we consider the case $n \geq 2$, since $\mathfrak{X}_T = \mathfrak{X}_T^{op}$ for $n = 1$ and there is nothing to prove. We denote by $\delta_i(a) \in GL_n$ the diagonal matrix whose j-th entry is 1 except the i-th which is $a \in GL_1$.

Lemma 3.4. *The condition $(\widetilde{C})$ is satisfied for $(x,y) \in \mathcal{S}$ for which $\det x_2 \neq 0$ or $\det y_2 \neq 0$.*

Proof. Let $\alpha = (x,y) \in \mathcal{S}$ and $\det x_2 \neq 0$. Then by the action of $P \times GL_n$, we may assume that $x_2 = 1_n$ and $x_1 = 0$, then $y_1 = 1_n$ since ${}^t x H_n y = 1_n$. Since $\alpha \in (P \times GL_n) \star \iota(\mathfrak{X})$, y_2 can be written as

$$y_2 = ph, \qquad (p \in B_n,\ h \in GL_n,\ {}^t\overline{h} = h),$$

and $0 = \prod_i \widetilde{f}_i(\alpha) = \prod_i d_i(y_2) = \prod_i d_i(h)$. For $q \in B_n$, we have

$$\begin{aligned}(\begin{pmatrix} {}^t q^{-1}\,{}^t p & 0 \\ 0 & q \end{pmatrix}, q) \star \alpha &= (\begin{pmatrix} {}^t q^{-1}\,{}^t p & 0 \\ 0 & q \end{pmatrix} \begin{pmatrix} 0 \\ 1_n \end{pmatrix} q^{-1}, \begin{pmatrix} {}^t q^{-1} & 0 \\ 0 & qp^{-1} \end{pmatrix} \begin{pmatrix} 1_n \\ y_2 \end{pmatrix} {}^t q) \\ &= (\begin{pmatrix} 0 \\ 1_n \end{pmatrix}, \begin{pmatrix} 1_n \\ qh\,{}^t q \end{pmatrix})\ (= \beta,\ \text{say}).\end{aligned}$$

Hence, by taking a suitable $q \in B_n$, we may make $qh\,{}^t q = 1_r \perp h_1,\ 0 \leq r < n$ such that h_1 is a hermitian matrix satisfying

- · the first row and column are zero, or
- · for some i, $(1 < i \leq n-r)$, each entry in the first row and column or in the i-th row and column is 0 except at $(1,i)$ or $(i,1)$ which are 1.

Then H_β contains the following elements, according to the above type of h_1,

$$(\left(\begin{array}{c|c} \delta_{r+1}(a) & \\ \hline & 1_n \end{array}\right), 1_n) \quad \text{or} \quad (\left(\begin{array}{c|c} \delta_{r+1}(a) & \\ \hline & \delta_{r+i}(a) \end{array}\right), \delta_{r+i}(a)) \quad (a \in GL_1),$$

and we see $\widetilde{\psi}_{r+1} \not\equiv 1$ on P_β.

The case $\alpha = (x,y) \in \mathcal{S}$ with $\det y_2 \neq 0$ is reduced to the case $\det x_2 \neq 0$, since $\beta = (y,x) \in \mathcal{S}$, $H_\beta = \{(\dot{p}, {}^t h^{-1}) \mid (p,h) \in H_\alpha\}$ and $\widetilde{\psi}_i(\dot{p}) = \widetilde{\psi}_i(p)^{-1}$. ∎

Now we have to consider for $(x,y) \in \mathcal{S}$ such that $\det x_2 = \det y_2 = 0$.

We set

$$\begin{aligned}
&\mathcal{S}_0 = \{\,(x,y)\in\mathcal{S} \mid \det x_2 = \det y_2 = 0\},\\
&J(i_1,i_2,\ldots,i_t)\in M_{nt}\,;\\
&\qquad\qquad 1\le i_1<i_2<\cdots<i_t\le n,\\
&\qquad\qquad \text{the entry at } (i_j,j) \text{ is 1, and all the other entries are 0.}
\end{aligned}$$

Lemma 3.5. *By the action of $P\times GL_n$, every element in $\mathcal{S}_0$ becomes the following type,*

$$\left(\left(\begin{array}{c|c} 0 & J_1\\ \hline J_2 & 0\end{array}\right), \left(\begin{array}{c|c} z_1 & 0\\ \hline z_2 & z_3\end{array}\right)\right), \qquad (J_1, z_3\in M_{n\ell},\ J_2, z_1, z_2\in M_{nk}),$$

where

$$J_1 = J(r_1,r_2,\ldots,r_\ell),\quad J_2 = J(e_1,e_2,\ldots,e_k),\quad 1\le \ell,k<n,\ \ \ell+k=n;$$

for z_1, the e_j-th row is the same as in J_2 and (i,j)-entry is 0 if $i<e_j$, $(1\le j\le k)$;
the r_j-th row of $(z_2\mid z_3)$ is the same as in $(0\mid J_1)$, and (i,j)-entry of z_3 is 0 if $i>r_j$, $(1\le j\le \ell)$.

Proof. Take an $\alpha=(x,y)\in\mathcal{S}_0$ and let $\operatorname{rank}(x_2)=k$. Then $1\le k<n$, and by the action of $P\times GL_n$, we make x into

$$\left(\begin{array}{c|c} 0 & x'\\ \hline J_2 & 0\end{array}\right).$$

Then, the rank of x' must be $\ell=n-k$, since $x\in\widetilde{\mathfrak{X}}$, and we may make x' into J_1, i.e. x into the required type. Further, the e_j-th rows in y_1 must be the same as in $(J_2\mid 0)$ and the r_j-th rows in y_2 must be the same as in $(0\mid J_1)$.

Multiplying y by a suitable element $p\in P$ from the left, we may make the latter ℓ columns of y_1 into 0 and $(i,k+j)$-entry of y_2 for $1\le j\le\ell,\ i>r_j$ into 0, while $\dot{p}x=x$. Since (e_j,r)-entry of y_1 is 0 unless $r=j$, we may make (i,j)-entry of y_1 for $1\le j\le k,\ i<r_j$ into 0 as keeping x. Thus we obtain a matrix of the form as in the statement. ∎

Lemma 3.6. *The condition $(\widetilde{C})$ is satisfied for elements in $\mathcal{S}_0$.*

Proof. We may assume $\alpha=(x,y)\in\mathcal{S}_0$ has the form as in Lemma 3.5.

Then, for any $a \in GL_1$,

$$\left(\left(\begin{array}{c|c} 1_n & 0 \\ \hline 0 & \delta_1(a) \end{array}\right), 1_n\right) \in H_\alpha \qquad \text{if } e_1 > 1,$$
$$\left(\left(\begin{array}{c|c} \delta_1(a) & 0 \\ \hline 0 & 1_n \end{array}\right), \delta_{k+1}(a)\right) \in H_\alpha \quad \text{if } r_1 = 1,$$
$$\left(\left(\begin{array}{c|c} a1_n & 0 \\ \hline 0 & 1_n \end{array}\right), a1_n\right) \in H_\alpha \qquad \text{if } z_2 = 0.$$

When $e_1 = 1$, $r_1 > 1$ and $z_2 \neq 0$, we modify α into $\beta = (x, y')$ by the $P \times GL_n$-action as below:

$$y' = \left(\begin{array}{c|c} z_1' & 0 \\ \hline z_2' & z_3' \end{array}\right), \; z_2' \neq 0,$$

the r_j-th row of z_3' is the j-th unit vector (the same as in J_1) for $1 \leq j \leq \ell$.
if the i-th row of z_2' is not 0, then the i-th row of z_3' is 0, $1 \leq i \leq n$. (3.10)

Then, for any $a \in GL_1$,

$$\left(\left(\begin{array}{c|c} D_n(a_i) & 0 \\ \hline 0 & 1_n \end{array}\right), \left(\begin{array}{c|c} 1_k & 0 \\ \hline 0 & a1_\ell \end{array}\right)\right) \in H_\beta,$$

where $D_n(a_i) = Diag(a_1, \ldots, a_n)$ with

$$a_i = \begin{cases} a \text{ if the } i\text{-th row of } z_2' \text{ is } 0 \\ 1 \text{ if the } i\text{-th row of } z_2' \text{ is not } 0, \end{cases}.$$

Hence, for any $\alpha \in \mathcal{S}_0$, $\widetilde{\psi}_n \not\equiv 1$ on P_α.

Now we explain how to obtain β as in (3.10) from α with $e_1 = 1$, $r_1 > 1$ and $z_2 \neq 0$. Let k' be the rank of x_2. Then for suitable $p_0 \in B_n$, we make $z_2' = p_0 z_2$ such that

> there exist integers $1 \leq s_1 < s_2 < \cdots < s_{k'} \leq n$ such that the i-th rows are 0 except for $i \in \{s_1, \ldots, s_{k'}\}$, and for each i, $1 \leq i \leq k'$, there exists distinct j_i for which
>
> 1 at (s_i, j_i)-entry,
>
> 0 at (s_i, j)-entry for $j < j_i$ and the (i', j_i)-entry for $i' > s_i$. (3.11)

Since every r_i-th row of z_2 is 0, we may assume each r_i-th row of p_0 is the r_i-th unit vector, hence ${}^t p_0^{-1} J_1 = J_1$ and the r_j-th row of $p_0 z_3$ remains to be the j-th unit vector. By a suitable matrix

$$h = \begin{pmatrix} 1_k & C \\ 0 & 1_\ell \end{pmatrix} \in GL_n,$$

we make each s_i-th row of $z_3' = z_2'C + p_0z_3$ into 0 for $1 \le i \le k'$ and remain the other rows as the same as in p_0z_3. Take the matrix $D \in M_n$ by putting the j-th row of z_1C into the r_j-th row for $1 \le j \le \ell$, and 0 at all other entries, then $z_1C = Dz_3'$. Setting

$$p_1 = \begin{pmatrix} 1_n & 0 \\ 0 & p_0 \end{pmatrix} \quad \text{and} \quad p = \begin{pmatrix} 1_n & -D \\ 0 & 1_n \end{pmatrix},$$

we obtain

$$pp_1yh = \begin{pmatrix} 1_n & -D \\ 0 & 1_n \end{pmatrix}\begin{pmatrix} z_1' & z_1C \\ z_2' & z_3' \end{pmatrix} = \begin{pmatrix} z_1' & 0 \\ z_2' & z_3' \end{pmatrix}, \qquad (z_1' = z_1 - Dz_2').$$

On the other hand, we have

$$\dot{p}\dot{p}_1x{}^th^{-1} = \begin{pmatrix} 1_n & {}^tD \\ 0 & 1_n \end{pmatrix}\begin{pmatrix} {}^tp_0^{-1} & 0 \\ 0 & 1_n \end{pmatrix}\begin{pmatrix} 0 & J_1 \\ J_2 & 0 \end{pmatrix}\begin{pmatrix} 1_k & 0 \\ -{}^tC & 1_\ell \end{pmatrix} = \begin{pmatrix} {}^tDJ_2 - J_1{}^tC & J_1 \\ J_2 & 0 \end{pmatrix},$$

and ${}^tDJ_2$ and $J_1{}^tC$ may have nonzero rows only at the r_i-th, $1 \le i \le \ell$, and

$$\begin{aligned}(r_i, j)\text{-entry of } {}^tDJ_2 &= (e_j, i)\text{-entry of } z_1C = (j, i)\text{-entry of } C \\ &= (r_i, j)\text{-entry of } J_1\,{}^tC.\end{aligned}$$

Thus we have the required element

$$\beta = (\dot{p}\dot{p}_1, {}^th) \star \alpha = (\begin{pmatrix} 0 & J_1 \\ J_2 & 0 \end{pmatrix}, \begin{pmatrix} z_1' & 0 \\ z_2' & z_3' \end{pmatrix}).$$

∎

Thus we have shown the condition $(\widetilde{C})$ is satisfied for every $(x, y) \in \mathcal{S}$, which shows that our $(\mathbb{X}, \mathbb{B})$ satisfies the condition $(A3)$ and Theorem 3.1 is established.

4. Spherical Fourier transform on $\mathcal{S}(K\backslash X_T)$

We consider the subspace $\mathcal{S}(K\backslash X_T)$ of $\mathcal{C}^\infty(K\backslash \mathfrak{X}_T/U(T))$ consisting of compactly supported functions modulo $U(T)$, which is an $\mathcal{H}(G, K)$-submodule (cf. (1.9)). We define the spherical Fourier transform F_T on $\mathcal{S}(K\backslash X_T)$, by setting

$$\begin{aligned} F_T : \mathcal{S}(K\backslash X_T) &\longrightarrow \mathbb{C}(q^{z_1}, \ldots, q^{z_n}), \\ \xi &\longmapsto F_T(\xi)(z) = \int_{X_T} \xi(x)\Psi_T(x; z)dx, \end{aligned} \tag{4.1}$$

where $\Psi_T(x;z) = G(z)\cdot\omega_T(x;z)$ and dx is the G-invariant measure on X_T. Since $\mathcal{S}(K\backslash X_T)$ is spanned by the characteristic functions of double cosets $KxU(T)$ in $K\backslash\mathfrak{X}_T/U(T) = K\backslash X_T$, the image of F_T is spanned by the set $\{\,\Psi_T(x;z) \mid x \in \mathfrak{X}_T\}$ over $\mathbb{C}$, and contained in

$$\mathcal{R} = \mathbb{C}[q^{\pm z_1}, \ldots, q^{\pm z_n}]^W$$

by Theorem 2.9. We decompose $\mathcal{R}$ in the following

$$\mathcal{R} = \bigoplus_{\mathbf{e}\in\{0,1\}^n} s_1^{e_1}\cdots s_n^{e_n}\,\mathcal{R}_0,$$

where

$$\mathcal{R}_0 = \mathbb{C}[q^{\pm 2z_1}, \ldots, q^{\pm 2z_n}]^W = \mathbb{C}[q^{2z_1}+q^{-2z_1}, \ldots, q^{2z_n}+q^{-2z_n}]^{S_n},$$

and $s_i = s_i(z)$ is the i-th fundamental symmetric polynomial of $\{\,q^{z_j}+q^{-z_j} \mid 1 \le j \le n\}$; $\mathcal{R}$ is a free $\mathcal{R}_0$-module of rank 2^n. We set

$$\mathcal{R}_{even} = \bigoplus_{\mathbf{e}:even} s_1^{e_1}\cdots s_n^{e_n}\,\mathcal{R}_0, \quad \mathcal{R}_{odd} = \bigoplus_{\mathbf{e}:odd} s_1^{e_1}\cdots s_n^{e_n}\,\mathcal{R}_0,$$

where $\mathbf{e} \in \{0,1\}^n$ is even (resp. odd) if $\sum_{i=1}^n ie_i$ is even (resp. odd). For each $T \in \mathcal{H}_n^{nd}$, we define

$$\mathcal{R}_{\langle T\rangle}$$

to be $\mathcal{R}_{even}$ or $\mathcal{R}_{odd}$ according to the parity of $v_\pi(\det(T))$.

Theorem 4.1. *For any $T \in \mathcal{H}_n^{nd}$, one has a surjective $\mathcal{H}(G,K)$-module homomorphism*

$$F_T : \mathcal{S}(K\backslash X_T) \longrightarrow \mathcal{R}_{\langle T\rangle},$$

and a commutative diagram

$$\begin{array}{ccccc} \mathcal{H}(G,K) & \times & \mathcal{S}(K\backslash X_T) & \stackrel{*}{\longrightarrow} & \mathcal{S}(K\backslash X_T) \\ \wr\downarrow & & \downarrow F_T & \circlearrowleft & \downarrow F_T \\ \mathcal{R}_0 & \times & \mathcal{R}_{\langle T\rangle} & \longrightarrow & \mathcal{R}_{\langle T\rangle}, \end{array} \tag{4.2}$$

where the upper horizontal arrow is given by the action of $\mathcal{H}(G,K)$ on $\mathcal{S}(K\backslash X_T)$, the left end vertical isomorphism is given by Satake isomorphism (1.12)

$$\mathcal{H}(G,K) \stackrel{\sim}{\longrightarrow} \mathcal{R}_0,\ \phi \longmapsto \lambda_z(\check{\phi}), \qquad (\check{\phi}(g) = \phi(g^{-1})),$$

and the lower horizontal arrow is given by the ordinary multiplication in $\mathcal{R}$.

Proof. For $\phi \in \mathcal{H}(G,K)$ and $\xi \in \mathcal{S}(K\backslash X_T)$, we have

$$\begin{aligned} F_T(\phi * \xi)(z) &= \int_X \int_G \phi(g)\xi(g^{-1}x)dg\Psi_T(x;z)dx \\ &= \int_X \xi(y)\int_G \phi(g)\Psi_T(gy;z)dgdy \\ &= \int_X \xi(y)(\check{\phi} * \Psi_T(\ ,z))(y)dy = \lambda_z(\check{\phi})F_T(\xi)(z), \end{aligned}$$

which gives the commutative diagram.

We recall the definition (1.8) of $\omega_T(x;z)$ and expand it in a region of absolute convergence. Then

$$\omega_T(x;z) = \sum_{\mu\in\mathbb{Z}^n} a_\mu q^{\langle \mu,\, z\rangle},$$

where $a_\mu = 0$ unless $|\mu|\ (=\sum_{i=1}^n \mu_i) \equiv v_\pi(\det(T)) \pmod 2$, since

$$v_\pi(f_{T,n}(x)) = v_\pi(\det(x_2T^{-1}x_2^*) \equiv v_\pi(\det(T)) \pmod 2, \text{ for any } x\in\mathfrak{X}_T^{op}$$

$$\begin{aligned} \langle \mu,\, z\rangle &= \sum_{i=1}^n \mu_i z_i = -\sum_{i=1}^n \mu_i(s_i+\cdots+s_n) \quad \text{(in } s\text{-variable)} \\ &= -\mu_1 s_1 - (\mu_1+\mu_2)s_2 - \cdots - (\mu_1+\cdots+\mu_n)s_n. \end{aligned}$$

Since

$$G(z) = \prod_{i<j}\left((1+q^{z_i-z_j}+q^{z_i+z_i}+q^{2z_i})\sum_{\ell,r\geq 0} q^{(\ell+r)z_i+(\ell-r)z_j-(\ell+r)}\right),$$

can be expanded only in terms $q^{\langle \nu,\, z\rangle}$ with $|\nu|$ is even, we may expand $\Psi_T(x;z) = \omega_T(x;z)G(z)$ in terms $q^{\langle \lambda,\, z\rangle}$ with $|\lambda| \equiv v_\pi(\det(T)) \pmod 2$, hence

$$\mathrm{Im}(F_T) \subset \mathcal{R}_{\langle T\rangle}. \tag{4.3}$$

On the other hand, by Remark 3.2 and Theorem 3.3 we see

$$\mathrm{Im}(F_T) \supset \left\{ H_\lambda(z) \;\middle|\; \lambda\in\Lambda_n^+,\ |\lambda| \equiv v_\pi(\det T) \pmod 2\right\},$$

and the image of F_T coincides with $\mathcal{R}_{\langle T\rangle}$. ∎

Remark 4.2. We expect that the spherical Fourier transform F_T is injective, which is equivalent to the identity

$$\mathfrak{X}_T = \bigcup_{\substack{\lambda\in\Lambda_n^+ \\ |\lambda|\equiv v_\pi(\det(T)) \pmod 2}} Kx_\lambda h_\lambda U(T), \tag{4.4}$$

where disjointness in the right hand side is known by Theorem 3.3. If it is true, then $\mathcal{S}(K\backslash X_T)$ would be a free $\mathcal{H}(G,K)$-module of rank 2^{n-1} and the set $\{\Psi_T(x; z+\widetilde{u}) \mid u \in \mathcal{U}\}$ would form a basis of spherical functions on X_T corresponding to $z \in \mathbb{C}^n$ through λ_z (cf. Proposition 1.3). This is true when $n = 1$ by Proposition 2.1, and we have the following.

Proposition 4.3. *Assume $n = 1$. Then the spherical transform F_T is injective and $\mathcal{S}(K\backslash X_T)$ is a free $\mathcal{H}(G,K)$-module of rank 1, in fact the image coincides with*

$$\mathbb{C}[q^{2z}+q^{-2z}] \text{ if } v_\pi(T) \text{ is even}, \quad (q^z+q^{-z})\mathbb{C}[q^{2z}+q^{-2z}] \text{ if } v_\pi(T) \text{ is odd.}$$

Any spherical function on X_T corresponding to $z \in \mathbb{C}$ through λ_z is a constant multiple of $\omega_T(x; z)$.

5. An application to hermitian Siegel series

We recall the hermitian Siegel series, and give an integral representation and a new proof of the functional equation as an application of spherical functions.

Let ψ be an additive character of k of conductor $\mathcal{O}_k$. For $T \in \mathcal{H}_n(k')$ and $t \in \mathbb{C}$, the hermitian Siegel series $b_\pi(T; s)$ is defined by

$$b_\pi(T; t) = \int_{\mathcal{H}_n(k')} \nu_\pi(R)^{-t}\psi(\mathrm{tr}(TR))dR, \tag{5.1}$$

where tr( ) is the trace of matrix and $\nu_\pi(R)$ is defined as follows: if the elementary divisors of R with negative π-powers are $\pi^{-e_1}, \ldots, \pi^{-e_r}$, then $\nu_\pi(R) = q^{e_1+\cdots+e_r}$, and $\nu_\pi(R) = 1$ otherwise (cf. [18] §13). The right hand side of (5.1) is absolutely convergent if $\mathrm{Re}(t)$ is sufficiently large.

In the following we assume that T is nondegenerate, since the properties of $b_\pi(T; t)$ can be reduced to the nondegenerate case. We give an integral expression of $b_\pi(T; t)$ in a similar argument for Siegel series in [8] §2.

We recall the set $\mathfrak{X}_T$ for $T \in \mathcal{H}_n^{nd}(k')$

$$\mathfrak{X}_T = \mathfrak{X}_T(k') = \{x \in M_{2n,n}(k') \mid H_n[x] = T\}$$

and take the measure $|\Theta_T|$ on $\mathfrak{X}_T$ simultaneously as the fibre space of T by the polynomial map $M_{2n,n}(k') \longrightarrow \mathcal{H}_n(k'), x \longmapsto H_n[x]$ defined over k.

Then the following identity holds (cf. [19], [8]-§2):

$$\begin{aligned}&\int_{\mathfrak{X}_T(k')} \phi(x)\,|\Theta_T|\,(x)\\&= \lim_{e\to\infty}\int_{\mathcal{H}_n(\pi^{-e})}\psi(-\mathrm{tr}(Ty))\int_{M_{2n,n}(k')}\phi(x)\psi(\mathrm{tr}(H_n[x]y))dxdy,\end{aligned}$$

where $\phi\in\mathcal{S}(M_{2n,n}(k'))$, a locally constant compactly supported function on $M_{2n,n}(k')$, and $\mathcal{H}_n(\pi^{-e})=\mathcal{H}_n(k')\cap M_n(\pi^{-e}\mathcal{O}_{k'})$.

The following lemma can be proved in the similar line to the case of symmetric matrices (cf. [8] §2).

Lemma 5.1. *If* $\mathrm{Re}(t)$ *is sufficiently large, one has*

$$\begin{aligned}&\int_{\mathfrak{X}_T(\mathcal{O}_{k'})}|N(\det x_2)|^{t-n}\,|\Theta_T|\,(x)\\&=\lim_{e\to\infty}\int_{\mathcal{H}_n(\pi^{-e}\mathcal{O}_{k'})}\psi(-\mathrm{tr}(Ty))\\&\qquad\times\int_{M_{2n,n}(\mathcal{O}_{k'})}|N(\det x_2)|^{t-n}\,\psi(\mathrm{tr}(H_n[x]y))dxdy.\qquad(5.2)\end{aligned}$$

Denote by $\zeta(k';t)$ the zeta function of the matrix algebra $M_n(k')$:

$$\zeta(k';t)=\int_{M_n(\mathcal{O}_{k'})}|\det x|_{k'}^{t-n}\,dx=\int_{M_n(\mathcal{O}_{k'})}|N(\det x)|^{t-n}\,dx,$$

whose explicit formula is well-known:

$$\zeta_n(k';t)=\prod_{i=1}^{n}\frac{1-q^{-2i}}{1-q^{-2(t-i+1)}}.$$

Then we have the following integral expression of hermitian Siegel series.

Theorem 5.2. *If* $\mathrm{Re}(t)>2n$*, we have*

$$b_\pi(T;t)=\zeta_n(k';\frac{t}{2})^{-1}\times\int_{\mathfrak{X}_T(\mathcal{O}_{k'})}|N(\det x_2)|^{\frac{t}{2}-n}\,|\Theta_T|\,(x).$$

Proof. We define the Fourier transform of $\phi\in\mathcal{S}(M_n(k'))$ by

$$\widehat{\phi}(z)=\int_{M_n(k')}\phi(w)\psi(T_{k'/k}(\mathrm{tr}(wz^*)))dw,$$

where $T_{k'/k}$ is the trace of the extension k'/k. Since we have

$$\begin{aligned}\mathrm{tr}(H_n[x]y) &= \mathrm{tr}(x_1^* x_2 y) + \mathrm{tr}(x_2^* x_1 y) = \mathrm{tr}(x_1^*(x_2 y)) + \mathrm{tr}((x_2 y)^* x_1)\\ &= T_{k'/k}(\mathrm{tr}(x_1(x_2 y)^*)),\end{aligned}$$

the second integral in the right hand side of (5.2) becomes

$$\begin{aligned}&\int_{M_n(\mathcal{O}_{k'})} |N(\det x_2)|^{t-n}\, \widehat{ch_{M_n(\mathcal{O}_{k'})}}(x_2 y) dx_2\\ &= \int_{M_n(\mathcal{O}_{k'})} |N(\det x_2)|^{t-n}\, ch_{M_n(\mathcal{O}_{k'})}(x_2 y) dx_2\\ &= \int_{M_n(\mathcal{O}_{k'})y^{-1}\cap M_n(\mathcal{O}_{k'})} |N(\det x_2)|^{t-n}\, dx_2\\ &= \int_{M_n(\mathcal{O}_{k'})D_y} |\det x_2|_{k'}^{t-n}\, dx_2,\end{aligned}$$

where $D_y = 1_n$ if $y \in M_n(\mathcal{O}_{k'})$, and $D_y = Diag(\pi^{e_1}, \ldots, \pi^{e_r}, 1, \ldots, 1)$ if the elementary divisors of y with negative π-powers are $\pi^{-e_1}, \ldots, \pi^{-e_r}$. Hence the second integral in the right hand side of (5.2) is equal to

$$|\det D_y|_{k'}^{t} \int_{M_n(\mathcal{O}_{k'})} |\det x_2|_{k'}^{t-n}\, dx_2 = \nu_\pi(y)^{-2t} \times \zeta_n(k'; t).$$

Now by Lemma 5.1, we obtain

$$\begin{aligned}&\int_{\mathfrak{X}_T(\mathcal{O}_{k'})} |N(\det x_2)|^{t-n}\, |\Theta_T|(x)\\ &= \zeta_n(k'; t) \times \lim_{e\to\infty} \int_{\mathcal{H}_n(\pi^{-e}\mathcal{O}_{k'})} \nu_\pi(y)^{-2t} \cdot \psi(-\mathrm{tr}(Ty)) dy\\ &= \zeta_n(k'; t) \times b_\pi(T; 2t),\end{aligned}$$

which gives the required identity. ∎

Setting, in s-variable,

$$s_t = \left(1 + \tfrac{\pi\sqrt{-1}}{\log q}, \ldots, 1 + \tfrac{\pi\sqrt{-1}}{\log q}\right) + \left(0, \ldots, 0, \frac{t}{2} - n - \frac{1}{2}\right) \in \mathbb{C}^n, \quad (5.3)$$

we see

$$\int_K |N(\det(kx)_2)|^{\frac{t}{2}-n}\, dk = |\det T|^{\frac{t}{2}-n}\, \omega_T(x; s_t). \quad (5.4)$$

Hence we may express $b_\pi(T; t)$ by using the spherical function $\omega_T(x; s)$.

Proposition 5.3. *Denote the K-orbit decomposition of $\mathfrak{X}_T(\mathcal{O}_{k'})$ as*

$$\mathfrak{X}_T(\mathcal{O}_{k'}) = \sqcup_{i=1}^{r} K x_i.$$

Then one has

$$b_\pi(T;t) = |\det T|^{\frac{t}{2}-n} \prod_{i=0}^{n-1}(1-q^{-t+2i}) \times \sum_{i=1}^{r} c_i \cdot \omega_T(x_i;s_t),$$

where $c_i = \left(\prod_{i=1}^{n}(1-q^{-2i})\right)^{-1} \cdot vol(Kx_i)$.

Proof. Since $\mathfrak{X}_T(\mathcal{O}_{k'})$ is compact, it is a finite union of K-orbits, which we write as above. By Theorem 5.2, we have

$$\begin{aligned} & b_\pi(T;t) \times \zeta_n(k';\frac{t}{2}) \\ &= \sum_{i=1}^{r} \int_{Kx_i} |N(\det y_2)|^{\frac{t}{2}-n} \, |\Theta_T|\,(y) \\ &= \sum_{i=1}^{r} \int_{Kx_i} \int_K |N(\det (ky)_2)|^{\frac{t}{2}-n} \, dk \, |\Theta_T|\,(y) \\ &= |\det T|^{\frac{t}{2}-n} \sum_{i=1}^{r} c_i' \cdot \omega_T(x_i;s_t), \end{aligned}$$

where $c_i' = vol(Kx_i)$. Substituting the explicit value of $\zeta_n(k';\frac{t}{2})$, we conclude the proof. ∎

By using Theorem 2.9, we have the following.

Corollary 5.4. *The function* $\{\prod_{i=0}^{n-1}(1-(-1)^i q^{-t+i})\}^{-1} \times b_\pi(T;t)$ *is holomorphic for any* t*, hence it is a polynomial in* q^t *and* q^{-t}.

Proof. We denote by z^* the corresponding value with s_t in z-variable. By Proposition 5.3 and Theorem 2.9, we see that

$$b_\pi(T;t) = \prod_{i=0}^{n-1}(1-q^{-t+2i}) \cdot \frac{1}{G(z^*)} \times (\text{a holomorphic function}).$$

By (5.3), (2.13), and the definition of $G(z)$, we obtain

$$\begin{aligned} G(z^*) &\equiv \prod_{i<j} \frac{1+(-1)^{j+i}q^{-t+i+j-1}}{1-(-1)^{i+j}q^{-t+i+j-2}} \pmod{\mathbb{C}^\times} \\ &= \prod_{i=1}^{n-1}\prod_{j=i+1}^{n} \frac{1-(-1)^{i+j-1}q^{-t+i+j-1}}{1-(-1)^{i+j-2}q^{-t+i+j-2}} \\ &= \prod_{i=1}^{n-1} \frac{1-(-1)^{n+i-1}q^{-t+n+i-1}}{1+q^{-t+2i-1}}, \end{aligned}$$

and

$$\prod_{i=0}^{n-1}(1-q^{-t+2i})\cdot\frac{1}{G(z^*)}\equiv\prod_{i=0}^{n-1}(1-(-1)^i q^{-t+i})\quad(\mathrm{mod}\ \mathbb{C}^\times),$$

which completes the proof. ∎

Remark 5.5. According to G. Shimura [18] Theorem 13.6, one may express $b_\pi(T;t)$ as follows (including ramified hermitian and split cases):

$$b_\pi(T;t)=f_T(q^{-t})\cdot g_T(q^{-t}), \tag{5.5}$$

where $f_T(X)$ is an explicitly given rational function of X, depending only on the type and size of T, and $g_T(X)$ is a (mysterious) polynomial with coefficients in $\mathbb{Z}$. For the unramified hermitian case, $f_T(X)$ is given for $T\in\mathcal{H}_n^{nd}$ by

$$f_T(X)=\prod_{i=0}^{n-1}(1-(-q)^i X),\qquad f_T(q^{-t})=\prod_{i=0}^{n-1}(1-(-1)^i q^{-t+i}).$$

In Corollary 5.4, we obtain the same factor $f_T(q^{-t})$ by using the spherical functions $\omega_T(x;z)$, and $f_T(q^{-t})^{-1}b_\pi(T;t)$ must be a polynomial in q^{-t} with coefficients in $\mathbb{Z}$, which we do not see from $\omega_T(x;z)$.

Now we give the functional equation of the hermitian Siegel series by using the results of functional equations of the spherical functions $\omega_T(x;s)$.

Theorem 5.6. *For any $T\in\mathcal{H}_n^{nd}$, one has*

$$\begin{aligned}&b_\pi(T;t)\\&=\chi_\pi(\det T)^{n-1}\,|\det T|^{t-n}\times\prod_{i=0}^{n-1}\frac{1-(-1)^i q^{-t+i}}{1-(-1)^i q^{-(2n-t)+i}}\times b_\pi(T;2n-t),\end{aligned}$$

where $\chi_\pi(a)=(-1)^{v_\pi(a)}$ for $a\in k^\times$.

Proof. Let us recall $\rho\in W$ given in Corollary 2.7. The value $s_t\in\mathbb{C}^n$ given by (5.3) corresponds to $z^*\in\mathbb{C}^n$ in z-variable where

$$z_i^*=-\frac{t}{2}+i-\frac{1}{2}-(n-i+1)\frac{\pi\sqrt{-1}}{\log q},\quad 1\le i\le n,$$

and $\rho(z^*)$ corresponds to

$$(1+\tfrac{\pi\sqrt{-1}}{\log q},\ldots,1\tfrac{\pi\sqrt{-1}}{\log q})+(0,\ldots,0,-\frac{t}{2}-\frac{1}{2}+(n-1)\tfrac{\pi\sqrt{-1}}{\log q})$$

in s-variable. By Corollary 2.7, we have

$$\omega_T(x;s_t) = \chi_\pi(\det T)^{n-1}\cdot \Gamma_\rho(z^*) \times \omega_T(x;s_{2n-t}).$$

Hence we obtain by Proposition 5.3,

$$b_\pi(T;t) = \chi_\pi(\det T)^{n-1}\,|\det T|^{t-n}\cdot \gamma_n(t)\times b_\pi(T;2n-t) \qquad (5.6)$$

where

$$\gamma_n(t) = \Gamma_\rho(z^*)\times \prod_{i=0}^{n-1}\frac{1-q^{-t+2i}}{1-q^{t-2(n-i)}} = \Gamma_\rho(z^*)\times(-1)^n q^{-nt+n(n+1)}\frac{1-q^{-t}}{1-q^{-t+2n}}.$$

Since we have

$$\Gamma_\rho(z^*) = \prod_{i<j}\frac{1-(-1)^{i+j}q^{-t+i+j-2}}{(-1)^{i+j}q^{-t+i+j-1}-q^{-1}} = (-q)^{\frac{n(n-1)}{2}}\prod_{i=1}^{n-1}\frac{1-(-1)^i q^{-t+i}}{1-(-1)^{n+i}q^{-t+n+i}},$$

we get

$$\begin{aligned}\gamma_n(t) &= (-1)^{\frac{n(n+1)}{2}}q^{-nt+\frac{n(3n+1)}{2}}\prod_{i=1}^{n}\frac{1-(-1)^{i-1}q^{-t+i-1}}{1-(-1)^{i+n}q^{-t+n+i}}\\ &= (-1)^{\frac{n(n+1)}{2}}q^{-nt+\frac{n(3n+1)}{2}}(-1)^{\frac{n(3n-1)}{2}}q^{ns-\frac{n(3n+1)}{2}}\\ &\qquad\times\prod_{i=1}^{n}\frac{1-(-1)^{i-1}q^{-t+i-1}}{1-(-1)^{i+n}q^{-(2n-t)+n-i}}\\ &= \prod_{i=0}^{n-1}\frac{1-(-1)^i q^{-t+i}}{1-(-1)^i q^{-(2n-t)+i}}, \end{aligned} \qquad (5.7)$$

hence we obtain the required functional equation of $b_\pi(T;t)$ by (5.6). ∎

Remark 5.7. Let us recall the decomposition (5.5) in Remark 5.5. Then by (5.7), we see

$$\gamma_n(t) = f_T(q^{-t})/f_T(q^{t-2n}),$$

and $\chi_\pi(\det T)^{n-1}\,|\det T|^{t-n}$ gives the Gamma factor for the functional equation of $g_T(q^{-t})$.

The above functional equation is related to an element of the Weyl group of $U(H_n)$, which is not the case for (symmetric) Siegel series when n is odd. F. Sato and the author have studied in a similar line for Siegel series, we needed some harmonic analysis on $O(H_n)$ to establish the functional equations, and employed some previous results on particular T's to determine the explicit Gamma factors. In the present case, we can obtain the explicit functional equations of hermitian Siegel series by a specialization of those of spherical functions $\omega_T(x;z)$.

Remark 5.8. The existence of the functional equation of $b_\pi(T;t)$ was known in an abstract form as functional equations of Whittaker functions of p-adic groups by M. L. Karel [10]. Recently T. Ikeda [9] has given explicit functional equations of $F_p(T;X) = g_T(X)$ on the basis of the results of S. S. Kudla and W. J. Sweet [12] for all quadratic extensions over $\mathbb{Q}_p$ containing split cases. There is a mistake in the range of i of the definition of $t_p(K/\mathbb{Q};X) = f_T(X)$ in [9] p.1112, and it is better to refer the original $f_T(X)$ in [18] Theorem 13.6; if $K/\mathbb{Q}$ is unramified at p, $t_p(K/\mathbb{Q};X)$ is the product of $1-(-p)^iX$ from $i=0$ to $n-1$ as in Remark 5.5.

References

1. A. Borel: *Linear Algebraic Groups, Second enlarged edition*, Graduate Texts in Mathematics **126**, Springer, 1991.
2. W. Casselman: The unramified principal series of $\mathfrak{p}$-adic groups I. The spherical functions, *Compositio Math.* **40**(1980), 387 – 406.
3. W. Casselman and J. Shalika: The unramified principal series of $\mathfrak{p}$-adic groups II. The Whitakker functions, *Compositio Math.* **41**(1980), 207 – 231.
4. Y. Hironaka: Spherical functions of hermitian and symmetric forms III, *Tôhoku Math. J.* **40**(1988), 651–671.
5. Y. Hironaka: Spherical functions and local densities on hermitian forms, *J. Math. Soc. Japan* **51**(1999), 553 – 581.
6. Y. Hironaka: Functional equations of spherical functions on p-adic homogeneous spaces, *Abh. Math. Sem. Univ. Hamburg* **75**(2005), 285 – 311.
7. Y. Hironaka: Spherical functions on p-adic homogeneous spaces, in *"Algebraic and Analytic Aspects of Zeta Functions and L-functions – Lectures at the French-Japanese Winter School (Miura, 2008)– ", MSJ Memoirs* **21**(2010), 50 – 72.
8. Y. Hironaka and F. Sato : The Siegel series and spherical functions on $O(2n)/(O(n)\times O(n))$, "Automorphic forms and zeta functions – Proceedings of the conference in memory of Tsuneo Arakawa –", World Scientific, 2006, p. 150 – 169.
9. T. Ikeda: On the lifting of hermitian modular forms, *Comp. Math.***144** (2008), 1107-1154.
10. M. L. Karel: Functional equations of Whittaker functions on p-adic groups, *Amer. J. Math.* **101**(1979), 1303 –1325.
11. S. Kato, A. Murase and T. Sugano: Whittaker-Shintani functions for orthogonal groups, *Tohoku Math. J.* **55**(2003), 1 – 64.
12. S. S .Kudla and W. J. Sweet: Degenerate principal series representations for $U(n,n)$, *Israel J. Math.* **98** (1997), 253 –306.
13. I. G. Macdonald: *Spherical functions on a group of p-adic type*, Univ. Madras, 1971.
14. I. G. Macdonald: Orthogonal polynomials associated with root systems, *Séminaire Lotharingien de Combinatoire* **45**(2000), Article B45a.

15. O. T. O'Meara: *Introduction to quadratic forms*, Grund. math. Wiss. **117**, Springer-Verlag, 1973.
16. O.Offen: Relative spherical functions on p-adic symmetric spaces, *Pacific J. Math.* **215**(2004), 97 – 149.
17. W. Scharlau: *Quadratic and hermitian forms*, Grund. math. Wiss. **270**, Springer-Verlag, 1985.
18. G. Shimura: *Euler products and Eisenstein series*, CBMS **93** (AMS), 1997.
19. T. Yamazaki: Integrals defining singular series, *Memoirs Fac. Sci. Kyushu Univ.* **37**(1983), 113 – 128.

Fourier transforms of weighted orbital integrals on the real symplectic group of rank two

Werner Hoffmann

Fakultät für Mathematik, Universität Bielefeld,
Postfach 100131, 33501 Bielefeld, Germany
E-mail: hoffmann@math.uni-bielefeld.de

We compute explicitly the Fourier transforms of the invariant distributions associated by Arthur to weighted orbital integrals on the real symplectic group of rank two. These distributions are the main terms on the geometric side of the invariant trace formula.

Keywords: weighted orbital integrals; Fourier transform; invariant distributions

0. Introduction

Weighted orbital integrals are distributions on a reductive group over a local field or an adèle ring. They appear on the geometric side of the local or global trace formula. Arthur has defined closely related invariant distributions which appear on the geometric side of the invariant trace formula. The corresponding objects on the spectral side of that trace formula are characters. Thus, it is desirable to express the geometric side in terms of characters as well. This amounts to the calculation of the Fourier transforms of those invariant distributions.

By a combination of several methods, one can determine those Fourier transforms for any given real reductive group. This has been done for $\mathrm{SL}_2(\mathbb{R})$ by Arthur/Sally/Herb,[1] and for groups of real rank one and the group $\mathrm{GL}_3(\mathbb{R})$ by the author.[12,14] In the present paper, we will complete this task for the group $\mathrm{Sp}_2(\mathbb{R})$. Gradually, more and more features of the general answer are emerging.

For the introduction, G could be any reductive Lie group in Harish-Chandra's class.[9] The Lie algebra of a group will be denoted by the corresponding lower-case gothic letter. We denote the intersection of the kernels of all continuous homomorphisms $G \to \mathbb{R}_+$ by G^1. Then G is the direct product of G^1 and a so-called split component A_G. The projection onto A_G composed with the logarithm yields a homomorphism $H_G : G \to \mathfrak{a}_G$. We fix a maximal compact subgroup K of G. Then, for any parabolic subgroup P with unipotent radical N and Levi component M, we can define $H_P : G \to \mathfrak{a}_M$ by setting $H_P(nmk) = H_M(m)$ for $m \in M$, $n \in N$ and $k \in K$.

Let us recall the definition of weighted orbital integrals. For a Levi subgroup M (i.e., a Levi component of a parabolic subgroup P) of G, an element $m \in M$ whose centraliser G_m is contained in M, and a test function f in Harish-Chandra's Schwarz space $\mathcal{C}(G)$, one defines

$$J_M(m,f) = |D(m)|^{1/2} \int_{G_m\backslash G} f(x^{-1}mx)v_M(x)\,dx,$$

where dx is a G-invariant measure and $v_M(x)$ is the convex hull in $\mathfrak{a}_M^G = \mathfrak{a}_M/\mathfrak{a}_G$ (assuming $A_G \subset A_M$) of all the points $-H_P(x)$ for parabolic subgroups P with Levi component M. In this paper, we will only consider semisimple elements m, and then

$$D(m) = \det(\mathrm{Id} - \mathrm{Ad}_{\mathfrak{g}_m\backslash\mathfrak{g}}\, m^{-1}).$$

If one applies the definition to a Levi subgroup $M' \supset M$ of G and a test function $f' \in \mathcal{C}(M')$, the resulting distribution will be denoted by $J_M^{M'}(m,f')$.

The Fourier transform of f is the function $\hat{f}$ on the set $\Pi_{\mathrm{temp}}(G)$ of equivalence classes of tempered representations of G defined by $\hat{f}(\pi) = \mathrm{tr}\,\pi(f)$, where $\pi(f) = \int_G f(g)\pi(g)\,dg$. The range of the Fourier transform is a space $\mathcal{I}(G)$ of functions on $\Pi_{\mathrm{temp}}(G)$ that can be descibed explicitly. The Fourier transform of an invariant distribution $I : \mathcal{C}(G) \to \mathbb{C}$ is then the map $\hat{I} : \mathcal{I}(G) \to \mathbb{C}$ defined by $\hat{I}(\hat{f}) = I(f)$.

For $M \neq G$, weighted orbital integrals do not have Fourier transforms in the aforementioned sense because they are non-invariant. Neither are the weighted characters, which appear on the spectral side of the trace formula and can be interpreted as continuous maps $\phi_M : \mathcal{C}(G) \to \mathcal{I}(M)$ generalising $\phi_G(f) = \hat{f}$. We will recall their definition in section 11. Arthur has proved that there exist unique invariant distributions $I_M(m,f)$ on G (and their analogues on its Levi subgroups) such that

$$J_M(m,f) = \sum_{M'\supset M} \hat{I}_M^{M'}(m, \phi_{M'}(f)), \tag{1}$$

where the sum is taken over all Levi subgroups containing M. It is the Fourier transforms of these invariant distributions that we are after.

First of all, one has to describe $\Pi_{\mathrm{temp}}(G)$. Let P be a parabolic subgroup with Levi component L, and let $\sigma \in \Pi_{\mathrm{temp}}(L)$. If the induced representation $\mathrm{Ind}_P^G(\sigma)$ is irreducible, its class does not depend on P and is denoted by $\sigma^G \in \Pi_{\mathrm{temp}}(G)$. This is the case for generic unramified twists $\sigma_\zeta(l) = e^{\zeta(H_L(l))}\sigma(l)$ but may fail for some twist parameters $\zeta \in i\mathfrak{a}_L^*$. For $\phi \in \mathcal{I}(G)$, $\phi(\sigma_\zeta^G)$ is then a Schwarz function of ζ invariant under the Weyl group $W_L = \mathrm{Norm}_G L/\mathrm{Cent}_G L$, but the compatibility conditions caused by reducibility are hard to describe.

To solve this problem, Arthur[5] has attached to every $\sigma \in \Pi_{\mathrm{temp}}(L)$ and every element r of the R-group of σ in G a virtual representation τ of G, which is a certain formal linear combination of the constituents of $\mathrm{Ind}_P^G(\sigma)$. These virtual tempered representations τ modulo equivalence make up a set $T(G)$. Those of its elements which cannot be induced from $T(L)$ for a proper Levi subgroup L make up a subset $T_{\mathrm{ell}}(G)$, which contains the set $\Pi_2(G)$ of classes of square-integrable (modulo centre) representations. Induction identifies $T(G)$ naturally with the disjoint union, over all Levi subgroups L up to conjugacy, of the W_L-orbits in $T_{\mathrm{ell}}(L)$, and $\mathcal{I}(G)$ can be simply described as the space of Weyl-invariant Schwarz functions on this set in the natural sense.

It turns out that the Fourier transform of the distribution $I_M(m)$ is a measure, whose atoms (modulo unramified twists) are always contained in a set $T_{\mathrm{disc}}(G)$ that can be larger than $T_{\mathrm{ell}}(G)$. In fact, Arthur[6] has proved that for each Levi subgroup M there exists a unique smooth function Φ_M on $(M \cap G_{\mathrm{reg}}) \times \bigsqcup_{[L]}(W_L\backslash T_{\mathrm{disc}}(L))$ such that

$$I_M(m, f) = \sum_{[L]} \int_{W_L\backslash T_{\mathrm{disc}}(L)} \Phi_M(m, \check{\tau})\Theta_{\tau^G}(f)\, d\tau,$$

where $\check{\tau}$ denotes the contragredient of τ and where $d\tau$ is induced from a Haar measure on $i\mathfrak{a}_L^*$. The function Φ_M can be regarded as the Fourier transform of I_M and is the object of the present paper.

Since Φ_M is invariant under conjugation as a function of m, it suffices to compute its restriction to any Cartan subgroup T of M. In view of the known local behaviour,[3] we may further restrict attention to the intersection of T with the set G_{reg} of regular semisimple elements.

In order to simplify the explicit formulas for an acceptable group G, it is preferable to fix a half-system Σ of roots for $\mathfrak{t}_{\mathbb{C}}$ in $\mathfrak{g}_{\mathbb{C}}$ (which amounts to the choice of a Borel subalgebra $\mathfrak{b} \supset \mathfrak{t}_{\mathbb{C}}$) and to replace $|D(t)|^{1/2}$ for $t \in T$

by the Weyl denominator

$$\Delta_\Sigma(t) = \xi_\Sigma(t) \prod_{\alpha\in\Sigma} (1 - t^{-\alpha}), \tag{2}$$

where ξ_Σ is the character of the centraliser of T in the complexification of G determined by the condition $\xi_\Sigma(t)^2 = \prod_{\alpha\in\Sigma} t^\alpha$.

One needs a further modification to ensure that the basic descent formula for invariant orbital integrals carries over, which reads

$$J_G(t, f) = J_{P/N}^{P/N}(t, f_P)$$

for any parabolic subgroup $P \supset T$ with unipotent radical N, where

$$f_P(t) = |\det \mathrm{Ad}_{\mathfrak{n}}\, t|^{1/2} \int_K \int_N f(ktnk^{-1})\, dn\, dk.$$

For this purpose, choose a minimal such P and a $\mathfrak{b} \subset \mathfrak{p}_{\mathbb{C}}$, and set

$$\tilde{\Delta}_\Sigma(t) = \Delta_\Sigma(t) \operatorname{sgn}\det(\mathrm{Id} - \mathrm{Ad}_{\mathfrak{n}}\, t^{-1}). \tag{3}$$

(The determinant on the right-hand side is just the partial product in (2) over the real roots in Σ.) The resulting distributions

$$J_{M,\Sigma}(t, f) = \tilde{\Delta}_\Sigma(t) \int_{G/T} f(xtx^{-1}) v_M(x)\, dx$$

lead via (1) to modified distributions $I_{M,\Sigma}(t, f)$. Their Fourier transforms will be denoted by $\Phi_{M,\Sigma}$, so that

$$\left.\frac{\Phi_M}{|D|^{1/2}}\right|_{(T\cap G_{\mathrm{reg}})\times T_{\mathrm{disc}}(L)} = \frac{\Phi_{M,\Sigma}}{\tilde{\Delta}_\Sigma}.$$

Thus everything comes down to explicit formulas for the functions $\Phi_{M,\Sigma}$. One should keep in mind that they depend implicitly on L and the Haar measures η_M chosen on the various groups A_M. We give total mass one to the maximal compact subgroup of T, so that $J_{M,\Sigma}$ is proportional to η_M and inversely proportional to η_G and η_T. The connected components of $T_{\mathrm{disc}}(L)$ are principal homogeneous spaces under the character group $\Pi(A_L)$. We endow them with the Plancherel measure of A_L, which is inversely proportional to η_L. In total, the function $\Phi_{M,\Sigma}$ is proportional to η_M and η_L and inversely proportional to η_G and η_T. The dependence on the Haar measure of G drops out. These remarks clarify how our formulas have to be modified if one deviates from the normalisation of Haar measures fixed at the end of section 1.

1. Parametrisation of Cartan subgroups

Let a symplectic form on $\mathbb{R}^4$ be given by the matrix

$$J = \begin{pmatrix} 0_2 & 1_2 \\ -1_2 & 0_2 \end{pmatrix},$$

where 0_2 and 1_2 denote the zero resp. unit matrix of size 2. We denote the real symplectic group of rank two by G. It consists of all real 4×4 matrices g such that $g^\top J g = J$.

We fix some Levi subgroups of G. Let H be the set of all diagonal matrices in G. Such matrices are of the form

$$\operatorname{diag}(a_1, a_2, a_1^{-1}, a_2^{-1}) \qquad \text{with } a_1,\ a_2 \in \mathbb{R}^\times,$$

where generally $\operatorname{diag}(a_1, a_2, \dots)$ denotes a matrix with the numbers (or blocks) a_1, a_2, $\dots$ along the diagonal and zeros elsewhere. Next, let M_1 resp. M_2 be the subgroup of all matrices in G of the form

$$\begin{pmatrix} * & & & \\ & * & & * \\ & & * & \\ & * & & * \end{pmatrix} \quad \text{resp.} \quad \begin{pmatrix} * & * & & \\ * & * & & \\ & & * & * \\ & & * & * \end{pmatrix},$$

where all empty places are filled with zeros. Then M_1 is isomorphic to $\mathrm{GL}_1 \times \mathrm{SL}_2$, while M_2 is isomorphic to GL_2. The groups H, M_1, M_2 and G form a system of representatives for the conjugacy classes of Levi subgroups of G.

Let us fix some Cartan subgroups of G. We denote by T_1 the set of all matrices of the form

$$\begin{pmatrix} a & & & \\ & \cos\theta & & \sin\theta \\ & & a^{-1} & \\ & -\sin\theta & & \cos\theta \end{pmatrix} \quad \text{with } a \in \mathbb{R}^\times,\ \theta \in \mathbb{R}.$$

Further, let T_2 be the set of all matrices of the form

$$\operatorname{diag}(c\tilde{r}, c^{-1}\tilde{r}) \qquad \text{with} \quad \tilde{r} = \begin{pmatrix} \cos\varphi & \sin\varphi \\ -\sin\varphi & \cos\varphi \end{pmatrix}, \quad \varphi \in \mathbb{R}, \quad c \in \mathbb{R}_+.$$

Finally, we denote by B the set of all matrices of the form

$$\begin{pmatrix} \cos\theta_1 & & \sin\theta_1 & \\ & \cos\theta_2 & & \sin\theta_2 \\ -\sin\theta_1 & & \cos\theta_1 & \\ & -\sin\theta_2 & & \cos\theta_2 \end{pmatrix} \quad \text{with } \theta_1,\ \theta_2 \in \mathbb{R}.$$

Then the groups H, T_1, T_2 and B form a system of representatives for the conjugacy classes of Cartan subgroups of G. Moreover, if i is 1 or 2, the groups H, T_i form a system of representatives for the conjugacy classes of Cartan subgroups of M_i.

We will mostly use the shorthands

$$u = e^{i\theta}, \quad r = e^{i\varphi}, \quad u_1 = e^{i\theta_1}, \quad u_2 = e^{i\theta_2}.$$

We have virtually defined functions

$$\begin{array}{ll} a_1,\, a_2 & \text{on } H, \\ a,\, u & \text{on } T_1, \\ c,\, r & \text{on } T_2, \\ u_1,\, u_2 & \text{on } B. \end{array}$$

Sometimes we will also need the functions $a_\pm = |a_1 a_2^{\pm 1}|^{1/2}$.

We can now describe the action of the real Weyl groups on the pair of parameters for each of the aforementioned Cartan subgroups.

group	order	action
W_H	8	permutation and inversion of either parameter
W_{T_1}	2	inversion of a
W_{T_2}	4	inversion of either parameter
W_B	2	permutation
$W_H^{M_1}$	2	inversion of a_2
$W_{T_1}^{M_1}$	1	trivial
$W_H^{M_2}$	2	permutation
$W_{T_2}^{M_2}$	2	inversion of r

For each of those Cartan subgroups, we choose a half-system of root characters and declare them to be positive:

$$\begin{aligned} \Pi &= \{a_1^2, a_2^2, a_1 a_2, a_1 a_2^{-1}\} && \text{for } H, \\ \Sigma_1 &= \{a^2, u^2, au, au^{-1}\} && \text{for } T_1, \\ \Sigma_2 &= \{c^2 r^2, c^2 r^{-2}, c^2, r^2\} && \text{for } T_2, \\ \Sigma &= \{u_1^2, u_2^2, u_1 u_2, u_1 u_2^{-1}\} && \text{for } B. \end{aligned}$$

The group G is acceptable, and the modified Weyl denominators from (3) are

$$\begin{aligned}\tilde{\Delta}_\Pi &= a_1^2 a_2 \left|1-a_1^{-2}\right| \left|1-a_2^{-2}\right| \left|1-a_1^{-1}a_2^{-1}\right| \left|1-a_1^{-1}a_2\right|,\\ \tilde{\Delta}_{\Sigma_1} &= a^2 u \left|1-a^{-2}\right| (1-u^{-2})(1-a^{-1}u^{-1})(1-a^{-1}u),\\ \tilde{\Delta}_{\Sigma_2} &= c^3 r(1-c^{-2}r^{-2})(1-c^{-2}r^2) \left|1-c^{-2}\right| (1-r^{-2}),\\ \tilde{\Delta}_\Sigma &= u_1^2 u_2 (1-u_1^{-2})(1-u_2^{-2})(1-u_1^{-1}u_2^{-1})(1-u_1^{-1}u_2).\end{aligned}$$

In the case of Σ, we may omit the tilde as there is no modification.

The positive root characters of T_i or H whose root spaces are contained in $\mathfrak{m}_{i,\mathbb{C}}$ make up half-systems

$$\begin{aligned}\Sigma_1^{M_1} &= \{u^2\}, &\qquad \Pi^{M_1} &= \{a_2^2\},\\ \Sigma_2^{M_2} &= \{r^2\}, &\qquad \Pi^{M_2} &= \{a_1 a_2^{-1}\}.\end{aligned}$$

To avoid clumsy notation, we index the corresponding modified Weyl denominators by the half-system for G as

$$\begin{aligned}\tilde{\Delta}_\Pi^{M_1} &= a_2 \left|1-a_2^{-2}\right|,\\ \tilde{\Delta}_{\Sigma_1}^{M_1} &= u(1-u^{-2}),\\ \tilde{\Delta}_{\Sigma_2}^{M_2} &= r(1-r^{-2}).\end{aligned}$$

We may omit the tilde on $\Delta_{\Sigma_i}^{M_i}$ as no modification takes place. Acceptability makes no sense for the non-connected groups M_2 and H. Here we set

$$\begin{aligned}\tilde{\Delta}_\Pi^{M_2} &= \operatorname{sgn}(a_2) a_- \left|1-a_1^{-1}a_2\right|,\\ \tilde{\Delta}_\Pi^{H} &= \operatorname{sgn} a_2.\end{aligned}$$

This seemingly absurd choice ensures that, for all our Levi subgroups M and Cartan subgroups $T \subset M$,

$$\tilde{\Delta}_\Pi(h) = \tilde{\Delta}_\Pi^M(h) \left|\det(\mathrm{Id} - \mathrm{Ad}_{\mathfrak{g}/\mathfrak{m}}\, h^{-1})\right|^{1/2}, \tag{4}$$

and that the analogous property is satisfied for Σ_i and also for G replaced by any one of our chosen Levi subgroups.

We choose Haar measures on the groups $H^0 = A_H$, $A_{T_1} = A_{M_1}$ and $A_{T_2} = A_{M_2}$ to be induced by the differential forms

$$\frac{da_1}{a_1} \wedge \frac{da_2}{a_2}, \qquad \frac{da}{a}, \qquad \frac{dc}{c},$$

respectively. The continuous characters of the three groups are of the form

$$a_1^{\lambda_1} a_2^{\lambda_2}, \qquad a^\lambda, \qquad c^\mu,$$

respectively. We can regard λ_1, λ_2 as coordinates on $\mathfrak{h}^*_{\mathbb{C}}$, λ as a coordinate on $\mathfrak{a}^*_{M_1,\mathbb{C}}$ and μ as a coordinate on $\mathfrak{a}^*_{M_2,\mathbb{C}}$. Unitary characters correspond to purely imaginary coordinates, and the Plancherel measures corresponding to the Haar measures chosen above are induced by the differential forms

$$\frac{d\lambda_1}{2\pi i} \wedge \frac{d\lambda_2}{2\pi i}, \quad \frac{d\lambda}{2\pi i}, \quad \frac{d\mu}{2\pi i}.$$

2. Parametrisation of representations

2.1. *Discrete series*

We will now parametrise the discrete series of G and its Levi subgroups chosen in section 1.

A square-integrable representation π of G is uniquely determined by the restriction of its global character Θ_π to the compact Cartan subgroup B, or else by the numerator

$$\Phi_{\Sigma,\pi} = \Delta_\Sigma \Theta_\pi.$$

Any such numerator is of the form

$$\Phi_{\Sigma,\pi} = \operatorname{sgn}(l_1 l_2 l_+ l_-)\big(u_1^{l_2} u_2^{l_1} - u_1^{l_1} u_2^{l_2}\big) \tag{5}$$

with l_1, $l_2 \in \mathbb{Z}$ such that the numbers l_1, l_2 and $l_\pm = l_1 \pm l_2$ are nonzero. Thus, $\Pi_2(G)$ is in bijection with the set of such pairs (l_1, l_2) modulo the action of the Weyl group W_B, which acts by permutation.

In the same fashion, if σ_i is a representation of M_i square-integrable modulo the centre, we consider the numerator

$$\Phi_{\Sigma_i,\sigma_i} = \Delta^{M_i}_{\Sigma_i} \Theta_{\sigma_i}$$

of the character Θ_{σ_i} on the elliptic Cartan subgroup T_i of M_i.

In the case of M_1, any such numerator is of the form

$$\Phi_{\Sigma_1,\sigma_1} = -\operatorname{sgn}(l) u^l \operatorname{sgn}(a)^N |a|^\lambda \tag{6}$$

with $l \in \mathbb{Z}$, $N \in \{0,1\}$ and $\lambda \in i\mathbb{R}$ such that $l \neq 0$. Thus, $\Pi_2(M_1)$ is in bijection with the set of such triples (l, N, λ), the group $W^{M_1}_{T_1}$ being trivial. The group W_{M_1} has order 2. It acts on $\Pi_2(M_1)$ by sign change of λ.

Similarly, any discrete series numerator for M_2 on T_2 is of the form

$$\Phi_{\Sigma_2,\sigma_2} = -\operatorname{sgn}(m)(r^m - r^{-m})c^\mu$$

with $m \in \mathbb{Z}$ and $\mu \in i\mathbb{R}$ such that $m \neq 0$. So $\Pi_2(M_2)$ is in bijection with the set of such pairs (m, μ) modulo the action of the the Weyl group $W^{M_2}_{T_2}$, which acts by sign change of m. The group W_{M_2} has order 2 and acts on $\Pi_2(M_2)$ by sign change of μ.

Finally, any unitary representation of the group H is one-dimensional and of the form

$$\chi = \operatorname{sgn}(a_1)^{N_1}|a_1|^{\lambda_1}\operatorname{sgn}(a_2)^{N_2}|a_2|^{\lambda_2}$$

with N_1, $N_2 \in \{0,1\}$ and λ_1, $\lambda_2 \in i\mathbb{R}$. Thus, $\Pi_2(H)$ is in bijection with the set of such quadruples $(N_1, \lambda_1, N_2, \lambda_2)$. The group W_H acts on this set by permutation of the indices and sign changes of the λ_i. The character numerator is in this case

$$\Phi_{\Pi,\chi} = \tilde{\Delta}_{\Pi}^{H}\chi.$$

If necessary, we will indicate the dependence of the representations π, σ_1, σ_2 and χ on the parameters and denote them by π_{l_1,l_2}, $\sigma_{l,N,\lambda}$, $\sigma_{m,\mu}$ and $\chi_{N_1,\lambda_1,N_2,\lambda_2}$, respectively. Passing to the contragredient representation amounts to sign changes in λ_i, l, λ, m, μ and l_i.

2.2. *Limits of discrete series*

Taking formally the one-sided limits as $l \to \pm 0$ on the right-hand side of (6), we obtain functions $\Phi_{\Sigma_1,\sigma_{\pm 0,N,\lambda}}$. These are the character numerators of certain tempered representations $\sigma_{\pm 0,N,\lambda}$ (obtained by Zuckerman tensoring) which, as N and λ vary, constitute the limit of discrete series of M_1. By Schmid's identity,[19] they appear as constituents of reducible induced representations, viz.

$$\chi^{M_1}_{N,\lambda,1,0} \cong \sigma_{+0,N,\lambda} \oplus \sigma_{-0,N,\lambda}.$$

We will need the elliptic virtual tempered representations

$$\tau_{N,\lambda} = \tfrac{1}{2}(\sigma_{+0,N,\lambda} - \sigma_{-0,N,\lambda}) \tag{7}$$

of M_1 with character numerators

$$\Phi_{\Sigma_1,\tau_{N,\lambda}} = -\operatorname{sgn}(a)^N|a|^\lambda.$$

(The factor $\frac{1}{2}$ is absent in Arthur's definition[5] but simplifies numerous formulas below.) This yields a parametrisation of the set $T_{\mathrm{ell}}(M_1) - \Pi_2(M_1)$ by pairs (N, λ), on which W_{M_1} acts by sign change of λ.

The group M_2 has no limit of discrete series, and $T_{\mathrm{ell}}(M_2) = \Pi_2(M_2)$.

In the case of G, passing formally to one-sided limits on the right-hand side of (5), we obtain character numerators of certain tempered representations $\pi_{l,\pm 0} \cong \pi_{\pm 0,l}$ and $\pi_{l\pm 0,-l\pm 0} \cong \pi_{-l\pm 0,l\pm 0}$, where the identifications come from the action of W_B. As l varies through the nonzero integers, they constitute the limits of discrete series of G. By Schmid's identity, we have

$$\sigma^G_{l,0,0} \cong \pi_{l,+0} \oplus \pi_{l,-0},$$
$$\sigma^G_{2l,0} \cong \pi_{l+0,-l+0} \oplus \pi_{l-0,-l-0}.$$

We need the elliptic virtual tempered representations of G

$$\tau'_l = \tfrac{1}{2}(\pi_{l,+0} - \pi_{l,-0}),$$
$$\tau''_l = \tfrac{1}{2}(\pi_{l+0,-l+0} - \pi_{l-0,-l-0})$$

with character numerators

$$\Phi_{\Sigma,\tau'_l} = \mathrm{sgn}(l)(u_2^l - u_1^l),$$
$$\Phi_{\Sigma,\tau''_l} = \mathrm{sgn}(l)(u_1^{-l}u_2^l - u_1^l u_2^{-l}).$$

In this way, we get a parametrisation of $T_{\mathrm{ell}}(G) - \Pi_2(G)$ by the set

$$(\mathbb{Z} - \{0\}) \sqcup (\mathbb{Z} - \{0\})/\{\pm 1\}.$$

The virtual representations $\tau_{N,\lambda}$, τ'_l and τ''_l will mostly be denoted by τ_1, τ' and τ'' for brevity.

3. Guide to the results

The following tables refer to the formulas for the components Φ_M of the Fourier transforms restricted to $(T \cap G_{\mathrm{reg}}) \times T_{\mathrm{ell}}(L)$. The Levi subgroup M is shown in the upper left corner, its Cartan subgroup T in the left column and the Levi subgroup L in the upper row.

G	G	M_1	M_2	H
H	(8)	(10)	(10)	(10)
T_1	(8)	(9)	(16)	(17)
T_2	(8)	(16)	(9)	(18)
B	(8)	(19)	(20)	(21)

M_1	G	M_1	M_2	H
H	(12)	(14)	(15)	12.1
T_1	(11)	13.1	9.1	12.2

M_2	G	M_1	M_2	H
H	(12)	(14)	(15)	12.1
T_2	(11)	9.1	13.1	12.2

H	G	M_1	M_2	H
H	(12)	13.2	13.2	12.3

Numbers in parentheses refer to displayed equations showing previously known formulas. The other numbers refer to theorems containing our main results.

We explain in section 14 how to obtain the values on $T_{\text{disc}}(L) - T_{\text{ell}}(L)$.

4. Global characters

In section 2, the elliptic tempered (virtual) characters were identified by giving their restriction to an elliptic Cartan subgroup. A full description requires their restriction to one Cartan subgroup from each conjugacy class.

As for the discrete series, we recall the formulas from Hirai's paper[11] and extend them to all chambers, taking into account that the numerator must exhibit the same equivariance as the denominator to make a fraction Weyl-group invariant. To simplify statements, we use the notation

$$\lfloor x \rfloor = \min{}'(x, x^{-1}), \quad \lceil x \rceil = \max{}'(x, x^{-1}) \quad \text{for } x \in \mathbb{R}^\times,$$

where we denote by $\min'(x, y, \dots)$ resp. $\max'(x, y, \dots)$ that number among x, y, ... which has the smallest resp. largest absolute value.

We start with the discrete series character numerators $\Phi_{\Pi,\sigma_i} = \tilde{\Delta}_{\Pi}^{M_i} \Theta_{\sigma_i}$ on $H \cap M_{i,\text{reg}}$. If σ_1 is the representation of M_1 with parameter (l, N, λ), then

$$\Phi_{\Pi,\sigma_1} = \operatorname{sgn}(a_1)^N |a_1|^\lambda \lfloor a_2^l \rfloor,$$

and if σ_2 is the representation M_2 with parameter (m, μ), then

$$\begin{aligned} \Phi_{\Pi,\sigma_2} &= (1 + \operatorname{sgn} a_1 a_2) a_+^\mu \lfloor a_1^m a_+^{-m} \rfloor \\ &= (1 + \operatorname{sgn} a_1 a_2) a_+^\mu \lfloor a_2^m a_+^{-m} \rfloor. \end{aligned}$$

Next we describe the character numerators $\Phi_{\Sigma_i,\pi} = \tilde{\Delta}_{\Sigma_i} \Theta_\pi$ of the representation π of G with parameter (l_1, l_2) on $T_i \cap G_{\text{reg}}$. We have

$$\begin{aligned} \Phi_{\Sigma_1,\pi} &= \operatorname{sgn}(l_+ l_-) \big(\operatorname{sgn}(l_2) \lfloor a^{l_1} \rfloor u^{l_2} - \operatorname{sgn}(l_1) \lfloor a^{l_2} \rfloor u^{l_1} \big), \\ \Phi_{\Sigma_2,\pi} &= \operatorname{sgn}(l_1 l_2 l_-) \lfloor c^{l_+} \rfloor \big(r^{l_-} - r^{-l_-} \big). \end{aligned}$$

In order to describe $\Phi_{\Pi,\pi} = \tilde{\Delta}_\Pi \Theta_\pi$ on $H \cap G_{\text{reg}}$, we distinguish two cases. If $l_1 l_2 > 0$ (holomorphic discrete series) or $a_1 a_2 < 0$, then

$$\Phi_{\Pi,\pi} = \operatorname{sgn}(l_+ l_-) \operatorname{sgn}((1 - a_1^{-1} a_2^{-1})(1 - a_1^{-1} a_2)) \big(\lfloor a_1^{l_2} \rfloor \lfloor a_2^{l_1} \rfloor - \lfloor a_1^{l_1} \rfloor \lfloor a_2^{l_2} \rfloor \big).$$

If $l_1 l_2 < 0$ (large discrete series) and $a_1 a_2 > 0$, then

$$\Phi_{\Pi,\pi} = \lfloor a_1^{l_2} \rfloor \lfloor a_2^{l_1} \rfloor + \lfloor a_1^{l_1} \rfloor \lfloor a_2^{l_2} \rfloor + 2\frac{\min'(\lfloor a_1^{l_1} \rfloor, \lfloor a_1^{l_2} \rfloor, \lfloor a_2^{l_1} \rfloor, \lfloor a_2^{l_2} \rfloor)}{\max'(\lfloor a_1^{l_1} \rfloor, \lfloor a_1^{l_2} \rfloor, \lfloor a_2^{l_1} \rfloor, \lfloor a_2^{l_2} \rfloor)}.$$

All of the formulas in this section remain valid for limits of discrete series if one takes the appropriate one-sided limits in a formal way. By forming their differences, we obtain the numerators of the virtual tempered characters introduced in subsection 2.2. If $\tau_1 \in T_{\mathrm{ell}}(M_1) - \Pi_2(M_1)$, then

$$\Phi_{\Pi,\tau_1} = 0.$$

If τ', $\tau'' \in T_{\mathrm{ell}}(G) - \Pi_2(G)$ in the notation of subsection 2.2, then

$$\begin{aligned} \Phi_{\Sigma_1,\tau'} &= \lfloor a^l \rfloor, \quad \Phi_{\Sigma_2,\tau'} = \lfloor c^l \rfloor (r^l - r^{-l}), \\ \Phi_{\Pi,\tau'} &= -(1 + \operatorname{sgn} a_1 a_2) \min'(\lfloor a_1^l \rfloor, \lfloor a_2^l \rfloor), \\ \Phi_{\Sigma_1,\tau''} &= -\lfloor a^l \rfloor (u^{-l} + u^l), \\ \Phi_{\Sigma_2,\tau''} &= \Phi_{\Pi,\tau''} = 0. \end{aligned}$$

5. Characters as Fourier transforms

The first results on Fourier transforms of orbital integrals are implicit in Harish-Chandra's work.[9] It follows from his results that $\Phi_{G,\Sigma}(\sigma)$ with $\sigma \in \Pi_2(M)$ and $T \subset M$ vanishes unless the Cartan subgroup T is elliptic in the Levi subgroup M, in which case it equals $\sum_{w \in W_M} \Phi^M_{\Sigma,w\sigma}$. Applied to our Levi subgroups, this implies

$$\Phi^{M_1}_{M_1,\Sigma_1}(\sigma_1) = \Phi_{\Sigma_1,\sigma_1}, \quad \Phi^{M_1}_{M_1,\Pi}(\sigma_1) = 0, \quad \Phi^{M_1}_{M_1,\Pi}(\chi) = \sum_{w \in W_H^{M_1}} \Phi_{\Pi,w\chi},$$

$$\Phi^{M_2}_{M_2,\Sigma_2}(\sigma_2) = \Phi_{\Sigma_2,\sigma_2}, \quad \Phi^{M_2}_{M_2,\Pi}(\sigma_2) = 0, \quad \Phi^{M_2}_{M_2,\Pi}(\chi) = \sum_{w \in W_H^{M_2}} \Phi_{\Pi,w\chi}.$$

Moreover, applied to our group G, it gives

$$\Phi_{G,\Sigma}(\pi) = \Phi_{\Sigma,\pi}, \quad \Phi_{G,\Sigma_1}(\pi) = \Phi_{G,\Sigma_2}(\pi) = \Phi_{G,\Pi}(\pi) = 0, \tag{8}$$

$$\Phi_{G,\Sigma_1}(\sigma_1) = \sum_{w \in W_{M_1}} \Phi_{\Sigma_1,w\sigma_1}, \quad \Phi_{G,\Sigma_2}(\sigma_2) = \sum_{w \in W_{M_2}} \Phi_{\Sigma_2,w\sigma_2}, \tag{9}$$

$$\Phi_{G,\Pi}(\chi) = \sum_{w \in W_H} \Phi_{\Pi,w\chi}, \quad \Phi_{G,\Pi}(\sigma_1) = \Phi_{G,\Pi}(\tau_1) = \Phi_{G,\Pi}(\sigma_2) = 0. \tag{10}$$

We can explicate the right-hand sides using the character formulas in section 2, e.g.

$$\Phi^{M_1}_{M_1,\Pi}(\chi) = \operatorname{sgn}(a_1)^{N_1}|a_1|^{\lambda_1} \operatorname{sgn}(a_2)^{N_2+1}(|a_2|^{\lambda_2} + |a_2|^{-\lambda_2}),$$
$$\Phi_{G,\Sigma_1}(\sigma_1) = -\operatorname{sgn}(l)u^l \operatorname{sgn}(a)^N(|a|^{\lambda} + |a|^{-\lambda}).$$

We leave the other cases to the reader.

The function $\Phi_{M,\Sigma}(\pi)$ on a Cartan subgroup T for $\pi \in \Pi_2(G)$ has been determined by Arthur.[2] It vanishes unless T is elliptic in M, in which case it equals $(-1)^{\dim A_M/A_G}\Phi_{\Sigma,\pi}$. These results have been extended by Arthur himself[5] to elliptic virtual tempered representations. With our modified normalisation (7) they read

$$\Phi^{M_1}_{H,\Pi}(\sigma_1) = -\Phi_{\Pi,\sigma_1}, \quad \Phi^{M_1}_{H,\Pi}(\tau_1) = -\Phi_{\Pi,\tau_1}, \quad \Phi^{M_2}_{H,\Pi}(\sigma_2) = -\Phi_{\Pi,\sigma_2}$$

and, for any $\tau \in T_{\mathrm{ell}}(G)$,

$$\Phi_{M_1,\Sigma_1}(\tau) = -\Phi_{\Sigma_1,\tau}, \quad \Phi_{M_2,\Sigma_2}(\tau) = -\Phi_{\Sigma_2,\tau}, \tag{11}$$
$$\Phi_{H,\Pi}(\tau) = \Phi_{\Pi,\tau}, \quad \Phi_{M_1,\Pi}(\tau) = \Phi_{M_2,\Pi}(\tau) = 0. \tag{12}$$

6. Geometric descent

Given a Levi subgroup M' of G and its Cartan subgroup T, we can simplify the calculation of the component $\Phi_{M',\Sigma}$ of our Fourier transform by descending from M' to any of its Levi subgroups M still containing T. Let L be a further Levi subgroup of G containing the minimal Levi subgroup H. Then Arthur's descent formula[6] reads, for $m \in M \cap G_{\mathrm{reg}}$ and $\tau \in T_{\mathrm{ell}}(L)$,

$$\Phi_{M'}(m,\tau) = \sum_{M''\supset M} d^G_M(M',M'') \sum_{\substack{w\in W_H^{M''}\backslash W_H \\ wL\subset M''}} \Phi^{M''}_M(m,w\tau). \tag{13}$$

The constant $d^G_M(M',M'')$ vanishes unless the natural map $\mathfrak{a}^{M''}_M \to \mathfrak{a}^G_M/\mathfrak{a}^{M'}_M \cong \mathfrak{a}^G_{M'}$ is an isomorphism, in which case $d^G_M(M',M'')$ times the image of the Haar measure on $\mathfrak{a}^{M''}_M$ equals the Haar measure on $\mathfrak{a}^G_{M'}$.

We can already apply this formula now in a special case. For $h \in H\cap G_{\mathrm{reg}}$ and $\tau \in T_{\mathrm{ell}}(M_j)$, it simplifies to

$$\Phi_{M_i}(h,\tau) = \sum_{w\in W_H/W_H^{M_j}} d^G_H(M_i, wM_j)\Phi^{M_j}_H(w^{-1}h,\tau).$$

Substituting the formulas of the last section, we get the following functions on $H \cap G_{\mathrm{reg}}$:

$$\begin{aligned}
\Phi_{M_1}(\sigma_1) &= -\operatorname{sgn}(a_2)^N(|a_2|^\lambda + |a_2|^{-\lambda})\operatorname{sgn}(a_1)\lfloor a_1^l \rfloor, \\
\Phi_{M_2}(\sigma_1) &= -\tfrac{1}{2}\operatorname{sgn}(a_1)^N(|a_1|^\lambda + |a_1|^{-\lambda})\operatorname{sgn}(a_2)\lfloor a_2^l \rfloor \\
&\quad - \tfrac{1}{2}\operatorname{sgn}(a_2)^N(|a_2|^\lambda + |a_2|^{-\lambda})\operatorname{sgn}(a_1)\lfloor a_1^l \rfloor,
\end{aligned} \tag{14}$$

$$\begin{aligned}
\Phi_{M_2}(\sigma_2) &= -\tfrac{1}{2}(\operatorname{sgn} a_1 + \operatorname{sgn} a_2)(a_-^\mu + a_-^{-\mu})\lfloor a_1^m a_-^{-m} \rfloor, \\
\Phi_{M_1}(\sigma_2) &= -\tfrac{1}{2}(\operatorname{sgn} a_1 + \operatorname{sgn} a_2)\big((a_+^\mu + a_+^{-\mu})\lfloor a_1^m a_+^{-m} \rfloor \\
&\qquad + (a_-^\mu + a_-^{-\mu})\lfloor a_1^m a_-^{-m} \rfloor\big),
\end{aligned} \tag{15}$$

$$\Phi_{M_1}(\tau_1) = \Phi_{M_2}(\tau_1) = 0.$$

In order to obtain the functions $\Phi_{M_i,\Pi}$, one has to multiply the right-hand sides by $\tilde{\Delta}_\Pi^H = \operatorname{sgn} a_2$.

7. Abelian Fourier inversion

The remaining components of the Fourier transforms of invariant orbital integrals can be determined by Fourier inversion on Cartan subgroups, where the character numerators satisfy certain orthogonality relations. General results have been obtained by Herb.[10] In the case of our Levi subgroups, the formulas had been known before and read

$$\Phi_{M_1,\Sigma_1}^{M_1}(\chi) = -\operatorname{sgn}(a)^{N_1}|a|^{\lambda_1} \sum_{n \equiv N_2+1} \left(\frac{1}{n+\lambda_2} + \frac{1}{n-\lambda_2}\right) u^n,$$

$$\Phi_{M_2,\Sigma_2}^{M_2}(\chi) = -c^{\lambda_+} \sum_{n \equiv N_+ +1} \left(\frac{1}{n+\lambda_-} + \frac{1}{n-\lambda_-}\right) r^n,$$

$$\Phi_{M_1,\Sigma_1}^{M_1}(\tau_1) = \Phi_{\Sigma_1,\tau_1},$$

where $\lambda_\pm = \lambda_1 \pm \lambda_2$, $N_+ \in \{0,1\}$ is determined by $N_+ \equiv N_1 + N_2 \pmod 2$, and the congruences in these and further summation conditions are taken modulo 2. If we observe that the function

$$S_\lambda(u) = \frac{\pi}{\sin \pi\lambda}(u^\lambda - u^{-\lambda})$$

(where we take the principal branch of the complex power) has the Fourier expansion

$$S_\lambda(u) = -\sum_n \left(\frac{1}{n+\lambda} + \frac{1}{n-\lambda}\right)(-u)^n,$$

we can write our formulas as

$$\Phi^{M_1}_{M_1,\Sigma_1}(\chi) = \tfrac{1}{2}\,\mathrm{sgn}(a)^{N_1}|a|^{\lambda_1}(S_{\lambda_2}(-u) - (-1)^{N_2}S_{\lambda_2}(u)),$$
$$\Phi^{M_2}_{M_2,\Sigma_2}(\chi) = \tfrac{1}{2}c^{\lambda_+}(S_{\lambda_-}(-r) - (-1)^{N_+}S_{\lambda_-}(r)).$$

Some components of the Fourier transforms on G can be reduced to the preceding results by the descent formula (13) specialised to $M' = G$, which can be modified to

$$\Phi_{G,\Sigma}(t,\tau) = \sum_{\substack{w\in W_H^M\backslash W_H\\ wL\subset M}} \Phi^M_{M,\Sigma}(t,w\tau)$$

due to the compatibility (4) of Weyl denominators. On the one hand, we get

$$\Phi_{G,\Sigma_1}(\sigma_2) = \Phi_{G,\Sigma_2}(\sigma_1) = \Phi_{G,\Sigma_2}(\tau_1) = 0 \tag{16}$$

because M_1 and M_2 are not conjugate. On the other hand,

$$\Phi_{G,\Sigma_i}(\chi) = \sum_{w\in W_H^{M_i}\backslash W_H} \Phi^{M_i}_{M_i,\Sigma_i}(w\chi).$$

If we plug in everything, this yields

$$\begin{aligned}
\Phi_{G,\Sigma_1}(\tau_1) &= -\,\mathrm{sgn}(a)^N(|a|^\lambda + |a|^{-\lambda}),\\
\Phi_{G,\Sigma_1}(\chi) &= -\,\mathrm{sgn}(a)^{N_1}(|a|^{\lambda_1} + |a|^{-\lambda_1}) \sum_{n\equiv N_2+1}\left(\frac{1}{n+\lambda_2} + \frac{1}{n-\lambda_2}\right)u^n\\
&\quad -\,\mathrm{sgn}(a)^{N_2}(|a|^{\lambda_2} + |a|^{-\lambda_2}) \sum_{n\equiv N_1+1}\left(\frac{1}{n+\lambda_1} + \frac{1}{n-\lambda_1}\right)u^n\\
&= \tfrac{1}{2}\,\mathrm{sgn}(a)^{N_1}(|a|^{\lambda_1} + |a|^{-\lambda_1})(S_{\lambda_2}(-u) - (-1)^{N_2}S_{\lambda_2}(u))\\
&\quad + \tfrac{1}{2}\,\mathrm{sgn}(a)^{N_2}(|a|^{\lambda_2} + |a|^{-\lambda_2})(S_{\lambda_1}(-u) - (-1)^{N_1}S_{\lambda_1}(u)),
\end{aligned} \tag{17}$$

$$\begin{aligned}
\Phi_{G,\Sigma_2}(\chi) &= -\,(c^{\lambda_+} + c^{-\lambda_+}) \sum_{n\equiv N_++1}\left(\frac{1}{n+\lambda_-} + \frac{1}{n-\lambda_-}\right)r^n\\
&\quad -\,(c^{\lambda_-} + c^{-\lambda_-}) \sum_{n\equiv N_++1}\left(\frac{1}{n+\lambda_+} + \frac{1}{n-\lambda_+}\right)r^n\\
&= \tfrac{1}{2}(c^{\lambda_+} + c^{-\lambda_+})(S_{\lambda_-}(-r) - (-1)^{N_+}S_{\lambda_-}(r))\\
&\quad + \tfrac{1}{2}(c^{\lambda_-} + c^{-\lambda_-})(S_{\lambda_+}(-r) - (-1)^{N_+}S_{\lambda_+}(r)).
\end{aligned} \tag{18}$$

We can also apply Herb's results[10] directly to our group G. Concerning the components of the Fourier transforms on $T_{\mathrm{ell}}(M_1)$, we get

$$\begin{aligned}\Phi_{G,\Sigma}(\sigma_1) &= \operatorname{sgn}(l) \sum_{n \equiv N} \left(\frac{1}{n+\lambda} + \frac{1}{n-\lambda} \right) (u_1^n u_2^l - u_1^l u_2^n) \\ &= \tfrac{1}{2} \operatorname{sgn}(l) \big(u_1^l (S_\lambda(-u_2) + (-1)^N S_\lambda(u_2)) \\ &\qquad - u_2^l (S_\lambda(-u_1) + (-1)^N S_\lambda(u_1)) \big), \\ \Phi_{G,\Sigma}(\tau_1) &= \sum_{n \equiv N} \left(\frac{1}{n+\lambda} + \frac{1}{n-\lambda} \right) (u_1^n - u_2^n) \\ &= \tfrac{1}{2} \big((S_\lambda(-u_2) + (-1)^N S_\lambda(u_2)) \\ &\qquad - (S_\lambda(-u_1) + (-1)^N S_\lambda(u_1)) \big). \end{aligned} \tag{19}$$

For the components on $\Pi_2(M_2)$ we obtain

$$\begin{aligned}\Phi_{G,\Sigma}(\sigma_2) &= 2 \sum_{\substack{n_1, n_2 \\ |n_-| = |m|}} \operatorname{sgn}(n_-) \left(\frac{1}{n_+ + \mu} + \frac{1}{n_+ - \mu} \right) u_1^{n_1} u_2^{n_2} \\ &= \operatorname{sgn}(m)(u_-^{-m} - u_-^m)(S_\mu(-u_+) + (-1)^m S_\mu(u_+)), \end{aligned} \tag{20}$$

where $u_\pm \in \mathrm{U}(1)$ are such that $u_+ u_- = u_1$, $u_+ u_-^{-1} = u_2$. (These numbers are determined up to simultaneous sign changes and parametrise a double covering of B.)

Finally, when writing down the components on $\Pi_2(H)$, we use averaging over the Weyl group in order to save space.

$$\begin{aligned}\Phi_{G,\Sigma}(\chi) = - \sum_{(N_1', \lambda_1', N_2', \lambda_2')} \sum_{\substack{n_1, n_2 \\ n_+ \equiv N_+ + 1}} &\left(\frac{1}{2} \cdot \frac{1}{n_+ - \lambda_+'} \left(\frac{1}{n_1 - \lambda_1'} - \frac{1}{n_2 - \lambda_2'} \right) \right. \\ &\left. + \frac{1}{2} \cdot \frac{(-1)^{n_1 + N_1'}}{(n_1 - \lambda_1')(n_2 - \lambda_2')} + \frac{1}{(n_+ - \lambda_+')(n_- - \lambda_-')} \right) u_1^{n_1} u_2^{n_2}. \end{aligned} \tag{21}$$

Here the exterior sum is over the eight quadruples obtained from

$(N_1, \lambda_1, N_2, \lambda_2)$ by the action of W_H. Alternatively,

$$\Phi_{G,\Sigma}(\chi) = -\frac{\pi^2}{4} \sum_{(N_1', \lambda_1', N_2', \lambda_2')} \left(\frac{(-u_1 u_2^{-1})^{\lambda_1'} \big((-u_2)^{\lambda_+'} - (-1)^{N_+'} u_2^{\lambda_+'}\big)}{\sin \pi\lambda_1' \sin \pi\lambda_+'} \right.$$
$$- \frac{(-u_1^{-1} u_2)^{\lambda_2'} \big((-u_1)^{\lambda_+'} - (-1)^{N_+'} u_1^{\lambda_+'}\big)}{\sin \pi\lambda_2' \sin \pi\lambda_+'}$$
$$+ \frac{(-1)^{N_1'} u_1^{\lambda_1'} (-u_2)^{\lambda_2'} - (-1)^{N_2'} (-u_1)^{\lambda_1'} u_2^{\lambda_2'}}{\sin \pi\lambda_1' \sin \pi\lambda_2'}$$
$$\left. + \frac{\big((-u_+)^{\lambda_+'} - (-1)^{N_+'} u_+^{\lambda_+'}\big)\big((-u_-)^{\lambda_-'} - (-1)^{N_+'} u_-^{\lambda_-'}\big)}{\sin \pi\lambda_+' \sin \pi\lambda_-'} \right).$$

8. Differential equations I

For each Cartan subgroup T of G, the orbital integrals $J_G(t, f)$ satisfy certain differential equations[8] as functions of $t \in T \cap G_{\mathrm{reg}}$. As a consequence, the Fourier transforms $\Phi_G(t, \tau)$ with fixed $\tau \in T_{\mathrm{ell}}(L)$ are eigenfunctions of the centre of the universal enveloping algebra of $\mathfrak{g}_{\mathbb{C}}$. The eigenvalues are given by the infinitesimal character of τ^G, which equals $W_{\mathfrak{t}_{\mathbb{C}}} O$, where the $W_{\mathfrak{t}_{\mathbb{C}}}^{\mathfrak{l}_{\mathbb{C}}}$-orbit $O \subset \mathfrak{t}_{\mathbb{C}}^*$ is the infinitesimal character of τ. If all $\nu \in W_{\mathfrak{t}_{\mathbb{C}}} O$ are regular, the pull-back of the corresponding eigenfunction under $\exp : \mathfrak{t} \to T$ is locally a linear combination of the functions $X \mapsto e^{\nu(X)}$.

As Arthur[3] has shown, the differential equations remain valid for $\Phi_M(t, \tau)$ with a Levi subgroup M containing T provided $\Phi_{M'}(t, \tau)$ vanishes identically for all Levi subgroups M' properly containing M. For our group G, this is the case for $M = M_1$, $L = M_2$ and also for $M = M_2$, $L = M_1$. Those differential equations are also valid for the functions $\Phi_{M,\Sigma}$, which differ by a locally constant factor only. We also know that these functions are W_M-equivariant and tempered as distributions on T.

We have to parametrise infinitesimal characters for our group G. Clearly, t_1 and t_2 can serve as coordinates on $\mathfrak{h}$, and θ_1, θ_2 as coordinates on $\mathfrak{b}$. If we write $|a| = e^t$ and $c = e^s$, the functions t, θ are coordinates on $\mathfrak{t}_1$, while s, φ are coordinates on $\mathfrak{t}_2$. Let us abbreviate the linear functional

$\lambda_1 t_1 + \lambda_2 t_2$	on $\mathfrak{h}_{\mathbb{C}}$ by	(λ_1, λ_2),
$li\theta + \lambda t$	on $\mathfrak{t}_{1,\mathbb{C}}$ by	(l, λ),
$mi\varphi + \mu s$	on $\mathfrak{t}_{2,\mathbb{C}}$ by	(m, μ),
$l_1 i\theta_1 + l_2 i\theta_2$	on $\mathfrak{b}_{\mathbb{C}}$ by	(l_1, l_2).

With the parametrisations of representations introduced in section 2, the infinitesimal character

$$\begin{aligned}
&\text{of } \chi \text{ is } \{(\lambda_1,\lambda_2)\},\\
&\text{of } \sigma_1 \text{ is } \{(\pm l,\lambda)\},\\
&\text{of } \sigma_2 \text{ is } \{(\pm m,\mu)\},\\
&\text{of } \pi \text{ is } \{(\pm l_1,\pm l_2),(\pm l_2,\pm l_1)\}.
\end{aligned}$$

The infinitesimal characters of τ_1, τ' and τ'' are obtained by setting the appropriate variables l or l_i equal to zero.

There is an element of $G_{\mathbb{C}}$ which conjugates $\mathfrak{t}_{1,\mathbb{C}}$ to $\mathfrak{t}_{2,\mathbb{C}}$ and identifies t with $s - i\varphi$ and $i\theta$ with $s + i\varphi$. This allows us to carry the infinitesimal character of σ_1^G from $\mathfrak{t}_{1,\mathbb{C}}^*$ to $\mathfrak{t}_{2,\mathbb{C}}^*$ and that of σ_2^G from $\mathfrak{t}_{2,\mathbb{C}}^*$ to $\mathfrak{t}_{1,\mathbb{C}}^*$. The upshot is that

$$\begin{aligned}
\Phi_{M_2,\Sigma_2}(\sigma_1) = {}& c_1 e^{(\lambda-|l|)|s|+(-\lambda-|l|)i\varphi} + c_2 e^{(-\lambda-|l|)|s|+(\lambda-|l|)i\varphi}\\
&+ c_3 e^{(\lambda-|l|)|s|+(\lambda+|l|)i\varphi} + c_4 e^{(-\lambda-|l|)|s|+(-\lambda+|l|)i\varphi}
\end{aligned}$$

with locally constant functions c_i on $\{t_2 \in T_2 : r \neq \pm 1\}$, the same formula with $l = 0$ being valid for $\Phi_{M_2,\Sigma_2}(\tau_1)$, and

$$\begin{aligned}
\Phi_{M_1,\Sigma_1}(\sigma_2) = {}& c_1' e^{(\mu-|m|)|t|/2+(\mu+|m|)i\theta/2} + c_2' e^{(-\mu-|m|)|t|/2+(\mu-|m|)i\theta/2}\\
&+ c_3' e^{(-\mu-|m|)|t|/2+(-\mu+|m|)i\theta/2} + c_4' e^{(\mu-|m|)|t|/2+(-\mu-|m|)i\theta/2}
\end{aligned}$$

with locally constant functions c_i' on $\{t_1 \in T_1 : u \neq \pm 1\}$.

9. Jump relations I

In order to determine the functions c_i and c_i' in the preceding formulas, we need the jump relations[2] satisfied by weighted orbital integrals at an intersection of two Cartan subgroups which has codimension one in both of them. In our situation, those intersections are characterised as follows.

$$\begin{aligned}
B \cap T_1 &: \quad u_1 = \operatorname{sgn} a, \quad u_2 = u.\\
B' \cap T_2 &: \quad u_1 = u_2^{-1} = r, \quad c = 1.\\
T_1 \cap H &: \quad a = a_1, \quad u = \operatorname{sgn} a_2.\\
T_2 \cap H &: \quad c = |a_1| = |a_2|, \quad r = \operatorname{sgn} a_1 = \operatorname{sgn} a_2.
\end{aligned}$$

Here $B' = gBg^{-1}$ is a Cartan subgroup to which we transport functions from B by conjugation with

$$g = g^{-1} = \frac{1}{\sqrt{2}} \begin{pmatrix} 1 & & & 1 \\ & 1 & 1 & \\ & 1 & -1 & \\ 1 & & & -1 \end{pmatrix}.$$

In each of the four cases, there is a Cayley transform, i.e. an inner automorphism of $\mathfrak{g}_{\mathbb{C}}$ mapping one complexified Cartan subalgebra to the other while fixing their intersection pointwise. In section 1, we have chosen positive roots for our Cartan subgroups, and we choose Σ' for B' compatibly. We require that under a Cayley transform those positive roots of the two Cartan subalgebras involved which vanish on their intersection correspond to each other. Then that transform is uniquely determined. Note that, under each of our four Cayley transforms, the half-systems of positive roots chosen on the two Cartan subalgebras correspond to each other.

The jump relations connect limits of derivatives of weighted orbital integrals on the two Cartan subgroups. We need only first derivatives obtained by applying the coroots corresponding to the positive roots singled out above. For our four Cayley transforms, these are

$$\partial_0^1 = \frac{\partial}{i\partial\theta_1} = u_1\frac{\partial}{\partial u_1} \text{ on } B \quad \text{and} \quad \partial_1^0 = \frac{\partial}{\partial t} = a\frac{\partial}{\partial a} \text{ on } T_1,$$

$$\partial_0^2 = \frac{\partial}{i\partial\theta_1} + \frac{\partial}{i\partial\theta_2} = u_1\frac{\partial}{\partial u_1} + u_2\frac{\partial}{\partial u_2} \text{ on } B \quad \text{and} \quad \partial_2^0 = \frac{\partial}{\partial s} = c\frac{\partial}{\partial c} \text{ on } T_2,$$

$$\partial_1 = \frac{\partial}{i\partial\theta} = u\frac{\partial}{\partial u} \text{ on } T_1 \quad \text{and} \quad \partial^1 = \frac{\partial}{\partial t_2} = a_2\frac{\partial}{\partial a_2} \text{ on } H,$$

$$\partial_2 = \frac{\partial}{i\partial\varphi} = r\frac{\partial}{\partial r} \text{ on } T_2 \quad \text{and} \quad \partial^2 = \frac{\partial}{\partial t_1} - \frac{\partial}{\partial t_2} = a_1\frac{\partial}{\partial a_1} - a_2\frac{\partial}{\partial a_2} \text{ on } H.$$

We interpret ∂_*^* as complex vector fields along the intersection of the two Cartan subgroups. When applying them to a function, we understand that the limit from above is taken with respect to the appropriate variable, viz.

$$\begin{aligned} &\theta_1 \text{ and } t \text{ at } B \cap T_1, \\ &\theta_1 + \theta_2 \text{ and } s \text{ at } B \cap T_2, \\ &\theta \text{ and } t_2 \text{ at } T_1 \cap H, \\ &\varphi \text{ and } t_1 - t_2 \text{ at } T_2 \cap H. \end{aligned}$$

The easier type of jump relations connects the values of a weighted orbital integral I_M on two Cartan subgroups T and T' contained in M, see Lemma 6.3 in Arthur's paper.[2] It implies a relation between the Fourier transforms $\Phi_M|_T$ and $\Phi_M|_{T'}$. This is useful for cross-checking our formulas but not for determining new instances of Fourier transforms.

The other type of jump relations, see Theorem 6.1 in Arthur's paper,[2] connects the values of I_M on a Cartan subgroup $T \subset M$ with those of $I_{M'}$ on a Cartan subgroup $T' \not\subset M$, where M' is the smallest Levi subgroup containing both M and T'. The corresponding relation for Fourier transforms on a component $T_{\text{disc}}(L)$ simplifies if $\Phi_{M'}|_{T\times T_{\text{disc}}(L)}$ vanishes. In our situation, because of (16), this yields the following relations:

$$\begin{aligned}
\partial_0^2\Phi_{G,\Sigma}(\sigma_1) &= \partial_2^0\Phi_{M_2,\Sigma_2}(\sigma_1),\\
\partial_0^2\Phi_{G,\Sigma}(\tau_1) &= \partial_2^0\Phi_{M_2,\Sigma_2}(\tau_1),\\
\partial_0^1\Phi_{G,\Sigma}(\sigma_2) &= \partial_1^0\Phi_{M_1,\Sigma_1}(\sigma_2).
\end{aligned}$$

They allow us to finish the calculation started in the preceding section.

Theorem 9.1. *On* $T_2\cap G_{\text{reg}}$, *we have*

$$\begin{aligned}
\Phi_{M_2,\Sigma_2}(\sigma_1) &= -\lfloor c^l\rfloor\sum_{n\equiv N}\Big(\frac{1}{n+\lambda}\big(r^{n+|l|}\lfloor c\rfloor^{\lambda}+r^{n-|l|}\lfloor c\rfloor^{-\lambda}\big)\\
&\qquad\qquad+\frac{1}{n-\lambda}\big(r^{n-|l|}\lfloor c\rfloor^{\lambda}+r^{n+|l|}\lfloor c\rfloor^{-\lambda}\big)\Big)\\
&= \frac{\pi}{2\sin\pi\lambda}\lfloor c^l\rfloor\Big(\big((-r)^{\lambda}+(-1)^N r^{\lambda}\big)\big(r^{|l|}\lfloor c\rfloor^{-\lambda}+r^{-|l|}\lfloor c\rfloor^{\lambda}\big)\\
&\qquad-\big((-r)^{-\lambda}+(-1)^N r^{-\lambda}\big)\big(r^{|l|}\lfloor c\rfloor^{\lambda}+r^{-|l|}\lfloor c\rfloor^{-\lambda}\big)\Big),
\end{aligned}$$

and $\Phi_{M_2,\Sigma_2}(\tau_1)=0$. *Moreover, on* $T_1\cap G_{\text{reg}}$, *we have*

$$\begin{aligned}
\Phi_{M_1,\Sigma_1}(\sigma_2) &= \sum_n\Bigg(\Big(\frac{2}{2n+|m|+\mu}+\frac{2}{2n-|m|-\mu}\Big)\lfloor|a|\rfloor^{(|m|-\mu)/2}\\
&\quad+\Big(\frac{2}{2n+|m|-\mu}+\frac{2}{2n-|m|+\mu}\Big)\lfloor|a|\rfloor^{(|m|+\mu)/2}\Bigg)(-u\,\mathrm{sgn}\,a)^n\\
&= S_{(|m|+\mu)/2}(u\,\mathrm{sgn}\,a)\lfloor|a|\rfloor^{(|m|-\mu)/2}+S_{(|m|-\mu)/2}(u\,\mathrm{sgn}\,a)\lfloor|a|\rfloor^{(|m|+\mu)/2}.
\end{aligned}$$

Proof. In the case of $\Phi_{M_2,\Sigma_2}(\sigma_1)$ and $\Phi_{M_2,\Sigma_2}(\tau_1)$, we deduce from the jump relations that

$$\begin{aligned}
c_1e^{-\lambda i\varphi} &= c_4e^{-\lambda i\varphi} = -\frac{\pi}{2\sin\pi\lambda}\big((-r)^{-\lambda}+(-1)^N r^{-\lambda}\big),\\
c_2e^{\lambda i\varphi} &= c_3e^{\lambda i\varphi} = \frac{\pi}{2\sin\pi\lambda}\big((-r)^{\lambda}+(-1)^N r^{\lambda}\big).
\end{aligned}$$

This yields the second expression given in the Theorem, while the first expression is its Fourier expansion in r.

In case of $\Phi_{M_1,\Sigma_1}(\sigma_2)$ we get a slightly more complicated expression on $B\cap T_1$ in terms of the variables $u_\pm$, which are determined only up to simultaneous sign changes by the conditions $u_+u_-=\mathrm{sgn}\,a$ and $u_+u_-^{-1}=u$. If one determines the nth Fourier coefficient by integrating against u^{-n}, the resulting series turns out to be expressible in terms of S again. In hindsight this could have been deduced from the identity

$$\left(z^m(-z)^\mu + (-z)^m z^\mu\right) \sin \pi \frac{m+\mu}{2} = \left(-z^2\right)^{\frac{m+\mu}{2}} \sin \pi\mu.$$

We omit the details of the computation. □

10. Differential equations II

For each Cartan subgroup T of G, the weighted orbital integrals $J_M(t, f)$ and the associated invariant distributions $I_M(t, f)$ for varying Levi subgroup $M \supset T$ satisfy systems of differential equations[3] as functions of $t \in T \cap G_{\mathrm{reg}}$, which generalise Harish-Chandra's differential equations for $J_G(t, f)$. They give rise to systems of differential equations satisfied by the Fourier transforms $\Phi_M(t, \tau)$ for $\tau \in T_{\mathrm{disc}}(L)$, which depend only on the infinitesimal character O of τ^G. By studying asymptotics at infinity, we[13] have singled out, for each parabolic subgroup P containing T, standard solutions of those differential equations pulled back under the exponential map to

$$\mathfrak{t}_P = \{X \in \mathfrak{t} : \operatorname{Re} \alpha(X) > 0 \text{ for all roots } \alpha \text{ of } \mathfrak{t}_{\mathbb{C}} \text{ in } \mathfrak{n}_{\mathbb{C}}\},$$

where N denotes the unipotent radical of P. They are of the form $e^{\nu(X)} F_M^{P,\nu}(\exp X)$ ($\nu \in O$ and M containing a Levi component of P), where $F_G^{P,\nu} = 1$, while $F_M^{P,\nu}$ is exponentially decreasing for $M \neq G$ as $\operatorname{Re} \alpha(X) \to \infty$ for all α as above. We have also seen that, for sufficiently regular O, any solution Φ_M on a connected component of the intersection of $T_P = \exp \mathfrak{t}_P$ with G_{reg} is of the form

$$\Phi_M = \sum_{\nu \in O} \sum_{M' \supset M} e_{M'}^{P,\nu} F_M^{P \cap M',\nu} \tag{22}$$

where $X e_{M'}^{P,\nu} = \nu(X) e_{M'}^{P,\nu}$ for all $X \in \mathfrak{t}_{\mathbb{C}}$ and where M' runs through the Levi subgroups of G containing M.

In our present group G, let P_1, P_2 and P be the smallest parabolic subgroups whose complexified Lie algebras contain the root spaces of Σ_1, Σ_2 and Π, respectively. Then M_i is a Levi component of P_i and H is a Levi component of P. In these cases, the aforementioned subsets of Cartan subgroups are characterised as follows:

H_{P_1}	by	$\lvert a_1 a_2\rvert > 1$, $\lvert a_1 a_2^{-1}\rvert > 1$,	T_{1,P_1}	by	$\lvert a\rvert > 1$,
H_{P_2}	by	$\lvert a_1\rvert > 1$, $\lvert a_2\rvert > 1$,	T_{2,P_2}	by	$c > 1$,
H_P	by	$\lvert a_1\rvert > \lvert a_2\rvert > 1$.			

When we apply this discussion to our Levi subgroups, we can characterise

$$H_{P\cap M_1} \quad \text{by} \quad |a_2| > 1,$$
$$H_{P\cap M_2} \quad \text{by} \quad |a_1| > |a_2|.$$

In order to describe the nontrivial standard solutions, we use the special function

$$b(s,z) = \sum_{n=1}^{\infty} \frac{z^n}{n+s} = z\int_0^1 \frac{t^s}{1-zt}\,dt$$

for $|z| < 1$ and $\operatorname{Re} s > -1$.

Let us start with the Levi subgroups. The infinitesimal character of $\chi \in \Pi(H)$ with respect to $\mathfrak{h}_{\mathbb{C}}$ is given by the pair (λ_1, λ_2), and we have

$$F_H^{P\cap M_1,(\lambda_1,\lambda_2)} = b(-\lambda_2, a_2^{-2}),$$
$$F_H^{P\cap M_2,(\lambda_1,\lambda_2)} = 2b(-\lambda_-, a_1^{-1}a_2)$$

on $H_{P\cap M_1}$ resp. $H_{P\cap M_2}$, where the factor 2 is the length of the coroot of $\mathfrak{h}$ in $\mathfrak{m}_2$ with respect to the quotient measure on H/A_2.

For H as a Cartan subgroup of G, we have the functions

$$F_{M_1}^{P_1,(\lambda_1,\lambda_2)} = b(-\lambda_1, a_1^{-2}) + b(-\lambda_+, a_1^{-1}a_2^{-1}) + b(-\lambda_-, a_1^{-1}a_2),$$
$$F_{M_2}^{P_2,(\lambda_1,\lambda_2)} = b(-\lambda_+, a_1^{-1}a_2^{-1}) + \tfrac{1}{2}b(-\lambda_1, a_1^{-2}) + \tfrac{1}{2}b(-\lambda_2, a_2^{-2})$$

on H_{P_1} resp. H_{P_2} (which will not be needed thanks to descent formulas), and $F_H^{P,(\lambda_1,\lambda_2)}$, whose bulky formula we will write down at the end of this section.

Next we come to the Cartan subgroups T_i. The infinitesimal character of $\sigma_1 \in \Pi_2(M_i)$ with respect to $\mathfrak{t}_{i,\mathbb{C}}$ is given by (l, λ) for $i = 1$ and by (m, μ) for $i = 2$, and we have

$$F_{M_1}^{P_1,(l,\lambda)} = b(-\lambda, a^{-2}) + b(-\lambda - l, u^{-1}a^{-1}) + b(-\lambda + l, ua^{-1}),$$
$$F_{M_2}^{P_2,(m,\mu)} = b(-\mu, c^{-2}) + \tfrac{1}{2}b(-\tfrac{\mu+m}{2}, r^{-2}c^{-2}) + \tfrac{1}{2}b(-\tfrac{\mu-m}{2}, r^2c^{-2})$$

on T_{1,P_1} resp. T_{2,P_2}.

When determining the functions $e_{M',\nu}^P$, we exploit their independence of M. E. g., in the case of the Levi subgroups M_i, we conclude from our knowledge of $\Phi_{M_i}^{M_i}(\chi)$ that

$$\begin{aligned}\Phi_H^{M_1}(\chi) &= \operatorname{sgn}(a_1)^{N_1}|a_1|^{\lambda_1}\operatorname{sgn}(a_2)^{N_2}|a_2|^{\lambda_2}\big(b(-\lambda_2, a_2^{-2}) + c_{M_1}^{P_1}(\chi)\big)\\ &\quad + \operatorname{sgn}(a_1)^{N_1}|a_1|^{\lambda_1}\operatorname{sgn}(a_2)^{N_2}|a_2|^{-\lambda_2}\big(b(\lambda_2, a_2^{-2}) + c_{M_1}^{P_1}(\chi_1)\big),\\ \Phi_H^{M_2}(\chi) &= 2\operatorname{sgn}(a_1)^{N_1}|a_1|^{\lambda_1}\operatorname{sgn}(a_2)^{N_2}|a_2|^{\lambda_2}\big(b(-\lambda_-, a_1^{-1}a_2) + c_{M_2}^{P_2}(\chi)\big)\\ &\quad + 2\operatorname{sgn}(a_1)^{N_2}|a_1|^{\lambda_2}\operatorname{sgn}(a_2)^{N_1}|a_2|^{\lambda_1}\big(b(\lambda_-, a_1^{-1}a_2) + c_{M_2}^{P_2}(\chi_2)\big)\end{aligned}$$

on $H_{P_1} \cap M_{1,\mathrm{reg}}$ resp. $H_{P_2} \cap M_{2,\mathrm{reg}}$, where χ_i is obtained by applying the nontrivial element of $W_H^{M_i}$ to χ.

Here is the promised formula, valid on H_P:

$$\begin{aligned} F_H^{P,(\lambda_1,\lambda_2)} = & \sum_{n_2 \ge n_1 \ge 1} \frac{(a_1 a_2^{-1})^{-n_1} (a_1 a_2)^{-n_2}}{(n_1 - \lambda_1)(n_2 - \lambda_2)} + \sum_{n_1 \ge n_2 \ge 1} \frac{2(a_1^2)^{-n_1} (a_2^2)^{-n_2}}{(2n_1 - \lambda_-)(2n_2 - \lambda_+)} \\ & + \sum_{n_1, n_2 \ge 1} \frac{(a_1 a_2^{-1})^{-n_1} (a_1^2)^{-n_2}}{(n_1 - \lambda_-)(n_2 - \lambda_2)} + \sum_{n_1 \ge 2n_2 \ge 1} \frac{(a_1 a_2^{-1})^{-n_1} (a_2^2)^{-n_2}}{(n_1 - \lambda_-)(n_2 - \lambda_1)} \\ & + \sum_{n_1, n_2 \ge 1} \frac{(a_1 a_2)^{-n_1} (a_2^2)^{-n_2}}{(n_1 - \lambda_-)(n_2 - \lambda_2)} + \sum_{n_2 \ge n_1 \ge 1} \frac{(a_1 a_2^{-1})^{-n_1} (a_2^2)^{-n_2}}{(n_1 - \lambda_+)(n_2 - \lambda_2)} \\ & + \sum_{n_1, n_2 \ge 1} \frac{2(a_1 a_2^{-1})^{-n_1} (a_1 a_2)^{-n_2}}{(n_1 - \lambda_-)(n_2 - \lambda_+)} + \sum_{n_1, n_2 \ge 1} \frac{(a_1^2)^{-n_1} (a_2^2)^{-n_2}}{(n_1 - \lambda_1)(n_2 - \lambda_2)}. \end{aligned}$$

11. Normalising factors

Given two parabolic subgroups P and P' of G with the same Levi component M, a representation $\sigma \in \Pi(M)$, and a complex-valued linear function ζ on $\mathfrak{a}_M$ whose real part is positive on $\mathfrak{a}_P^+ \cap \mathfrak{a}_{P'}^-$, one defines the Knapp-Stein intertwining operator $J_{P'|P}(\sigma_\zeta)$ from the space of smooth vectors of $\mathrm{Ind}_P^G(\sigma_\zeta)$ to that of $\mathrm{Ind}_{P'}^G(\sigma_\zeta)$ by a convergent integral. It can be analytically continued to a meromorphic function of ζ. There are scalars $r_{P'|P}(\sigma_\zeta)$, which depend meromorphically on ζ, such that the the normalised intertwining operators

$$R_{P'|P}(\sigma_\zeta) = r_{P'|P}(\sigma_\zeta)^{-1} J_{P'|P}(\sigma_\zeta)$$

are holomorphic at $\zeta = 0$ and have the properties

$$R_{P''|P'}(\sigma) R_{P'|P}(\sigma) = R_{P''|P}(\sigma), \qquad R_{P'|P}(\sigma)^* = R_{P|P'}(\check{\sigma}).$$

Let θ_P be the product of the fundamental coroots of $\mathfrak{a}_M$ in the unipotent radical of P, considered as a polynomial function on $\mathfrak{a}_M^*$ and so normalised that

$$v_M(x) = \lim_{\zeta \to 0} \sum_P \frac{e^{-\zeta(H_P(x))}}{\theta_P(\zeta)},$$

where the sum is over all parabolic subgroups with Levi component M. One can then consider the unbounded operator

$$R_M(\sigma, P_0) = \lim_{\zeta \to 0} \sum_P \frac{R_{P|P_0}(\sigma)^{-1} R_{P|P_0}(\sigma_\zeta)}{\theta_P(\zeta)}$$

in the space of $\mathrm{Ind}_{P_0}^G(\sigma_\zeta)$, realised as a space of functions on K independent of ζ. For $f \in \mathcal{C}(G)$, the weighted character[4]

$$\phi_M(f,\sigma) = \mathrm{tr}(R_M(\sigma, P_0)\,\mathrm{Ind}_{P_0}^G(\sigma, f)),$$

which is used in (1), does not depend on P_0. The function

$$r_M(\sigma, P_0) = \lim_{\zeta\to 0} \sum_P \frac{r_{P|P_0}(\sigma)^{-1} r_{P|P_0}(\sigma_\zeta)}{\theta_P(\zeta)} \tag{23}$$

will also play a role in our results.

There are several choices for the normalising factors $r_{P'|P}$, which affect the definition of the distributions I_M. We choose the normalising factors that are given as quotients of local L-factors as suggested by Langlands.[17] (One may pass to different normalising factors with the aid of the product formula for (G,M)-families.) The L-factor of a representation is given in terms of its Langlands parameter. To describe the latter, we realise the L-group of our rank two split symplectic group G as ${}^LG = \check{G} \times W_{\mathbb{R}}$, where $W_{\mathbb{R}}$ is the Weil group of the field $\mathbb{R}$ and $\check{G} \cong \mathrm{SO}_5$ is the orthogonal group consisting of all complex 5×5 matrices g such that $g^\top \check{J} g = \check{J}$, for

$$\check{J} = \begin{pmatrix} & & & & 1 \\ & & & 1 & \\ & & 1 & & \\ & 1 & & & \\ 1 & & & & \end{pmatrix}.$$

The maximal torus $\check{T}$ consisting of all diagonal matrices in $\check{G}$ can be identified with each of the groups $\check{H}$, $\check{T}_1$, $\check{T}_2$ and $\check{B}$ in such a way that their respective half-system Π, Σ_1, Σ_2, Σ of positive roots is identified with the half-system

$$\{\mathrm{diag}(2,0,0,0,-2), \mathrm{diag}(0,2,0,-2,0), \\ \mathrm{diag}(1,1,0,-1,-1), \mathrm{diag}(1,-1,0,1,-1)\} \subset \check{\mathfrak{t}}$$

of positive coroots for $\check{T}$. The duals of our chosen Levi subgroups are then ${}^LH = \check{T} \times W_{\mathbb{R}}$ and ${}^LM_i = \check{M}_i \times W_{\mathbb{R}}$, where $\check{M}_1 \cong \mathrm{GL}_1 \times \mathrm{SO}_3 \cong \mathrm{GL}_1 \times \mathrm{PSL}_2$ and $\check{M}_2 \cong \mathrm{GL}_2 \times \mathrm{SO}_1 \cong \mathrm{GL}_2$ are the subgroups of $\check{G}$ consisting of all matrices of the form

$$\begin{pmatrix} * & & & & \\ & * & * & * & \\ & * & * & * & \\ & * & * & * & \\ & & & & * \end{pmatrix} \quad \text{resp.} \quad \begin{pmatrix} * & * & & & \\ * & * & & & \\ & & * & & \\ & & & * & * \\ & & & * & * \end{pmatrix}.$$

Given a packet of square-integrable representations of one of our Levi subgroups M, we can choose its Langlands parameter $\phi : W_{\mathbb{R}} \to {}^L M$ such that its image is contained in the dual of a fixed Cartan subgroup of M. We identify the trivial connected component of $W_{\mathbb{R}}$ with $\mathbb{C}^\times$ and fix an element ς of the other connected component such that $\varsigma^2 = -1$. Then ϕ is determined by the values $\phi(\varsigma)$ and $\phi(z)$ for $z \in \mathbb{C}^\times$. Since the composition of ϕ with the natural projection ${}^L M \to W_{\mathbb{R}}$ must be the identity and since our Cartan subgroups are split, it suffices to give the composition $\tilde{\phi}$ of ϕ with the projection ${}^L M \cong \check{M} \times W_{\mathbb{R}} \to \check{M}$.

We denote the Langlands parameters of $\chi \in \Pi(H)$, $\sigma_1 \in \Pi_2(M_1)$ and $\sigma_2 \in \Pi_2(M_2)$ by

$$\phi : W_{\mathbb{R}} \to {}^L H, \quad \phi_1 : W_{\mathbb{R}} \to {}^L T_1 \quad \text{resp.} \quad \phi_2 : W_{\mathbb{R}} \to {}^L T_2,$$

while we ignore that of $\pi \in \Pi_2(G)$ as it will be of no importance to us. According to Langlands' classification,[18] we can take

$$\begin{aligned}
\tilde{\phi}(z) &= \operatorname{diag}\big((z\bar{z})^{\lambda_1}, (z\bar{z})^{\lambda_2}, 1, (z\bar{z})^{\lambda_2}, (z\bar{z})^{\lambda_1}\big),\\
\tilde{\phi}_1(z) &= \operatorname{diag}\big((z\bar{z})^{\lambda}, (z/\bar{z})^{l}, 1, (z/\bar{z})^{-l}, (z\bar{z})^{-\lambda}\big),\\
\tilde{\phi}_2(z) &= \operatorname{diag}\big(|z|^{\mu+m}\bar{z}^{-m}, |z|^{\mu-m}\bar{z}^{m}, 1, |z|^{-\mu+m}\bar{z}^{-m}, |z|^{-\mu-m}\bar{z}^{m}\big),\\
\tilde{\phi}(\varsigma) &= \operatorname{diag}\big((-1)^{N_1}, (-1)^{N_2}, 1, (-1)^{N_2}, (-1)^{N_1}\big),
\end{aligned}$$

$$\tilde{\phi}_1(\varsigma) = \begin{pmatrix} (-1)^N & & & & \\ & 0 & & 1 & \\ & & -1 & & \\ & 1 & & 0 & \\ & & & & (-1)^N \end{pmatrix}, \quad \tilde{\phi}_2(\varsigma) = \begin{pmatrix} 0 & 1 & & & \\ (-1)^m & 0 & & & \\ & & 1 & & \\ & & & 0 & (-1)^m \\ & & & 1 & 0 \end{pmatrix}.$$

In Arthur's notation[4] we have to identify, for two parabolic subgroups P, P' with the same Levi component M, the adjoint representation $\rho_{P'|P}$ of ${}^L M$ on ${}^L\mathfrak{n}_{P'} \cap {}^L\mathfrak{n}_P \backslash {}^L\mathfrak{n}_{P'}$, where ${}^L\mathfrak{n}_P$ is the Lie algebra of the unipotent radical of ${}^L P$. Actually we need the contragredient representation $\tilde{\rho}_{P'|P}$, which is the adjoint representation of ${}^L M$ on $\mathfrak{n}_P \cap {}^L\mathfrak{n}_{P'} \backslash {}^L\mathfrak{n}_P$. We write the elements of $\check{T}$ in the form

$$\operatorname{diag}(t_1, t_2, 1, t_2^{-1}, t_1^{-1}),$$

thus defining functions t_1 and t_2 on $\check{T}$. Here are the root characters of $\check{T}$ in the unipotent radical of the dual of each of the parabolic subgroups fixed in section 10:

t_2	for $P \cap M_1$,
$t_1 t_2^{-1}$	for $P \cap M_2$,
$t_1,\ t_1t_2,\ t_1t_2^{-1}$	for P_1,
$t_1,\ t_2,\ t_1t_2$	for P_2,
$t_1,\ t_2,\ t_1t_2,\ t_1t_2^{-1}$	for P.

This allows us, for each of the above parabolics P, to decompose $\tilde{\rho}_{\bar{P}|P} \circ \phi$ into irreducible representations of $W_{\mathbb{R}}$:

		z	ς
ϕ	$P \cap M_1$	$(z\bar{z})^{\lambda_2}$	$(-1)^{N_2}$
ϕ	$P \cap M_2$	$(z\bar{z})^{\lambda_-}$	$(-1)^{N_+}$
ϕ	P_1	$(z\bar{z})^{\lambda_1}$	$(-1)^{N_1}$
		$(z\bar{z})^{\lambda_+}$	$(-1)^{N_+}$
		$(z\bar{z})^{\lambda_-}$	$(-1)^{N_+}$
ϕ	P_2	$(z\bar{z})^{\lambda_1}$	$(-1)^{N_1}$
		$(z\bar{z})^{\lambda_2}$	$(-1)^{N_2}$
		$(z\bar{z})^{\lambda_+}$	$(-1)^{N_+}$
ϕ_1	P_1	$(z\bar{z})^{\lambda}$	$(-1)^{1-N}$
		$(z\bar{z})^{\lambda}(z/\bar{z})^{\pm l}$	transp.
ϕ_2	P_2	$(z\bar{z})^{\mu}$	$(-1)^{1-M}$
		$\lvert z\rvert^{\mu\pm m}\bar{z}^{\mp m}$	transp.

Here $M \in \{0,1\}$ is determined by $M \equiv m \pmod 2$. Each row corresponds to an irreducible representation. Numbers are eigenvalues of the group element in the column header, while the specification "transp." indicates that ς transposes the two eigenspaces of z.

With this information, we can finally write down the normalising factors in terms of the functions

$$\Gamma_{\mathbb{R}}(s) = \pi^{-s/2}\Gamma(s/2), \qquad \Gamma_{\mathbb{C}}(s) = 2(2\pi)^{-s}\Gamma(s),$$

namely

$$r_{\overline{P\cap M_1}|P\cap M_1}(\chi) = \frac{\Gamma_{\mathbb{R}}(\lambda_2 + N_2)}{\Gamma_{\mathbb{R}}(\lambda_2 + N_2 + 1)},$$

$$r_{\overline{P\cap M_2}|P\cap M_2}(\chi) = \frac{\Gamma_{\mathbb{R}}(\lambda_- + N_+)}{\Gamma_{\mathbb{R}}(\lambda_- + N_+ + 1)}$$

in case of the Levi subgroups and

$$r_{\bar{P}_1|P_1}(\chi^{M_1}) = \frac{\Gamma_{\mathbb{R}}(\lambda_1+N_1)\Gamma_{\mathbb{R}}(\lambda_+ + N_+)\Gamma_{\mathbb{R}}(\lambda_- + N_+)}{\Gamma_{\mathbb{R}}(\lambda_1+N_1+1)\Gamma_{\mathbb{R}}(\lambda_+ + N_+ + 1)\Gamma_{\mathbb{R}}(\lambda_- + N_+ + 1)},$$
$$r_{\bar{P}_2|P_2}(\chi^{M_2}) = \frac{\Gamma_{\mathbb{R}}(\lambda_1+N_1)\Gamma_{\mathbb{R}}(\lambda_2 + N_2)\Gamma_{\mathbb{R}}(\lambda_+ + N_+)}{\Gamma_{\mathbb{R}}(\lambda_1+N_1+1)\Gamma_{\mathbb{R}}(\lambda_2 + N_2 + 1)\Gamma_{\mathbb{R}}(\lambda_+ + N_+ + 1)},$$
$$r_{\bar{P}_1|P_1}(\sigma_1) = \frac{\Gamma_{\mathbb{R}}(\lambda+1-N)\Gamma_{\mathbb{C}}(\lambda+|l|)}{\Gamma_{\mathbb{R}}(\lambda+2-N)\Gamma_{\mathbb{C}}(\lambda+|l|+1)},$$
$$r_{\bar{P}_2|P_2}(\sigma_2) = \frac{\Gamma_{\mathbb{R}}(\mu+1-M)\Gamma_{\mathbb{C}}(\frac{\mu+|m|}{2})}{\Gamma_{\mathbb{R}}(\mu+2-M)\Gamma_{\mathbb{C}}(\frac{\mu+|m|}{2}+1)},$$

in case of G. Now we can compute the limit (23) in various cases. With the notation

$$v(N,\lambda) = \frac{1}{2}\left(\psi\left(\frac{\lambda+N+1}{2}\right) - \psi\left(\frac{\lambda+N}{2}\right)\right) = \sum_{n\geq 0} \frac{(-1)^n}{n+N+\lambda},$$

where $\psi(s) = \Gamma'(s)/\Gamma(s)$, we have

$$\begin{aligned}
r_H^{M_1}(\chi, P\cap M_1) &= v(N_2,\lambda_2),\\
r_H^{M_2}(\chi, P\cap M_2) &= 2v(N_+,\lambda_-),\\
r_{M_1}(\chi^{M_1}, P_1) &= v(N_1,\lambda_1) + v(N_+,\lambda_+) + v(N_+,\lambda_-),\\
r_{M_2}(\chi^{M_2}, P_2) &= v(N_+,\lambda_+) + \tfrac{1}{2}v(N_1,\lambda_1) + \tfrac{1}{2}v(N_2,\lambda_2),\\
r_{M_1}(\sigma_1, P_1) &= v(1-N,\lambda) + \tfrac{1}{\lambda+|l|},\\
r_{M_1}(\tau_1, P_1) &= v(1-N,\lambda) + \tfrac{1}{\lambda},\\
r_{M_2}(\sigma_2, P_2) &= v(1-M,\mu) + \tfrac{1}{\mu+|m|}.
\end{aligned}$$

Let M_i' be the Levi subgroup of G conjugate to but different from M_i. Then

$$\begin{aligned}
r_H^{M_1'}(\chi, P\cap M_1') &= v(N_1,\lambda_1),\\
r_H^{M_2'}(\chi, P\cap M_2') &= 2v(N_+,\lambda_+).
\end{aligned}$$

Finally we state, without writing out the (G,H)-family, that

$$\begin{aligned}
r_H(\chi,P) = v(N_1,\lambda_1)v(N_2,\lambda_2) &+ v(N_1,\lambda_1)v(N_+,\lambda_+)\\
&+ v(N_1,\lambda_1)v(N_+,\lambda_-) + v(N_2,\lambda_2)v(N_+,\lambda_+)\\
&\qquad + v(N_2,\lambda_2)v(N_+,\lambda_-) + 2v(N_+,\lambda_+)v(N_+,\lambda_-).
\end{aligned}$$

12. The asymptotic formula

Arthur has proved an asymptotic formula[7] for weighted orbital integrals and for the associated invariant distributions I_M. We have deduced an asymptotic formula[15] for the Fourier transforms $\Phi_M(\tau)$, $\tau \in T_{\mathrm{disc}}(L)$, of those distributions on the subset T_P of a Cartan subgroup T. Now we consider the case that $\tau = \sigma \in \Pi_2(L)$. In the notation of (22), we have $e_{M'}^{P,\nu}(\sigma) = 0$ if $\mathrm{Re}\,\nu$ is positive at some point of $\mathfrak{a}_P^+$, and

$$\sum_{\substack{\nu \in O \\ \mathrm{Re}\,\nu|_{\mathfrak{a}_M}=0}} e_M^{P,\nu}(\sigma) = \sum_{[L',\sigma']_M} r_M(\check{\sigma}'^M, P)\Phi_M^M(\sigma'), \tag{24}$$

where the sum on the right-hand side is taken over all M-conjugacy classes of pairs (L', σ') which are G-conjugate to (L, σ) and such that $L' \subset M$.

If we apply this to the groups M_i in place of G, we can compute the missing values in the formulas for $\Phi_H^{M_i}(\chi)$ in section 10, namely

$$c_{M_1}^{P_1}(\chi) = r_H^{M_1}(\check{\chi}, P \cap M_1), \qquad c_{M_2}^{P_2}(\chi) = r_H^{M_2}(\check{\chi}, P \cap M_2).$$

This special case has already been treated by Arthur, Herb and Sally.[1] We can easily extend $\Phi_H^{M_i}(\chi)$ to $H \cap M_{i,\mathrm{reg}}$ using its $W_H^{M_i}$-invariance. In order to state the result, it is useful to introduce

$$\varepsilon_i = \mathrm{sgn}\log|a_i|, \qquad \varepsilon_\pm = \mathrm{sgn}\log a_\pm.$$

In fact, $\lfloor a_i \rfloor = a_i^{-\varepsilon_i}$ and $\lfloor a_1 a_2^{\pm 1} \rfloor = (a_1 a_2^{\pm 1})^{-\varepsilon_\pm}$. E. g., the $W_H^{M_1}$-invariant extension of

$$\sum_{\lambda_2'=\pm\lambda_2} \mathrm{sgn}(a_2)^{N_2}|a_2|^{\lambda_2'}\big(b(-\lambda_2', a_2^{-2}) + v(N_2, -\lambda_2')\big)$$

from H_{P_1} to $H \cap M_{1,\mathrm{reg}}$ is

$$\sum_{\lambda_2'=\pm\lambda_2} \mathrm{sgn}(a_2)^{N_2}|a_2|^{\lambda_2'}\big(b(-\varepsilon_2\lambda_2', \lfloor a_2^2 \rfloor) + v(N_2, -\varepsilon_2\lambda_2')\big).$$

We go on and deduce the following result with the aid of the descent identity (13).

Theorem 12.1. *As functions on $H \cap G_{\mathrm{reg}}$,*

$$\begin{aligned}\Phi_{M_1}(\chi) = \sum_{(N_1',\lambda_1',N_2',\lambda_2')} & \operatorname{sgn}(a_1)^{N_1'}|a_1|^{\lambda_1'}\operatorname{sgn}(a_2)^{N_2'}|a_2|^{\lambda_2'} \\ & \times\Bigg(\sum_{n\geq 1}\frac{\lfloor a_1^2\rfloor^n}{n-\varepsilon_1\lambda_1'}+\sum_{n\geq 0}\frac{(-1)^n}{n+N_1'-\varepsilon_1\lambda_1'} \\ & \quad+\sum_{n\geq 1}\frac{\lfloor a_1a_2\rfloor^n}{n-\varepsilon_+\lambda_+'}+\sum_{n\geq 0}\frac{(-1)^n}{n+N_+'-\varepsilon_+\lambda_+'} \\ & \quad+\sum_{n\geq 1}\frac{\lfloor a_1a_2^{-1}\rfloor^n}{n-\varepsilon_-\lambda_-'}+\sum_{n\geq 0}\frac{(-1)^n}{n+N_+'-\varepsilon_-\lambda_-'}\Bigg)\end{aligned}$$

and

$$\begin{aligned}\Phi_{M_2}(\chi) = \sum_{(N_1',\lambda_1',N_2',\lambda_2')} & \operatorname{sgn}(a_1)^{N_1'}|a_1|^{\lambda_1'}\operatorname{sgn}(a_2)^{N_2'}|a_2|^{\lambda_2'} \\ & \times\Bigg(\sum_{n\geq 1}\frac{\lfloor a_1a_2\rfloor^n}{n-\varepsilon_+\lambda_+'}+\sum_{n\geq 0}\frac{(-1)^n}{n+N_+'-\varepsilon_+\lambda_+'} \\ & \quad+\frac{1}{2}\Bigg(\sum_{n\geq 1}\frac{\lfloor a_1^2\rfloor^n}{n-\varepsilon_1\lambda_1'}+\sum_{n\geq 0}\frac{(-1)^n}{n+N_1'-\varepsilon_1\lambda_1'} \\ & \quad\quad+\sum_{n\geq 1}\frac{\lfloor a_2^2\rfloor^n}{n-\varepsilon_2\lambda_2'}+\sum_{n\geq 0}\frac{(-1)^n}{n+N_2'-\varepsilon_2\lambda_2'}\Bigg)\Bigg).\end{aligned}$$

In either case the exterior sum is over the eight quadruples that are obtained from $(N_1, \lambda_1, N_2, \lambda_2)$ by the action of W_H.

Note that Φ_{M_i} carries no subscript Π as it is more convenient here not to use (modified) Weyl denominators.

We can apply the asymptotic formula directly to our group G. Again, we need some sign functions:

$$\varepsilon = \operatorname{sgn}\log|a|, \qquad \delta = \operatorname{sgn}\log c.$$

Theorem 12.2.

$$\Phi_{M_1,\Sigma_1}(\chi) = -\sum_{(N_1',\lambda_1',N_2',\lambda_2')} \mathrm{sgn}(a)^{N_1'}|a|^{\lambda_1'} \sum_{n_2\equiv N_2'+1} \frac{u^{n_2}}{n_2-\lambda_2'}$$
$$\times\Bigg(\sum_{n_1\geq 1}\frac{\lfloor a^2\rfloor^{n_1}}{n_1-\varepsilon\lambda_1'}+\sum_{n_1\geq 0}\frac{(-1)^{n_1}}{n_1+N_1'-\varepsilon\lambda_1'}$$
$$+\sum_{n_+\geq 1}\frac{(\lfloor a\rfloor u^{-\varepsilon})^{n_+}}{n_+-\varepsilon\lambda_+'}+\sum_{n_+\geq 0}\frac{(-1)^{n_+}}{n_++N_+'-\varepsilon\lambda_+'}$$
$$+\sum_{n_-\geq 1}\frac{(\lfloor a\rfloor u^{\varepsilon})^{n_-}}{n_--\varepsilon\lambda_-'}+\sum_{n_-\geq 0}\frac{(-1)^{n_-}}{n_-+N_+'-\varepsilon\lambda_-'}\Bigg).$$

$$\Phi_{M_2,\Sigma_2}(\chi) = -\sum_{(N_1',\lambda_1',N_2',\lambda_2')} c^{\lambda_+'} \sum_{n_-\equiv N_+'+1} \frac{r^{n_-}}{n_--\lambda_-'}$$
$$\times\Bigg(\sum_{n_+\geq 1}\frac{\lfloor c^2\rfloor^{n_+}}{n_+-\delta\lambda_+'}+\sum_{n_+\geq 0}\frac{(-1)^{n_+}}{n_++N_+'-\delta\lambda_+'}$$
$$+\frac{1}{2}\Bigg(\sum_{n_1\geq 1}\frac{(\lfloor c\rfloor r^{-\delta})^{2n_1}}{n_1-\delta\lambda_1'}+\sum_{n_1\geq 0}\frac{(-1)^{n_1}}{n_1+N_1'-\delta\lambda_1'}$$
$$+\sum_{n_2\geq 1}\frac{(\lfloor c\rfloor r^{\delta})^{2n_2}}{n_2-\delta\lambda_2'}+\sum_{n_2\geq 0}\frac{(-1)^{n_2}}{n_2+N_2'-\delta\lambda_2'}\Bigg)\Bigg).$$

In either case the exterior sum is over the eight quadruples that are obtained from $(N_1,\lambda_1,N_2,\lambda_2)$ *by the action of* W_H.

Proof. In the presentation (22), we indicate the dependence on χ by writing $\Phi_{M_1}^{M_1}(\chi)$ and $e_{M'}^{P_1,\nu}(\chi)$ and modify both sides by replacing $|D|^{1/2}$ with $\tilde{\Delta}_{\Sigma_1}$, which we indicate by a subscript Σ_1. The functions $e_{G,\Sigma_1}^{P_1,\nu}(\chi)$ can be read off from our formula for $\Phi_{G,\Sigma_1}(\chi)$. This shows that, on $T_{1,P_1}\cap G_{\mathrm{reg}}$,

$$\Phi_{M_1,\Sigma_1}(\chi) = \sum_{(N_1',\lambda_1',N_2',\lambda_2')} \mathrm{sgn}(a)^{N_1'}|a|^{\lambda_1'}\frac{\pi\big((-u)^{\lambda_2'}-(-1)^{N_2'}u^{\lambda_2'}\big)}{2\sin\pi\lambda_2'}F_{M_1}^{P_1,(\lambda_1',\lambda_2')}$$
$$+\sum_{\nu\in O} e_{M_1,\Sigma_1}^{P_1,\nu}(\chi).$$

All of the elements of O have purely imaginary restrictions to $\mathfrak{a}_{M_1}$, hence the contribution of the functions $e_{M_1,\Sigma_1}^{P_1,\nu}(\chi)$ is given by the asymptotic formula (24) in terms of $\Phi_{M_1,\Sigma_1}^{M_1}(\chi)$ and $r_{M_1}(\check{\chi}^{M_1},P_1)$.

The proof for $\Phi_{M_2,\Sigma_2}(\chi)$ is analogous. □

By combining the asymptotic formula with descent identities for the functions $e_M^{P,\nu}$, we have deduced[15] that, as a function on H_P,

$$\Phi_H(\chi) = \sum_{w\in W_H} w\chi \sum_{M\supset H} r_H^M(w\check{\chi}^M, P\cap M)F_M^{P,w(\lambda_1,\lambda_2)}.$$

Thus, we obtain the following result.

Theorem 12.3. *As a function on* H_P,

$$\begin{aligned}\Phi_H(\chi) = \sum_{w\in W_H} w\chi\Big(F_H^{P,w\nu} &+ r_H^{M_1}(w\check{\chi}, P\cap M_1)F_{M_1}^{P,w\nu} \\ &+ r_H^{M_1'}(w\check{\chi}, P\cap M_1')F_{M_1'}^{P,w\nu} + r_H^{M_2}(w\check{\chi}, P\cap M_2)F_{M_2}^{P,w\nu} \\ &+ r_H^{M_2'}(w\check{\chi}, P\cap M_2')F_{M_2'}^{P,w\nu} + r_H(w\check{\chi}, P)\Big),\end{aligned}$$

where $\nu = (\lambda_1, \lambda_2)$ *is the infinitesimal character of* χ.

We leave to the reader the task of plugging in the bulky formulas.

13. Jump relations II

Now we come back to the jump relations that connect the values of the distributions I_M on two Cartan subgroups T and T' of G related by a Cayley transform. As in section 9, we assume that only T is contained in M, and denote by M' the smallest Levi subgroup of G containing both M and T'. In order that there be a limit as $t \in T$ approaches a semiregular point of $T_1 \cap T_2$, Arthur[2] has modified $J_M(t)$ by a multiple of $J_{M'}(t)$. In our case, this leads to the following modifications of the Fourier transforms:

$$\begin{aligned}\Phi^0_{M_1,\Sigma_1} &= \Phi_{M_1,\Sigma_1} + \Phi_{G,\Sigma_1}\log\left|a - a^{-1}\right|, \\ \Phi^0_{M_2,\Sigma_2} &= \Phi_{M_2,\Sigma_2} + \Phi_{G,\Sigma_2}\log\left|c - c^{-1}\right|, \\ \Phi^1_{H,\Pi} &= \Phi_{H,\Pi} + \Phi_{M_1,\Pi}\log\left|a_2 - a_2^{-1}\right|, \\ \Phi^2_{H,\Pi} &= \Phi_{H,\Pi} + 2\Phi_{M_2,\Pi}\log\left|a_- - a_-^{-1}\right|.\end{aligned}$$

In the notation of section 9, we have the additional jump relations

$$\partial_0^1\Phi_{G,\Sigma} = \partial_1^0\Phi^0_{M_1,\Sigma_1}, \qquad \partial_1\Phi_{M_1,\Sigma_1} = \partial^1\Phi^1_{H,\Pi},$$
$$\partial_0^2\Phi_{G,\Sigma} = \partial_2^0\Phi^0_{M_2,\Sigma_2}, \qquad \partial_2\Phi_{M_2,\Sigma_2} = \partial^2\Phi^2_{H,\Pi}.$$

Combined with the previous methods, they enable us to determine the following components of the Fourier transforms.

Theorem 13.1. *As a function on* $T_1 \cap G_{\text{reg}}$,

$$\Phi_{M_1,\Sigma_1}(\sigma_1) = -\sum_{\lambda'=\pm\lambda} \Bigg(\operatorname{sgn}(a) \lfloor a^l \rfloor \sum_{n \equiv N+1} \frac{u^n}{n-\lambda'}$$
$$+ \operatorname{sgn}(l) u^l \operatorname{sgn}(a)^N |a|^{\lambda'} \Bigg(\sum_{n\geq 1} \frac{\lfloor a \rfloor^{2n}}{n - \varepsilon\lambda'} + \sum_{n \geq 0} \frac{(-1)^n}{n+1-N-\varepsilon\lambda'}$$
$$+ \sum_{n\geq 1} \frac{(\lfloor a \rfloor u)^n}{n-l-\varepsilon\lambda'} + \sum_{n\geq 1} \frac{(\lfloor a \rfloor u^{-1})^n}{n+l-\varepsilon\lambda'} + \frac{1}{|l|-\lambda'} \Bigg)\Bigg),$$

and $\Phi_{M_1,\Sigma_1}(\tau_1)$ *is obtained by taking one half of the difference of the limits as* $l \to \pm 0$ *formally. As a function on* $T_2 \cap G_{\text{reg}}$,

$$\Phi_{M_2,\Sigma_2}(\sigma_2) = -\sum_{\substack{m'=\pm m \\ \mu'=\pm\mu}} \Bigg(\lfloor c^m \rfloor \sum_n \frac{r^{2n}}{2n-m'-\mu'}$$
$$+ \operatorname{sgn}(m') r^{m'} c^{\mu'} \Bigg(\sum_{n\geq 1} \frac{\lfloor c \rfloor^{2n}}{n-\delta\mu'} + \sum_{n\geq 0} \frac{(-1)^n}{n+1-M-\delta\mu'}$$
$$+ \sum_{n\geq 1} \frac{(\lfloor c \rfloor r)^{2n}}{2n-m'-\delta\mu'} + \sum_{n\geq 1} \frac{(\lfloor c \rfloor r^{-1})^{2n}}{2n+m'-\delta\mu'} + \frac{1}{|m|-\delta\mu'} \Bigg)\Bigg),$$

where $M \in \{0,1\}$ *is determined by* $M \equiv m \pmod 2$.

Proof. We present $\Phi_{M_1,\Sigma_1}(\sigma_1)$ in the form (22), indicating the dependence on Σ_1 and σ_1 on the right-hand side, too. The functions $e^{P_1,\nu}_{G,\Sigma_1}(\sigma_1)$ can be read off from our formula for $\Phi_{G,\Sigma_1}(\sigma_1)$. The functions $e^{P_1,\nu}_{M_1,\Sigma_1}(\sigma_1)$ with $\nu = (\pm l, \pm\lambda)$ can be determined with the aid of the asymptotic formula (24), while the exponents $(\pm\lambda, |l|)$ are excluded by the temperedness condition. We see that $\Phi_{M_1,\Sigma_1}(\sigma_1)$ restricted to $T_{1,P_1} \cap G_{\text{reg}}$ equals

$$\sum_{\lambda'=\pm\lambda} \Big(-\operatorname{sgn}(l) u^l \operatorname{sgn}(a)^N |a|^{\lambda'} \big(F^{P_1,(l,\lambda')}_{M_1} + r_{M_1}(\check\sigma_1', P_1)\big) + e^{P_1,(\lambda',-|l|)}_{M_1,\Sigma_1}(\sigma_1) \Big),$$

where σ_1' is the representation with parameter (l, λ'). We compute the derivatives occurring in $\partial_1^0 \Phi^0_{M_1,\Sigma_1}$ using the formula

$$z \frac{\partial}{\partial z} b(s,z) = \frac{z}{1-z} - s b(s,z),$$

take the limit as $t \to +0$ using

$$\lim_{z\to 1} \big(b(s_1,z) - b(s_2,z)\big) = \psi(s_2+1) - \psi(s_1+1)$$

and simplify the result using

$$b(s,z)-b(-s,z^{-1})=\frac{\pi(-z)^{-s}}{\sin\pi s}-\frac{1}{s},$$
$$\psi(s)-\psi(1-s)=-\pi\cot\pi s.$$

In particular, since $b(0,z)=-\log(1-z)$,

$$\lim_{|a|\to 1+0} a\frac{\partial}{\partial a}\sum_{\lambda'=\pm\lambda}|a|^{\lambda'}\big(b(-\lambda',a^{-2})+\log|a-a^{-1}|\big)=\pi\lambda\cot\pi\lambda+1.$$

We also compute $\partial_0^1\Phi_{G,\Sigma}(\sigma_1)$ using the expression in terms of S_λ. Now the jump relation implies

$$\sum_{\lambda'=\pm\lambda} e_{M_1,\Sigma_1}^{P_1,(\lambda',|l|)}(\sigma_1)\big|_{a=\pm 1}=-\tfrac{1}{2}a^{l+1}\big(S_\lambda(-u)-(-1)^N S_\lambda(u)\big).$$

We obtain the value of the left-hand side for general $|a|>1$ by multiplying with $a^{-|l|}$ and extend $\Phi_{M_1,\Sigma_1}(\sigma_1)$ to $T_1\cap G_{\mathrm{reg}}$ using its W_{M_1}-invariance.

The proof for $\Phi_{M_1,\Sigma_1}(\tau_1)$ and $\Phi_{M_2,\Sigma_2}(\sigma_2)$ is similar. □

Theorem 13.2. *As functions on $H\cap G_{\mathrm{reg}}$,*

$$\begin{aligned}\Phi_H(\sigma_1)=-\sum_{\lambda'=\pm\lambda}\Bigg(&\operatorname{sgn}(a_1)^N|a_1|^{\lambda'}\operatorname{sgn}(a_2)\lfloor a_2^l\rfloor\\
&\times\Bigg(\sum_{n\ge 1}\frac{\lfloor a_1^2\rfloor^n}{n-\varepsilon_1\lambda'}+\sum_{n\ge 0}\frac{(-1)^n}{n+1-N-\varepsilon_1\lambda'}\\
&\quad+\sum_{n\ge 1}\frac{\lfloor a_1a_2^{-1}\rfloor^n}{n-|l|-\varepsilon_1\lambda'}+\sum_{n\ge 0}\frac{\lfloor a_1a_2\rfloor^n}{n+|l|-\varepsilon_1\lambda'}\Bigg)\\
&+\operatorname{sgn}(a_2)^N|a_2|^{\lambda'}\operatorname{sgn}(a_1)\lfloor a_1^l\rfloor\\
&\times\Bigg(\sum_{n\ge 1}\frac{\lfloor a_2^2\rfloor^n}{n-\varepsilon_2\lambda'}+\sum_{n\ge 0}\frac{(-1)^n}{n+1-N-\varepsilon_2\lambda'}\\
&\quad+\sum_{n\ge 1}\frac{\lfloor a_1a_2^{-1}\rfloor^n}{n+|l|+\varepsilon_2\lambda'}+\sum_{n\ge 0}\frac{\lfloor a_1a_2\rfloor^n}{n+|l|-\varepsilon_2\lambda'}\Bigg)\Bigg),\end{aligned}$$

$$\Phi_H(\tau_1)=0,$$

$$\Phi_H(\sigma_2) = -(\operatorname{sgn} a_1 + \operatorname{sgn} a_2)$$
$$\times \sum_{\mu'=\pm\mu} \Bigg(a_+^{\mu'} \lfloor a_1^m a_+^{-m} \rfloor \Bigg(\sum_{n\geq 1} \frac{\lfloor a_1 a_2 \rfloor^n}{n - \varepsilon_+ \mu'} + \sum_{n \geq 0} \frac{(-1)^n}{n+1-M-\varepsilon_+\mu'}$$
$$+ \sum_{n\geq 0} \frac{\lfloor a_1^2 \rfloor^n}{2n + |m| - \varepsilon_+\mu'} + \sum_{n\geq 1} \frac{\lfloor a_2^2 \rfloor^n}{2n - |m| - \varepsilon_+\mu'} \Bigg)$$
$$+ a_-^{\mu'} \lfloor a_1^m a_-^{-m} \rfloor \Bigg(\sum_{n\geq 1} \frac{\lfloor a_1 a_2^{-1} \rfloor^n}{n - \varepsilon_- \mu'} + \sum_{n \geq 0} \frac{(-1)^n}{n+1-M-\varepsilon_-\mu'}$$
$$+ \sum_{n\geq 0} \frac{\lfloor a_1^2 \rfloor^n}{2n + |m| - \varepsilon_-\mu'} + \sum_{n\geq 1} \frac{\lfloor a_2^2 \rfloor^n}{2n + |m| + \varepsilon_-\mu'} \Bigg)\Bigg).$$

Proof. We present $\Phi_H(\sigma_1)$ in the form (22). The functions $e_G^{P_1,\nu}(\sigma_1)$ vanish because $\Phi_{G,\Pi}(\sigma_1)$ does. The functions $e_{M_i}^{P_1,\nu}(\sigma_1)$ can be read off from our formulas for $\Phi_{M_i,\Pi}(\sigma_1)$. We see that, on $H_P \cap G_{\mathrm{reg}}$,

$$\Phi_H(\sigma_1) = -\sum_{\lambda'=\pm\lambda} \Bigg(\operatorname{sgn}(a_1)^N |a_1|^{\lambda'} \operatorname{sgn}(a_2) a_2^{-|l|} \Big(b(-\lambda', a_1^{-2})$$
$$+ b(-\lambda' - l, a_1^{-1} a_2) + b(-\lambda' + l, a_1^{-1} a_2^{-1}) \Big) + e_H^{P,(\lambda',-l)}(\sigma_1)$$
$$+ \operatorname{sgn}(a_2)^N |a_2|^{\lambda'} \operatorname{sgn}(a_1) a_1^{-|l|} \Big(b(-\lambda', a_2^{-2})$$
$$+ b(\lambda' + l, a_1^{-1} a_2) + b(-\lambda' + l, a_1^{-1} a_2^{-1}) \Big) + e_H^{P,(-l,\lambda')}(\sigma_1) \Bigg).$$

Now we compute $\partial^1 \Phi^1_{H,\Pi}(\sigma_1)$ and $\partial_1 \Phi_{M_1,\Sigma_1}(\sigma_1)$ using the properties of $b(s,z)$ mentioned in the proof of Theorem 13.1. The jump relation implies that

$$e_H^{P,(\lambda,-l)}(\sigma_1)\Big|_{a_2=\pm 1} = \operatorname{sgn}(a_1)^N |a_1|^{\lambda'} a_2^{l+1} r_{M_1}(\check{\sigma}_1, P_1).$$

Next we compute $\partial^2 \Phi^2_{H,\Pi}(\sigma_1)$ and $\partial_2 \Phi_{M_2,\Sigma_2}(\sigma_1)$. Here the jump relation implies that

$$\Big(e_H^{P,(\lambda,-l)}(\sigma_1) - e_H^{P,(-l,\lambda)}(\sigma_1) \Big)\Big|_{a_1=a_2} = 0.$$

Together with the W_H-invariance of $\Phi_H(\sigma_1)$, this determines our functions on the subset of $H \cap G_{\mathrm{reg}}$ where $\operatorname{sgn} a_1 = \operatorname{sgn} a_2$. We conclude from the smoothness of $\Phi_H(\sigma_1)$ at $a_1 = -a_2$ that the same formula must be true on the components of H where $\operatorname{sgn} a_1 = -\operatorname{sgn} a_2$.

The proof for $\Phi_H(\tau_1)$ and $\Phi_H(\sigma_2)$ is similar. □

14. Spectral descent

For each Levi subgroup L, the set $T_{\text{disc}}(L) - T_{\text{ell}}(L)$ consists of certain[5] (classes of) virtual tempered representations τ^L induced from $\tau \in T_{\text{ell}}(L_1)$ for a proper Levi subgroup L_1 of L. Here we list them for our group G together with the parameters of τ in the notation of section 2.

L	L_1	τ	param.
M_1	H	χ	$(N, \lambda, 0, 0)$
M_2	H	χ	(N, λ, N, λ)
	H	χ	$(0, 0, 0, 0)$
G	M_1	σ_1	$(l, 1, 0)$
	M_1	τ_1	$(0, N, 0)$
	M_2	σ_2	$(2l+1, 0), (0, 0)$

The value of $\Phi_M(\tau^L)$ is given by Arthur's[6] descent formula (4.7) in terms of functions $\Phi_M^{M'}(\tau')$ for pairs (L_1', τ') conjugate to (L_1, τ) and such that $L_1' \subset M'$ and $\dim \mathfrak{a}_{M'}^G = \dim \mathfrak{a}_{L_1}^L$.

References

1. J. Arthur, R. Herb, P. Sally Jr., The Fourier transform of weighted orbital integrals on SL(2, $\mathbb{R}$), in: The Selberg trace formula and related topics, *Contemp. Math.*, vol. 53, pp. 17–37 (Amer. Math. Soc., Providence, RI, 1986).
2. J. Arthur, The characters of discrete series as orbital integrals, *Invent. Math.* **32** (1976), 205–261.
3. J. Arthur, The local behaviour of weighted orbital integrals, *Duke Math. J.* **56** (1988), 223–293.
4. J. Arthur, Intertwining operators and residues I: Weighted characters, *J. Funct. Anal.* **84** (1989), 19–84.
5. J. Arthur, On elliptic tempered characters, *Acta Math.* **171** (1993), 73–138.
6. J. Arthur, On the Fourier transforms of weighted orbital integrals, *J. reine angew. Math.* **452** (1994), 163–217.
7. J. Arthur, An asymptotic formula for real groups, *J. reine angew. Math.* **601** (2006), 163–230.
8. Harish-Chandra, A formula for semisimple Lie groups, *Trans. Amer. Math. Soc.* **79** (1957), 733–760.
9. Harish-Chandra, Harmonic analysis on real reductive groups I: The theory of the constant term, *J. Funct. Anal.* **19** (1975), 104–204.
10. R. Herb, Discrete series characters and Fourier inversion on semisimple real Lie groups, *Trans. Amer. Math. Soc.* **277** (1983), 241–262.
11. T. Hirai, Explicit form of the characters of discrete series representations of semisimple Lie groups, in: Harmonic analysis on homogeneous spaces, *Proc. Sympos. Pure Math.*, Vol. XXVI, pp. 281–287 (Amer. Math. Soc., Providence, R. I., 1973).

12. W. Hoffmann, The Fourier transforms of weighted orbital integrals on semisimple groups of real rank one, *J. reine angew. Math.* **489** (1997), 53–97.
13. W. Hoffmann, On the differential equations satisfied by weighted orbital integrals, *Annals of Math.* **155** (2002), 759–802.
14. W. Hoffmann, Weighted orbital integrals, in: Proceedings of the conference on Representation Theory of Real Reductive Lie Groups, Snowbird 2006. *Contemporary Mathematics*, vol. 472, pp. 153–173 (Amer. Math. Soc., Providence, RI, 2008).
15. W. Hoffmann, Asymptotic and descent formulas for weighted orbital integrals, preprint, `http://arxiv.org/abs/0808.4144`
16. A. Knapp, *Representation theory of semisimple Groups* (Princeton University Press, Princeton, 1986).
17. R. Langlands, On the functional equations satisfied by Eisenstein series, *Lecture Notes in Math.*, Vol. 544 (Springer Verlag, New York/Berlin, 1976).
18. R. Langlands, The classification of representations of real reductive groups, in: *Math. Surveys and Monographs*, vol. 31 (Amer. Math. Soc., Providence, RI, 1988).
19. W. Schmid, On the characters of discrete series: the Hermitian symmetric case, *Invent. Math.* **30** (1975), 47–144.

An Atkin-Lehner type theorem on Siegel modular forms and primitive Fourier coefficients

T. Ibukiyama

Department of Mathematics, Graduate School of Science, Osaka University, Machikaneyama 1-1, Toyonaka, Osaka 560-0043 Japan
E-mail: ibukiyam@math.sci.osaka-u.ac.jp

H. Katsurada

Department of Mathematics, Muroran Institute of Technology, Mizumoto 27-1, Muroran, 050-8585 Japan
E-mail: hidenori@mmm.muroran-it.ac.jp

We give a generalization of the Atkin-Lehner theory to Siegel modular forms of level N and define new forms in this sense. As an application, we show that any non-zero new form in the above sense has a non-zero Fourier coefficient at a primitive half-integral symmetric matrix.

Keywords: Siegel modular forms; Atkin-Lehner theory; Fourier coefficient

1. Introduction

There are several tries to define new forms in case of Siegel modular forms. One of the oldest attempts would be [7] and [8] on parahoric subgroups focusing a Langlands type correspondence with the compact twist. Recent local theories in R. Schmidt [12] or in B. Roberts and R. Schimidt [5], [6] gave a local picture on this subject, giving local new vectors for each local representation. On the other hand, the classical theory of Atkin-Lehner described new forms of one variables by their Fourier coefficients. In the one variable case, the Fourier coefficients and eigenvalues of the Hecke operators are almost the same objects, but for Siegel modular forms, they are quite different. Sometimes we are interested in the Fourier coefficients themselves and it would be of some use to generalize a part of Atkin-Lehner theory to the case of Siegel modular forms. An aim of this paper is to give such generalization. We define old forms and new forms in Atkin-Lehner like way. One application is to show that, when the degree is greater than one, any non-zero new form has a non-zero Fourier coefficient at some primitive half-integral matrix (cf. Theorem 2.2). This result is a refinement of a part of Yamana's result in [13] (cf. also Theorem 2.1), where he claims that any non-zero Siegel modular form has a non-zero Fourier coefficient at some half-integral matrix whose *content* divides a certain divisor of the level. Historically, his result is a generalization of the results in [3], [4], [9], [10] on level one to the congruence subgroups of general levels. We also show that the space of new forms is invariant by Hecke operators $T(m)$ when m is coprime to the level. We state our main results in section 2 and prove them in section 3.

2. Main results

We fix a natural number n throughout the paper. We denote by H_n the Siegel upper half space of degree n and by $Sp(n,\mathbb{R})$ the symplectic group of matrix size $2n$. We put

$$GSp^+(n,\mathbb{R}) = \left\{g \in GL_{2n}(\mathbb{R});\ {}^t gJg = n(g)J \text{ for some real number } n(g) > 0\right\},$$

where $J = \begin{pmatrix} 0 & 1_n \\ -1_n & 0 \end{pmatrix}$. For any natural number k, any $g = \begin{pmatrix} A & B \\ C & D \end{pmatrix} \in GSp^+(n,\mathbb{R})$ and any function $F(Z)$ on H_n, we write

$$(F|_k[g])(Z) = n(g)^{nk/2} \det(CZ+D)^{-k} F(gZ).$$

Let Γ be an arithmetic discrete subgroup of $Sp(n,\mathbb{R})$ and χ a character of Γ. We denote by $A_k(\Gamma,\chi)$ the space of Siegel modular forms $F(Z)$ of weight k of Γ with character χ, i.e. holomorphic functions on H_n such that

$$F|_k[\gamma] = \chi(\gamma)F$$

for any $\gamma \in \Gamma$ (with finiteness condition at cusps when $n = 1$.) A cusp form is defined as usual as a Siegel modular form which vanishes at every cusp of Γ. Let N be any natural number. We fix a primitive Dirichlet character χ with conductor m_χ and assume that $m_\chi | N$. We put

$$\Gamma_0^{(n)}(N) = \left\{g = \begin{pmatrix} A & B \\ C & D \end{pmatrix} \in Sp(n,\mathbb{Z}); C \equiv 0 \bmod N\right\}.$$

We regard χ as a character of $\Gamma_0^{(n)}(N)$ by putting $\chi(\gamma) = \chi(\det(D))$. Hereafter we write $\Gamma_0(N) = \Gamma_0^{(n)}(N)$, since no confusion is likely. We write $A_k(N,\chi) = A_k(\Gamma_0(N),\chi)$ and denote by $S_k(N,\chi)$ the space of cusp forms in $A_k(N,\chi)$. When χ is trivial, we write $A_k(N) = A_k(N,\chi)$ and

$S_k(N) = S_k(N, \chi)$. Any element $F \in A_k(N, \chi)$ has the unique Fourier expansion

$$F(Z) = \sum_T a(T)e(\mathrm{tr}(TZ)),$$

where we put $e(x) = e^{2\pi ix}$ and T runs over positive semi-definite half integral symmetric matrices. For a half integral symmetric matrix $T = \left(\frac{\delta_{ij}+1}{2}t_{ij}\right)$, where δ_{ij} is Kronecker's delta and $t_{ij} \in \mathbb{Z}$, we denote by $\mathrm{cont}(T)$ the greatest common divisor of all t_{ij}. We say that T is primitive if $\mathrm{cont}(T) = 1$.

We quote first a part of Yamana's theorem in [13] on tube domains.

Theorem 2.1 (Yamana). *Assume that the degree $n \geq 2$. For $F \in A_k(N, \chi)$, assume that its Fourier coefficient $a(T) = 0$ for every T with $cont(T)|(N/m_\chi)$. Then $F = 0$.*

He used the Fourier Jacobi expansion of F and the Taylor expansion of the Jacobi forms, then reduced everything to one variable case. Here we give a refinement of his result by restricting Siegel modular forms to *new forms* (cf. Theorem 2.2 below). We define now the space of new cusp forms. If d and M are natural numbers such that $dM|N$ and $m_\chi|M$, then for any $f(Z) \in S_k(M, \chi)$, we have $f(dZ) \in S_k(N, \chi)$. We denote by $S_k^1(N, \chi)$ the linear subspace of $S_k(N, \chi)$ spanned by the set

$$\{f(dZ); f(Z) \in S_k(M, \chi), m_\chi|M, dM|N, M \neq N\}.$$

We denote by $S_k^0(N, \chi)$ the orthogonal complement of $S_k^1(N, \chi)$ in $S_k(N, \chi)$ with respect to the Petersson inner product. In this paper, we call an element of $S_k^0(N, \chi)$ a new form.

Theorem 2.2. *Assume that $n \geq 2$. Assume that $F \in S_k^0(N, \chi)$ and its Fourier coefficients $a(T) = 0$ for all the primitive symmetric half-integral matrix T. Then we have $F = 0$.*

To prove this, we need the following key theorem, which is interesting as itself.

Theorem 2.3. *We fix $F(Z) = \sum_T a(T)e(\mathrm{tr}(TZ)) \in A_k(N, \chi)$ and assume that there exists a natural number l such that $a(T) = 0$ for any T with $(\mathrm{cont}(T), l) = 1$. Then we have*
(1) If $\gcd(l, N/m_\chi) = 1$, then $F = 0$.

(2) If $\gcd(l, N/m_\chi) \neq 1$*, then there exists* $f_p(Z) \in A_k(N/p, \chi)$ *for each prime factor* p *of* $\gcd(l, N/m_\chi)$ *such that*

$$F(Z) = \sum_{\substack{p|\gcd(l,N/m_\chi) \\ p:prime}} f_p(pZ).$$

If F *is a cusp form, we can take* f_p *to be cusp forms.*

The above claim (1) is an easy consequence of Theorem 2.1, but we shall give here a different proof.

The following corollary to Theorem 2.3 is trivial.

Corollary 2.1. *For* $F \in S_k^0(N, \chi)$*, if there exists a natural number* l *such that* $a(T) = 0$ *for any* T *with* $(\mathrm{cont}(T), l) = 1$*, then* $F = 0$.

Now finally we give a comment on Hecke theory. We put

$$\Delta_0(N) = \left\{ \begin{pmatrix} A & B \\ NC & D \end{pmatrix} \in GSp^+(n, \mathbb{Q}) \cap M_{2n}(\mathbb{Z}); \begin{array}{c} C \equiv 0 \bmod N \\ \text{and } (\det(A), N) = 1 \end{array} \right\}.$$

We denote by $\mathcal{H}(N) = \mathcal{H}(\Delta_0(N), \Gamma_0(N))$ the Hecke algebra generated by $\Gamma_0(N)g\Gamma_0(N)$ with $g \in \Delta_0(N)$ and by $\mathcal{H}_N(N)$ the subalgebra of $\mathcal{H}(N)$ generated by $\Gamma_0(N)g\Gamma_0(N)$ with $g \in \Delta_0(N)$ such that $(n(g), N) = 1$. We follow the explanation in Andrianov [1] on the structure of $\mathcal{H}_N(N)$. For a_i, $d_i \in \mathbb{Z}_{\geq 1}$ with $a_i | a_{i+1}$, $a_i d_i = m$ with $1 \leq i \leq n$, we put

$$T(a_1, \ldots, a_n; d_1, \ldots, d_n) = \Gamma_0(N)\mathrm{diag}(a_1, \ldots, a_n, d_1, \ldots, d_n)\Gamma_0(N),$$

where $\mathrm{diag}(x_1, \ldots, x_{2n})$ is the diagonal matrix whose diagonal components are x_i. The Hecke algebra $\mathcal{H}_N(N)$ is generated by $\mathbb{T}(p) = T(\underbrace{1, \ldots, 1}_{n}; \underbrace{p, \ldots, p}_{n})$ and $\mathbb{T}_i(p^2) = (\underbrace{1, \ldots, 1}_{i}, \underbrace{p, \ldots, p}_{n-i}; \underbrace{p^2, \ldots, p^2}_{i}, \underbrace{p, \ldots, p}_{n-i})$ with p coprime to N with $0 \leq i \leq n-1$.

Proposition 2.1. *The space* $S_k^0(N, \chi)$ *is stable under the action of the Hecke operators in* $\mathcal{H}_N(N)$.

Remark. In the above Atkin-Lehner-like theory, there is a very different point from the case of modular forms of one variable. Even if a Siegel cusp form $F \in S_k^0(N, \chi)$ is a common Hecke eigen form of all the Hecke operators $T(m) = \sum_{n(g)=m} \Gamma_0(N)g\Gamma_0(N)$ with $(m, N) = 1$ (where $g \in M_{2n}(\mathbb{Z}) \cap GSp(n, \mathbb{Q})$ runs over the representatives of such double cosets), it is not necessarily true that F is uniquely determined by all the eigenvalues $\lambda(m, F)$ of F of $T(m)$ with $(m, N) = 1$. The counter example is found in [7] p.607. It is as follows. Notation being the same as in [7], if we put $F = Y^2Z + 5XYK - 1024YZ^2 - 5120XZK$, $G = 13XYK - 2X^2YZ + X^3K - 4608YZ^2 + 5760K^2 - 9728XZK$, then we have F, $G \in S_{12}(2)$ and we have $\lambda(3, F) = \lambda(3, G) = -88488$ (cf. loc.cit. p.607). These are obviously new forms in the above sense. Besides the two-dimensional space V spanned by F and G is a common eigenspace of all $T(m)$ with odd m and $\lambda(m, F) = \lambda(m, G)$ for any odd m, as was explained in [7] p.612. This comes from the fact that the action of double cosets $\Gamma_0(2)g\Gamma_0(2)$ with $n(g) = 2$ on V cannot be diagonalized simultaneously.

3. Proof

Although the contents of Theorem 2.3 are analogues of the paper [2] of Atkin and Lehner, the method of the proofs largely consist of analogues of Miyake's book [11].

3.1. *Action of Hecke operators*

We review the action of Hecke operators here. We define $\Delta_0(N)$ as before. For a (not necessarily primitive) Dirichlet character χ modulo N and any $g \in \Delta_0(N)$, we write $\chi(g) = \overline{\chi(\det(A))}$ by abuse of notation. Then χ is a character of the semi-group $\Delta_0(N)$. If $g \in \Gamma_0(N)$, then this means $\chi(g) = \chi(\det(D))$. Now for any natural numbers N and M, take $g \in \Delta_0(N) \cap \Delta_0(M)$ with $n(g) = m$. We take a Dirichlet character χ such that the conductor $m_\chi | \gcd(N, M)$. If $T := \Gamma_0(N)g\Gamma_0(M) = \bigcup \Gamma_0(N)g_i$ (disjoint), then for $F \in A_k(\Gamma_0(N), \chi)$, we define

$$F|_k[\Gamma_0(N)g\Gamma_0(M)] = m^{(nk-n(n+1))/2} \sum_i \overline{\chi(g_i)} F|_k[g_i].$$

This does not depend on the choice of g_i and we have $F|_kT \in A_k(M, \chi)$. In particular, when $N = M$, the above double coset is in $\mathcal{H}(N) = \mathcal{H}(\Delta_0(N), \Gamma_0(N))$.

For any prime q and any $F(Z) = \sum_T a(T)e(\mathrm{tr}(TZ)) \in A_k(N, \chi)$, we define an operator $U(q)$ by

$$F\big|U(q) = \sum_T a(qT)e(\mathrm{tr}(TZ)).$$

If q divides N, then we have

$$\Gamma_0(N) \begin{pmatrix} 1_n & 0 \\ 0 & q1_n \end{pmatrix} \Gamma_0(N) = \sum_{S={}^tS \in M_n(\mathbb{Z}/q\mathbb{Z})} \Gamma_0(N) \begin{pmatrix} 1_n & S \\ 0 & q1_n \end{pmatrix},$$

where S runs over a complete set of representatives of integral symmetric matrices modulo q. In this case $F|U(q)$ is the same as the usual action of this double coset on F and we have $F|U(q) \in A_k(N,\chi)$. For any q, we can regard $F \in A_k(N,\chi)$ as an element in $A_k(qN,\chi)$, since $A_k(N,\chi) \subset A_k(qN,\chi)$. So, even in the case when q is coprime to N, we can define $F|_kU(q)$ as the action of the double coset with respect to $\Gamma_0(qN)$. In this case we have $F|U(q) \in A_k(qN,\chi)$ and $(F|U(q))(qZ) \in A_k(q^2N,\chi)$. In particular, $F(Z) - (F|U(q))(qZ)$ is an element of $A_k(q^2N,\chi)$ such that $a(T) = 0$ for any T with $q|\mathrm{cont}(T)$. We sometimes write $F|U(q) = F|_kU(q)$ when we emphasize that the weight of F is k.

3.2. *Some group theory*

First we give several technical lemmas. For any natural integers N, M, we define a subgroup $\Gamma_0(N,M)$ of $Sp(n,\mathbb{Z})$ by

$$\Gamma_0(N,M) = Sp(n,\mathbb{Z}) \cap \left\{ \begin{pmatrix} A & MB \\ NC & D \end{pmatrix} ; A,B,C,D \in M_n(\mathbb{Z}) \right\}.$$

For any symmetric matrix $S \in M_n(\mathbb{R})$, we put $u(S) = \begin{pmatrix} 1_n & S \\ 0 & 1_n \end{pmatrix}$ and for any $U \in GL_n(\mathbb{R})$, we put $t(U) = \begin{pmatrix} U & 0 \\ 0 & {}^tU^{-1} \end{pmatrix}$. We put

$$U_n(\mathbb{Z}) = \left\{ u(S); S = {}^tS \in M_n(\mathbb{Z}) \right\},$$
$$T_n(\mathbb{Z}) = \{ t(U); U \in GL_n(\mathbb{Z}) \}.$$

In the next lemma, we do not assume any relation between N and p. We have $T_n(\mathbb{Z}) \subset \Gamma_0(N,M)$ for any N, M.

Lemma 3.1. *Let N be any natural number and let p be any prime. Then the group $\Gamma_0(N)$ is generated by $\Gamma_0(N,p)$ and $U_n(\mathbb{Z})$.*

Proof. We take $\gamma = \begin{pmatrix} A & B \\ NC & D \end{pmatrix} \in \Gamma_0(N)$. Put $r = \mathrm{rank}(A \bmod p)$. Then by the elementary divisor theorem, there exists U, $V \in GL_n(\mathbb{Z})$ such that $UAV \equiv \begin{pmatrix} A_{11} & 0 \\ 0 & 0 \end{pmatrix} \bmod p$ for some $A_{11} \in M_r(\mathbb{Z})$ such that $\det(A_{11})$ is coprime to p. Put $UB\,{}^tV^{-1} = \begin{pmatrix} B_{11} & B_{12} \\ B_{21} & B_{22} \end{pmatrix}$, where $B_{11} \in M_r(\mathbb{Z})$, $B_{22} \in M_{n-r}(\mathbb{Z})$, B_{12}, ${}^tB_{21} \in M_{r,n-r}(\mathbb{Z})$. If we write $A_1 = UAV$ and

$B_1 = UB^tV^{-1}$, then the matrix $A_1\,{}^tB_1$ is symmetric. We have

$$A_1\,{}^tB_1 \equiv \begin{pmatrix} A_{11}\,{}^tB_{11} & A_{11}\,{}^tB_{21} \\ 0 & 0 \end{pmatrix} \bmod p.$$

Hence $A_{11}\,{}^tB_{21} \equiv 0 \bmod p$ and $B_{21} \equiv 0 \bmod p$. Therefore $\det(B_{22})$ is coprime to p, since (A_1, B_1) is a coprime symmetric pair and the rank of $(A_1, B_1) \bmod p$ is n. Now put

$$S_1 = \begin{pmatrix} 0 & 0 \\ 0 & N1_{n-r} \end{pmatrix}$$

and write

$$t(U)\gamma\, t(V)\,{}^t(u(S_1)) = \begin{pmatrix} A_2 & B_2 \\ NC_2 & D_2 \end{pmatrix}.$$

Then we have

$$A_2 \equiv \begin{pmatrix} A_{11} & B_{12} \\ 0 & NB_{22} \end{pmatrix} \bmod p.$$

If $p \nmid N$, then $\det(A_2) \equiv \det(A_{11})\det(NB_{22}) \bmod p$ is coprime to p. If $p|N$, then $\det(A)$ is coprime to p, so $r = n$ and $\det(A_2) \equiv \det(A_{11}) \bmod p$. Hence we see $\det(A_2)$ is coprime to p in any case. Since $A_2^{-1}B_2$ is symmetric, there exists an integral symmetric matrix S such that $A_2S + B_2 \equiv 0 \bmod p$. So we see

$$\begin{pmatrix} A_2 & B_2 \\ NC_2 & D_2 \end{pmatrix} \times \begin{pmatrix} 1_n & S \\ 0 & 1_n \end{pmatrix} = \begin{pmatrix} A_2 & A_2S + B_2 \\ NC_2 & C_2S + D_2 \end{pmatrix} \in \Gamma_0(N, p).$$

This implies that γ is contained in the group generated by $\Gamma_0(N, p)$ and $U_n(\mathbb{Z})$. q.e.d.

3.3. *Level change*

For any prime p, we put

$$\delta_p = \begin{pmatrix} p1_n & 0 \\ 0 & 1_n \end{pmatrix}.$$

Proposition 3.1. *Let F be any holomorphic function on H_n such that $F(Z+S) = F(Z)$ for any symmetric matrices $S \in M_n(\mathbb{Z})$. We assume that $F(pZ) \in A_k(N, \chi)$ for some prime number p.*
(1) If $p|(N/m_\chi)$, then $F(Z) \in A_k(N/p, \chi)$.
(2) If $p \nmid (N/m_\chi)$, then $F(Z) = 0$.

Proof. First we show (1). Since $\Gamma_0(N) = \delta_p^{-1}\Gamma_0(N/p,p)\delta_p$, and $F|_k u(S) = F$ for any $S = {}^tS \in M_n(\mathbb{Z})$ by assumption, we see $F \in A_k(\Gamma,\chi)$ where Γ is a group generated by $\Gamma_0(N/p,p)$ and $U_n(\mathbb{Z})$. Here χ is well-defined on Γ since we assumed $m_\chi|(N/p)$. By Lemma 3.1, we see $\Gamma = \Gamma_0(N/p)$. Hence we prove (1). Next, we prove (2). First we assume that $p \nmid N$. Put $G(Z) = F(pZ)$. Then since $G(Z+S/p) = G(Z)$, $G(Z)$ is a modular form with respect to the group Γ' generated by $\Gamma_0(N)$ and $\begin{pmatrix} 1_n & S/p \\ 0 & 1_n \end{pmatrix}$ with $S = {}^tS \in M_n(\mathbb{Z})$. We show that Γ' is too big. We put $H = G|_k[J]$, where $J = \begin{pmatrix} 0 & -1_n \\ 1_n & 0 \end{pmatrix}$ as before. Let S be any integral symmetric matrix. We put

$$\gamma_1 = J^{-1}\begin{pmatrix} 1_n & 0 \\ -NS & 1_n \end{pmatrix} J = \begin{pmatrix} 1_n & NS \\ 0 & 1_n \end{pmatrix}.$$

Then since $\gamma_1 \in J^{-1}\Gamma_0(N)J$, we have $H|_k[\gamma_1] = H$. This means $H(Z+NS) = H(Z)$ for any integral symmetric matrices S. So $H(Z)$ has a Fourier expansion $H(Z) = \sum_T a(T)e(\mathrm{tr}(TZ/N))$, where T runs over positive semi-definite half-integral symmetric matrices. On the other hand, we take c, $d \in \mathbb{Z}$ so that $pd - Nc = 1$. We see

$$\begin{pmatrix} p1_n & 1_n \\ Nc1_n & d1_n \end{pmatrix}\begin{pmatrix} 1_n & -p^{-1}1_n \\ 0 & 1_n \end{pmatrix} = \begin{pmatrix} p1_n & 0 \\ Nc1_n & p^{-1}1_n \end{pmatrix} \in \Gamma'.$$

We put

$$\gamma_2 = J^{-1}\begin{pmatrix} p1_n & 0 \\ Nc1_n & p^{-1}1_n \end{pmatrix} J = \begin{pmatrix} p^{-1}1_n & -Nc1_n \\ 0 & p1_n \end{pmatrix}.$$

Since $H|_k[\gamma_2] = \chi(d)^n H$, we have $H(p^{-2}Z - p^{-1}Nc1_n) = p^{nk}\chi(p)^{-n}H(Z)$. So we have $H(Z) = p^{-nk}\chi(p)^{-n}\sum_T a(T)e(\mathrm{tr}((T/Np^2)Z - cT/p))$. Comparing this with the original Fourier expansion, we see that $a(T) = 0$ unless $p^2|\mathrm{cont}(T)$. Inductively, we see that $a(T) = 0$ for all T and that $H = 0$. Now we assume $p|N$. We denote by e_{11} the $n \times n$ matrix such that $(1,1)$ component is 1 and the other components are zero. Then since $F(pZ) \in A_k(N,\chi)$ by our assumption and $\delta_p^{-1}\begin{pmatrix} 1_n & 0 \\ (N/p)e_{11} & 1_n \end{pmatrix}\delta_p \in \Gamma_0(N)$, we have

$$F|_k\begin{bmatrix} 1_n & 0 \\ (N/p)e_{11} & 1_n \end{bmatrix} = F.$$

So, if we put

$$\gamma_3 = \begin{bmatrix} 1_n & S_2 \\ 0 & 1_n \end{bmatrix}\begin{bmatrix} 1_n & 0 \\ (N/p)e_{11} & 1_n \end{bmatrix}\begin{bmatrix} 1_n & S_1 \\ 0 & 1_n \end{bmatrix}$$

for integral symmetric matrices S_1, S_2, we have $F|_k[\gamma_3] = F$. We take an integer m_1 such that $m_1N/p + 1 \not\equiv 0 \bmod p$. We also take an integer m_2 such that $m_1(1 + m_2N/p) + m_2 = m_1 + m_2(1 + m_1N/p) \equiv 0 \bmod p$. We put $S_1 = m_1e_{11}$ and $S_2 = m_2e_{11}$. Then we have

$$\delta_p^{-1}\gamma_3\delta_p = \begin{pmatrix} 1 + m_2N/pe_{11} & p^{-1}(m_1(1 + m_2(N/p)) + m_2)e_{11} \\ Ne_{11} & 1 + m_1(N/p)e_{11} \end{pmatrix} \in \Gamma_0(N).$$

Hence, we have $F|_k[\gamma_3] = \chi(1+m_1N/p)F$. So if $F \neq 0$, then $\chi(1+m_1N/p) = 1$ for any m_1 such that $(m_1N/p+1, p) = 1$. This means that χ is a Dirichlet character defined modulo N/p and hence $m_\chi|(N/p)$, which is a contradiction. This proves (2). q.e.d.

Lemma 3.2. *Let N be a natural number, $M = lN$ a multiple of N, and p be a prime with $p \nmid l$.*
(1) We have $[\Gamma_0(N) : \Gamma_0(N,p)] = [\Gamma_0(M) : \Gamma_0(M,p)]$.
(2) For $F \in A_k(pN, \chi)$, we have

$$F\big|_k\left[\Gamma_0(pN)\begin{pmatrix} 1_n & 0 \\ 0 & p1_n \end{pmatrix}\Gamma_0(N)\right] = F\big|_k\left[\Gamma_0(pM)\begin{pmatrix} 1_n & 0 \\ 0 & p1_n \end{pmatrix}\Gamma_0(M)\right].$$

(3) For $F \in A_k(pN, \chi)$, we have

$$F\big|_k\left[\Gamma_0(pN)\begin{pmatrix} 1_n & 0 \\ 0 & p1_n \end{pmatrix}\Gamma_0(N)\right][\delta_l] = F\big|_k[\delta_l]\left[\Gamma_0(pM)\begin{pmatrix} 1_n & 0 \\ 0 & p1_n \end{pmatrix}\Gamma_0(M)\right].$$

Proof. We show (1). Since $\Gamma_0(M,p) \subset \Gamma_0(M) \subset \Gamma_0(N)$ and $\Gamma_0(M,p) \subset \Gamma_0(N,p) \subset \Gamma_0(N)$, the assertion is equivalent to

$$[\Gamma_0(N) : \Gamma_0(M)] = [\Gamma_0(N,p) : \Gamma_0(M,p)].$$

For any natural number m, we have

$$\begin{pmatrix} 1_n & 0 \\ 0 & p1_n \end{pmatrix}\Gamma_0(m,p)\begin{pmatrix} 1_n & 0 \\ 0 & p^{-1}1_n \end{pmatrix} = \Gamma_0(mp),$$

so we have

$$[\Gamma_0(N,p) : \Gamma_0(M,p)] = [\Gamma_0(Np) : \Gamma_0(Mp)].$$

Now, for any natural number m, we denote by $\Gamma(m)$ the principal congruence subgroup of $Sp(n,\mathbb{Z})$ of level m. We see

$$\begin{aligned}[\Gamma_0(N), \Gamma_0(M)] &= [\Gamma_0(N) : \Gamma(M)]/[\Gamma_0(M) : \Gamma(M)] \\ &= [\Gamma_0(N) : \Gamma(N)][\Gamma(N) : \Gamma(M)]/[\Gamma_0(M) : \Gamma(M)].\end{aligned}$$

The index $[Sp(n,\mathbb{Z}):\Gamma(N)]$ is well known (e.g. Séminaire Cartan 1957/58) and from that we see

$$[\Gamma(N):\Gamma(M)] = l^{2n^2+n}\prod_{\substack{q|l,q\nmid N\\ q:\text{prime}}}\prod_{i=1}^{n}\left(1-\frac{1}{q^{2i}}\right),$$

$$[\Gamma_0(N):\Gamma(N)] = \#(GL_n(\mathbb{Z}/N\mathbb{Z}))N^{n(n+1)/2},$$
$$[\Gamma_0(M):\Gamma(M)] = \#(GL_n(\mathbb{Z}/M\mathbb{Z}))M^{n(n+1)/2}.$$

Since we have

$$\#(GL_n(\mathbb{Z}/N\mathbb{Z})) = N^{n^2}\prod_{\substack{q|N\\ q:\text{prime}}}\prod_{i=1}^{n}\left(1-\frac{1}{q^{i}}\right),$$

$$\#(GL_n(\mathbb{Z}/M\mathbb{Z})) = N^{n^2}l^{n^2}\prod_{\substack{q|Nl\\ q:\text{prime}}}\prod_{i=1}^{n}\left(1-\frac{1}{q^{i}}\right),$$

we get

$$\#(GL_2(\mathbb{Z}/N\mathbb{Z}))/\#(GL_2(\mathbb{Z}/M\mathbb{Z})) = \frac{1}{l^{n^2}\prod_{\substack{q|l,q\nmid N\\ q:\text{prime}}}\prod_{i=1}^{n}(1-q^{-i})}.$$

Hence we have

$$[\Gamma_0(N):\Gamma_0(M)] = l^{n(n+1)/2}\prod_{\substack{q|l,q\nmid N\\ q:\text{prime}}}\prod_{i=1}^{n}\left(1+\frac{1}{q^{i}}\right).$$

Since $p\nmid l$, the above formula means that it is equal to $[\Gamma_0(Np):\Gamma_0(Mp)]$. Now we prove (2). We may write

$$\Gamma_0(pM)\begin{pmatrix}1_n & 0\\ 0 & p1_n\end{pmatrix}\Gamma_0(M) = \bigcup_i \Gamma_0(pM)\begin{pmatrix}1_n & 0\\ 0 & p1_n\end{pmatrix}\gamma_i \quad \text{(disjoint)},$$

where γ_i are a complete set of representatives of $\Gamma_0(M,p)\backslash\Gamma_0(M)$. Here of course we have $\gamma_i\gamma_j^{-1}\in\Gamma_0(M)$. So if $\gamma_i\gamma_j^{-1}\in\Gamma_0(N,p)$ besides, then we have $\gamma_i\gamma_j^{-1}\in\Gamma_0(M)\cap\Gamma_0(N,p)=\Gamma_0(M,p)$, hence $\gamma_i=\gamma_j$. This means that cosets $\Gamma_0(pN)\begin{pmatrix}1_n & 0\\ 0 & p1_n\end{pmatrix}\gamma_i$ are disjoint for different i. The number of $\Gamma_0(pN)$ cosets in the double cosets $\Gamma_0(pN)\begin{pmatrix}1_n & 0\\ 0 & p1_n\end{pmatrix}\Gamma_0(N)$ is equal to $[\Gamma_0(N);\Gamma_0(N,p)]$, and by (1), this is exactly the number of γ_i, so we prove (2). Next we prove (3). We note that for any m, $|_k[\delta_l]$ is an operator from $A_k(m,\chi)$ to $A_k(lm,\chi)$ (or more precisely, from $A_k(\Gamma_0(m,l),\chi)$ to $A_k(\Gamma_0(lm),\chi)$). We take γ_i as before. We take $F\in A_k(pN,\chi)$ and put $g=F|_k[\delta_l]$. Then we have $g\in A_k(pM,\chi)$ and

$$\begin{aligned} &F\big|_k[\delta_l]\left[\Gamma_0(pM)\begin{pmatrix}1_n & 0\\ 0 & p1_n\end{pmatrix}\Gamma_0(M)\right]\\ &= g\big|_k\left[\Gamma_0(pM)\begin{pmatrix}1_n & 0\\ 0 & p1_n\end{pmatrix}\Gamma_0(M)\right]\\ &= p^{(nk-n(n+1))/2}\sum_i\overline{\chi(\gamma_i)}g\big|_k\left[\begin{pmatrix}1_n & 0\\ 0 & p1_n\end{pmatrix}\gamma_i\right]\\ &= p^{(nk-n(n+1))/2}\sum_i\overline{\chi(\gamma_i)}F\big|_k[\delta_l]\left[\begin{pmatrix}1_n & 0\\ 0 & p1_n\end{pmatrix}\gamma_i\right]\\ &= p^{(nk-n(n+1))/2}\overline{\chi(\delta_l\gamma_i\delta_l^{-1})}F\big|_k\left[\begin{pmatrix}1_n & 0\\ 0 & p1_n\end{pmatrix}\delta_l\gamma_i\delta_l^{-1}\right][\delta_l]. \end{aligned}$$

Here we used the fact that $\chi(\gamma_i)=\chi(\delta_l\gamma_i\delta_l^{-1})$. We shall show that

$$\Gamma_0(pN)\begin{pmatrix}1_n & 0\\ 0 & p1_n\end{pmatrix}\Gamma_0(N)=\bigcup_i\Gamma_0(pN)\begin{pmatrix}1_n & 0\\ 0 & p1_n\end{pmatrix}\delta_l\gamma_i\delta_l^{-1}.$$

First of all, we see $\delta_l\gamma_i\delta_l^{-1}\in\Gamma_0(N,l)\subset\Gamma_0(N)$. So for any i, j, we have $(\delta_l\gamma_i\delta_l^{-1})(\delta_l\gamma_j\delta_l^{-1})^{-1}\in\Gamma_0(N,p)$ and this element belongs to $\Gamma_0(pN)$ only if $\gamma_i\gamma_j^{-1}\in\Gamma_0(Nl,p)=\Gamma_0(M,p)$ since $p\nmid l$. By our definition of γ_i, this means $i=j$ and $\gamma_i=\gamma_j$. Hence $\delta_l\gamma_i\delta_l^{-1}$ gives a different coset for each i. The claim that these exhaust cosets can be proved again by (1).

3.4. *Proof of main results*

First we prove Theorem 2.3. We may assume that l is squarefree. We prove by induction on the number of prime divisors of l. If l is a prime, this is nothing but an assertion of Proposition 3.1. We assume that Theorem 2.3 is true for any proper divisor of l. We take a prime $p|l$ and put $l_1=l/p$. For $F(Z)=\sum_T a(T)e(\mathrm{tr}(TZ))\in A_k(N,\chi)$, we put

$$\begin{aligned} h(Z) &= \sum_{(\mathrm{cont}(T),l_1)\neq 1} a(T)e(\mathrm{tr}(TZ)),\\ g(Z) &= F(Z)-h(Z)=\sum_{(\mathrm{cont}(T),l_1)=1} a(T)e(\mathrm{tr}(TZ)). \end{aligned}$$

We can obtain $g(Z)$ by repeating the mapping $F\to F(Z)-(F|_kU(q))(qZ)$ for every $q|l_1$, and so we have $g(Z)\in A_k(Nl_1^2,\chi)$. Hence we have also $h(Z)\in$

$A_k(Nl_1^2, \chi)$. If we write $g(Z) = \sum_T b(T)e(\mathrm{tr}(TZ))$, then by our assumption on F in Theorem 2.3, we have $b(T) \neq 0$ only when $p|\mathrm{cont}(T)$. So if we put $g_p(Z) = g(Z/p)$, then $g_p(Z+S) = g_p(Z)$ for any $S = {}^tS \in M_n(\mathbb{Z})$. If p is coprime to N/m_χ, then by Proposition 3.1 (2), we have $g(Z) = 0$. Hence we can reduce the case l to the case l_1 and by the inductive assumption, Theorem 2.3 is proved. Now assume that $p|(N/m_\chi)$. Then by Proposition 3.1, we have $g_p \in A_k(Nl_1^2/p, \chi)$. We put

$$\Gamma_0(Nl_1^2)\begin{pmatrix} 1_n & 0 \\ 0 & p1_n \end{pmatrix}\Gamma_0(Nl_1^2/p) = \bigcup_i \Gamma_0(Nl_1^2)\begin{pmatrix} 1_n & 0 \\ 0 & p1_n \end{pmatrix}\gamma_i, \qquad \text{(disjoint)}\ ,$$

where $\gamma_i \in \Gamma_0(Nl_1^2/p)$. Then we get

$$\begin{aligned} g|_k\left[\Gamma_0(Nl_1^2)\begin{pmatrix} 1_n & 0 \\ 0 & p1_n \end{pmatrix}\Gamma_0(Nl_1^2/p)\right] &= p^{(nk-n(n+1))/2}\sum_i \overline{\chi(\gamma_i)}g|_k\begin{pmatrix} 1_n & 0 \\ 0 & p1_n \end{pmatrix}\gamma_i \\ &= p^{-n(n+1)/2}\sum_i \overline{\chi(\gamma_i)}g_p|_k\gamma_i = c_p g_p \end{aligned}$$

with some positive constant c_p. Hence we have

$$g(Z/p) = g_p(Z) = c_p^{-1}\left(g|_k\left[\Gamma_0(Nl_1^2)\begin{pmatrix} 1_n & 0 \\ 0 & p1_n \end{pmatrix}\Gamma_0(Nl_1^2/p)\right]\right).$$

Now we put

$$f_p(Z) = c_p^{-1}\left(F|_k\left[\Gamma_0(N)\begin{pmatrix} 1_n & 0 \\ 0 & p1_n \end{pmatrix}\Gamma_0(N/p)\right]\right).$$

Then of course we have $f_p \in A_k(N/p, \chi)$ and $f_p(pZ) \in A_k(N, \chi)$. If we can show that $F(Z) - f_p(pZ)$ satisfies the condition of the theorem for l_1, then by inductive assumption, we are done. By using the above expression of g_p and Lemma 3.2 (1), we have

$$\begin{aligned} &F(Z) - f_p(pZ) \\ &= F(Z) - f_p(pZ) - (g(Z) - g_p(pZ)) \\ &= F(Z) - g(Z) - c_p^{-1}\left((F-g)|_k\left[\Gamma_0(Nl_1^2)\begin{pmatrix} 1_n & 0 \\ 0 & p1_n \end{pmatrix}\Gamma_0(Nl_1^2/p)\right]\right)(pZ) \\ &= h(Z) - c_p^{-1}\left(h|_k\left[\Gamma_0(Nl_1^2)\begin{pmatrix} 1_n & 0 \\ 0 & p1_n \end{pmatrix}\Gamma_0(Nl_1^2/p)\right]\right)(pZ). \end{aligned}$$

By repeating the annihilating operator $h(Z) - (h|_kU(q))(qZ)$ for $q|l_1$, we can show that $h(Z) = \sum_{q|l_1} h_q(qZ)$ for some $h_q \in A_k(Nl_1^3, \chi)$. We put

$\delta_q = \begin{pmatrix} q1_n & 0 \\ 0 & 1_n \end{pmatrix}$ as before. Then we have

$$\begin{aligned}
&\left(h\big|_k\left[\Gamma_0(Nl_1^2)\begin{pmatrix} 1_n & 0 \\ 0 & p1_n \end{pmatrix}\Gamma_0(Nl_1^2/p)\right]\right)(Z) \\
&= \left(h\big|_k\left[\Gamma_0(Nl_1^3)\begin{pmatrix} 1_n & 0 \\ 0 & p1_n \end{pmatrix}\Gamma_0(Nl_1^3/p)\right]\right)(Z) \\
&= \sum_{q|l_1}\left(q^{-nk/2}h_q\big|_k[\delta_q]\big|_k\left[\Gamma_0(Nl_1^3)\begin{pmatrix} 1_n & 0 \\ 0 & p1_n \end{pmatrix}\Gamma_0(Nl_1^3/p)\right]\right)(Z) \\
&= \sum_{q|l_1}\left(h_q\big|_k\left[\Gamma_0(Nl_1^3)\begin{pmatrix} 1_n & 0 \\ 0 & p1_n \end{pmatrix}\Gamma_0(Nl_1^3/p)\right]\right)(qZ).
\end{aligned}$$

Since $h_q \in A_k(Nl_1^3, \chi)$, we have

$$h_q\big|_k\left[\Gamma_0(Nl_1^3)\begin{pmatrix} 1_n & 0 \\ 0 & p1_n \end{pmatrix}\Gamma_0(Nl_1^3/p)\right] \in A_k(\Gamma_0(Nl_1^3/p), \chi)$$

and hence the coefficient of $\left(h|_k\left[\Gamma_0(Nl_1^2)\begin{pmatrix} 1_n & 0 \\ 0 & p1_n \end{pmatrix}\Gamma_0(Nl_1^2/p)\right]\right)(Z)$ at T vanishes unless $\mathrm{cont}(T)$ is divisible by some $q|l_1$. This is the same for h. Hence the coefficient of $F(Z) - f_p(pZ)$ at T vanishes if $\mathrm{cont}(T)$ is coprime to l_1. Hence by inductive assumption, we prove the theorem. When F is a cusp form, all the forms which appear in the above proof are cusp forms and the assertion on the cuspidality follows easily. q.e.d.

Now we denote by $S_k^1(N, \chi)$ the subspace of $S_k(N, \chi)$ spanned by all $f(dZ)$ where $f \in S_k(\Gamma_0(M), \chi)$ with $m_\chi|M$, $dM|N$ and $M \neq N$. We denote by $S_k^0(N, \chi)$ the orthogonal complement of $S_k^1(N, \chi)$ in $S_k(N, \chi)$ with respect to the Petersson inner product.

Proof of Theorem 2.2. Assume that $F \in S_k^0(N, \chi)$ and $a(T) = 0$ for any T with $\mathrm{cont}(T) = 1$. We put

$$G(Z) = \sum_{d|(N/m_\chi)} \mu(d)(F|_kU(d))(dZ),$$

where μ is the Möbius function and $U(d) = \prod_{p|d,p:prime} U(p)$ for square-free d. Since $d|(N/m_\chi)$, we have $F|_kU(d) \in S_k(N, \chi)$ and $(F|_kU(d))(dZ) \in S_k(dN, \chi)$, hence $G \in S_k(N^2/m_\chi, \chi)$. We write $F(Z) = \sum_T a(T)e(\mathrm{tr}(TZ))$ and $G(Z) = \sum_T b(T)e(\mathrm{tr}(TZ))$. Fix a half integral symmetric matrix T_1. We write $\mathrm{cont}(T_1) = m$ and put $m_0 = (m, N/m_\chi)$. Then, since $(F|_kU(d))(dZ) = \sum_T a(dT)e(\mathrm{tr}(dTZ))$, the coefficient at $e(\mathrm{tr}(T_1Z))$ of

$G(Z)$ vanishes unless $d|m_0$. We have

$$b(T_1) = a(T_1) \sum_{d|m_0} \mu(d) = \begin{cases} a(T_1) & \text{if } m_0 = 1, \\ 0 & \text{otherwise.} \end{cases}$$

So in general we have $b(T) = 0$ if $(\mathrm{cont}(T), N/m_\chi) \neq 1$. Now assume that $\mathrm{cont}(T)|(N^2/m_\chi^2)$. Then for any prime p with $p|\mathrm{cont}(T)$, we have $p|(N/m_\chi)$, so if $\mathrm{cont}(T) \neq 1$, then $1 < (\mathrm{cont(T)}, N/m_\chi)$. In this case we have $b(T) = 0$ as shown above. If $\mathrm{cont}(T) = 1$, then we have $b(T) = a(T)$, and by our assumption, we have $a(T) = 0$ for any primitive T. Hence by applying Yamana's theorem in [13] for $G \in S_k(N^2/m_\chi, \chi)$, we have $G = 0$. So we have

$$F(Z) = -\sum_{d|(N/m_\chi), d\neq 1} \mu(d)(F|_k U(d))(dZ).$$

But obviously the Fourier coefficients at T of the right hand side vanish if $(\mathrm{cont}(T), N/m_\chi) = 1$. So taking $l = N/m_\chi$ in Theorem 2.3, we have $F = \sum_{p|l, p:prime} f_p(pZ)$ for $f_p \in A_k(N/p, \chi)$ and this means that $F \in S_k^1(N, \chi)$. Since $S_k^0(N, \chi) \cap S_k^1(N, \chi) = \{0\}$, we have $F = 0$. q.e.d.

Proof of Proposition 2.1. It is sufficient to show that $S_k^1(N, \chi)$ is invariant by $T(p)$ and $T_i(p^2)$ $(0 \le i \le n-1)$ for any p coprime to N, since the dual of $T(p)$ or $T_i(p^2)$ with respect to the Petersson metric is a constant multiple of $T(p)$ or $T_i(p^2)$. We take $F(Z) = \sum_\nu F_\nu(d_\nu Z)$, where $F_\nu(Z) \in S_k(M_\nu, \chi)$, $d_\nu M_\nu | N$, $M_\nu \neq N$, $m_\chi | M_\nu$. We put $T = T(p)$ or $T_i(p^2)$. Since d_ν is coprime to p, for each T, we have

$$T = \bigcup_{A,B,D} \Gamma_0(N) \begin{pmatrix} A & B \\ 0 & D \end{pmatrix}$$

for some matrices A, B, $D \in M_n(\mathbb{Z})$ such that the matrices $\begin{pmatrix} A & d_\nu B \\ 0 & D \end{pmatrix}$ are also the complete set of representatives of the left cosets in T. Here the set of A, B, D can be taken independent of N as far as p is coprime to N. (cf. Andrianov [1]). So if $m_\chi | M_\nu | N$ then we have $S_k(M_\nu, \chi) \subset S_k(N, \chi)$, and the operator T on $S_k(M_\nu, \chi)$ is just the restriction of the operator T on $S_k(N, \chi)$. Also, as operators on $S_k(M_\nu, \chi)$, we have $[\delta_{d_\nu}]T = T[\delta_{d_\nu}]$ when p is coprime to $d_\nu M_\nu$, so we have the following commutative diagram.

$$\begin{array}{ccc} S_k(M_\nu, \chi) & \xrightarrow{T} & S_k(M_\nu, \chi) \\ \downarrow [\delta_{d_\nu}] & & \downarrow [\delta_{d_\nu}] \\ S_k(d_\nu M_\nu, \chi) & \xrightarrow{T} & S_k(d_\nu M_\nu, \chi) \end{array}$$

This is because we have

$$\begin{aligned} F\big|_k[\delta_{d_\nu}]\big|_k T &= const. \sum_{A,B,D} F\big|_k[\delta_{d_\nu}]\big|_k \begin{bmatrix} A & B \\ 0 & D \end{bmatrix} \\ &= const. \sum_{A,B,D} F\big|_k \begin{bmatrix} A & d_\nu B \\ 0 & D \end{bmatrix} \big|_k[\delta_{d_\nu}] \\ &= F\big|_k T\big|_k[\delta_{d_\nu}] \end{aligned}$$

Hence $S_k^1(N,\chi)$ is invariant by $\mathcal{H}_N(N)$. q.e.d.

References

1. A. N. Andrianov, *Quadratic Forms and Hecke Operators,* Springer Verlag, Berlin Heidelberg New York London Paris Tokyo, (1987), pp.374+xii.
2. A. O. Atkin and J. Lehner, Hecke operators on $\Gamma_0(m)$, Math. Ann. **185**(1970), 134–160.
3. S. Breulmann and W. Kohnen, Twisted Maass-Koecher series and spinor zeta functions, Nagoya Math. J. **155**(1999),153–160.
4. S. Breulmann and W. Kohnen, On Hecke eigenforms of degree n, Abh. Math. Sem. Univ. Hamburg **70**(2000), 119–123.
5. B. Roberts and R. Schmidt, On modular forms for the paramodular groups. Automorphic forms and zeta functions, 334–364, World Sci. Publ., Hackensack, NJ, 2006.
6. B. Roberts and R. Schmidt, *Local Newforms for GSp(4)*, Lecture Notes in Mathematics **1918**, Springer Berlin Heidelberg (2007) pp.307+viii.
7. T. Ibukiyama, On symplectic Euler factors of genus two, J. Fac. Sci. Univ. Tokyo Sect. IA Math. **30** (1984), no. 3, 587–614.
8. T. Ibukiyama, On relations of dimensions of automorphic forms of $\mathrm{Sp}(2,R)$ and its compact twist Sp(2) (I). Automorphic forms and number theory (Sendai, 1983), 7–30, Adv. Stud. Pure Math. **7**, North-Holland, Amsterdam, 1985.
9. H. Katsurada, On the coincidence of Hecke-Eigenforms, Abh. Math. Sem. Univ. Hamburg **70**(2000),77-83.
10. H. Katsurada, A remark on the coincidence of Hecke-eigenforms. Comment. Math. Univ. St. Pauli Vol. **52** No. 1 (2003), 47–54.
11. T. Miyake, *Modular Forms*, (1989) pp.335+ viii, Springer-Verlag Berlin Heidelberg New York.
12. R. Schmidt, Iwahori-spherical representations of GSp(4) and Siegel modular forms of degree 2 with square-free level. J. Math. Soc. Japan **57** (2005), no. 1, 259–293.
13. S. Yamana, Determination of holomorphic modular forms by primitive Fourier coefficients, Math. Ann. **344** (2009), 853–862.

Cohomology of Siegel modular varieties of genus 2 and corresponding automorphic forms

Takayuki Oda

Graduate School of Mathematical Sciences, University of Tokyo,
3-8-1 Komaba, Meguro, Tokyo, 153-8914 Japan
E-mail: takayuki@ms.u-tokyo.ac.jp

1. Introduction

This is a short survey on the papers of myself directly related to the cohomology groups of the Siegel modular varieties of genus two, and on the other hand the papers on the spherical functions on the real symplectic group $G = Sp(2, \mathbf{R})$, written by me and by the people around me. These two themes are strongly related to each other. There are two different groups of mathematicians, one is working in algebro-geometric method, and the other in representation-theoretic method. But either group does not seem to understand the other group, because either is already a rather sophisticated theme, and many mathematicians are neither naive nor honest to admit that they are ignorant of the other side of this important theme.

The article consists of five parts. The former three parts are rather analytic, though they have an obvious connection with the geometry of the complex analytic or algebraic varieties: in the first part (§1), we recall some general facts on cohomology groups of the uniform lattices in Lie groups (alias the cocompact discrete subgroups in Lie groups) including the Matsushima isomorphism; in the second part (§2 and §3), we specialize the results of §1 to the group $G = Sp(2, \mathbf{R})$, and a down-to-earth description of the automorphic realization of the (large) discrete series representations of G is given; in the third part (§4), the known results on the explicit formulas of various spherical functions on G are reviewed in short. The last two parts are geometric; in the part four (§5) we recall the basic facts on the mixed Hodge structures of the cohomology groups of the Siegel modular varieties of genus two, and in the last part (§6) some attempt to construct non-trivial cohomology classes of degree two on the Siegel modular varieties of genus two will be reviewed.

This short paper is never intended to be a general survey of the status quo of the theme of the title, but I hope it is still helpful for some readers. I want to thank the reviewer(s) for the very careful reading of the manuscript.

2. The Matsushima isomorphism and related results

The Matsushima isomorphism is the generalization of the Eichler isomorphism to a higher-dimensional locally symmetric space $\Gamma\backslash G/K$, where G is a semisimple real Lie group, K a maximal compact subgroup, and Γ a cocompact discrete subgroup of G. It was established around 1960.

I had a chance to hear the historical review by Haefliger on the time around its discovery, at the occasion of the memorial workshop for Armand Borel. This result appeared suddenly; before this, there are only some results on cocompact discrete subgroups in nilpotent Lie groups.

As shown in the proceeding of the conference at Boulder, Colorado ('67), this activated the investigation on the cohomology groups of various vector-bundles on locally symmetric spaces (see a series of paper by Matsushima-Murakami).

We know that in 70's Borel developed the theory of "stable cohomology", partly motivated by an investigation of K-groups of algebraic number fields. Utilizing the Borel-Serre bordification, he can show that the cohomology groups with low degrees compared with the rank of G, the cohomology class are square-integrable. Moreover for some typical series G_n (say, $G_n = SL(n, \mathbf{R})$ or $G = Sp(n, \mathbf{R})$) of classical groups of the same type, the stable part of cohomology groups of fixed degree that is independent of n, is generated by the cocycles coming from the compact dual of G, hence is independent of Γ_n ($\Gamma_n = SL(n, \mathbf{Z})$, or $\Gamma_n = Sp(n, \mathbf{Z})$)

When Γ is non-cocompact but with covolume finite, there are still many things to be done. This will be discussed in Part 3-5 (§4-6).

2.1. *Cohomology of discrete subgroups*

In this section, we recall basic facts on cohomology groups of discrete subgroups in real semisimple Lie groups. A good reference is a survey article by A. Borel [B1]. The book [BW] of Borel-Wallach also has been a very important reference, though this is a bit difficult to read through since it is not written as a textbook. The original papers ([Ma1,Ma2]) about this theme are due to Matsushima. In the case when G/K is Hermitian, there is a paper by the author which is more specialized to the Hodge theoretic aspect of the problem [O85].

2.2. *Shift to the relative Lie algebra cohomology groups*

Given an arithmetic discrete subgroup Γ in a semisimple real Lie group G, we consider its Eilenberg-Maclane cohomology group $\mathrm{H}^i(\Gamma, \mathbf{C})$. Or more generally if a finite dimensional rational representation $r : G \to GL(V)$ of G is given, we may regard it as a Γ-module by restriction, and we can form the cohomology group $\mathrm{H}^i(\Gamma, V)$.

Fix a maximal compact subgroup K of G to get a Riemannian symmetric space $X = G/K$. For simplicity assume that we have no elements of finite order in Γ, then Γ acts on X from the right side without fixed point, and the quotient $\Gamma\backslash X$ becomes a manifold. In this case Γ is isomorphic to the fundamental group of this manifold (X is contractible to a point), and the Γ-modules V defines a local system $\tilde{V}$ on this quotient manifold. Then we have an isomorphism of cohomology groups: $\mathrm{H}^*(\Gamma, V) \cong \mathrm{H}^*(\Gamma\backslash X, \tilde{V})$.

Let $\sigma : X \to \Gamma\backslash X$ be the canonical map, by pulling back differential forms with respect to σ we have a monomorphism of de Rham complexes $\sigma^0 : \Omega^*_{\Gamma\backslash X} \to \Omega^*_X$. Moreover on the target complex, Γ acts naturally. Let $(\Omega^*_X)^\Gamma$ be the invariant subcomplex. Then it coincides with the image of σ^0, and by the de Rham theorem, we have an isomorphism of cohomology groups:

$$\mathrm{H}^*(\Gamma, V) = \mathrm{H}^*(\Gamma\backslash X, \tilde{V}) = \mathrm{H}^*(\Omega_X(V)^\Gamma).$$

The last complex in the above isomorphim is identified with the complex of differential forms on G as follows.

Let π^0 be the pull-back homomorphism of differential forms with respect to the canonical map $\pi : \Gamma\backslash G \to \Gamma\backslash X$. Then a form $\omega \in \Omega_X(V)^\Gamma$ defines a differential form ω^0 on G by

$$x \in G \mapsto r(x)^{-1}\pi^0(\omega)(x),$$

and denote by $A^*_0(G, \Gamma, V)$ the image of this homomorphism. Then since σ^0 is a monomorphism, we have an isomorphism of complexes $\Omega^*_X(V)^\Gamma \cong A^*_0(G, \Gamma, V)$. Here r is the restriction of G on V.

Here the last complex is identified with the right K-invariant subcomplex of the de Rham complex $\Omega^*_{\Gamma\backslash G}(V) = \Omega^*_{\Gamma\backslash G} \otimes V$, which is defined over $\Gamma\backslash G$ and takes values in V.

Since the tangent space at each point of G/K is identified with the orthogonal complement $\mathfrak{p}$ of $\mathfrak{k}$ in $\mathfrak{g}$ with respect to the Killing form, the module of the i-th cochains becomes $\mathrm{Hom}_K(\wedge^i\mathfrak{p}, C^\infty(\Gamma\backslash G)\otimes V)$. Therefore, the cochain complex defined in this manner gives the relative Lie algebra cohomology groups. We have an isomorphism:

$$\mathrm{H}^*(\Gamma, V) \cong \mathrm{H}^*(\mathfrak{g}, K; C^\infty(\Gamma\backslash G) \otimes V).$$

2.3. *Matsushima isomorphism*

Let G be a connected semisimple real Lie group with finite center. Assume that the discrete subgroup Γ is cocompact, i.e. the quotient $\Gamma\backslash G$ is compact.

Let $L^2(\Gamma\backslash G)$ be the space of L^2-functions on $\Gamma\backslash G$ with respect the Haar measure on G, on which G acts unitarily by the right action. By assumption the space $C^\infty(\Gamma\backslash G)$ is a subspace of this space.

Proposition 2.1. *(Gelfand-Graev,Piateskii-Shapiro) If Γ is cocompact, we have the direct sum decomposition of the unitary representation $L^2(\Gamma\backslash G)$ into irreducible components:*

$$L^2(\Gamma\backslash G) = \tilde{\oplus}_{\pi\in\hat{G}} m(\pi, \Gamma) M_\pi,$$

with finite multiplicities $m(\pi, \Gamma)$. Here $\hat{G}$ is the unitary dual of G, i.e., the set of the unitary equivalence classes of irreducible unitary representations of G, and M_π denotes the representation space of π.

By the above proposition, one has a decomposition

$$C^\infty(\Gamma\backslash G) = \tilde{\oplus}_{\pi\in\hat{G}} m(\pi, \Gamma) M_\pi^\infty$$

of topological linear spaces. Here M_π^∞ is the subspace consisting of C^∞-vectors in the representation space M_π.

Theorem 2.2. *For a finite dimensional rational G-module V over the complex number field, we have an isomorphism:*

$$\mathrm{H}^*(\Gamma, V) = \oplus_{\pi\in\hat{G}} m(\pi, \Gamma) \mathrm{H}^*(\mathfrak{g}, K, M_\pi^\infty \otimes V)$$

$$= \oplus_{\pi\in\hat{G}} \{\mathrm{Hom}_G(M_\pi, L^2(\Gamma\backslash G)) \otimes_{\mathbf{C}} \mathrm{H}^*(\mathfrak{g}, K, M_\pi^\infty \otimes V)\}$$

The key point of the proof here is the complete direct sum $\tilde{\oplus}$ is replaced by a simple algebraic direct sum by passing to the cohomology (*cf.* Borel [B1]).

We can equip a K-invariant inner product on V. By using this, we may regard $\mathrm{H}^m(\mathfrak{g}, K, M_\pi^\infty \otimes V)$ as a space of a kind of harmonic forms, i.e. the totality of cochains annihilated by the Laplace operator. Thus we have the following.

Proposition 2.3. *Let (r, V) be an irreducible G-module of finite dimension. For any $\pi \in \hat{G}$ we have the following.*

(1) If $\chi_r(C) = \chi_\pi(C)$, there is an isomorphism

$$\mathrm{Hom}_K(\wedge^m \mathfrak{p}, M_\pi^\infty \otimes V) = C^m(\mathfrak{g}, K, M_\pi^\infty \otimes V) \cong \mathrm{H}^*(\mathfrak{g}, K, M_\pi^\infty \otimes V).$$

(2) If $\chi_r(C) \neq \chi_\pi(C)$, there is an isomorphism

$$\mathrm{H}^m(\mathfrak{g}, K, M_\pi^\infty \otimes V) = \{0\}.$$

Here χ_r and χ_π are the infinitesimal characters of r and π, respectively, and C denotes the Casimir operator.

Remark 2.1. In particular, when V is the trivial G-module $\mathbf{C}$, the above theorem is no other than the original formula of Betti numbers of $\Gamma\backslash X$ by Matsushima [Ma1].

Also there is a variant of this type of vanishing theorem shown by D. Wigner. But it is omitted here.

We recall here that the relative Lie algebra cohomology group $\mathrm{H}^m(\mathfrak{g}, K, M_\pi^\infty \otimes V)$ is isomorphic to the continuous cohomology group $\mathrm{H}^m_{ct}(G, M_\pi^\infty \otimes V)$ if G is connected. This is shown by using differential cohomology and van Est spectral sequence.

2.4. *"Classical" vanishing theorems*

A number of vanishing theorems were found in '60's: Calabi-Vesentini, Weil etc. In their proof, the same type of computation of "curvature forms" is done, which is similar to a proof of Kodaira vanishing theorem.

Firstly, Matsushima's vanishing theorem of the 1-st Betti number of $\Gamma\backslash X$ was also proved by such a method [Ma1].

This type of vanishing theorem is vastly improved by a representation theoretic method. Probably the best result of this category is the following result by Zuckerman [Z] (see also [BW], Chapter V, §2—§3 (p.150–155)).

Theorem 2.4. *Let G be a simple real algebraic group, (π, M_π) a nontrivial irreducible unitary representation of G, and (r, V) a finite dimensional representation of G. Then for $k < \mathrm{rank}_{\mathbf{R}} G$,*

$$\mathrm{H}^k_{ct}(G, M_\pi^\infty \otimes V) = \{0\}.$$

Corollary 2.5. *Given a cocompact discrete subgroup Γ of G, for $k < \mathrm{rank}_{\mathbf{R}} G$, the restriction homomorphism*

$$\mathrm{H}^k_{ct}(G, V) \to \mathrm{H}^k(\Gamma, V)$$

is an isomorphism.

2.5. *Enumeration and construction of unitary cohomological representations*

We shortly review the state of arts on the cohomological representations defined below.

Definition 2.1. An irreducible (unitary) representation $(\pi, M_\pi) \in \hat{G}$ is called *cohomological*, if there is a finite dimensional G-module F such that

$$\mathrm{H}^i(\mathfrak{g}, K; M_\pi^\infty \otimes_{\mathbf{C}} F) \neq \{0\}$$

for some $i \in \mathbf{N}$. The set of equivalent classes of cohomological representations is denoted by $\hat{G}_{coh}$.

Enumeration of such cohomological representations was done for the case. trivial $F = \mathbf{C}$ by Kumaresan [Km], and for general case by Vogan-Zuckerman [VZ]. This was originally described by using the cohomological induction functor $\mathcal{A}_\lambda(\mathfrak{q})$ first. And later a global realization of this Zuckerman module was obtained by H.-W. Wong [Wg].

2.6. *Example, the classical case* $SL(2, \mathbf{R})$

Let $G = SL(2, \mathbf{R})$. Then it has discrete series representations D_k^+ and D_k^- with the Blattner parameters $k \geq 2$. But the naming, holomorphic and anti-holomorphic discrete series are sometimes confused. The cohomological representations are exhausted by the set:

$$\{\mathbf{1}\} \cup \{D_k^\pm \mid k = 2, 3, \cdots\}.$$

These satisfy

$$\mathrm{H}^i(\mathfrak{g}, K; \mathbf{1}) = \begin{cases} \mathbf{C}, & i = 0 \text{ or } = 2; \\ 0, & i = 1 \text{ or } \geq 3. \end{cases}$$

and

$$\mathrm{H}^i(\mathfrak{g}, K; D_{k+2}^\pm \otimes Sym^k) = \begin{cases} \mathbf{C}, & i = 1 \\ 0, & i \neq 1. \end{cases}$$

Here Sym^k is the degree k symmetric tensor product of the standard representation of $SL(2, \mathbf{R})$ on $\mathbf{C}^2$.

Hence for a cocompact $\Gamma \subset G$,

$$\mathrm{H}^1(X_\Gamma, Sym^k) \cong \mathrm{Hom}_G(D_{k+2}^-, L^2(\Gamma\backslash G)) \oplus \mathrm{Hom}_G(D_{k+2}^+, L^2(\Gamma\backslash G)).$$

and the spaces in the right hand side are identified with the spaces of holomorphic and anti-holomorphic automorphic forms of weight $k+2$ with respect to Γ, respectively. This is a paraphrase of the Eichler isomorphism. The modification by Shimura for a non-cocompact Γ is to replace :
(a) $\mathrm{H}^1(X_\Gamma, Sym^k)$ by the image $\mathrm{H}^1_!(X_\Gamma, Sym^k)$ of the canonical homomorphism $\mathrm{H}^1_c(X_\Gamma, Sym^k) \to \mathrm{H}^1(X_\Gamma, Sym^k)$;
(b) the space $L^2(\Gamma\backslash G)$ by $L^2(\Gamma\backslash G)_{dis} = L^2(\Gamma\backslash G)_{cusp} \oplus L^2(\Gamma\backslash G)_{res}$.
Here $L^2(\Gamma\backslash G)_{dis}$ is the sum of all the irreducible subrepresentations and $L^2(\Gamma\backslash G)_{res} = \mathbf{C1}$ is the residual spectrum.

References

B1. Borel, A.: *Cohomologie des sous-groupes discrete et répresentations de groupes* . Astérisque **32-33** (1976), 73–112.

BW. Borel, A. and Wallach, N.: Continuous cohomology, discrete subgroups, and representations of reductive groups. Annals of Math. Studies No.**94**, Princeton University Press. 1980

Km. Kumaresan, S.: *On the canonical k-types in the irreducible unitary $\mathfrak{g}$-modules with nonzero relative cohomology.* Invent. Math. **59** (1980), 1–11.

Ma1. Matsushima, Y.: *On the first Betti number of compact quotient of higher dimensional symmetric spaces.* Annals of Math. **75** (1962), 312–30.

Ma2. Matsushima, Y.: *On Betti numbers of compact, locally symmetric Riemannian manifolds.* Osaka Math. J. **14** (1962), 1–20.

VZ. Vogan, D. and Zuckerman, G.: *Unitary representations with non-zero cohomology.* Comp. Math. **53** (1984), 51–90.

W1. Wallach, N.R.: *Real Reductive Groups I*, Academic Press, 1988.

Wg. Wong, H.-W.: *Dolbeault cohomological realization of Zuckerman modules associated with finite rank representations.* Journ. of Funct. Analy. **129** (1995), 428–454.

Z. Zuckerman, G.: *Continuous cohomology and unitary representations of real reductive groups.* Ann. of Math. **107** (1978), 495–516.

3. Basic rudiments on Siegel modular varieties

In this section, we recall some basic facts on the Lie group G and the homogeneous space G/K with a maximal compact subgroup K of G in the first subsection. The book [Sa] is readable for a reference on Hermitian symmetric spaces.

3.1. *The Lie group and the associated homogeneous space, etc.*

A standard realization of the symplectic group of rank n is

$$Sp(n, A) := \left\{ g \in M_{2n}(A) \mid {}^t g J_n g = J_n,\ J_n = \begin{pmatrix} 0 & 1_n \\ -1_n & 0 \end{pmatrix} \right\},$$

for a subring A of $\mathbf{C}$. This group is a split semisimple algebraic group of rank n with root system of type C_n.

On the real simple group $G = Sp(n, \mathbf{R})$, we can define a Cartan involution $\theta(g) = {}^t g^{-1}$. Then the fixed part $K := G^\theta$ is a maximal compact subgroup of G.

When $n = 1$, $G = SL(2, \mathbf{R})$ and $K = SO(2)$, and the homogeneous space G/K is identified with the complex upper half plane or with the complex unit disk. For general n, corresponding to these two realizations, we have the Siegel upper half space:

$$H_n := \{ Z = {}^t Z \mid Im(Z) >> 0 \}$$

and the other realization:

$$D_n := \{ Z = {}^t Z \mid 1_n - {}^t \bar{Z} Z >> 0 \}.$$

In either case, the action of $g = \begin{pmatrix} A & B \\ C & D \end{pmatrix} \in G$ is given by the "fractional linear transformation"

$$Z \mapsto g(Z) := (AZ + B)(CZ + D)^{-1}.$$

The subgroup K is the isotropy subgroup (or the stabilizer) of $\sqrt{-1}1_n \in H_n$ or $0 \in D_n$. The map $Z \mapsto (Z - \sqrt{-1}1_n)(Z + \sqrt{-1}1_n)^{-1} \in D_n$ is a biholomorphic map from H_n to D_n.

The Cartan involution θ defines the Cartan decomposition:

$$\mathfrak{g} = \mathfrak{k} + \mathfrak{p},$$

where $\mathfrak{g}$, $\mathfrak{k}$ are the corresponding Lie algebras of G, K, and $\mathfrak{p} = \mathfrak{g}^{-\theta}$. Since the group G is a split group, we can choose a maximal abelian subalgebra $\mathfrak{a}$ in $\mathfrak{p}$ of real dimension n.

3.2. *The discrete subgroups*

The group $Sp(n, \mathbf{Z})$ is a discrete subgroup of G, which has finite covolume $vol(Sp(n, \mathbf{Z}) \backslash G) < \infty$, as proven by Reduction Theory. More generally, we may consider a subgroup Γ in $Sp(n, \mathbf{Q})$ commensurable with $Sp(n, \mathbf{Z})$.

Then it also has a finite covolume. The variety $X_\Gamma := \Gamma\backslash H_n$ is a Siegel modular variety which parametrizes certain complex abelian varieties of dimension n with principal polarization and some level structure. And the theory of Satake compactification (later generalized by Baily-Borel), or a chronologically earlier result of Siegel on the algebraic dimension of the quotient X_Γ implies that this quotient is a complex algebraic variety, which is quasi-projective that is compactified as a projective algebraic variety. And this compactification, which has terrible singularities, so-called cusp singularities at infinity, is resolved by toroidal resolution. This is one way to grasp the quotient X_Γ, the view-points of algebraic geometry.

The other way is to consider X_Γ as a quotient of the symmetric space H_n under the simple Lie group G, whose canonical Kähler structure is inherited to the quotient X_Γ. Then we can use the tool and the results of the Kähler geometry, and even the representation theory of the group of symmetry G of H_n. This is the view-points of transcendental algebraic geometry, or/and the view-points of complex and real analysis.

The combination of these two view-points, if adequately utilized, brings many deep results and beautiful methods to investigate X_Γ.

4. Example: The Matsushima isomorphism for compact quotients of $Sp(2, \mathbf{R})$

From now on we set $n = 2$. We are going to see the Matsushima isomorphism in this case rather closely. The symmetric domain H_2 has complex dimension 3. In the first place we fix a discrete subgroup Γ such that $X_\Gamma = \Gamma\backslash H_2$ is compact.

For such a cocompact group, the Matsushima isomorphism is already explained. We see what will happen for this special case when $G = Sp(2, \mathbf{R})$.

Remark 4.1. Examples of such groups that X_Γ is compact are found in the monograph of Siegel "Symplectic Geometry" in the classical literature, which are also the sources of Shimura varieties having "strange model" in the saying of Deligne. But Shimura himself seems to be proud to find this kind of models, which utilizes various results of modular embeddings to get a canonical field of definition.

4.1. *Finite-dimensional representations of G*

To have natural local systems over the quotient X_Γ, we firstly recall the irreducible finite-dimensional rational representations of G. The result is algebraic and does not depend on the special field $\mathbf{R}$. The group G is split of rank 2, and fix a split Cartan subgroup

$$S = \{\mathrm{diag}(t_1, t_2, t_1^{-1}, t_2^{-1}) \mid t_i \in \mathbf{G}_m = \mathbf{R}^\times\}.$$

Then the root system of G with respect to S is of type C_2:

$$\Phi = \Phi(\mathfrak{g}, \mathfrak{s}) = \{\pm 2e_1,\ \pm 2e_2, \pm(e_1 \pm e_2)\}.$$

Here $\{e_1, e_2\}$ is the standard orthonormal basis in the Euclidean plane. A positive system in Φ is chosen: $\Phi_+ = \{2e_1, 2e_2, e_1 - e_2, e_1 + e_2\}$, with the half-sum of positive roots $\rho = (2, 1)$. Then the set of dominant weights is $\{l = (l_1, l_2) \mid l_1 \geq l_2 \geq 0\}$.

Let (ρ_l, V_l) be the irreducible rational representation G associated with the dominant weight l with the representation space V_l, then the Weyl dimension formula tells us that

$$\dim \rho_l = \dim V_l = \frac{\prod_{\alpha \in \Phi_+} < l + \rho, \alpha >}{\prod_{\alpha \in \Phi_+} < \rho, \alpha >}$$
$$= \frac{1}{6}(l_1 + 2)(l_2 + 1)(l_1 - l_2 + 1)(l_1 + l_2 + 3).$$

If we restrict this representation of G to the subgroup Γ, then we have a representation of the fundamental group $\pi_1(X_\Gamma, *) \cong \Gamma$ for Γ torsion-free. Hence this defines a local (coefficient) system $\tilde{V}_l$ on X_Γ. If the intersection $\Gamma \cap Z(G)$ acts trivially on V_l, we can associate a constructible sheaf $\tilde{V}_l$ on X_Γ similarly. The extended version of the Matsushima isomorphism gives the description of $\mathrm{H}^i(X_\Gamma, \tilde{V}_l)$ in terms of cohomological automorphic forms. In particular when $i = 3$, it is a direct sum of spaces of automorphic forms belonging to the discrete series representations of G.

4.2. *The representations of discrete series of $Sp(2, \mathbf{R})$*

Among cohomological representations, there are rather "big" representations which normally contribute to the cohomology groups of the middle degree. These are discrete series representations.

These representations make up a large part of the "main series" (tempered) of unitary representations of semisimple Lie groups, and historically speaking, the systematic research was started by the colossal effort of Harish-Chandra, up to mid 60's.

4.2.1. *Discrete series, or square-integrable representations*

Definition 4.1. An irreducible unitary representation $\pi \in \hat{G}$ is called a *discrete series* representation or belongs to the discrete series, if it satisfies

one of the following equivalent conditions:
(C1) The equivalence class of π is isolated in $\hat{G}$ with respect to the dual topology;
(C2) π is square-integrable, i.e., all the matrix coefficients of π belong to $L^2(G)$, or equivalently a non-zero matrix coefficient of π belongs to $L^2(G)$ (this part is due to Godement);
(C3) π is realized as a closed (irreducible) subspace in $L^2(G)$. (In effect, in the $G\times G$-modules $L^2(G)$ it occurs in the multiplicity-free subspace $\pi \boxtimes \pi^*$, where π^* is the contragradient (discrete series) representation of G.)

A general criterion for a semisimple Lie group to have discrete series representations, is that it has a compact Cartan subgroup.

4.2.2. *Harish-Chandra's parametrization of discrete series for $Sp(2,\mathbf{R})$*

In our case, we know rank$G$ = rank$K$ = 2 for G. Therefore we can choose a compact Cartan subgroup $T \subset K \subset G$. For the standard choice of K, we may take

$$T := \left\{ \begin{pmatrix} \cos\theta_1 & 0 & \sin\theta_1 & 0 \\ 0 & \cos\theta_2 & 0 & \sin\theta_2 \\ -\sin\theta_1 & 0 & \cos\theta_1 & 0 \\ 0 & -\sin\theta_2 & 0 & \cos\theta_2 \end{pmatrix} \mid \theta_1, \theta_2 \in \mathbf{R} \right\}.$$

Then the root system $\Phi(\mathfrak{g},\mathfrak{t})$ is given by

$$\Phi(\mathfrak{g},\mathfrak{t}) = \{\pm 2e_1, \pm 2e_2, \pm(e_1 \pm e_2)\}$$

with the Euclid basis $e_1 = (1,0), e_2 = (0,1)$. Then in the root space decomposition: $\mathfrak{g}_{\mathbf{C}} = \mathfrak{t}_C \oplus \bigoplus_{\alpha\in\Phi} \mathfrak{g}_\alpha$, the compact part is given by $\mathfrak{k}_C = \mathfrak{t}_{\mathbf{C}} \oplus \mathfrak{g}_{e_1-e_2} \oplus \mathfrak{g}_{-e_1+e_2}$. Therefore $\pm(e_1-e_2)$ are called compact roots, and the others non-compact roots.

Since the group G has a compact Cartan subgroup T in K, a weight l for the pair $(\mathfrak{g}_{\mathbf{C}}, \mathfrak{t}_{\mathbf{C}})$ is naturally identified with a weight for the pair $(\mathfrak{k}_{\mathbf{C}}, \mathfrak{t}_{\mathbf{C}})$. The dominant integral weights for the latter pair in turn parametrize the equivalence classes of continuous finite-dimensional representations of the connected compact Lie group $K \cong U(2)$, and also the equivalence classes of rational irreducible representations of the algebraic group $K_{\mathbf{C}} \cong GL(2,\mathbf{C})$. We denote the irreducible representation of $K \cong U(2)$ or of $K_{\mathbf{C}} \cong GL(2,\mathbf{C})$ with highest weight l by τ_l.

We fix a set of simple roots as $\{e_1 - e_2, 2e_2\}$. Then the half sum ρ of positive roots is given by $\rho = (2,1) \in \sqrt{-1}\mathfrak{t}^*$, which is integral, i.e., it is a derivation of a unitary character of T. Then for the integral weights $L_T = \sqrt{-1}\log\hat{T} \subset \mathfrak{t}^*_{\mathbf{C}}$, we have $L_T = L_T + \rho$. Let L_T^{reg} be the subset of the regular weights defined as follows.

Let Ξ be the subset of L_T consisting of integral weights satisfying the conditions below:

(i) The weight $n = (n_1, n_2) \in L_T$ be regular, i.e., $< n, \alpha > \neq 0$ for any $\alpha \in \Phi(\mathfrak{g}, \mathfrak{t})$, or equivalently

$$n_1 \neq 0, \quad n_2 \neq 0, \quad n_1 - n_2 \neq 0, \quad \text{and } n_1 + n_2 \neq 0.$$

(ii) Moreover n is positive with respect to the compact positive root $e_1 - e_2$, i.e., $n_1 - n_2 > 0$.

Then Ξ is a representative of the orbit space L_T^{reg}/W_K with W_K the Weyl group of K. This set parametrizes the subset $\hat{G}_{DS}$ consisting of discrete series in the unitary dual $\hat{G}$ of G. Let (D_n, H_{D_n}) be the discrete series representation corresponding to $n \in \Xi$. Historically speaking, each D_n for $n \in \Xi$ was firstly characterized by the values of its character (distribution) at T.

The set Ξ has a natural partition: $\Xi = \Xi_I \cup \Xi_{II} \cup \Xi_{III} \cup \Xi_{IV}$ with

$$\Xi_I := \{n = (n_1, n_2) \in \Xi \mid n_1 > n_2\}, \quad \Xi_{II} := \{n_1 > -n_2 > 0\},$$
$$\Xi_{III} := \{-n_2 > n_1 > 0\}, \quad \Xi_{IV} := \{-n_2 > -n_1 > 0\}.$$

The Weyl chambers containing these subset has the adjacency: $I \leftrightarrow II \leftrightarrow III \leftrightarrow IV$.

(A) Those discrete series D_n with parameters in $\Xi_I \cup \Xi_{IV}$ are called either holomorphic discrete series or anti-holomorphic discrete series, which have the Gel'fand-Kirillov dimension $3 = \dim_{\mathbf{C}} G/K$.

(B) The others with parameters $n \in \Xi_{II} \cup \Xi_{III}$ are called large, because they have the largest possible Gel'fand-Kirillov dimension $4 = \dim_{\mathbf{R}} N_{min}$ with N_{min} the unipotent radical of the minimal parabolic subgroup. These representations have Whittaker models.

4.2.3. *The discrete series with the same infinitesimal characters*

The reflection s_{2e_2} with respect to the root $2e_2$ maps an element $n = (n_1, n_2)$ in Ξ_I to $n' = s_{2e_2}(n) = (n_1, -n_2)$ in Ξ_{II}. Further by the reflection $s_{e_1+e_2}$ we can map n' to $n'' = (n_2, -n_1)$ in Ξ_{III}. Last, via the reflection s_{2e_1} we can map n'' to $\bar{n} = (-n_2, -n_1)$ in Ξ_{IV}. These four discrete series have the same infinitesimal characters. We introduce a symbol.

Notation 4.2. We denote by W^1 a subset of the Weyl group $W(G,T)$:

$W^1 = \{w_0 := e,\ w_1 := s_{2e_2},\ w_2 = s_{e_1+e_2} \circ s_{2e_2},\ w_3 = s_{2e_1} \circ s_{e_1+e_2} \circ s_{2e_2}\}.$

Then we have the following.

Proposition 4.1. *Let $l = (l_1, l_2)$ be a dominant weight in L_T $(l_1 \geq l_2 \geq 0)$. Further let*

$$n_I = l + \rho \in \Xi_I; \quad n_{II} = w_1(l+\rho) \in \Xi_{II};$$
$$n_{III} = w_2(l+\rho) \in \Xi_{III}; \quad n_{IV} = w_3(l+\rho) \in \Xi_{IV}.$$

Then the four discrete series associated to these four Harish-Chandra parameters have the same infinitesimal characters $\chi_{D_*} : Z(\mathfrak{g}) \to \mathbf{C}$. Moreover these exhaust all the discrete series representation with the same infinitesimal characters as that of the finite dimensional representation ρ_l.

4.2.4. *K-types of a discrete series*

Given a discrete series D_n with parameter n, then we can describe the K-types $D_n|K$. For discrete series of a general simple Lie group satisfying the condition $\mathrm{rank}G = \mathrm{rank}K$, this was conjectured by Blattner, and proved by Hecht-Schmid in general in '76. Among the K-types, there is one $\tau_{min}(D_n)$ called the *minimal K-type*, that occurs with multiplicity one. The parameter of $\tau_{min}(D_n)$ is sometimes called the *Blattner parameter* $\beta(n)$ which is obtained from the Harish-Chandra parameter n adding some offset. We describe this deviations directly for $Sp(2,\mathbf{R})$. General case is found in, say, Borel-Wallach [BW], Chapter II, §5.

Proposition 4.2.

$$\begin{cases} \beta(n) = n + (1,2), & \text{\textit{if} } n \in \Xi_I; \\ \beta(n) = n + (1,0), & \text{\textit{if} } n \in \Xi_{II}; \\ \beta(n) = n + (0,-1), & \text{\textit{if} } n \in \Xi_{III}; \\ \beta(n) = n + (-2,-1), & \text{\textit{if} } n \in \Xi_{IV}. \end{cases}$$

The Blattner parameters have geometric meaning in the cohomological realization of discrete series, and cohomological automorphic forms.

4.3. *Cohomological representations*

Discrete series representations are firstly investigated, probably because they are parts of those representations which contribute to the Plancherel formula. This means that they are a part of the tempered representaions. But discrete series are very important representation also from geometric view-points because they have non-trivial $(\mathfrak{g}, K)$-cohomology group $\mathrm{H}^*(\mathfrak{g}, K; H_{\pi,K} \otimes V_l) \neq \{0\}$. There are other non-tempered representations, which also contribute to $(\mathfrak{g}, K)$-cohomology groups. Those representations appear in the complicated composition series of (generalized) principal series with regular (Harish-Chandra) parameters; "complicated" because Yoneda interpretation of cohomology classes of the relative Lie algebra cohomology and the information derived from the composition series of generalized principal series are not directly connected in the case of Lie groups of higher rank.

4.4. *The list of cohomological representations with trivial infinitesimal characters*

We consider the trivial representation $\rho_{(0,0)} = \mathbf{1}$ of G. It is finite-dimensional, but also an irreducible unitary representations. Let us enumerate the unitary representations $\pi \in \hat{G}$ so that $\mathrm{H}^i(\mathfrak{g}, K; H_{\pi,K}) \neq \{0\}$ for some i.

4.4.1. *Discrete series*

There are four discrete series $\pi \in \{D_{(2,\pm1)}, D_{(\pm1,-2)}\}$ such that $\mathrm{H}^3(\mathfrak{g}, K; H_{\pi,K}) \neq \{0\}$, but $\mathrm{H}^i(\mathfrak{g}, K; H_{\pi,K} \otimes V_l) = \{0\}$ for $i \neq 3$. This is discussed already in the previous section.

4.4.2. *Non-tempered cohomological representations*

There are the other four cohomological representations $\{\mathbf{1}, J_{2,0}, J_{1,1}, J_{0,2}\}$. All these contribute to

$$\mathrm{H}^2(\mathfrak{g}, K; H_{\pi,K}) \cong \mathbf{C} \text{ and } \mathrm{H}^4(\mathfrak{g}, K; H_{\pi,K}) \cong \mathbf{C}.$$

When $\pi = \mathbf{1}$ it also contributes $\mathrm{H}^i(\mathfrak{g}, K; H_{\pi,K}) \cong \mathbf{C}$ for $i = 0, 6$. And for other degrees, cohomology groups vanish. The representation $J_{2,0}$, $J_{1,1}$ and $J_{0,2}$ have the minimal K-types $\tau_{(3,1)}$, $\tau_{(2,-2)}$, and $\tau_{(-1,-3)}$, respectively. Here $\tau_{(l_1,l_2)}$ is the simple K-module with highest weight (l_1, l_2).

Remark 4.2. We have no unitary representations, contributing to $\mathrm{H}^1(\mathfrak{g}, K; M_\pi)$. But if it were just $(\mathfrak{g}, K)$-modules, there are such ones. But these are not unitarizable. Note that $\mathbf{1}$ is unitary, but the other finite-dimensional representations are not.

When $l = (l_1, l_2)$ is regular, i.e., $l_1 > l_2 > 0$, only the discrete series with $\chi_\pi = \chi_{\rho_l^*}$ are cohomological, and only the degree 3 part $\mathrm{H}^3(\mathfrak{g}, K; H_{\pi,K} \otimes V_l)$ is non-vanishing.

The cases $l = (l_1, 0)$ $(l_1 > 0)$ and $l = (l_1, l_2)$ $(l_1 = l_2)$ will be omitted to discuss.

4.5. *The Matsushima isomorphism for compact quotients*

Let $l = (l_1, l_2)$ be a dominant integral weight, and let (ρ_l, V_l) be the finite-dimensional representation of $G = Sp(2, \mathbf{R})$ with highest weight $l = (l_1, l_2)$. Given a discrete subgroup Γ in G, then we have Γ-module V_l by restriction. If Γ is torsion-free, Γ is canonically isomorphic to the fundamental group of the quotient $\Gamma \backslash H_2$. Hence V_l defines a local (coefficient) system $\tilde{V}_l$ on $X_\Gamma = \Gamma \backslash H_2$. When $\Gamma \cap Z(G)$ acts trivially V_l, even if $\Gamma/(\Gamma \cap Z(G))$ has torsion, we can define similarly a constructible sheaf $\tilde{V}_l$ associated with V_l. Thus we consider the cohomology groups with twisted coefficients: $\mathrm{H}^i(X_\Gamma, \tilde{V}_l)$.

For the reason of dimension, $\mathrm{H}^i(X_\Gamma, \tilde{V}_l) = \{0\}$ for $i > 6$. If ρ_l is not trivial, $i = 0$ -th cohomology vanishes and for $i = 6$, it also vanishes by Poincaré duality. Other parts are described by the following result.

Theorem 4.3. *Let Γ be a cocompact subgroup of $Sp(2, \mathbf{R})$.*
(a) When $i = 3$,

$$\mathrm{H}^3(X_\Gamma, \tilde{V}_l) \cong \oplus_{i=0}^3 \mathrm{Hom}_G(D_{w_i(l+\rho)}, L^2(\Gamma \backslash G))$$

with w_i in the subset W^1 of the Weyl group.
(b) The groups H^1 and H^5 vanish.
(c) When l is regular, both for H^2 and H^4 vanish. When $l = (0, 0)$ we have

$$\mathrm{H}^2(X_\Gamma, \mathbf{C}) \cong \{\oplus_{i=0}^2 \mathrm{Hom}_G(J_{(i,2-i)}, L^2(\Gamma \backslash G))\} \oplus \mathrm{H}^{(1,1)}(\mathfrak{g}, K; \mathbf{C}),$$

and a similar isomorphism for H^4 (by Poincaré duality).

Remark 4.3. When l is not regular and $l \neq (0, 0)$, we have similar results for H^2 and H^4.

Remark 4.4. For the discrete series $D_{w_i(l+\rho)}$, we have

$$\mathrm{H}^3(\mathfrak{g}, K; D_{w_i(l+\rho)} \otimes V_l) \cong \mathbf{C}.$$

Remark 4.5. The reason of vanishing $\mathrm{H}^1(X_\Gamma, \tilde{V}_l) = \{0\}$ is non-existence of cohomological unitary representation $[\pi]$ of G contributing $\mathrm{H}^1(\mathfrak{g}, K; M_\pi \otimes V_l)$. The same is valid for H^5 by the Poincaré duality.

Remark 4.6. We have another proof of the above theorem from the context of complex analytic geometry, utilizing variation of Hodge structures etc. We refer to the latter part of [O85]. Comparing with the formulation of there, you can tell the Hodge type of each direct factor of the above theorem (compare with the paper of Zucker quoted there). We want to refer to yet another paper of Faltings [Fa], which gives another proof of $(\mathfrak{g}, K)$-cohomology groups in the case when G/K is of Hermitian type. On this paper, there is a review MR732471(85k:22028) by S. Murakami about the relation with the former results in the literature.

4.6. *Scholia for the Matsushima isomorphism for* $Sp(2, \mathbf{R})$

Let us consider the discrete series $D_{(-l_2-1,-l_1-2)}$ which appears in the Matsushima isomorphism of $\mathrm{H}^3(X_\Gamma, \tilde{V}_l)$. Its minimal K-type is $\tau_{(-l_2-3,-l_1-3)} \in \hat{K}$.

Let $j(\gamma, z) = (cz + d) \in GL(2, \mathbf{C})$ be the canonical automorphy factor with $\gamma = \begin{pmatrix} a & b \\ c & d \end{pmatrix}$, $z \in H_2$. Let $M_l(\Gamma)$ be the space of vector-valued holomorphic automorphic forms F with values in V_l, satisfying

$$F(\gamma(z)) = \tau_l(j(\gamma, z))F(z).$$

We note that we embed the maximal compact subgroup K of G into $GL(2, \mathbf{C})$ by $k \mapsto j(k, \sqrt{-1}I_2)$.

Proposition 4.4. *We have the canonical identification :*

$$\mathrm{Hom}_G(D_{(-l_2-1,-l_1-2)}, L^2(\Gamma\backslash G)) \cong M_{(l_1+3,l_2+3)}(\Gamma)$$

unique up to scalar multiple.

Proof. The non-zero K-homomorphism

$$j : \tau_{min} = \tau_{(-l_2-3,-l_1-3)} \hookrightarrow D_{(-l_2-1,-l_1-2)}$$

unique up to scalar multiple gives the associated restriction map:

$$res : \mathrm{Hom}_G(D_{(-l_2-1,-l_1-2)}, L^2(\Gamma\backslash G)) \to \mathrm{Hom}_K(\tau_{(-l_2-3,-l_1-3)}, L^2(\Gamma\backslash G)).$$

Moreover we have the natural identification

$$\mathrm{Hom}_G(D_{(-l_2-1,-l_1-2)}, L^2(\Gamma\backslash G)) = \{L^2(\Gamma\backslash G) \otimes \tau_{(l_1+3,l_2+3)}\}^K.$$

Therefore the image $res(\varphi) = \varphi \cdot j$ of an intertwining operator φ in $\mathrm{Hom}_G(D_{(-l_2-1,-l_1-2)}, L^2(\Gamma\backslash G))$ is an element of $\{L^2(\Gamma\backslash G)\otimes\tau_{(l_1+3,l_2+3)}\}^K$.

On the other side, the minimality of K-type implies that the image of j is annihilated by the elements of $\mathfrak{p}_-$. Since φ is an intertwiner, the image $res(\varphi)$ is also annihilated by the right quasi-regular action R_X $(X \in \mathfrak{p}_-)$ on $L^2(\Gamma\backslash G)\otimes\tau_{(l_1+3,l_2+3)}$. We can find readily that these are Cauchy-Riemann operators, which ensure that an element in $res(\varphi)$ corresponds to a vector-valued holomorphic automorphic forms belonging to $M_{(l_1+3,l_2+3)}(\Gamma)$. □

Similarly we have the canonical identification :

$$\mathrm{Hom}_G(D_{(l_1+2,l_2+1)}, L^2(\Gamma\backslash G)) \cong \overline{M_{(l_1+3,l_2+3)}(\Gamma)}.$$

4.6.1. *The middle components*

The two spaces

$$\mathrm{Hom}_G(D_{(l_2+1,-l_1-2)}, L^2(\Gamma\backslash G)) \text{ and } \mathrm{Hom}_G(D_{(l_1+2,-l_2-1)}, L^2(\Gamma\backslash G)),$$

which are mutually complex conjugate, appeared only after 60's, and we do not understand these spaces well.

In the former case, the minimal K-type of $D_{(l_2+1,-l_1-2)}$ is $\tau_{(l_2+1,-l_1-3)}$. Therefore the target of the restriction map

$$\begin{aligned} res : \mathrm{Hom}_G(D_{(l_2+1,-l_1-2)}, L^2(\Gamma\backslash G)) &\to \\ \mathrm{Hom}_K(\tau_{(l_2+1,-l_1-3)}, L^2(\Gamma\backslash G)) &\cong \{L^2(\Gamma\backslash G)\otimes\tau^*_{(l_2+1,-l_1-3)}\}^K \end{aligned}$$

specifies the K-type of the vector-valued (harmonic) automorphic forms corresponding to the source intertwining space $\mathrm{Hom}_G(D_{(l_2+1,-l_1-2)}, L^2(\Gamma\backslash G))$. However in this case "harmonicity" does not mean that the automorphic forms in the image are holomorphic automorphic forms. There is another system of partial differential operators, which characterize the images, that we call the *Dirac-Schmid* operator.

4.6.2. *The Dirac-Schmid operator*

We have to analyze the space generated by the gradient operators on G/K from the minimal K-type. The complexification of the real tangent space of G/K at the origin $eK \in G/K$ is canonically identified with $\mathfrak{p}_{\mathbf{C}}$, which is a K-module by the adjoint representation.

Consider the tensor product of K-modules $\mathfrak{p}_{\mathbf{C}}\otimes\tau_{(l_2+1,-l_1-2)}$, and the natural map $X\otimes v \to Xv \in M_\pi$ into the representation space M_π of $\pi = D_{(l_2+1,-l_1-2)}$. Recall here the Clebsh-Gordan decomposition of the K-module

$$\mathfrak{p}_{\mathbf{C}}\otimes\tau_\beta = \{\mathfrak{p}_+\otimes\tau_\beta\}\oplus\{\mathfrak{p}_-\otimes\tau_\beta\}$$

with the Blattner parameter $\beta = (l_2 + 1, -l_1 - 2)$:

$$\mathfrak{p}_+ \otimes \tau_\beta \cong \tau_{\beta+(2,0)} \oplus \tau_{\beta+(1,1)} \oplus \tau_{\beta+(0,2)} \text{ and}$$
$$\mathfrak{p}_- \otimes \tau_\beta \cong \tau_{\beta+(0,-2)} \oplus \tau_{\beta+(-1,-1)} \oplus \tau_{\beta+(-2,0)}.$$

Let ι_{+2}, ι_0 and ι_{-2} be the K-homomorphism of each irreducible component in the right hand side into the tensor product of the left hand side in the above Clebsh-Gordan decomposition. Then the fact that β is the minimal K-type means that there is no K-types $\tau_{\beta+(1,1)}$, $\tau_{\beta+(0,2)}$ and $\tau_{\beta+(-2,0)}$ in our representation $D_{(l_2+1,-l_1-2)}$. This is a part of deep results on the K-type of the discrete series representations, which was proved earlier than more sophisticated results. Anyway this fact implies that the restriction maps

$$res : \mathrm{Hom}_G(D_{(l_2+1,-l_1-2)}, L^2(\Gamma\backslash G)) \to$$
$$\mathrm{Hom}_K(\tau_{\beta+(1,1)}, L^2(\Gamma\backslash G)) \cong \{L^2(\Gamma\backslash G) \otimes \tau^*_{\beta+(1,1)}\}^K$$

and other two similar restriction maps are all zero-maps. This can be interpreted in the dual picture, i.e., in terms of the gradient operators:

$$\nabla_{\mathfrak{p}_\pm} : \{L^2(\Gamma\backslash G) \otimes \tau^*_\beta\}^K \to \{L^2(\Gamma\backslash G) \otimes \tau^*_\beta \otimes \mathfrak{p}^*_\pm\}^K.$$

Note here that $\mathfrak{p}^*_\pm$ is the cotangent spaces at the origin of G/K

Let pr_{+2}, pr_0 and pr_{-2} be the dual projectors to the K-homomorphism ι_{+2} etc. Then the image F_ϕ of an intertwiner

$\phi \in \mathrm{Hom}_G(D_{(l_2+1,-l_1-2)}, L^2(\Gamma\backslash G))$ with respect to the above restriction map *res* is annihilated by a system of operators:

$$pr_{-2} \circ \nabla_{\mathfrak{p}_+} F_\phi = 0, \quad pr_0 \circ \nabla_{\mathfrak{p}_+} F_\phi = 0, \text{ and}$$
$$pr_{-2} \circ \nabla_{\mathfrak{p}_-} F_\phi = 0.$$

This system of operators is called *Schmid operator* and the equation *Schmid* equation.

Because the Schmid equation is the fundamental tool to have explicit formulas for various spherical functions belonging to the discrete series representations, in the next subsection we give a down-to-earth description of this equation for the case of the large discrete series representation with trivial infinitesimal character.

4.7. *Explicit description of the automorphic realization of the Schmid equation*

Set $l = (0, 0)$ and consider

$$res : \mathrm{Hom}_G(D_{(+1,-2)}, L^2(\Gamma\backslash G)) \to$$
$$\{F \in \{L^2(\Gamma\backslash G) \otimes \tau^*_{\beta+(1,-3)}\}^K \mid F \text{ satisfies the Schmid equation}\}.$$

We try to give a down-to-earth description of the target space.

4.7.1. *The natural trivialization of the bundle* $\Omega^2_{H_2} \otimes \bar{\Omega}^1_{H_2}$

Given an element $g = \begin{pmatrix} A & B \\ C & D \end{pmatrix}$ of $G = Sp(2, \mathbf{R})$ and $Z \in H_2$ we set $Z^* = g(Z)$ with

$$Z^* = (AZ + B) \cdot (CZ + D)^{-1} = {}^t(CZ + D)^{-1} \cdot {}^t(AZ + B).$$

Choose another point $W \in Z$ and $W^* = g(W)$. Then the following formulas are well-known.

Formula 4.5. Let $\bar{W}$ be the complex conjugate of the matrix W. Then

$$Z^* - \bar{W}^* = {}^t(C\bar{W} + D)^{-1} \cdot (Z - \bar{W}) \cdot (CZ + D)^{-1}.$$

In particular if we set $W = Z$, $Y = \mathrm{Im}(Z) = \frac{1}{2\sqrt{-1}}(Z - \bar{Z})$ and $Y^* = \mathrm{Im}(Z^*)$, we have

$$Y^* = {}^t(C\bar{Z} + D)^{-1} \cdot Y \cdot (CZ + D)^{-1}.$$

Formula 4.6. We have $Z^* - W^* = {}^t(CW + D)^{-1} \cdot (Z - W) \cdot (CZ + D)^{-1}$. In particular, let $W = Z + dZ$ with the matrix of differentials $dZ = \begin{pmatrix} dz_{11} & dz_{12} \\ dz_{12} & dz_{22} \end{pmatrix}$, then

$$dZ^* = {}^t(CZ + D)^{-1} \cdot dZ \cdot (CZ + D)^{-1}.$$

4.7.2. *The G-invariant Kähler metric and the associated metric form*

The unique G-invariant Kähler metric is given by

$$\begin{aligned}
&\mathrm{tr}(Y^{-1}dZY^{-1}d\bar{Z}) = \mathrm{tr}\{(Y^{-\frac{1}{2}}dZY^{-\frac{1}{2}}) \cdot (Y^{-\frac{1}{2}}d\bar{Z}Y^{-\frac{1}{2}})\} \\
&= \det(Y)^{-2}\{y_{22}^2 dz_{11} \otimes d\bar{z}_{11} + 2(y_{11}y_{22} + y_{12}^2)dz_{12} \otimes d\bar{z}_{12} + y_{11}^2 dz_{22} \otimes d\bar{z}_{22} \\
&+ y_{12}^2(dz_{11} \otimes d\bar{z}_{22} + dz_{22} \otimes d\bar{z}_{11}) - 2y_{12}y_{22}(dz_{12} \otimes d\bar{z}_{22}) - 2y_{12}y_{11}(dz_{22} \otimes d\bar{z}_{12})\}
\end{aligned}$$

If one replaces the symbol $\otimes$ of tensor product by the symbol $\wedge$ of wedge product in the above formula, then the obtained (metric) form is of $(1,1)$-type and d-closed. On passing to the quotient $X_\Gamma = \Gamma\backslash H_2$ with a discrete cocompact subgroup Γ, this defines a "universal" class in $\mathrm{H}^2(X_\Gamma, \mathbf{C})$, corresponding to the canonical element $\kappa \otimes 1$ in $\mathrm{H}^2(\mathfrak{g}, K; \mathbf{C}) \otimes \mathrm{Hom}_G(\mathbf{C}, L^2(\Gamma\backslash G))$.

4.7.3. *The natural trivialization of* $\Omega^1_{H_2}$ *and* $\Omega^2_{H_2}$

For $(g, Z) \in G \times H_2$, the canonical automorphy factor $j(g,Z)$ is given by

$$j(g,Z) = CZ + D \text{ with } g = \begin{pmatrix} A & B \\ C & D \end{pmatrix}.$$

Then the formula

$$dZ^* = {}^t j(g,Z)^{-1} \cdot dZ \cdot j(g,Z)^{-1} \text{ for } Z^* = g(Z)$$

gives a natural way to trivialize the locally free sheaf $\Omega^1_{H_2}$: each local section of this sheaf is written as

$$\omega_f = f_{11}(Z)dz_{11} + f_{12}(Z)dz_{12} + f_{22}(Z)dz_{22} = \mathrm{tr}\left\{ \begin{pmatrix} f_{11}(Z) & \frac{1}{2}f_{12}(Z) \\ \frac{1}{2}f_{12}(Z) & f_{22}(Z) \end{pmatrix} dZ \right\}$$

via holomorphic functions $f_{ij}(Z)$ defined on an open subset of H_2. Put

$$F_f(Z) = \begin{pmatrix} f_{11}(Z) & \frac{1}{2}f_{12}(Z) \\ \frac{1}{2}f_{12}(Z) & f_{22}(Z) \end{pmatrix},$$

then the coefficient matrix $F_f(Z)$ of the form ω_f has to satisfy the automorphy condition

$$F_f(\gamma(Z)) = j(\gamma, Z) \cdot F_f(Z) \cdot {}^t j(\gamma, Z)$$

for any $\gamma \in \Gamma$ in order that ω_f defines a holomorphic 1-form on the quotient X_Γ, i.e., $\omega_f(\gamma(Z)) = \omega_f(Z)$.

Now let us consider the case of holomorphic 2-forms. A natural trivialization of $\Omega^2_{H_2}$ is derived from the above trivialization of $\Omega^2_{H_2}$ as follows.

Notation 4.3. We define a matrix of 2-forms by

$$d^{(2)}Z = \begin{pmatrix} dz_{11} \wedge dz_{12} & \frac{1}{2}dz_{11} \wedge dz_{22} \\ \frac{1}{2}dz_{11} \wedge dz_{22} & dz_{12} \wedge dz_{22} \end{pmatrix}.$$

Then we have

Formula 4.7. Let $Z^* = g(Z)$. Then we have

$$d^{(2)}Z^* = \{\det j(g,Z)\}^{-1} j(g,Z)^{-1} \cdot d^{(2)}Z \cdot j(g,Z)^{-1}.$$

A proof of this is obtained by direct computation. Then an expression of a local section η_h of $\Omega^2_{H_2}$ is written as

$$\begin{aligned}\eta_h &= \operatorname{tr}\left\{\begin{pmatrix} h_{11}(Z) & h_{12}(Z) \\ h_{12}(Z) & h_{22}(Z)\end{pmatrix} d^{(2)}Z\right\} \\ &= h_{11}(Z)dz_{11}\wedge dz_{12} + h_{12}(Z)dz_{11}\wedge dz_{22} + h_{22}(Z)dz_{12}\wedge dz_{22}.\end{aligned}$$

In order that this η_h defines a holomorphic 2-form on X_Γ we have to impose the automorphy condition

$$H(\gamma(Z)) = \{\det(j(\gamma, Z)\}\cdot j(\gamma, Z)H(Z)\cdot {}^t j(\gamma, Z)$$

for the coefficient matrix $H(Z) = \begin{pmatrix} h_{11}(Z) & h_{12}(Z) \\ h_{21}(Z) & h_{22}(Z)\end{pmatrix}$.

4.7.4. *Decomposition of the trivialization on $\Omega^2_{H_2}\otimes\bar{\Omega}^1_{H_2}$*

The fiber at the origin of $G/K = H_2$ of the vector bundle $\Omega^2_{H_2}\otimes\bar{\Omega}^1_{H_2}$ has natural structure of a K-module obtained from the G-linearization (or the G-action) on $\Omega^2_{H_2}\otimes\bar{\Omega}^1_{H_2}$. Then the Clebsh-Gordan formula gives natural decomposition of this bundle into rank 5, rank 3, and rank 1 subbundles with induced trivializations. Here is the explicit computation of these induced trivializations.

A natural system of global generators to trivial this tensor product is given by the Kronecker product

$$(Y^{-1}d\bar{Z}Y^{-1})\otimes d^{(2)}Z = \begin{pmatrix} a_{11} & a_{12} & a_{13} & a_{14} \\ a_{21} & a_{22} & a_{23} & a_{24} \\ a_{31} & a_{32} & a_{33} & a_{34} \\ a_{41} & a_{42} & a_{43} & a_{44}\end{pmatrix},$$

where if we write

$$Y^{-1}d\bar{Z}Y^{-1} = \begin{pmatrix} \alpha & \beta \\ \gamma & \delta\end{pmatrix}, \text{ and } d^{(2)}Z = \begin{pmatrix} A & B \\ B & C\end{pmatrix},$$

we have

$$\begin{aligned} a_{11} &= \alpha A,\ a_{12} = \beta A,\ a_{22} = \gamma A \\ a_{13} &= \alpha B,\ a_{14} = \beta B,\ a_{24} = \gamma B \\ a_{23} &= \alpha C,\ a_{34} = \beta C,\ a_{44} = \gamma C\end{aligned}$$

by the definition of the Kronecker product

$$\begin{pmatrix} \begin{pmatrix} \alpha & \beta \\ \beta & \gamma \end{pmatrix} A & \begin{pmatrix} \alpha & \beta \\ \beta & \gamma \end{pmatrix} B \\ \begin{pmatrix} \alpha & \beta \\ \beta & \gamma \end{pmatrix} B & \begin{pmatrix} \alpha & \beta \\ \beta & \gamma \end{pmatrix} C \end{pmatrix}$$

of matrices. These nine 3-forms $a_{ij} = a_{ij}(Z)$ are real analytic global section of $\Omega^2_{H_2} \otimes \bar{\Omega}^1_{H_2}$. which have the following explicit formulas. Since the expression of A, B, C are simple, it suffices to give the explicit formulas of α, β, γ.

Formula 4.8. The entries of the matrix $Y^{-1} d\bar{Z} Y^{-1} = \begin{pmatrix} \alpha & \beta \\ \gamma & \delta \end{pmatrix}$ are given by

$$\begin{pmatrix} \alpha \\ \beta \\ \gamma \end{pmatrix} = (\det Y)^{-2} \begin{pmatrix} y_{22}^2 & -2y_{12}y_{22} & y_{12}^2 \\ -y_{12}y_{22} & y_{11}y_{22} + y_{12}^2 & -y_{11}y_{12} \\ y_{12}^2 & -2y_{11}y_{12} & y_{11}^2 \end{pmatrix} \begin{pmatrix} d\bar{z}_{11} \\ d\bar{z}_{12} \\ d\bar{z}_{22} \end{pmatrix}.$$

On the other side, by definition we have

$$A = dz_{11} \wedge dz_{12}, \quad B = \frac{1}{2} dz_{11} \wedge dz_{22}, \quad C = dz_{12} \wedge dz_{12}.$$

Hence we have

$$a_{11}(Z) = \alpha A = (\det Y)^{-2} \{y_{22}^2 d\bar{z}_{11} - 2y_{12}y_{22} d\bar{z}_{12} + y_{12}^2 d\bar{z}_{22}\} \otimes (dz_{11} \wedge dz_{12}),$$

and similar formulas for the other eight elements $a_{ij}(Z)$.

Now we have the decomposition of $\Omega^2_{H_2} \otimes \bar{\Omega}^1_{H_2}$ according to the Clebsh-Gordan formula: $\tau_{(3,1)} \otimes \tau_{(0,-2)} = \tau_{(3,-1)} \oplus \tau_{(2,0)} \oplus \tau_{(1,1)}$.

Notation 4.4. Put

$$\mathbf{V}_{(3,-1)}(Z) = \begin{pmatrix} a_{22}(Z) \\ \frac{1}{2}\{-a_{12}(Z) + a_{24}(Z)\} \\ \frac{1}{6}\{a_{11}(Z) + a_{44}(Z) - 4a_{14}(Z)\} \\ \frac{1}{2}\{-a_{34}(Z) + a_{13}(Z)\} \\ a_{33}(Z) \end{pmatrix},$$

$$\mathbf{V}_{(2,0)}(Z) = \begin{pmatrix} \frac{1}{2}\{a_{11}(Z) - a_{44}(Z)\} & a_{13}(Z) + a_{34}(Z) \\ a_{12}(Z) + a_{24}(Z) & \frac{1}{2}\{a_{44}(Z) - a_{11}(Z)\} \end{pmatrix},$$

and

$$\begin{aligned} \mathbf{v}_{(1,1)} &= a_{11}(Z) + 2a_{14}(Z) + a_{44}(Z) \\ &= (\det Y)^{-2}[(dz_{11} \wedge dz_{12}) \otimes \{y_{22}^2 d\bar{z}_{11} - 2y_{12}y_{22} d\bar{z}_{12} + y_{12}^2 d\bar{z}_{22}\} \\ &+ (dz_{11} \wedge dz_{22}) \otimes \{-y_{12}y_{22} d\bar{z}_{11} + (y_{11}y_{22} + y_{12}^2) d\bar{z}_{12} - y_{11}y_{12} d\bar{z}_{22}\} \\ &+ (dz_{12} \wedge dz_{22}) \otimes \{y_{12}^2 d\bar{z}_{11} - 2y_{11}y_{12} d\bar{z}_{12} + y_{11}^2 d\bar{z}_{22}\}]. \end{aligned}$$

Remark 4.7. If we take the "matrix product" of 2×2 matrices with entries in 1-forms and 2-forms on H_2:

$$H(Z) := (Y^{-1} d\bar{Z} Y^{-1}) \cdot d^{(2)} Z = \begin{pmatrix} a_{11}(Z) + a_{14}(Z) & a_{13}(Z) + a_{34}(Z) \\ a_{12}(Z) + a_{24}(Z) & a_{14}(Z) + a_{44}(Z) \end{pmatrix},$$

then $\mathbf{v}_{(1,1)}(Z) = \mathrm{tr}\{H(Z)\}$ and

$$\mathbf{V}_{(2,0)}(Z) = H(Z) - \frac{1}{2} \mathrm{tr}\{H(Z)\} \cdot 1_2.$$

These are the twice contraction and one-time contraction for the tensor $(Y^{-1} d\bar{Z} Y^{-1}) \otimes d^{(2)} Z$.

We have to confirm that these are corresponding to some direct factors of rank 1, rank 3, and rank 5 of $\Omega^2_{H_2} \otimes \bar{\Omega}^1_{H_2}$ with the restricted G-linearization by the following formulas.

Lemma 4.1. *Let $Z^* = g(Z)$ with $g \in G$ and $Z,\ Z^* \in H_2$. Then we have*

$$\begin{aligned} \mathbf{v}_{(1,1)}(Z^*) &= \det(j(g,Z)) \cdot \mathbf{v}_{(1,1)}(Z), \\ \mathbf{V}_{(2,0)}(Z^*) &= \det(j(g,Z)) j(g,Z) \cdot \mathbf{V}_{(2,0)}(Z) \cdot j(g,Z)^{-1}. \end{aligned}$$

Moreover given $M = \begin{pmatrix} m_{11} & m_{12} \\ m_{21} & m_{22} \end{pmatrix} \in GL(2, \mathbf{C})$, we define a 5×5 matrix $Sym^4(M)$ by

$$\begin{pmatrix} x'^4 \\ x'^3 y' \\ x'^2 y'^2 \\ x' y'^3 \\ y'^4 \end{pmatrix} = Sym^4(M) \begin{pmatrix} x^4 \\ x^3 y \\ x^2 y^2 \\ x y^3 \\ y^4 \end{pmatrix} \quad \text{for} \quad \begin{pmatrix} x' \\ y' \end{pmatrix} = \begin{pmatrix} m_{11} & m_{12} \\ m_{21} & m_{22} \end{pmatrix} \begin{pmatrix} x \\ y \end{pmatrix},$$

utilizing symbolic variables x, y, x', y'. Then

$$\mathbf{V}_{(3,-1)}(Z^*) = \det(j(g,Z)) Sym^4({}^t j(g,Z))^{-1} \mathbf{V}_{(3,-1)}(Z).$$

Remark 4.8. We write only the left three columns of the matrix 5×5 matrix $Sym^4(M)$ for confirmation:

$$\begin{pmatrix} m_{11}^4 & 4m_{11}^3 m_{12} & 6m_{11}^2 m_{12}^2 & * & * \\ m_{11}^3 m_{21} & 3m_{11}^2 m_{12} m_{21} + m_{11}^3 m_{22} & 3m_{11} m_{12}^2 m_{21} + 3m_{11}^2 m_{12} m_{22} & * & * \\ m_{11}^2 m_{21}^2 & 2m_{11} m_{12} m_{21}^2 & m_{12}^2 m_{21}^2 + 4m_{11} m_{12} m_{21} m_{22} + m_{11}^2 m_{22}^2 & * & * \\ m_{11} m_{21}^3 & m_{21}^2 (3m_{11} m_{22} + m_{12} m_{21}) & 3m_{21} m_{22} (m_{11} m_{22} + m_{12} m_{21}) & * & * \\ m_{21}^4 & 4m_{21}^3 m_{22} & 6m_{21}^2 m_{22}^2 & * & * \end{pmatrix}.$$

The readers will be able to guess by symmetry or confirm directly the suppressed entries denoted by $*$'s.

Now recall the restriction map

$$res : \varphi \in \mathrm{Hom}_G(D_{(1,-2)}, L^2(\Gamma\backslash G)) \\ \to F_\varphi \in \{L^2(\Gamma\backslash G) \otimes \tau_{(1,-3)}\}^K$$

which we have to describe. Then in general F_φ is written as an inner product of a row vector of five C^∞ functions

$$(f_0^{(3,-1)}(Z), f_1^{(3,-1)}(Z), f_2^{(3,-1)}(Z), f_3^{(3,-1)}(Z), f_4^{(3,-1)}(Z))$$

and the column vector $\mathbf{V}_{(3,-1)}(Z)$, which is a C^∞-section of the real analytic vector bundle $\Omega^2_{H_2} \otimes \bar{\Omega}^1_{H_2}$ on the smooth locus of X_Γ.

Before finishing this subsubsection, we note that the bundle $\Omega^2 \otimes \bar{\Omega}^2$ has a similar decomposition into a rank 5, a rank 3, and rank 1 real analytic bundles, corresponding to the fact the fiber at the origin $[K] \in G/K$ has the decomposition of K-modules : $\tau_{(3,1)} \otimes \tau_{(-1,-3)} = \tau_{(2,-2)} \oplus \tau_{(1,-1)} \oplus \tau_{(0,0)}$. The author recommends the readers to try the explicit computation of this decomposition as an exercise.

4.7.5. *Automorphic realization of the Schmid operator*

Note that the Schmid operator is the gradient operator $d = \nabla = \partial + \bar{\partial}$ ($\partial = \nabla_+$ and $\bar{\partial} = \nabla_-$), composed with the projectors corresponding to the decomposition of the G-linearized bundle, which is the target of the gradient operator, into indecomposable G-linearlized subbundles. Among others we have to take notice on the $\bar{\partial}$-operator

$$\bar{\partial} : \Omega^2_{H_2} \otimes \bar{\Omega}^1_{H_2} \to \Omega^2_{H_2} \otimes \bar{\Omega}^2_{H_2}$$

of the holomorphic vector bundle $\Omega^2_{H_2}$. Note also that the C^ω-bundle $\Omega^2_{H_2} \otimes \bar{\Omega}^2_{H_2}$ also decomposes as a direct sum of subbundles of rank 5, 3 and 1, similarly as the case of $\Omega^2_{H_2} \otimes \bar{\Omega}^1_{H_2}$, though we do not write the decomposition explicitly here.

For a scalar valued C^∞ function u, the operator $\bar{\partial} u$ is

$$\frac{\partial u}{\partial \bar{z}_{11}} \otimes d\bar{z}_{11} + \frac{\partial u}{\partial \bar{z}_{12}} \otimes d\bar{z}_{12} + \frac{\partial u}{\partial \bar{z}_{22}} \otimes d\bar{z}_{22} = \mathrm{tr}(\frac{\partial u}{\partial \bar{Z}} \cdot d\bar{Z}),$$

where the matrix of derivations is defined by

$$\frac{\partial u}{\partial \bar{Z}} = \begin{pmatrix} \frac{\partial u}{\partial \bar{z}_{11}} & \frac{1}{2}\frac{\partial u}{\partial \bar{z}_{12}} \\ \frac{1}{2}\frac{\partial u}{\partial \bar{z}_{12}} & \frac{\partial u}{\partial \bar{z}_{22}} \end{pmatrix}.$$

We can extend this to the space of forms with values in $\Omega^2_{H_2} \otimes \bar{\Omega}^1_{H_2}$. Then we have the canonical morphism of G-linearlized bundles

$$\bar{\partial} : \Omega^2_{H_2} \otimes \bar{\Omega}^1_{H_2} \to \Omega^2_{H_2} \otimes \bar{\Omega}^1_{H_2} \otimes \bar{\Omega}^1_{H_2},$$

and composed with the canonical projection $\bar{\Omega}^1_{H_2} \otimes \bar{\Omega}^1_{H_2} \to \bar{\Omega}^2_{H_2}$ to get $\bar{\partial}$ as an exterior derivation.

Passing to the five dimensional subbundle in the source, and to the one dimensional and three dimensional subbundles in the target side of

$$\bar{\partial} : \Omega^2_{H_2} \otimes \bar{\Omega}^1_{H_2} \to \Omega^2_{H_2} \otimes \bar{\Omega}^2_{H_2},$$

we have the first two equations of the Schmid operator. The other remaining equation comes from the the formal adjoint operator with respect to the Petersson metrics (i.e., the canonical G-equivariant Hermitian metrics) on $\Omega^i_{H_2}$, $\bar{\Omega}^i_{H_2}$ and the induced metrics on their tensor products

$$\bar{\partial}^* : \Omega^2_{H_2} \otimes \bar{\Omega}^i_{H_2} \to \Omega^2_{H_2} \otimes \bar{\Omega}^{i-1}_{H_2}.$$

The common kernel is also the kernel of $\square_{\bar{\partial}} = \bar{\partial}\bar{\partial}^* + \bar{\partial}^*\bar{\partial}$. We refer to the papers Narasimhan-Okamoto [NO], and Schmid [Sd1,Sd2] for their origin.

Remark 4.9. The reader might notice that the subbundles of rank 3 and 1 in $\Omega^2_{H_2} \otimes \bar{\Omega}^1_{H_2}$ play no role. The fact that the kernel of $\square_{\bar{\partial}}$ has no components belonging to these subbundles of rank 3 and 1. This is a non-trivial fact reflecting that H_2 is a homogeneous space under G. Say, when you consider the quotient $SU(3,1)/K$ the situation is rather different, though you have the corresponding Schmid equations for the large discrete series representations of $SU(3,1)$. Thus X_Γ is not a general threefold in any mean.

References

Fa. Faltings, G.: *On the cohomology of locally symmetric Hermitian spaces.* Paul Dubriel and Marie-Paule Malliavin algebra seminar, 35th year (Paris, 1982), 55-98, Lecture Notes in Math., **1029**, Springer, 1983 (MR732471).

NO. Narasimhan, M. S. and Okamoto, K.: *An analogue of the Borel-Weil-Bott theorem for hermitian symmetric pairs on non-compact type.* Annals of Math. (2) **91** (1970), 486–511.

O85. Oda, T.: *Hodge structures attached to geometric automorphic forms.* Adv. Stud. in pure Math. **7** (1985), 223-276.

Sa. Satake, I.: Algebraic structures of symmetric domains. Kanô Memorial Lectures, 4. Iwanami Shoten, Tokyo; Princeton University Press, Princeton, N.J., 1980. xvi+321 pp.

Sd1. Schmid, W.: *On the realization of the discrete series of a semisimple Lie group.* Rice Univ. Studies **56** 1970 *no. 2* (1971) 99–108.

Sd2. Schmid, W.: *L^2-cohomology and the discrete series.* Ann. of Math. (2) **103** (1976), no. 2, 375–394.

5. Analytic aspect: Spherical functions

Because Siegel modular varieties are not general algebraic varieties, but are a more special kind of algebraic varieties which are locally symmetric, for deeper study of them we have to utilize this property "locally symmetric". Say the Matsushima isomorphism is valid because the varieties are locally symmetric.

This means that we have to use the information of the harmonic analysis of the Lie group $Sp(2, \mathbf{R})$, i.e., various spherical functions on this group belonging to some representations. Generally speaking a typical problem of the spherical functions on a semisimple Lie group G occurs in the following setting.

We have some closed subgroup R of G, and some admissible (irreducible) representation π of G with representation space M_π, and a unitary irreducible representation η of R with representation space η. We consider the smoothly induced representation $\mathrm{Ind}_R^G(\eta)$ of G, and the intertwining space

$$I_{\eta,\pi} := \mathrm{Hom}_G(\pi, \mathrm{Ind}_R^G(\eta))$$

or its algebraic analogue

$$I_{\eta,\pi}^{alg} := \mathrm{Hom}_{(\mathfrak{g}\ ,K)}(\pi, \mathrm{Ind}_R^G(\eta))$$

between the associated $(\mathfrak{g}, K)$-modules. For example if R is a maximal unipotent subgroup of G and η is a unitary character, then the element of $I_{\eta,\pi}$ is called *Whittaker vector* or *Whittaker functional.*

This problem is an old problem originating from the dawn of the representation theory of Lie groups, and we have a long list of literature. But for application to number theory, sometimes quite important and basic problems are not solved.

In the subsequent subsections, we shall give some introductory remarks for our relatively recent results of the various spherical functions on $G = Sp(2, \mathbf{R})$.

5.1. *Fourier expansion*

When an arithmetic subgroup Γ has no cusps, we can get rid of many troubles in the theory of automorphic forms. But also this means that we lose a fundamental tool to investigate automorphic forms much deeper: Fourier expansion of automorphic forms.

When the discrete subgroup Γ has cusps, we can consider some kind of Fourier expansion of automorphic forms. But situation is not so simple as the case of elliptic modular case where the meaning of "Fourier expansion" is clear and unique.

When Γ is a subgroup of $G_{\mathbf{Q}} := Sp(2, \mathbf{Q})$ that has three different parabolic subgroup the minimal parabolic subgroup P_{min}, a maximal parabolic subgroup P_S (the Siegel parabolic subgroup) with abelian unipotent radical, and the other maximal parabolic subgroup P_J, which we refer as *Jacobi parabolic subgroup*, that some people call *Klingen parabolic subgroup*.

Basically we should have a Fourier expansion associated to each type of parabolic subgroup, but the author does not know even what is the best definition of Fourier expansion for each case. We can state only the state of arts for the local aspect of this.

5.1.1. *Whittaker functions*

When we consider the Fourier expansion with respect to P_{min}, we have to choose an element c in the finite set $\Gamma\backslash G_{\mathbf{Q}}/P_{min}$. Let us denote the conjugate class of P_{min} corresponds to c by the symbol $P_{min,c}$. Let $N_{min,c}$ be the unipotent radical of $P_{min,c}$. Then an automorphic form $\varphi(g)$ is a left $\Gamma_{min,c} := \Gamma \cap N_{min,c}$ invariant function, hence the function

$$(n, g) \in \Gamma_{min,c}\backslash N_{min,c}(\mathbf{R}) \times G \to \varphi(ng)$$

is a continuous function in the first variable n on the compact quotient (note here that $N_{min,c}$ is a unipotent algebraic group). If we decompose the space $L^2(\Gamma_{min,c}\backslash N_{min,c}(\mathbf{R}))$ into isotypic components, we can consider the associated decomposition of $\varphi(ng) = \sum_{\eta \in \hat{N_{min}(\mathbf{R})}} \varphi^{\eta}(n, g)$. When η is one-dimensional representation, the component associated to this has an intertwining property: $\varphi_\eta(n, g) = \eta(n)\varphi_\eta(1, g)$. Here the factor $W_\varphi(g) := \varphi_\eta(1, g)$ is a function in $g \in G$ which has intertwining property $W_\varphi(ng) = \eta(n)W_\varphi(g)$ $(n \in N_{min}(\mathbf{R}),\ g \in G)$.

If the function $\varphi(g)$ generates a representation (π, H_φ) of G by the right quasi-regular action, then the function W_φ belongs to the image of some intertwining space $I_{\eta,\pi}$ with $R = N_{min}(\mathbf{R})$. This kind of functions which belong to the image of a Whittaker functional are called Whittaker functions.

Let $W(g)$ be a Whittaker function on G belonging to a K-type τ of a representation π of G, then because of the Iwasawa decomposition $G = N_{min}(\mathbf{R})AK$ with a split Cartan subgroup A of G, the function W is determined by its values on A, in view of its intertwining property with respect to the left action of N and the right action of K.

We had some results on the explicit formulas of the A-radial part $W|A$ of W when W belongs to a good K-type. Note that $\dim_{\mathbf{R}} A = 2$, hence $W|A$ is a function of two variables.

Chronologically speaking, we first had a result for large discrete series representation π with the minimal K-type τ in [O94]: an integral expression of $W|A$, in terms of the classical Whittaker function (one variable). In retrospect, we can say that this was much easier, compared with the case of principal series, later handled by Ishii, and the reason is that the degree of the associated variety of our π is 4, and in the case of the principal series it is 8. The author was helped much by the papers of Hiroshi Yamashita [Y1,Y2] which handled the case of group $SU(2,2)$.

The method to obtain the explicit formula of $W|A$ is basically as follows: Step 1) We have the Schmid operator as shown in the previous section; Step 2) We can compute the radial part of this operator; Step 3) We find that this system of vector-valued partial differential equations is a holonomic system of rank 4, and a nice solution with rapid decreasing property is found by inspection.

The similar method worked for generalized principal series with respect to the Jacobi parabolic subgroup, because the representation has similar K-types as the case of large discrete series (*cf.* Miyazaki-Oda [M-O2]).

The more difficult case when π is principal series representation, i.e., the case of the parabolic induction, were handled by Niwa [N2] and Ishii [Ishi2]. The method of Niwa utilizes Weil representation, and Ishii utilized the holonomic system of the radial part of the Whittaker functions with minimal K-types belonging to the principal series representations, which was obtained by Miyazaki-Oda [M-O1].

Hasegawa [Hase] generalized the result of Ishii for the case of 1-dimensional K to have M-Whittaker functions (or the secondary Whittaker functions in our terminology, see the example below for definition) belonging to the generalized principal series induced via the Siegel parabolic subgroup. This case seems to be most difficult to have tractable formulas, because the Whittaker functions are vector-valued and the rank of the holonomic system is 8.

Example

It is better to write some formulas to see how these Whittaker functions are complicated. Given an element $\mathrm{diag}(a_1, a_2, a_1^{-1}, a_2^{-1})$ of the split Cartan subgroup A, we define the coordinates $y_1 = a_1/a_2$ and $y_2 = a_2^2$ associated with a basis $(1,-1),(0,2)$ of the root system of type C_2. Let $\partial_i = y_i(\partial\ /\partial y_i)$ be the Euler operator for y_i $(i = 1,2)$. Then up to a simple multiplier corresponding to the half-sum of positive roots, the radial part $\psi(y)$ of class one Whittaker function belonging to a class one principal series satisfies a system of partial differential equations :

$$[{\partial_1}^2 + 2{\partial_2}^2 - 2\partial_1\partial_2 - (2\pi y_1)^2 - 2(2\pi y_2)^2]\psi = \frac{1}{2}(\nu_1^2+\nu_2^2)\psi,$$

and

$$\begin{aligned}[(\partial_1^2-1)\{(\partial_1-2\partial_2)^2-1\} + (2\pi y_1)^4 - 2(2\pi y_1)^2(\partial_1+1)(\partial_1-2\partial_2+1)\\ - 4(2\pi y_2)^2(\partial_1^2-1)]\psi = (\nu_1^2-1)(\nu_2^2-1)\psi,\end{aligned}$$

(a part of Theorems in [M-O1]. This system is a holonomic system of rank 8 with regular singularities along $y_1 = 0$ and $y_2 = 0$. And we have 8 different power series solutions $M_{(\nu_1',\nu_2')}(y)$, $(\nu_1',\nu_2') = w(\nu_1,\nu_2)$ $(w \in W)$ with distinct characteristic exponents, whose coefficients are expressed by the values of the generalized hypergeometric function ${}_3F_2$ (Theorem 2.1 of [Ishi2]). Here W is the Weyl group of G.

The unique solution $W^{(0,0)}_{(\nu_1,\nu_2)}$ of the holonomic system which is of moderate growth, i.e., of rapid decrease at infinity, has an expansion in terms of these 8 solutions :

$$\begin{aligned}&W^{(0,0)}_{(\nu_1,\nu_2)}(y) = 2^{-4}\sum_{w\in W}\\ &w\Big[\Gamma(-\frac{\nu_1-1}{2})\Gamma(-\frac{\nu_2-1}{2})\Gamma(-\frac{\nu_1+\nu_2}{2})\Gamma(-\frac{\nu_1-\nu_2}{2})M^{(0,0)}_{(\nu_1,\nu_2)}(y)\Big],\end{aligned}$$

(Theorem 4.1 of [Ishi2]). We refer to these $M^{(0,0)}_{(\nu_1',\nu_2')}$ as M-Whittaker functions or *the secondary Whittaker functions.* In the original paper [Ishi2] similar results are shown for other principal series, not only the class one case.

Remark 5.1. Recently Tadashi Miyazaki obtained explicit integral expression of (general) principal series Whittaker functions with minimal K-type on the complex group $Sp(2,\mathbf{C})$. We have to remark that Proskurin [Pr], Part 2 has similar expression by direct evaluation of the Jacquet integral for class one case.

5.1.2. *Application to automorphic L-functions*

Hori [Hori] seems to be the first to handle the problem of the archimedean places for the Andrianov' L-function, or by alias the spinor L-function. He utilized the generalized Whittaker function obtained by Niwa, referred in the next subsubsection. Miyazaki [Mi3] extends this kind of results for the generalized principal series with respect to the Jacobi parabolic subgroup and the large discrete series. He also utilized generalized Whittaker functions, because the method to construct the L-function is that of Andrianov.

Moriyama handled the case of automorphic representations whose archimedean component is a large discrete series, utilizing the formula of Whittaker functions in [O94] and Moriyama [M4]. And Ishii-Moriyama [I-M] fixed the case of principal series representations.

Remark 5.2. Some retrospect: We are not yet in a circumstance to have enough library to handle archimedean computation freely. But for Whittaker functions, the situation is much better compared with several years ago. If one see the monograph of Proskurin [Pr], the large part is devoted to the discussion of Whittaker functions on $Sp(2, \mathbf{C})$. Now we have better understanding of many ingenious computations in [Pr], though we cannot yet manipulate them as a kind of routine work.

Problem One possibility of a more direct application is to consider the Poincaré series associated to M-Whittaker functions, or alias "the secondary Whittaker functions". In the case of semisimple Lie group with real rank one, Miatello-Wallach [M-W1]. Their result has an application to generalized Kloosterman sums. There are interesting attempts to generalize such a construction to a higher rank Lie group by Stade [Std] and by Stevens [Stv]. The problem is obviously not yet cultivated enough. Say, one has to make sure the convergence of the series, and after that there is another problem of analytic continuation.

5.1.3. *Siegel-Whittaker functions*

The Fourier expansion with respect to the Siegel parabolic subgroup has a longer history compared with other parabolic subgroups. In the study of Siegel on definite quadratic forms, these were basic tool. Because what he considered were holomorphic modular forms, the representations of $Sp(2, \mathbf{R})$ generated by such modular forms were holomorphic or anti-holomorphic discrete series. These representations have smaller Gel'fand-Kirillov dimensions. Heuristically speaking, this means that the annihilator ideal of the representation π in the universal enveloping algebra $U(\mathfrak{g})$ is much larger than the case of the large discrete series or the case of the principal series, hence they have no Whittaker model. And the Fourier expansion with respect to the Siegel parabolic subgroup P_S also has very simple form, found in the classical literature. Therefore when π is holomorphic or anti-holomorphic discrete series, you can take the unipotent radical N_S of P_S as the group R and η as a non-degenerate character. Then the intertwining space $I_{\eta,\pi}$ is finite-dimensional.

But this is not the case when π is a principal series representation or a large discrete series representation. As R we take a semidirect product of N_S and a stabilizer S_ξ in the Levi part L_S of P_S which stabilizes a non-degenerate character ξ in the dual action of the conjugate action of L_S on the radical N_S. Depending on the signature of the character, S_ξ is isomorphic to either $SO(2)$ or $SO(1,1)$. Then we set $R = N_S \times S_\xi$ (semi-direct) and η is a semi-direct of ξ and a character of S_ξ.

Now the functions belonging to the intertwining operator in $I_{\eta,\pi}$ are sometimes called *generalized Whittaker functions* or *Siegel-Whittaker functions*. Also some people refer to the same thing as *Bessel functions*.

Anyway, for the case of class-one principal series of $Sp(2,\mathbf{R})$ and for a definite character ξ, an integral expression was obtained by Niwa [N1], evaluating certain integral that appears in the Weil representation for $Sp(2,\mathbf{R}) \times SO(2,2)$.

Our method is to investigate the image $U(v)$ of non-vanishing intertwining operator $U \in I_{\eta,\pi}$ for $v \in M_\pi$ with a good K-type, which occurs in (π, M_π) with multiplicity one. Since we want to know the nature of $U(v)(g)$ on $Sp(2,\mathbf{R})$ precisely, we utilize th double coset decomposition $G = RAK$ to have a holonomic system on A of the A-radial part $U(v)|A$ of $U(v)$.

For generalized principal series representations and for the large discrete series representations, the nature of the the function $U(v)|A$ was firstly analyzed by Takuya Miyazaki [Mi3] for definite character ξ. Among others he showed that $I_{\eta,\pi}^{alg,mod}$ has dimension one, where the superscript *mod* means that it is the subspace of $I_{\eta,\pi}^{alg}$ consisting of moderate growth functionals.

For general principal series π (i.e., the parabolic induction with respect $P_{min}(\mathbf{R})$) and when ξ is definite, Ishii [Ishi1] analyzed the holonomic system on A when $v \in M_\pi$ belongs to the minimal K-type of the principal series representation π. An interesting point is here Meijer's G-functions are employed, which also appeared in the papers [Hira1,Hira2] of Hirano. And this result was extended by Moriyama [M6] to the case where ξ is indefinite.

Cohomological representations of $Sp(2,\mathbf{R})$ which are not discrete series have no Whittaker model because they have smaller Gel'fand-Kirillov dimensions, but may have other generalized Whittaker model. We refer to Miyazaki [Mi2] for such a problem.

Remark 5.3. We have to say that there are still many problems for Siegel-Whittaker functions. Among others, we do not yet have satisfactory results for integral expression of Siegel-Whittaker functions. And if one try to have a global Fourier expansion, some term of Fourier expansion are not expressed as discrete sums of coefficients times special functions. We find "continuous spectrum" in some case, which is quite similar to a difficulty of construction of some prehomogeneous zeta functions.

Problem 5.1. *In the case of the Siegel-Whittaker function, we know little about the integral expression, except for the case of class one representation. Probably it would be useful to have (asymptotic) expansion of the integral in terms of power series defined for regular singularities at infinity. Corresponding problem for Whittaker functions is studied already for various cases.*

5.1.4. *Fourier-Jacobi functions*

We can consider a kind of Fourier expansion along the other maximal parabolic subgroup P_J, the Jacobi parabolic subgroup in our terminology. For some classical domains including the Siegel upper half space of genus n, such Fourier expansion for holomorphic automorphic forms has been called *Fourier-Jacobi expansion.*

When we consider more general automorphic forms beyond the holomorphic ones, we need some basic facts on the nature of functions that should appear in such expansions.

Hirano [Hira1,Hira2,Hira3] investigates this problem rather systematically. But, among the standard representations obtained by parabolic induction, the case of generalized principal series with respect to the Siegel parabolic subgroup P_S is not yet handled. In this case the spherical subgroup R is a subgroup of P_J which is a semi-direct product of a subgroup of the Levi part of P_J isomorphic to $SL(2,\mathbf{R})$ and the unipotent radial N_J of P_J. We refer to [Hira1,Hira2,Hira3] for details of η. We only note here that we use a double coset decomposition $G = RA_1K$ with a 1-dimensional subgroup A_1 of A.

Problem If one has reasonable theory for global Fourier-Jacobi expansion of an automorphic form belonging to a large discrete series, then one can extend the method of Kohnen-Skoruppa [K-S] to have a spinor L-function, for automorphic forms belonging to the large discrete series.

5.2. *Spherical functions with respect to reductive subgroups*

We can consider spherical functions when R is a reductive subgroup of G and the pair (G, R) is symmetric.

5.2.1. *Matrix coefficients*

When $R = K$ is a maximal compact subgroup and η is a finite-dimensional representation (τ, W_τ) of K with a basis $\{w_i\}$, so that for a unitary representation (π, M_π) of G which has a K-type τ^* contragradient to τ with the dual basis $\{w_j^*\}$, we can consider the map

$$v \in H \mapsto \Phi(v)(g) := \sum_i (\pi(g)v, w_i^*)w_i \in W_\tau.$$

Then for any $g' \in G$ we have $\Phi(v)(gg') = \Phi(\pi(g')v)(g)$, i.e., this is an intertwining operator in $\mathrm{Hom}_G(\pi, \mathrm{Ind}_K^G(\tau))$.

Iida [Iida] investigates the A-radial part of the matrix coefficients with special K-types for generalized principal series induced via P_J. Among others he showed that they are represented by Apell's hypergeometric series of two variables of F_2-type. Oda [O11] have a similar result for large discrete series representations.

The case of class one principal series, the A-radial part of the matrix coefficients with the trivial K-type are BC_2-type hypergeometric series investigated by Koornwinder [Krn], Debiard-Gaveau [DG], and Heckman-Opdam [H-Op]. We do not try to give an exhaustive reference here.

5.2.2. *The cases of other spherical subgroups*

The group G is large enough to have other nice reductive subgroups $R = H$. Moriyama [M1,M2] investigated the spherical functions for the pairs $(G, H) = (Sp(2, \mathbf{R}), SL(2, \mathbf{R}) \times SL(2, \mathbf{R})$ and $(Sp(2, \mathbf{R}), SL(2, \mathbf{C}))$.

5.3. *Intertwining operators between different spherical models*

Let (R_1, η_1) and (R_2, η_2) be two subgroups of G and η_i be a unitary irreducible representation of each group R_i, respectively. When we have non-zero $T_i \in I_{\pi,\eta_i}$ for both i, for a given admissible irreducible representation π, then the images $Im(T_i)$ are isomorphic G-modules. Then there is a problem whether we have a description of the intertwining operators $i : Im(T_1) \to Im(T_2)$. If one can express i in terms of some integral kernel (which should be a hyperfunction on G), we have applications to automorphic forms. But we have few results on this problem.

As a case study, we consider a different but related problem. In our paper [HIO1], the deformation of the pair (R_t, η_t) is considered. And we showed that the Siegel-Whittaker functions are deformed to the Whittaker functions for a special case. This idea is very simple, and should work for more general situation.

5.4. *Postscript for this section*

There are some points which are not clear in the general theory. The first question is for what quadruple $(G, \pi; R, \eta)$, our intertwining space $I_{\pi,\eta}$ is of finite dimension. When (G, R) is a pair of a semisimple Lie group G and a reductive subgroup R in G, the notion of spherical subgroup in the sense of M. Brion might be helpful. Also the experience tells us that if R is reductive and $I_{\pi,\eta}$ is of finite dimension, then the radial part of the image of of an intertwining operator T of a minimal K-type τ in π satisfies a holonomic system. Moreover this holonomic system has regular singularities if R is reductive. And if R is a subgroup of a parabolic subgroup, the corresponding holonomic system has some irregular singularities. Is there a good way to handle this kind of problem?

Because to avoid the too long reference list, we did not refer to the papers on spherical functions on other groups than $Sp(2, \mathbf{R})$. There are papers by T. Hayata, Y. Gon, and G. Bayarmagnai on $SU(2,2)$ (including the joint paper by T. Hayata, H. Koseki and Oda), and since the group $SU(2,2)$ has the same restricted root system as $Sp(2, \mathbf{R})$, the obtained holonomic systems are similar to those of $Sp(2, \mathbf{R})$. Also T. Ishii has many results on explicit formulas of Whittaker functions on the classical groups.

We refer to our papers which handle certain Whittaker functions on $Sp(3, \mathbf{R})$ that is the group next to $Sp(2, \mathbf{R})$: [HO1,HIO2].

References

DG. Debiard, A. and Gaveau, B.: *Représentation intégrale de certaines séries de fonctions sphériques d'un système de racines BC*. J. Func. Analy. **96** (1991) 256–296.

Hase. Hasegawa, Y.: *Principal series and generalized principal series Whittaker functions with peripheral K-types on the real symplectic group of rank 2*, manuscripta math., (2010),

H-Op. Heckman, G.J. and Opdam, E. M.: *Root system and hypergeometric functions I* . Comp. Math. **64** (1987) 329–352.

Hira1. Hirano, M.: *Fourier-Jacobi type spherical functions for principal series representations of* $Sp(2, \mathbf{R})$, Indag. Math. **15** (2004), 43–53.

Hira2. Hirano, M.: *Fourier-Jacobi type spherical functions for* P_J*-principal series representations of* $Sp(2, \mathbf{R})$, J. London Math. Soc. (2) **65** (2002), 524–546.

Hira3. Hirano, M.: *Fourier-Jacobi type spherical functions for discrete series representations of* $Sp(2, \mathbf{R})$, Comp. Math. **128** (2001), 177–216.

HO1. Hirano, M. and Oda, T.: *Secondary Whittaker functions for* P_J*-principal series representations of* $\mathrm{Sp}(3, \mathbf{R})$. Proc. Japan Acad. Ser. A Math. Sci. **81** (2005) 105–109.

HIO1. Hirano, M., Ishii, T. and Oda, T.: *Confluence from Siegel-Whittaker functions to Whittaker functions on* $\mathrm{Sp}(2, \mathbf{R})$. Math. Proc. Cambridge Philos. Soc. **141** (2006) 15–31.

HIO2. Hirano, M., Ishii, T. and Oda, T.: *Whittaker functions for* P_J*-principal series representations of* $\mathrm{Sp}(3, \mathbf{R})$. Adv. Math. **215** (2007) 734–765.

Hori. Hori, A.: *Andrianov's L-functions associated to Siegel wave forms of degree 2*, Math. Ann. **303** (1995), 195-226.

Iida. Iida, M.: *Spherical functions of the principal series representations of* $\mathrm{Sp}(2, \mathbf{R})$ *as hypergeometric functions of* C_2*-type*. Publ. Res. Inst. Math. Sci. 32 (1996), no. 4, 689–727; Errata: Publ. Res. Inst. Math. Sci. 43 (2007), no. 2, 521–524.

Id-O. Iida, M. and Oda, T.: *Exact power series in the asymptotic expansion of the matrix coefficients with the corner* K*-type of* P_J*-principal series representations of* $\mathrm{Sp}(2, \mathbf{R})$. J. Math. Sci. Univ. Tokyo **15** (2008) 521–543.

Ishi1. Ishii, T.: *Siegel-Whitakker functions on* $Sp(2, \mathbf{R})$ *for principal series representations*, J. Math. Sci. Univ. of Tokyo **9** (2002), 303–346.

Ishi2. Ishii, T.: *On principal series Whittaker functions on* $Sp(2, \mathbf{R})$, J. of Funct. Analysis **225** (2005), 1–32.

Krn. Koornwinder, T.: *Orthogonal polynomials in two variables which are eigenfunctions of two algebraically independent partial differential operators I, II, III, IV.* Indag. math. **A77** (1975) 48–65, 357–381.

I-M. Ishii, T. and Moriyama, T.: *Spinor L-functions for generic cusp forms on GSp(2) belonging to principal series representations.* Trans. Amer. Math. Soc. **360** (2008), 5683–5709.

K-S. Kohnen, W. and Skoruppa, N.-P.: *A certain Dirichlet series attached to Siegel modular forms of degree two.* Invent. Math. **95** (1989) 541–558.

M-W1. Miatello, R. and Wallach, N. R.: *Automorphic forms constructed from Whittaker vectors.* J. Funct. Anal. **86** (1989) 411–487.

M-W2. Miatello, R. and Wallach, N. R.: *The resolvent of the Laplacian on locally symmetric spaces.* J. Differential Geom. **36** (1992) 663–698.

Mi2. Miyazaki, Takuya: *Nilpotent orbits and Whittaker functions for derived functor modules of* $Sp(2, \mathbf{R})$. Canad. J. Math. **54** (2002) 769–794.

Mi3. Miyazaki, Takuya: *The generalized Whittaker functions for* $Sp(2, \mathbf{R})$ *and the gamma factor of the Andrianov L-function*, J. Math. Sci. Univ. of Tokyo **7** (2000), 241–295.

MTd. Miyazaki, Tadashi: *Principal series Whittaker functions on* $Sp(2, \mathbf{C})$, Preprint Dec. 2010.

M-O1. Miyazaki, Takuya and Oda, T.: *Principal series Whittaker functions on* $Sp(2, \mathbf{R})$, *– explicit formulae of differential equations –*, Proc. of the 1993 Workshop "Automorphic forms and related topics", (1993), 59–92.

M-O2. Miyazaki, Takuya and Oda, T.: *Principal series Whittaker functions on* $Sp(2, \mathbf{R})$, Tohoku Math. J. **2-50** (1998), 244–260; Errata **2-54** (2002), 161–162.

M1. Moriyama, T.: *Spherical functions with respect to the semisimple symmetric pair* $(\mathrm{Sp}(2, \mathbf{R}), \mathrm{SL}(2, \mathbf{R}) \times \mathrm{SL}(2, \mathbf{R}))$. J. Math. Sci. Univ. Tokyo **6** (1999) 127–179.

M2. Moriyama, T.: *Spherical functions for the semisimple symmetry pair* $(\mathrm{Sp}(2, \mathbf{R}), \mathrm{SL}(2, \mathbf{C}))$. Canad. J. Math. **54** (2002) 828–865.

M3. Moriyama, T.: *A remark on Whittaker functions on* $\mathrm{Sp}(2, \mathbf{R})$. J. Math. Sci. Univ. Tokyo **9** (2002) 627–635.

M4. Moriyama, T.: *Entireness of the spinor L-functions for certain generic cusp forms on* $\mathrm{GSp}(2)$. Amer. J. Math. **126** (2004) 899–920.

M5. Moriyama, T.: *L-functions for* $\mathrm{GSp}(2) \times \mathrm{GL}(2)$*: Archimedean theory and applications.* Canad. J. Math. **61** (2009), no. 2, 395–426.

M6. Moriyama, T.: *Generalized Whittaker functions on* $GSp(2, \mathbf{R})$ *associated with indefinite characters.* In Press, to appear in J. Math. Soc. Japan.

N1. Niwa, S.: *On generalized Whittaker functions on Siegel's upper half space of degree 2*, Nagoya Math. J., **121** (1991), 171–184.

N2. Niwa, S: *Commutation relations of differential operators and Whittaker functions on* $Sp_2(\mathbf{R})$. Proc. Japan Acad., **71**. Ser. A (1995), 189–191.

O94. Oda, T.: *An explicit integral representation of Whittaker functions on* $Sp(2, \mathbf{R})$ *for the large discrete series representations*, Tohoku Math. J. **2-46** (1994), 261–279.

O11. Oda, T.: *Matrix coefficients of the large discrete series representations of* $Sp(2, \mathbf{R})$ *as hypergeometric series of two variables.* Preprint, 2011

Pr. Proskurin, N.: *Cubic metaplectic forms and theta functions*, Lecture Notes in Math. **1677**, Springer, 1998.

Std. Stade, E.: *Poincaré series for* $GL(3, \mathbf{R})$*-Whittaker functions.* Duke Math. J. **58** (1989), no. 3, 695–729.

Stv. Stevens, G.: *Poincaré series on* $GL(r)$ *and Kloostermann sums.* Math. Ann. **277** (1987), no. 1, 25–51.

Y1. Yamashita, H.: *Embedding of discrete series into induced representations of semisimple Lie groups I, – General theory and the case for* $SU(2, 2)$ –, Japan J. Math. **16** (1990), 31–95.

Y2. Yamashita, H.: *Embedding of discrete series into induced representations of semisimple Lie groups, II, – Generalized Whittaker models for* $SU(2, 2)$ –. J. Math., Kyoto Univ. **31** (1991), 543–571.

6. Cohomology groups of Siegel modular groups of genus two

6.1. *Definition of variants of cohomology groups*

Assume that Γ is a discrete subgroup in $G = Sp(2, \mathbf{R})$, which is commensurable with $Sp(2, \mathbf{Z})$. Then it is of covolume finite, i.e., $vol(\Gamma\backslash G) < \infty$. Since $Sp(2, \mathbf{Q})$ has split rank two, the cusps, i.e., the structure of the rational boundary components with respect Γ is rather complicated. Anyway the quotient $X_\Gamma = \Gamma\backslash H_2$ is a non-compact quasi-projective algebraic variety of complex dimension three. We review first some results on the cohomology groups of such quotients $\mathrm{H}^i(X_\Gamma, \mathbf{C})$ or $\mathrm{H}^i(X_\Gamma, \tilde{V}_l)$. We note that X_Γ is *rational homology manifold* or *rationally smooth*, i.e., though it has singularities, these are mild enough so that the local cohomology groups $\mathrm{H}^i_{\{p\}}(X_\Gamma, \mathbf{Q}) = \mathrm{H}^i(X_\Gamma, X_\Gamma - \{p\}, \mathbf{Q})$ for each point p are the same as smooth topological manifolds. Hence we can define the fundamental class in $\mathrm{H}^6_{\{p\}}(X_\Gamma, \mathbf{Q})$ for each point p, and can show the Poincaré duality: $\mathrm{H}^i(X_\Gamma, \mathbf{Q}) \times \mathrm{H}^{6-i}_c(X_\Gamma, \mathbf{Q}) \to \mathbf{Q}$.

Firstly we recall various variants of the cohomology groups.

Notation 6.1. We define the following symbols:
(a) the interior cohomology group:

$$\mathrm{H}^i_!(X_\Gamma, \mathbf{Q}) := \mathrm{Im}\{\mathrm{H}^i_c(X_\Gamma, \mathbf{Q}) \to \mathrm{H}^i(X_\Gamma, \mathbf{Q})\}.$$

The open variety X_Γ has various compactifications: one is the Borel-Serre bordification $\bar{X}_\Gamma$ which is a topological compactification as manifold with corners, that is homotopically equivalent to the original space X_Γ; another is the Satake compactification X_Γ^{min}, and the other is the toroidal compactification X_Γ^{tor}. The last one is the oldest compactification constructed by Satake for genus 2 , historically speaking. The latter two compactifications are utilized to define the following:
(b1) the mixed Hodge structure on $\mathrm{H}^i(X_\Gamma, \mathbf{Q})$ and the homogeneous part of weight i:

$$W_i\mathrm{H}^i(X_\Gamma, \mathbf{Q}) = \mathrm{Im}\{\mathrm{H}^i(X_\Gamma^{tor}, \mathbf{Q}) \to \mathrm{H}^i(X_\Gamma, \mathbf{Q})\},$$

(b2) and the intersection cohomology with middle perversity on X_Γ^{min} :

$$\mathrm{IH}^i(X_\Gamma^{min}, \mathbf{Q}),$$

which has a homogeneous Hodge structure of weight i.

There exists the canonical G-invariant Kähler metric on H_2, which induces a metric on X_Γ.

(c1) We can define the square-integrable de Rham complex and the associated cohomology $\mathrm{H}^i_{(2)}(X_\Gamma, \mathbf{C})$, that is the L^2-cohomology group in the sense of Borel, which satisfies :

$$\mathrm{H}^i_{(2)}(X_\Gamma, \mathbf{C}) \cong \mathrm{H}^i(\mathfrak{g}, K; L^2(\Gamma\backslash G)).$$

(c2) We can also consider the image :

$$\mathrm{H}^i_{[2]}(X_\Gamma, \mathbf{C}) := \mathrm{Im}\{\mathrm{H}^i_{(2)}(X_\Gamma, \mathbf{C}) \to \mathrm{H}^i_{dR}(X_\Gamma)\}$$

which is the subspace spanned by square-integrable classes.
(c3) The cuspidal cohomology group is yet another subspace in $\mathrm{H}^i_{dR}(X_\Gamma)$ defined by

$$\mathrm{H}^i_{cusp}(X_\Gamma, \mathbf{C}) := \mathrm{H}^i(\mathfrak{g}, K; L^2_{cusp}(\Gamma\backslash G)).$$

Here $L^2_{cusp}(\Gamma\backslash G)$ is the closed subspace generated by cusp forms in $L^2(\Gamma\backslash G)$.

The cohomology groups defined above are related. The following result is conjectured by S. Zucker and proved by Saper-Stern, and Looijenga by different methods, which tells us that the intersection cohomology groups with the middle perversity are isomorphic to square-integrable cohomology groups for general arithmetic quotients X_Γ of a general bounded symmetric domain. The case up to rank two was fixed by Borel-Casselmann somewhat earlier. The key-point in the proof of Saper-Stern is to show that the local L^2 complex satisfies the "axiom" of the intersection cohomology.

Theorem 6.1. *Let $\tilde{V}$ be the constructible sheaf obtained by a finite-dimensional representation of G. Then*

$$\mathrm{IH}^i(X_\Gamma, \tilde{V}) \otimes_{\mathbf{Q}} \mathbf{C} \cong \mathrm{H}^i_{(2)}(X_\Gamma, \tilde{V}).$$

This implies

$$W_i\mathrm{H}^i(X_\Gamma, \mathbf{Q}) \otimes \mathbf{C} = \mathrm{H}^i_{[2]}(X_\Gamma, \mathbf{C})$$

automatically.

Here are some more general facts added.

(1) $\mathrm{H}^i(X_\Gamma, \mathbf{Q})$ is a sub Hodge structure of $W_i\mathrm{H}^i(X_\Gamma, \mathbf{Q})$.
(2) We have the spectral decomposition: $L^2(\Gamma\backslash G) = L^2_{cusp}(\Gamma\backslash G) \oplus L^2_{res}(\Gamma\backslash G)$. Hence if we denote $\mathrm{H}^i(\mathfrak{g}, K; L^2_{res}(\Gamma\backslash G))$ by $\mathrm{H}^i_{res}(X_\Gamma, \tilde{V})$, we have

$$\mathrm{H}^i(X_\Gamma, \tilde{V}) = \mathrm{H}^i_{cusp}(X_\Gamma, \tilde{V}) \oplus \mathrm{H}^i_{res}(X_\Gamma, \tilde{V}).$$

Remark 6.1. It seems to be a difficult problem to understand the image of the natural map $\mathrm{H}^i_{res}(X_\Gamma, \tilde{V}) \to \mathrm{H}^i(X_\Gamma, \tilde{V})$, especially when the local coefficient system $\tilde{V}$ is trivial. We note that the information on the poles and zeros of the L-functions of elliptic modular forms is important here, through the Eisenstein cohomology classes.

6.2. *Cohomology groups of degree $i \neq 3$*

The cohomology groups of degree $i = 0$, 6 are almost trivial. But let me collect the facts:

Proposition 6.2. *(a) By definition and by the Poincaré duality,* $\mathrm{H}^0(X_\Gamma, \mathbf{Q}) \cong \mathbf{Q}$ *and* $\mathrm{H}^6_c(X_\Gamma, \mathbf{Q}) \cong \mathbf{Q}[-3]$. *Here* $[-3]$ *signifies the Tate twist.*
(b) Since X_Γ *is non-compact, and also by the Poincaré duality,* $\mathrm{H}^6(X_\Gamma, \mathbf{Q})$ *and* $\mathrm{H}^0_c(X_\Gamma, \mathbf{Q}) = \{0\}$.
(c) If $\tilde{V}$ *contains non constant part, the cohomology groups of degree* 0 *or* 6 *with coefficients* $\tilde{V}$ *vanish.*

Other cohomology groups $\mathrm{IH}^0, W_0\mathrm{H}^0, \mathrm{IH}^6, W_6\mathrm{H}^6$ *etc. are immediately determined from these basic facts.*

Proposition 6.3. *(a)* $\mathrm{H}^1_{(2)}(X_\Gamma, \tilde{V}) = \{0\}$ *and* $\mathrm{H}^5_{(2)}(X_\Gamma, \tilde{V}) = \{0\}$.
(b) $\mathrm{H}^1(X_\Gamma, \tilde{V}) = \{0\}$ *and* $\mathrm{H}^5_c(X_\Gamma, \tilde{V}) = \{0\}$.

The idea of the proof. It is enough to consider the cohomology of degree 1. The claim for the degree 5 follows from this by Poincaré duality.
The step (a) $\Rightarrow$ (b): The estimate of inner product of closed 1-cycles on X_Γ utilizing the Borel-Serre bordification implies that the natural map

$$\mathrm{H}^i_{(2)}(X_\Gamma, \tilde{V}) \to \mathrm{H}^i(X_\Gamma, \tilde{V}) \quad (i \leq 2)$$

is surjective (this kind of result was crucial in the stable cohomology theory of Armand Borel). Thus it is enough to show that the left side (i.e., the source side) of the above map is zero.
To show (a), we use the Matsushima isomorphism enhanced by Borel

$$\mathrm{H}^i_{(2)}(X_\Gamma, \tilde{V}) \cong \oplus_{\pi \in \hat{G}, \chi_\pi = \chi_{V^*}} \; m(\pi, \Gamma)\mathrm{H}^i(\mathfrak{g}, K; M_\pi \otimes V_l).$$

Here V^* is the contragradient representation of V.

However we have no unitary irreducible representation π of G with infinitesimal character $\chi_\pi = \chi_{V^*}$, so that its first $(\mathfrak{g}, K)$-cohomology group $\mathrm{H}^1(\mathfrak{g}, K; M_\pi \otimes V)$ does not vanish (see Kumaresan [Km], Vogan-Zuckerman [VZ]).

Remark 6.2. In this case, there is yet other proofs:
(A) Because of "Congruence Subgroup Property" of $Sp(2,\mathbf{Z})$ (*cf.* the works of Bass, Serre in the old days, and the almost "terminating" work of G. Prasad), the abelianization $\Gamma/[\Gamma,\Gamma]$ is finite;
(B) The other proof is elementary and topological argument utilizing the van Kampen theorem, to show the fundamental group $\pi_1(X_{\Gamma'},*)$ is finite, for a congruence subgroup Γ' of $Sp(2,\mathbf{Z})\cap\Gamma$.

Remark 6.3. We have $\mathrm{IH}^1(X_\Gamma,\tilde{V}_l)=\mathrm{H}^1_{(2)}=\{0\}$. Hence when V_l is trivial, we also have $W_1\mathrm{H}^1(X_\Gamma,\mathbf{Q})=\{0\}=\mathrm{H}^1_{[2]}(X_\Gamma,\mathbf{C})=\mathrm{H}^1_{cusp}(X_\Gamma,\mathbf{C})$. This is not enough to control $\mathrm{H}^1_c(X_\Gamma,\mathbf{Q})$ or dually $\mathrm{H}^5(X_\Gamma,\mathbf{Q})$.

In the above proof, we have the surjectivity of $\mathrm{H}^2_{(2)}(X_\Gamma,\mathbf{C})\to\mathrm{H}^2(X_\Gamma,\mathbf{C})$ which is equivalent to the surjectivity of $\mathrm{IH}^2(X_\Gamma,\mathbf{Q})\to\mathrm{H}^2(X_\Gamma,\mathbf{Q})$. This shows the following.

Proposition 6.4. $\mathrm{H}^2(X_\Gamma,\mathbf{C})=\mathrm{H}^2_{[2]}(X_\Gamma,\mathbf{C})=W_2\mathrm{H}^2(X_\Gamma,\mathbf{C})$.

We have $\mathrm{H}^2_{cusp}(X_\Gamma,\mathbf{C})\subset\mathrm{H}^2_{[2]}(X_\Gamma,\mathbf{C})$ in general, and the complement of the cuspidal part is given by $\mathrm{H}^2_{res}(X_\Gamma,\mathbf{C})=\mathrm{H}^2(\mathfrak{g},K;L^2_{res}(\Gamma\backslash G))$, i.e., the cohomology classes which come from residual "Eisenstein cohomology classes" in the sense of Harder.

We discuss the problem of H^2 in the next section

Remark 6.4. It is difficult to describe the weight filtration for cohomology of the Siegel modular varieties with general genus n. But for the extremal weight one can get some result in a rather naive way (see, T. Miyazaki [M96]).

7. The second cohomology group of Siegel modular varieties

The second cohomology group $\mathrm{H}^2(X_\Gamma,\mathbf{C})$ consisting of square-integrable cohomology classes, as shown by Borel employing the theory of bordification (of Borel-Serre), hence it has a homogeneous Hodge structure of weight two. We give another proof for this via Igusa desingularization (*cf.* [OS1], Theorem 4.1). Moreover non-zero elements with $(2,0)$ Hodge type were constructed by theta series of the norm form of imaginary quadratic fields. These are expected to appear in the residual part of the Eisenstein series associated to P_J. Heuristically speaking, the holomorphic 2-forms obtained in this way correspond to Grössencharacter ([OS1], §6).

7.1. *Modular divisors*

There is a way to have elements in $\mathrm{H}^{(1,1)} \cap \mathrm{H}^2(X_\Gamma, \mathbf{Q})$. By Lefschetz $(1,1)$ criterion, these elements should come from divisors on X_Γ.

As found in the first paper of Hecke's "Werke", the algebraic cycles on X_Γ are obtained as the locus of "singular period" in the sense of G. Humbert. This means that there are integers A, B, C, D, E and $Z = \begin{pmatrix} z_1 & z_3 \\ z_3 & z_2 \end{pmatrix}$ satisfies

$$Az_1 + Bz_3 + Cz_2 + D(z_3^2 - z_1 z_2) + E = 0.$$

This means that if A_Z be the 2-dimensional abelian variety with period modulus Z, then the Néron-Severi group (or the Picard group) of A has an element independent of the canonical polarization class of A_Z. The vector $l = (A, B, C, D, E)$ is identified with an element of $\mathrm{H}^2(A_Z, \mathbf{Z})$, corresponding to an algebraic cycle on A_Z, and by Hodge's index theorem $\Delta = B^2 - 4AC - 4DE$ is necessarily positive (this number is (-1)-times the self-intersection number of l). The group $G_\mathbf{Q} = Sp(2, \mathbf{Q})$ acts canonically on $\mathrm{H}^1(A_Z, \mathbf{Q})$, hence also on $\mathrm{H}^2(A_Z, \mathbf{Q})$ by the induced action. Let H_l be the algebraic subgroup stabilizing the vector l. Then it is isomorphic to $SL(2) \times SL(2)$ over the extension $\mathbf{Q}(\sqrt{\Delta})$.

The locus $Y_l := \{Z \mid Az_1 + Bz_3 + Cz_2 + D(z_3^2 - z_1 z_2) + E = 0\}$ is biholomorphically identified with a product $H_1 \times H_1$ of two copies of the complex upper half-plane H_1. The group $\Gamma_l := \Gamma \cap H_l(\mathbf{Q})$ acts on Y_l properly discontinuously. Then a holomorphic morphism $\Gamma_l \backslash Y_l \to X_\Gamma$ of algebraic varieties are well-defined, and its image is a divisor on X_Γ.

7.2. *Green function associated to modular divisors*

We have a method to construct the Green current associated with the above modular divisors. This is a generalization of the classical situation for points divisors on modular curves. In [OT1], we considered an affine symmetric pair (G, H) of rank 1, so that both G/K and $H/(H \cap K)$ have invariant Hermitian structures. The pair $(Sp(2, \mathbf{R}), SL(2, \mathbf{R}) \times SL(2, \mathbf{R}))$ is a special case. Then we can consider the eigenvalue problem $\Delta\phi_s = (n^2 - s^2)\phi_s$ for an H-invariant function ϕ on G/K for the invariant Laplacian Δ. Here n is a constant determined naturally by the pair (G, H) and s is a complex parameter. This ϕ is the right K-invariant function in $L^2(H\backslash G)$ if s is chosen adequately.

Here we are interested in the *secondary spherical function* $\phi_s^{(2)}$ which is a distribution on G/K with singular support only the H-orbit $Y := HK/K$ in G/K which satisfies the same Poisson equation as ϕ_s. If we specify the behavior of $\phi^{(2)}$ at infinity, then it is determined uniquely. Given an arithmetic discrete subgroup Γ, form the Poincaré series

$$G_s(z) = \sum_{\gamma \in \Gamma \cap H \backslash \Gamma} \phi^{(2)}(\gamma z), \qquad z \in \Gamma \backslash G/K,$$

which converges when $Re(s) > n$. Then G_s has the Laurent expansion :

$$\frac{\kappa}{s-n} + \mathcal{G} + O(s-n)$$

with a constant κ as the residue at $s = n$, the constant term $\mathcal{G} = \mathcal{G}(z)$, and the Landau symbol $O(s-n)$, and we have the Green formula:

$$-\sqrt{-1}\partial\bar{\partial}\mathcal{G} - \pi\delta_Y = \kappa \cdot \omega_{\Gamma\backslash G/K} + 4\sqrt{-1}\Psi$$

with the Dirac current δ_Y of the divisor $(H \cap \Gamma)\backslash Y$, the Kähler form $\omega_{\Gamma\backslash G/K}$ and a harmonic $(1,1)$-type form belonging to a certain unitary cohomological representation $J_{(1,1)}$ of G with non-trivial $\mathrm{H}^{1,1}(\mathfrak{g}, K; J_{1,1}) = \mathrm{H}^2(\mathfrak{g}, K; J_{1,1})$ ([OT1], §7). Here (-1) times of the right hand side of this formula is the Chern form of the divisor $(H \cap \Gamma)\backslash Y$.

Bruinier [Bru] also discusses some related problems. There are many papers by Tsuzuki and his collaborators for similar Poincaré series and their applications.

Remark 7.1. We, optimists, have a **hope** that the Chern classes of modular divisors generate the whole algebraic cycles in $\mathrm{H}^2(X_\Gamma, \mathbf{Q})$. In the case of complex threefolds associated to an inner twist of $Sp(2, \mathbf{Q})$, by a simple argument originating from [O81], we can show the vanishing of $(2,0)$-type Hodge component [OS2]. By Lefschetz criterion, this means that the second cohomology classes of such threefolds are all algebraic. Compare with some papers of Venkataramana (e.g. [CV,Ven]).

Remark 7.2. We can extend the above results to modular cycles of higher codimension ([OT2A,OT2B]). In the announcement [OT2A], the readers might find some helpful remarks which are not found in the full paper [OT2B].

Remark 7.3. The fundamental work by T. Kobayashi on the branching rule of the cohomological representations, when they are restricted to a large reductive subgroup H of G ([K1–K3]) is related to the above construction of Green currents. Actually the theme of the joint paper [KO1] is the restriction of the cohomology classes of a modular quotient to a modular cycle, i.e., the dual situation to the construction of Green currents.

References

Bru. Bruinier, J.H.: *Borcherds products on* $O(2,l)$ *and Chern classes of Heegner divisors*. Springer Lecture Notes in Mathematics, **1780**. Springer, New York (2002).

CV. Clozel, L. and Venkataramana, T. N.: *Restriction of the holomorphic cohomology of a Shimura variety to a smaller Shimura variety*. Duke Math. J. **95** (1998) 51–106.

K1. Kobayashi, T.: *Discrete decomposability of the restriction of* $A_q(\lambda)$ *with respect to reductive subgroups and its applications*. Invent. math. **117** (1994), 181–205.

K2. Kobayashi, T.: *Discrete decomposability of the restriction of* $A_q(\lambda)$*, Part II. Micro-local analysis and asymptotic* K*-support*. Annals of Math. **147** (1998), 709–726.

K3. Kobayashi, T.: *Discrete decomposability of the restriction of* $A_q(\lambda)$*, Part III. -Restriction of Harish-Chandra modules and associated varieties* . Invent. math. **131** (1998), 229–256.

KO1. Kobayashi, T. and Oda, T.: *A vanishing theorem for modular symbols on locally symmetric spaces*. Comment. Math. Helv. **73** (1998), 45–70.

M96. Miyazaki, Takuya: *Mixed Hodge structures of Siegel modular varieties and Siegel Eisenstein series*. J. Math. Sci. Univ. Tokyo **3** (1996) 297–330.

O81. Oda, T.: *A note on the Albanese variety of an arithmetic quotient of the complex hyperball*. J. Fac. Sci. Univ. Tokyo Sect. IA Math. **28** (1981) 481–486.

OS1. Oda, T., and Schwermer, J: *Mixed Hodge structures and automorphic forms for Siegel modular varieties of degree two*. Math. Annalen, **286** (1990), 481–509.

OS2. Oda, T. and Schwermer, J.: *On mixed Hodge structures of Shimura varieties attached to inner forms of the symplectic group of degree two*. Tohoku Math. J. (2) **61** (2009) 83–113.

OT1. Oda, T. and Tsuzuki, M.: *Automorphic Green functions associated with the secondary spherical functions*. Publ. Res. Inst. Math. Sci. **39** (2003), no. 3, 451–533.

OT2A. Oda, T. and Tsuzuki, M.: *The secondary spherical functions and Green currents associated with certain symmetric pairs*. "Representation theory and analysis on homogeneous spaces", 121–135, RIMS Kokyuroku Bessatsu, **B7**, Res. Inst. Math. Sci. (RIMS), Kyoto, 2008.

OT2B. Oda, T. and Tsuzuki, M.: *The secondary spherical functions and automorphic Green currents for certain symmetric pairs*. Pure Appl. Math. Q. **5** (2009) 977–1028.

Ven. Venkataramana, T. N.: *Abelianness of Mumford-Tate groups associated to some unitary groups*. Compositio Math. **122** (2000) 223–242.

Reducibility and discrete series in the case of classical p-adic groups; an approach based on examples

Marko Tadić*

Department of Mathematics, University of Zagreb,
Bijenička 30, 10000 Zagreb, Croatia
Email: tadic@math.hr

Dedicated to Professor Takayuki Oda on the occasion of his 60th birthday

In this paper, we first present the basic ideas of the method of determining reducibility or irreducibility of parabolically induced representations of classical p-adic groups using Jacquet modules. After that we explain the construction of irreducible square integrable representations by considering characteristic examples. We end with a brief presentation of the classification of irreducible square integrable representations of these groups modulo cuspidal data, which was obtained jointly with C. Mœglin.

Keywords: p-adic fields; classical groups; induced representations; reducibility; square integrable representations

Introduction

Professor Oda invited me in 2007 to the University of Tokyo, to give several lectures presenting the main ideas of my work on reducibility of parabolically induced representations and discrete series of classical p-adic groups. This paper is based on the notes of lectures given during that visit. We are very thankful to Professor Oda for the interest that he expressed in our work, and for explaining to us the relationship of our work to his.

This paper has two objectives. The first objective of this paper is to explain how Jacquet modules can be used for analyzing reducibil-

---

2010 Mathematics Subject Classification: Primary 22E50, 22E55; Secondary 11F70, 11S37

*The author is partly supported by Croatian Ministry of Science, Education and Sports grant #037-0372794-2804.

ity of parabolically induced representations of classical groups over non-archimedean local fields, based on examples. The second objective is to explain, also based on examples, ideas in the classification of irreducible square integrable representations of these groups modulo cuspidal data. This classification has been obtained jointly with C. Mœglin.

Although this paper deals with problems from the representation theory of p-adic groups, the questions that we consider have at least some number-theoretic relevance. Namely, non-archimedean local fields show up as completions of number fields (in characteristic zero), and we shall say something regarding this a little bit later. In general, it is well known that harmonic analysis on the groups that we consider in this paper have a very strong relationship to some aspects of number theory. For example, Plancherel measures are expressed by the local L-functions etc.

The full power of the methods and results that we are presenting in this paper can be found in a number of already published ones. The method of applying Jacquet modules for determining reducibility of parabolically induced representations and their composition series was introduced in [39]. This method has been significantly improved by C. Jantzen (he studied the degenerate principal series; see [14] and several of his other papers) and by G. Muić (he studied generalized principal series in [27] and standard representations in [28]). The classification of irreducible square integrable representations of classical p-adic groups modulo cuspidal data is covered by [21] and [24] (we also have several other papers giving more precise information about some classes of square integrable representations). In general, these papers are very technical. Rather then trying to present these technically very complicated aspects of results in general case, we put an emphasis on important characteristic examples, which are relatively simple in comparison with the general case, but still illustrate the methods and results, sometimes better than the final results. Our experience is that we often understand new topics much more easily in this way. This is the reason that this paper may be useful not only to those working in the area of representation theory of reductive p-adic groups, but also to those whose primary interest is automorphic forms.

For examining reducibility questions (in particular related to the tempered setting) and square integrability, there are other (mainly analytic) methods. We shall not study them in this paper, but shall briefly comment on some of them on a few occasions. These methods are less elementary than the methods that we present in this paper. They are based on standard intertwining operators (and their normalizations), Plancherel measures (and

L-functions), R-groups etc. (see [12], [25], [26], [31], [32] and [33] among others). They play a very important role in C. Mœglin's paper [21] on the exhaustion of square integrable representations, which is closely related to the topic of our paper.

This paper has three main parts. In the first part "Representations of classical p-adic groups and structures built on them" (sections 1. to 14.) we present general facts related to the study of reducibility of parabolically induced representations of classical p-adic groups, and to the construction of irreducible square integrable representations of these groups. We start the second part "Jacquet module method of determining reducibility of parabolically induced representations" (sections 15. to 27.) with relatively simple examples of determining reducibility or irreducibility of specific parabolically induced representations. Later on, we consider more complicated cases, and study their composition series. In this way we obtain some square integrable subquotients. We further study parabolically induced representations related to the square integrable representations that we have obtained. In this way we notice some phenomena which are used in the third part "On irreducible square integrable representations and their parameters" (sections 28. to 35.), where invariants (defined by C. Mœglin) are attached to irreducible square integrable representations which classify them modulo cuspidal data (and a natural assumption). In this part of the paper we present characteristic examples of irreducible square integrable representations (we emphasize that these invariants classify them modulo cuspidal data and a natural assumption). We end the paper with a general formulation of the classification of irreducible square integrable representations modulo cuspidal data.

The reader familiar with general facts of representation theory of reductive p-adic groups can skip the first 8 sections, and go directly to section 9. Those who are also familiar with the Bernstein-Zelevinsky theory can go directly to section 14. Very basic ideas regarding studying reducibility of parabolically induced representations are contained in sections 15. - 22. The reader more interested in reducibility questions can read the remaining sections 23. - 27. for additional information and examples. The reader more interested in square integrability can skip these five sections and go directly to section 28.

We are thankful to C. Jantzen for a very careful reading of the paper, and a number of suggestions which have significantly improved the style of the paper. I. Matić has also passed through the whole paper, and made a numerous corrections. The referee gave a number of useful comments.

Discussions with M. Hanzer, A. Moy and G. Muić on some aspects of this paper, helped us to better understand some topics and explain them in the paper. We are thankful to all of them. The first version of this paper was written when we were visiting the University of Tokyo. We are thankful to the University for its hospitality.

Representations of classical p-adic groups and structures built on them

1. Historical observations

Induced representation are the main object of the study in this paper. Let us recall that they were present in representation theory from the very beginning in the development of the subject. Namely, the notion of a representation of a finite group was introduced by Frobenius in 1897 (the notion of a character of a representation had already existed). The following year Frobenius introduced induced representations (and Frobenius reciprocity, which plays a very important role in our paper).

A representation of a group G is a homomorphism π of G into the group of linear isomorphisms of a complex vector space V (we shall consider only complex representations here; representations over other fields are also very important). For a representation (σ, U) of a subgroup P of a finite group G, the representation of G **induced** by σ is the representation of G on the space $\text{Ind}_P^G(\sigma)$ of all functions $f : G \to U$ which satisfy

$$f(pg) = \sigma(p)f(g), \qquad \forall p \in P, \forall g \in G. \tag{1.1}$$

The group G acts by the right translations $R_g f(x) = f(xg)$. The most basic fact regarding induced representations is **Frobenius reciprocity**:

$$\text{Hom}_G(\pi, \text{Ind}_P^G(\sigma)) \cong \text{Hom}_P(\pi|_P, \sigma), \tag{1.2}$$

where π is a representation of G ($\pi|_P$ denotes the restriction to P of the representation π).

Induced representations play a very important role in the construction of irreducible representations ((π, V) is irreducible if $\{0\}$ and V are the only subspaces of V invariant for all $\pi(g)$, $g \in G$). For example, one can get from (simple) one-dimensional representations of a subgroup, such as the trivial representation, interesting non-trivial representations of the group.

It is interesting to note that induction did not play a role in the construction of irreducible representations of compact Lie groups (induced representations here are too big). Also, it did not play a role in the case of commutative groups, since irreducible representations are one-dimensional in this case. We note that in both cases induced representations were studied (at least induced representations by trivial representations of subgroups).

Induction played a very important role in the next phase of development - harmonic analysis on groups, especially harmonic analysis on reductive groups over local fields. What is particularly important here is a special case of induction, parabolic induction. I. M. Gelfand and M. A. Naimark were the ones who were first aware of the importance of these representations for harmonic analysis in this case. They obtained the first important results about them, which were later extended to more general cases by Harish-Chandra and others.

2. Smooth representations and the unitary dual

The fundamental object for harmonic analysis on a locally compact group G is the **(unitary) dual** $\hat{G}$ of G, i.e., the set of equivalence classes of (topologically) irreducible unitary representations (on Hilbert spaces) of G (a unitary representation is a continuous homomorphism π of G into the group of unitary operators on a Hilbert space H; a unitary representation is irreducible if $\{0\}$ and H are the only closed subspaces of H invariant for all $\pi(g)$, $g \in G$). See [38] for more discussion regarding harmonic analysis on groups.

We shall consider (linear) reductive groups over non-archimedean local fields. Thanks (mostly) to the work of Harish-Chandra and J. Bernstein, the problem of studying the irreducible unitary representations can be algebraized (and we can forget Hilbert space representations and continuity conditions): it reduces to smooth representations. We shall now recall the definition.

Reductive groups over non-archimedean local fields are locally compact and totally disconnected. A locally compact totally disconnected group G has a neighborhood of the identity consisting of open compact subgroups. A representation (π, V) of G is called **smooth** if the stabilizer of v in G is open, for each $v \in V$. The set of all equivalence classes of irreducible smooth representations of G will be denoted by $\tilde{G}$, and called the **non-unitary dual** of G (the term non-unitary that we have used here is to stress that we do not require unitarity in the definition of $\tilde{G}$; it does not mean that $\tilde{G}$ consists of non-unitary representations only). Another (maybe better) choice would be to call $\tilde{G}$ smooth dual (or admissible) dual of G.

In what follows, by a representation, we shall mean a smooth representation.

A smooth representation is called **unitarizable** if there exists a G-invariant inner product on V. If π is irreducible, such an inner product is unique up to a positive scalar (this follows by Schur's lemma). Denote by $\hat{G}$ the subset of all unitarizable classes in $\tilde{G}$. Then $\hat{G}$, defined in this way is in a canonical bijection with the unitary dual of G defined above.

The above reduction is enabled by the following fundamental fact proved in [2] by J. Bernstein: each irreducible unitary representation of G on a Hilbert space H is admissible, i.e. for each open compact subgroup K of G, the K-invariants H^K form a finite dimensional (complex) vector space.

All representations in $\tilde{G}$ are also admissible (this fact was proved by H. Jacquet). A smooth representation which is admissible will be called an admissible representation. It is easy to show that unitarizable admissible representations are always semisimple.

3. The non-unitary dual and unitary dual

Attempts to get the classification of $\hat{G}$ directly did not produce results (for general G as above) and it was much easier to construct elements of $\tilde{G}$. These facts motivated the following strategy for classifying $\hat{G}$, due to Harish-Chandra. The strategy has two steps:

- classify $\tilde{G}$ (the **problem of non-unitary dual**);
- identify unitarizable classes in $\tilde{G}$, i.e., determine the subset $\hat{G} \subset \tilde{G}$ (the **unitarizability problem**).

In these lectures we shall talk only about the first problem, the problem of the non-unitary dual.

A general framework for handling the first problem is offered by the Langlands classification. Some definitions are necessary to explain this classification. Square integrable representations and parabolic induction play crucial role in it.

4. Square integrable representations

Let G be a reductive group over a local non-archimedean field F. One important example is $G = GL(n, F)$, and $F = \mathbb{Q}_p$, i.e., the field of p-adic numbers.

Let (π, V) be a smooth representation of G. On the space V' of linear forms on V there exists a natural representation: $(\pi'(g)v')(v) = v'(\pi(g^{-1})v)$. The set of all linear forms with open stabilizer is a (smooth) subrepresentation, which we denote by $(\tilde{\pi}, \tilde{V})$, and call the **contragredi-**

ent of (π, V). We denote by $\bar{\pi}$ the complex conjugate representation of a representation π.

Functions of the form $g \mapsto \tilde{v}(\pi(g)v), G \to \mathbb{C}$, for $v \in V, \tilde{v} \in \tilde{V}$, are called **matrix coefficients** of the representation (π, V).

Denote by $Z(G)$ the center of G. The group $G/Z(G)$ (as well as G) is unimodular. When we talk about integrability, we mean with respect to the invariant measure. Schur's lemma implies that for each smooth irreducible representation (π, V) there is a character ω_π of $Z(G)$ such that $\pi(z) = \omega_\pi(z)\,\mathrm{id}_V$ for all $z \in Z(G)$. The character ω_π is called the **central character** of π.

Definition. *An irreducible representation (π, V) of G is called* ***square integrable modulo center*** *if*

(1) the central character of π is a unitary character;
(2) the absolute values of all the matrix coefficients of π are square integrable functions on $G/Z(G)$.

If the center of G is compact, square integrable modulo center representations will be simply called square integrable. Sometimes, if the center is not compact, we also abbreviate square integrable modulo center representations by square integrable.

It is easy to show that these representations are unitarizable (the G-invariant inner product is given by

$$(v_1, v_2) \mapsto \int_{G/Z(G)} \tilde{v}(\pi(g)v_1)\overline{\tilde{v}(\pi(g)v_2)}dg,$$

where $\tilde{v} \in \tilde{V}\backslash\{0\}$ is fixed). We denote by $D_u(G)$ the set of all such classes (of representations) in $\hat{G}$.

A smooth representation (π, V) of G is called **essentially square integrable** if it becomes square integrable modulo center after twisting by some character of G. The set of all such representations will be denoted by $D(G)$.

Square integrable representations play a very important role in a number of questions, from several points of view (not only in harmonic analysis). For example, irreducible square integrable representations are very distinguished representations from the functional-analytic point: if the center of G is compact, they can be characterized as irreducible subrepresentations of $L^2(G)$ (the representation of G on $L^2(G)$ defined by right translations).

On the other side, irreducible essentially square integrable representations of reductive groups are also particularly interesting to understand from the point of view of the modern theory of automorphic forms. Namely, the **local Langlands correspondence** for a split semi simple group G predicts the existence of a natural partition of $\tilde{G}$ into finite sets of representations

$$\tilde{G} = \underset{\varphi}{\cup}\, \Pi_\varphi,$$

where φ runs over all conjugacy classes of continuous homomorphisms of $W_F \times SL(2,\mathbb{C})$, where W_F denotes the **Weil group** of F †, into the complex **dual Langlands** L**-group** ${}^LG^0$ (see below), such that φ maps elements of the first factor to semi simple elements and is algebraic on the second factor (such homomorphisms will be called **admissible homomorphisms**). Elements of Π_φ are called L**-packets** (since all representations in an L-packet should have the same L-functions). The mapping which sends $\pi \in \Pi_\varphi$ to φ will be denoted by Φ_G, or simply by Φ when it is evident which correspondence we consider. This mapping is called the **local Langlands correspondence** for G.

For establishing local Langlands correspondences, the case of square integrable representations is crucial. For them, there should exist a partition

$$D(G) = \cup_\varphi \Pi_\varphi,$$

where φ runs over all admissible homomorphisms whose images are not contained in any proper Levi subgroup (such homomorphisms will be called **discrete admissible homomorphisms**).

Further, elements in a square integrable L-packet Π_φ should be parameterized (roughly) by the equivalence classes of irreducible representations of the **component group**

$$\mathrm{Cent}_{{}^LG^0}(\varphi(W_F \times SL(2,\mathbb{C})))/\mathrm{Cent}_{{}^LG^0}(\varphi(W_F \times SL(2,\mathbb{C}))^0\ Z({}^LG^0).$$

Above, $\mathrm{Cent}_{{}^LG^0}X$ denotes the centralizer of $X \subseteq {}^LG^0$, while $(\mathrm{Cent}_{{}^LG^0}X)^0$ denotes the connected component of the identity.

---

†The Weil group is a dense subgroup of the absolute Galois group G_F of F (i.e. of the Galois group over F of its separable algebraic closure). More precisely, denote by $\mathcal{O}_F$ the maximal compact subring of F and by $\mathfrak{p}_F$ its unique maximal ideal. Consider the natural homomorphism $r : G_F \to G_{\mathcal{O}_F/\mathfrak{p}_F}$. Let $<$ Frob $>$ be the subgroup generated by the Frobenius automorphism Frob $\in G_{\mathcal{O}_F/\mathfrak{p}_F}$. Then $W_F = r^{-1}(<\text{Frob}>)$. The topology on W_F is that for which (the inertia group) Ker(r) gets the topology from G_F, and Ker(r) is open in W_F.

Instead of going into the general definition of ${}^LG^0$ (which involves root data and dual root data), let us indicate what ${}^LG^0$ is in the cases of interest of us:

$$
\begin{aligned}
{}^LGL(n,F)^0 &= GL(n,\mathbb{C}),\\
{}^LSp(2n,F)^0 &= SO(2n+1,\mathbb{C}),\\
{}^LSO(2n+1,F)^0 &= Sp(2n,\mathbb{C}).
\end{aligned}
$$

In the first case, the component groups are trivial, while in the remaining two cases the component groups are isomorphic to direct sums of copies of $\mathbb{Z}/2\mathbb{Z}$ (therefore, L-packets should have cardinalities which are always a power of 2). In the case of $GL(n,F)$, discrete admissible homomorphisms are just continuous irreducible representations of $W_F \times SL(2,\mathbb{C})$ which are semi simple on the first factor and algebraic on the second factor.

Later on, we shall also see the importance of square integrable representations for the classification of non-unitary duals.

5. Parabolic subgroups

We shall fix a **maximal split torus** $A_\emptyset$ in G and a **minimal parabolic subgroup** $P_\emptyset$ in G containing $A_\emptyset$. Subgroups containing $P_\emptyset$ will be called **standard parabolic subgroups**. Let P be a standard parabolic subgroup. The Levi decomposition $P = MN$ (with N unipotent) will be called a **standard Levi decomposition** if M contains $A_\emptyset$.

In the case of $G = GL(n,F)$, we shall take for $P_\emptyset$ (resp., $A_\emptyset$) the subgroup of upper triangular (resp., diagonal) matrices in G. Let $\alpha = (n_1,\ldots,n_l)$ be an ordered partition of n. Denote by M_α^{GL} (resp., N_α^{GL}) block-diagonal (resp., unipotent block-upper triangular) matrices of type $(n_1,\ldots,n_l)$ in G. Then $P_\alpha^{GL} := M_\alpha^{GL}N_\alpha^{GL}$ is a standard parabolic subgroup, and this is its standard Levi decomposition. In this way we obtain a bijection of ordered partitions of n onto standard parabolic subgroups of G. The Levi subgroup M_α^{GL} is isomorphic to $GL(n_1,F) \times \cdots \times GL(n_l,F)$ in a natural way.

When we consider the classical groups $Sp(2n,F)$ or $SO(n,F)$, we fix maximal split tori and minimal parabolic subgroups obtained by intersection with the corresponding subgroups from general linear groups. A similar situation occurs with standard parabolic subgroups and standard Levi decompositions.

6. Parabolic induction, tempered representations and the Langlands classification

Let $P = MN$ be a parabolic subgroup of G. Denote by δ_P the **modular character** of P. Fix a smooth representation (σ, U) of M. On the space $\text{Ind}_P^G(\sigma)$ of all locally constant functions $f : G \to U$ satisfying

$$f(mng) = \delta_P(m)^{1/2}\sigma(m)f(g), \quad \forall m \in M, \forall n \in N, \forall g \in G, \tag{6.1}$$

the group G acts by right translations:

$$(R_g f)(x) = f(xg).$$

This defines a smooth representation of G, which is called the representation of G **parabolically induced** by σ from P. Parabolic induction carries unitarizable representations to unitarizable ones (this is the reason to have the modular character $\delta_P^{1/2}$ in the requirement (6.1)).

Definition. *An irreducible smooth representation τ of G is called* ***tempered*** *if there exists a parabolic subgroup $P = MN$ and an irreducible square integrable modulo center representation δ of M such that*

$$\tau \hookrightarrow Ind_P^G(\delta).$$

The **Langlands classification** (or maybe more appropriately, the Langlands parameterization), starts with a triple (P, τ, χ), where $P = MN$ is a standard parabolic subgroup of G, τ is an irreducible tempered representation of M, where $P = MN$ is standard Levi decomposition of P, and χ is a positive valued character of M satisfying a certain positivity condition (which we shall illustrate in several examples). To such a triple one attaches the irreducible quotient of $\text{Ind}_P^G(\chi\tau)$ (the irreducible quotient is unique and it has multiplicity one in the whole induced representation). In this way $\tilde{G}$ is parameterized by triples (P, τ, χ) as above.

Example - The case of general linear groups. *We shall fix on F the absolute value satisfying $d(ax) = |a|_F dx$ (i.e. the* ***modulus character****), where dx is an invariant measure on the additive group of F. We shall denote by ν the character*

$$\nu : g \mapsto |det(g)|_F$$

of $GL(n, F)$.

Define

$$D = \cup_{n=1}^{\infty} D(GL(n, F)), \quad D_u = \cup_{n=1}^{\infty} D_u(GL(n, F)).$$

Let $M(D)$ be the set of all finite multisets in D (in multisets, multiplicities greater than 1 are allowed). For $\delta \in D$ there exists a unique $e(\delta) \in \mathbb{R}$ and $\delta^u \in D_u$ such that

$$\delta \cong \nu^{e(\delta)} \delta^u.$$

For

$$d = (\delta_1, \dots, \delta_l) \in M(D),$$

take a permutation p of $\{1, \dots, l\}$ such that $e(\delta_{p(1)}) \geq \dots \geq e(\delta_{p(l)})$. Then the representation

$$Ind(\delta_{p(1)} \otimes \dots \otimes \delta_{p(l)})$$

(of corresponding general linear group, parabolically induced from the corresponding standard parabolic subgroup) has a unique irreducible quotient, which we denote by

$$L(d).$$

This is the Langlands classification for general linear groups:

$$d \mapsto L(d) \text{ is a bijection of } M(D) \text{ onto } \cup_{n=0}^{\infty} (GL(n,F))\tilde{}.$$

The key fact that enables the above description of the Langlands classification, is the irreducibility of tempered parabolic induction for general linear groups (proved by H. Jacquet, among others). The above description of the Langlands classification is very much in the spirit of the local Langlands correspondences for general linear groups.

We have a similar descriptions of the Langlands classification for other classical groups (we shall give that description later).

Note that the Langlands classification reduces the problem of classifying the non-unitary dual to the following two problems:

(*i*) classification of irreducible square integrable modulo center representations of Levi factors;
(*ii*) understanding reducibility of representations parabolically induced by irreducible square integrable modulo center representations of Levi factors.

Having in mind (the formulation of) the Langlands classification, it is important to understand reducibility of much more general (and non-unitarizable) parabolically induced representations then in (ii) (at least in some very important situations).

These two problems - classification of irreducible square integrable representations of classical groups and understanding of irreducibility/reducibility of parabolically induced representations - will be the main topic of the rest of our paper.

Remark. *We shall not discuss the unitarizability problem in this paper. Let us note that having a good understanding of both of the above problems is very important for unitarizability. For working on unitarizability one needs to have a more explicit understanding of non-unitary dual than the general form of the Langlands classification. Reducibility of parabolically induced representations is related to the size of the unitary dual. For one thing, it is related to the existence of complementary series.*

Before we discuss how one can deal with the above problems, we recall some basic facts that we shall need for dealing with that problems.

7. Parabolic induction - basic facts

We take a moment to recall some general facts regarding parabolic induction (one can find more information in [3] and [4]):

(1) Fix a parabolic subgroup P of G and its Levi subgroup M. The process of attaching the parabolically induced representation $\mathrm{Ind}_P^G(\sigma)$ of G to a representation σ of M lifts in a natural way to a functor, which is called the **functor of parabolic induction**. This functor is **exact**, and it carries representations of finite length of M to the representations of finite length of G.

(2) Parabolic induction commutes with contragredients, i.e.,

$$\mathrm{Ind}_P^G(\sigma)^{\tilde{}} \cong \mathrm{Ind}_P^G(\tilde{\sigma}).$$

(3) **Induction in stages**: If $P_1 = M_1N_1$ and $P_2 = M_2N_2$ are standard parabolic subgroups with standard Levi decompositions such that $P_1 \subseteq P_2$, and if σ is a smooth representation of M_1, then for parabolically induced representations we have

$$\mathrm{Ind}_{P_1}^G(\sigma) \cong \mathrm{Ind}_{P_2}^G(\mathrm{Ind}_{P_1\cap M_2}^{M_2}(\sigma)).$$

(4) **Induction from associate representations and associate parabolic subgroups**: If $P_1 = MN_1$ and $P_2 = MN_2$ are two (not necessarily standard) parabolic subgroups and σ is a smooth representation of M of finite length, then $\mathrm{Ind}_{P_1}^G(\sigma)$ and $\mathrm{Ind}_{P_2}^G(\sigma)$ have the same composition series.

Two parabolic subgroups $P_1 = M_1N_1$ and $P_2 = M_2N_2$ are called **associate** if M_1 is a conjugate of M_2.

8. Jacquet modules

Frobenius reciprocity for parabolically induced representations holds in the same form as for the finite groups. However, there is a big problem: that the restriction is in general a very big representation of the subgroup (usually of infinite length). From the other side, parabolic induction is induction from P, but with the representation **trivial on** N. Using this fact, we easily get the form of Frobenius reciprocity below.

For a representation (π, V) of a group G and a parabolic subgroup $P = MN$ of G, let

$$V(N) = \text{span}_{\mathbb{C}}\{\pi(n)v - v; v \in V, n \in N\}.$$

Since M normalizes N, this space is M-invariant. Denote by

$$r_M^G(\pi)$$

the quotient representation of M on $V/V(N)$, twisted by $(\delta_P|_M)^{-1/2}$. Then $r_M^G(\pi)$ is called **Jacquet module** of π with respect to $P = MN$.

Now, **Frobenius reciprocity** becomes

$$\text{Hom}_G(\pi, \text{Ind}_P^G(\sigma)) \cong \text{Hom}_M(r_M^G(\pi), \sigma) \tag{8.1}$$

for representations π and σ of G and M respectively.

The fundamental properties of Jacquet modules are listed below (see [3] and [4] for more information):

(1) Jacquet modules lift in a natural way to a functor, which is called the **Jacquet module functor**. This functor is **exact** and it carries representations of finite length of G into representations of finite length of M.

(2) **Transitivity of Jacquet modules**: If $P_1 = M_1N_1$ and $P_2 = M_2N_2$ are standard parabolic subgroups with standard Levi decompositions, such that $P_1 \subseteq P_2$, and π a smooth representation of G, we have

$$r_{M_1}^G(\pi) \cong r_{M_1}^{M_2}(r_{M_2}^G(\pi)).$$

(3) **Jacquet module of the contragredient representation**: Let π be a smooth representation of G of finite length, and let $P = MN$ be a standard parabolic subgroup (with standard Levi decomposition). Denote by $\bar{P} = M\bar{N}$ the opposite parabolic subgroup (this is a parabolic subgroup satisfying $P \cap \bar{P} = M$). Then $r_M^G(\tilde{\pi})$ is isomorphic to the

contragredient representation of the Jacquet module of π with respect to $\bar{P} = M\bar{N}$.

Suppose that for an irreducible representation π of G we have $r_M^G(\pi) \neq \{0\}$ for some (standard) proper parabolic subgroup $P = MN$ of G. Then Frobenius reciprocity implies that π embeds into $\mathrm{Ind}_P^G(\sigma)$ for some irreducible representation σ of M. Taking P minimal with property $r_M^G(\pi) \neq \{0\}$, we would get that π embeds into $\mathrm{Ind}_P^G(\sigma)$, where σ satisfies

$$r_{M'}^M(\sigma) = \{0\}$$

for all proper parabolic subgroups $P' = M'N'$ of M.

Definition. *An irreducible representation π of G for which*

$$r_M^G(\pi) = \{0\}$$

holds for all proper parabolic subgroups $P = MN$ of G is called ***cuspidal****. Otherwise, we say that representation is non-cuspidal.*

This definition obviously opens the following strategy for classifying $\tilde{G}$ in two stages:

(1) first classify the irreducible cuspidal representations of Levi subgroups;

(2) then classify the non-cuspidal irreducible representations, i.e., the irreducible subquotients (or subrepresentations) of representations of G parabolically induced by irreducible cuspidal representations from proper (standard) parabolic subgroups.

Langlands classification may be viewed as a refinement of this strategy: it reduces the second problem to the problem of classifying of all irreducible tempered subquotients. As we know, this question splits into the following problems:

(a) classifying the irreducible square integrable modulo center subquotients of representations parabolically induced by irreducible cuspidal ones;
(b) understanding representations parabolically induced by square integrable modulo center representations.

The matrix coefficients of irreducible cuspidal representations are compactly supported modulo center (and the converse also holds; [4]). Therefore, irreducible cuspidal representations are essentially square integrable.

In this paper we shall deal with the problem of constructing and classifying all the non-cuspidal irreducible square integrable modulo center subquotients of parabolically induced representations. We shall not go into the problem of constructing irreducible cuspidal representations.

9. The geometric lemma

For a non-cuspidal representation, it is very useful to know its Jacquet modules explicitly. Since we are interested in irreducible subquotients of parabolically induced representations, it is important for us to know the Jacquet modules of induced representations, at least to a certain level. The complete structure of these Jacquet modules is usually hard to understand, at least for the most interesting cases (already for $SL(2,F)$: for example it is not that easy to tell if the Jacquet module of $\mathrm{Ind}^{SL(2,F)}(\chi)$ from the minimal parabolic subgroup is semi simple when $\chi^2 \equiv 1$, which is in the unitary situation; see [4]). However, there is a very useful result in that direction. It does not completely describe the Jacquet modules of parabolically induced representations, but rather some filtrations. Moreover, these subquotients of Jacquet modules of parabolically induced representations are described as representations parabolically induced by certain Jacquet modules of the inducing representation. This (technical) description is very useful, and for the sake of completeness, we shall give it here.

Fix a parabolic subgroup $P = MN$ of G and a smooth representation σ of M. For $x \in G$, $x^{-1}\sigma$ will denote the representation of $x^{-1}Mx$ given by $(x^{-1}\sigma)(x^{-1}mx) = \sigma(m)$, $m \in M$. In other words, $(x^{-1}\sigma)(m') = \sigma(xm'x^{-1})$, $m' \in x^{-1}Mx$.

Denote by W the Weyl group of $A_\emptyset$ (the quotient group of the normalizer of $A_\emptyset$ by the centralizer). Let $P = MN$ and $Q = LU$ be standard parabolic subgroups, with standard Levi decompositions. Define

$$W^{M,L} = \{w \in W; w(M \cap P_\emptyset)w^{-1} \subseteq P_\emptyset, w^{-1}(L \cap P_\emptyset)w \subseteq P_\emptyset\}.$$

Now by J. Bernstein, A.V. Zelevinsky ([3]), and by W. Casselman ([4]), we have:

Geometric lemma. *Let σ be a smooth admissible representation of M. Then one can enumerate the elements $w_1, w_2, \ldots, w_m$ of $W^{M,L}$ in such a way that there exists a filtration*

$$\{0\} = \tau_0 \subseteq \tau_1 \subseteq \ldots \subseteq \tau_m = r_L^G(Ind_P^G(\sigma))$$

of $r_L^G(Ind_P^G(\sigma))$ such that for $1 \leq i \leq m$,

$$\tau_i/\tau_{i-1} \cong Ind^L_{wMw^{-1}\cap L}(w(r^M_{M\cap wLw^{-1}}(\sigma))).$$

For more details one should consult the original papers (the Geometric lemma in terms of roots may be more suitable for some calculations; see [4]).

As we shall soon see, already for the analysis of a single induced representation, we need to consider various Jacquet modules of several induced representations and their filtrations. This requires (among other things) frequent computations of $W^{M,L}$ and the groups $M \cap w^{-1}Lw$, $N' = wMw^{-1} \cap L$. These computations may take a considerable time even for a single representation. Therefore, it would be convenient to have a simple way (or algorithm) for obtaining the above filtrations.

Surprisingly, for classical groups there is very elegant solution to this problem. In a way, this problem is solved by "algebraization" of the Geometric lemma (in this case). Before we go on to explain it, let us derive some direct consequences of the Geometric lemma.

10. Some general consequences

Suppose below that σ is an irreducible cuspidal representation of M.

(1) A simple case of the Geometric lemma implies that the composition series of the Jacquet module $r_M^G(\mathrm{Ind}_P^G(\sigma))$ consist of all $w\sigma$ such that w runs over all representatives of $W_M \backslash W_G / W_M$ which normalize M (W_G and W_M denote the Weyl groups of G and M respectively with respect to $A_\emptyset$). From this, one can get a number of interesting general facts.

Conjugating an induced representation and applying induction from associate parabolic subgroups (see (4) of section 7.), for a standard parabolic subgroup $P' = M'N'$ associate to P, one gets irreducible subquotients of $r_{M'}^G(\mathrm{Ind}_P^G(\sigma))$ by conjugating the irreducible subquotients of $r_M^G(\mathrm{Ind}_P^G(\sigma))$ by the corresponding element (thus, they are all cuspidal).

Further, if a standard parabolic subgroup P' does not contain parabolic subgroup associate to P, then $r_{M'}^G(\mathrm{Ind}_P^G(\sigma))$ is trivial (i.e., equal to $\{0\}$). In the case that the parabolic subgroup strictly contains a parabolic subgroup associate to P, the Jacquet modules are non-trivial, and all irreducible subquotients are non-cuspidal.

(2) The description of composition series implies that if $r_{M'}^G(\mathrm{Ind}_P^G(\sigma))$ has a cuspidal subquotient for a standard parabolic subgroup $P' = M'N'$, then P and P' are associate. Further, if σ' is an irreducible cuspidal subquotient, then σ and σ' are conjugated.

More generally, let τ be an irreducible admissible representation of M and ρ', ρ'' irreducible cuspidal subquotients of $r^G_{M'}\left(\mathrm{Ind}^G_P(\tau)\right)$ and $r^G_{M''}\left(\mathrm{Ind}^G_P(\tau)\right)$ respectively, where $P' = M'N'$ and $P'' = M''N''$ are standard parabolic subgroups (with standard Levi decompositions). Then P' and P'' are associate and ρ', ρ'' are conjugate.
Similarly, if $P' = M'N'$, $P'' = M''N''$ denote two parabolic subgroups of G, and σ', σ'' denote irreducible cuspidal representations of M' and M'' respectively, such that $\mathrm{Ind}^G_P(\sigma')$ and $\mathrm{Ind}^G_{P''}(\sigma'')$ have a common composition factor, then P' and P'' must be associate, while σ' and σ'' are conjugate.

(3) The description of Jacquet modules in (1), together with some simple representation theory, implies that if π is an irreducible subquotient of $\mathrm{Ind}^G_P(\sigma)$ (as above, σ is irreducible cuspidal), then there exists $w \in G$ which normalizes M such that π is isomorphic to a subrepresentation of $\mathrm{Ind}^G_P(w\sigma)$. Now Frobenius reciprocity implies that $r^G_M(\pi)$ is non-zero (and each irreducible subquotient is cuspidal). Further, if $P' = M'N'$ is a parabolic subgroup such that $r^G_{M'}\left(\mathrm{Ind}^G_P(\sigma)\right)$ has an irreducible cuspidal subquotient, then $r^G_{M'}(\pi) \neq \{0\}$.
If τ is an irreducible admissible representation of M, and π an irreducible subquotient of $\mathrm{Ind}^G_P(\tau)$ such that $r^G_{M'}\left(\mathrm{Ind}^G_P(\tau)\right) \neq \{0\}$ for some standard parabolic subgroup $P' = M'N'$ of G, then from the transitivity of Jacquet modules, we get $r^G_{M'}(\pi) \neq \{0\}$.

In the above considerations, we have used only a small portion of the information coming from the Geometric lemma, namely, only the information coming from the calculation of the minimal non-trivial Jacquet modules. Much more information is contained in the Jacquet modules with respect to the other parabolic subgroups. We shall soon explain how one can use this information to understand irreducibility/reducibility of parabolically induced representations in some important cases. Before we explain this, we shall show how one can get good control of these other Jacquet modules for series of classical groups. But first, we shall say a few words about

11. The case of maximal parabolic subgroups

In this section, we shall see what the Geometric lemma gives in the simplest case, when we induce from an irreducible cuspidal representation of a maximal parabolic subgroup, like $P_\emptyset$ in $SL(2,F)$. We start with the case $G = SL(2,F)$ and $P = P_\emptyset$.

We identify the subgroup of diagonal matrices $A_{\emptyset}$ in $G = SL(2,F)$ with $F^{\times}$ using the identification $\text{diag}(a, a^{-1}) \leftrightarrow a$. Let χ be a character of $F^{\times}$. Now the Geometric lemma implies that χ, χ^{-1} is the composition series of $r_M^G\left(\text{Ind}_P^G(\chi)\right)$. This implies that the length of $\text{Ind}_P^G(\chi)$ is at most two.

Suppose that χ is unitary. Then if $\chi \neq \bar{\chi}$, the unitarizability of $\text{Ind}_P^G(\chi)$ and Frobenius reciprocity imply the irreducibility of $\text{Ind}_P^G(\chi)$.

Suppose that χ is not unitary, and that $\text{Ind}_P^G(\chi)$ reduces. Then one irreducible subquotient is a square integrable representation by Casselman's square integrability criterion (which will be discussed later). Since square integrable representations are unitarizable, which implies that they are equivalent to their Hermitian contragredients (i.e., the complex conjugates of the contragredients), $\text{Ind}_P^G(\chi)$ and $\text{Ind}_P^G((\tilde{\chi})^{-})$ have an irreducible subquotient in common. Now, general properties regarding principal series (discussed in section 10.) imply that $\chi \cong (\tilde{\chi})^{-}$ or $\tilde{\chi} \cong (\tilde{\chi})^{-}$. Since χ is not unitary, we must have $\chi = \bar{\chi}$.

Thus, here the Geometric lemma gives irreducibility in the case of non-real valued characters, but does not tell us anything about reducibility points (in the most interesting case) of real valued characters (observe that we know here that $\text{Ind}_P^G(|\ |_F^{1/2})$ reduces since 1_G is subquotient; here 1_G denotes the representation of G on one-dimensional vector space, with the trivial action of G). Summing up, Jacquet modules provide us only qualitative information.

Quantitative information (which is the most interesting), i.e. where exactly reducibility points are, is obtained using analytic methods (from which one can also deduce the above qualitative information).

Therefore, it is not surprising that for a long time Jacquet modules were not used much in treating more general reducibility problems (and construction of square integrable representations). In the case of general linear groups, Bernstein and Zelevinsky used something much stronger - Gelfand-Kazhdan derivatives - and got complete answers modulo cuspidal representations (we shall describe briefly these answers later).

For other (classical) groups, the same problems are much, much harder, while the analogous powerful tool does not exist in this case. Therefore, it was natural to try to use analytic methods there. But analytic methods, which are not even very simple for $SL(2)$, became quite involved in more general cases.

For example, R. Gustafson needed considerable work to settle (degenerate principal series) representations of $Sp(2n, F)$ parabolically induced by unramified characters from the Siegel parabolic subgroup (see [8]). S.S.

Kudla and S. Rallis needed additional significant work to settle the case of remaining ramified characters (see [18]). The case of Siegel parabolic subgroup fits well with their methods since the unipotent radical is commutative (which is not the case with other maximal parabolic subgroups). After these two relatively simple cases, the question which arises is what should we expect in more complicated cases, and what kind of methods should we use to handle these problems. It seems that the best is a combination of both methods. When one induces from an irreducible cuspidal representation of a maximal parabolic subgroup, one is forced to use the analytic tools (simply put, Jacquet modules do not give much). In other cases, Jacquet module methods happen to be surprisingly powerful. For example, although not very useful in the simple case of $SL(2,F) = Sp(2,F)$, the use of Jacquet module methods gives complete answers (i.e. reducibility points, lengths of representations and Langlands parameters of irreducible subquotients) for all degenerate principal series of all $Sp(2n,F)$ or $SO(2n+1,F)$, $n \geq 2$, induced from maximal parabolic subgroups (with the help of the Bernstein-Zelevinsky theory; see [14]). Roughly, the power of Jacquet module methods for bigger groups lies in the fact that in this case we have more standard parabolic subgroups, and by comparing information coming from different parabolic subgroups we can obtain surprising amount in this case.

The same type of analysis which we have described for $SL(2,F)$ can be applied to a maximal parabolic subgroup $P = MN$ of a connected reductive F-group and an irreducible cuspidal representation σ of M. For simplicity, we shall assume that the center of G is compact. Suppose that $\mathrm{Ind}_P^G(\sigma)$ reduces. Then general facts on induced representations (in section 7.) imply that there exists $w \in W$ which is not in M, but normalizes M. Now $\{\sigma, w\sigma\}$ is the composition series of $r_M^G\left(\mathrm{Ind}_P^G(\sigma)\right)$. If σ is unitarizable, then Frobenius reciprocity (and semi simplicity) implies irreducibility if $\sigma \not\cong w\sigma$ (regular case).

Suppose that σ is not unitarizable, and that $\mathrm{Ind}_P^G(\sigma)$ reduces. Then the length of $\mathrm{Ind}_P^G(\sigma)$ is two and one irreducible subquotient is again square integrable by the Casselman criterion. In the same way as above, we get that $\mathrm{Ind}_P^G(\sigma)$ and $\mathrm{Ind}_P^G((\tilde{\sigma})^-)$ have an irreducible subquotient in common. Now section 10. implies $\sigma \cong (\tilde{\sigma})^-$ or $w\sigma \cong (\tilde{\sigma})^-$, and further non-unitarity implies $w\sigma \cong (\tilde{\sigma})^-$.

The hard part here is to understand when we have reducibility in the case $w\sigma \cong (\tilde{\sigma})^-$. This is a very hard problem, and it requires analytic methods (see [31] and [32]). We shall say few words about this later.

12. Hopf algebras in the case of general linear groups

In this section we recall the notation and structures that have been introduced by J. Bernstein and A.V. Zelevinsky. They have introduced an algebraic structure in which they incorporated some very basic properties of the representation theory of general linear groups (like induction in stages, transitivity of Jacquet modules, the Geometric lemma etc.). It is interesting that they did not use this structure much in their work, but rather pointed it out as an interesting curiosity (later Zelevinsky used it in the finite field case; we have used this structure in a substantial way in our work on representation theory of general linear groups over p-adic division algebras in [35]).

For a reductive F-group G, we denote by

$$\mathcal{R}(G)$$

the free $\mathbb{Z}$-module with basis $\tilde{G}$ (this is the Grothendieck group of the category of all smooth finite length representations of G). The multiplicity of $\tau \in \tilde{G}$ in a finite length representation π will be denoted by $m(\tau : \pi)$. We denote the semi simplification of π by

$$\text{s.s.}(\pi) = \sum_{\tau \in \tilde{G}} m(\tau : \pi)\tau \in \mathcal{R}(G).$$

There is a natural partial ordering $\leq$ on $\mathcal{R}(G)$ (the positive cone consists of finite (formal) sums of elements of $\tilde{G}$). When we write $\leq$ between representations (of finite length), this will actually mean inequality among their semi simplifications.

For a parabolic subgroup $P = MN$, the functors r_M^G and Ind_P^G factor in a natural way to homomorphisms of Grothendieck groups (from $\mathcal{R}(G)$ into $\mathcal{R}(M)$ and from $\mathcal{R}(M)$ into $\mathcal{R}(G)$ respectively). These homomorphisms are also denoted by the same symbol as the functors. They are homomorphisms of ordered groups (i.e., they also respect orderings).

In what follows we shall often use the fact that for two reductive groups G_1 and G_2 we have

$$\mathcal{R}(G_1 \times G_2) \cong \mathcal{R}(G_1) \otimes \mathcal{R}(G_2)$$

(this is a consequence of the fact that representations in $\tilde{G}_i$ are admissible; see [4]).

Now we return to the general linear groups and recall the Bernstein-Zelevinsky notation. If π_1 and π_2 are representations of $GL(n_1, F)$ and

$GL(n_2, F)$, let

$$\pi_1 \times \pi_2 = \mathrm{Ind}_{P^{GL}_{(n_1,n_2)}}^{GL(n_1+n_2,F)}(\pi_1 \otimes \pi_2).$$

Then induction by stages implies

$$\pi_1 \times (\pi_2 \times \pi_3) \cong (\pi_1 \times \pi_2) \times \pi_3.$$

Set

$$R_n = \mathcal{R}(GL(n,F)) \quad \text{and} \quad R = \bigoplus_{n \in \mathbb{Z}_{\geq 0}} R_n.$$

Lift $\times$ to a multiplication on R, and factor $\times : R \times R \to R$ through a mapping

$$m : R \otimes R \to R.$$

In this way R becomes a **commutative ring** (the commutativity follows from induction from associate parabolic subgroups: $\pi_1 \times \pi_2$ and $\pi_2 \times \pi_1$ have the same composition series).

For $\pi \in GL(n,F)\tilde{}$ consider s.s. $\left(r_{M^{GL}_{(k,n-k)}}^{GL(n,F)}(\pi)\right) \in R_k \otimes R_{n-k}$ and define

$$m^*(\pi) = \sum_{k=0}^{n} \text{s.s.} \left(r_{M^{GL}_{(k,n-k)}}^{GL(n,F)}(\pi)\right) \in \sum_{k=0}^{n} R_k \otimes R_{n-k} \hookrightarrow R \otimes R\ .$$

Lift m^* to additive mapping $m^* \colon R \to R \otimes R$, which will be called **comultiplication**. In this way R becomes **coalgebra** (coassociativity follows from the transitivity of Jacquet modules). Moreover, with m and m^*, R is a **Hopf algebra** over $\mathbb{Z}$, i.e., the comultiplication is multiplicative:

$$m^*(\pi_1 \times \pi_2) = m^*(\pi_1) \times m^*(\pi_2), \tag{12.1}$$

which follows from the Geometric lemma.

Observe that the above formula enables us to get, in a very simple and direct way, composition factors of Jacquet modules (for maximal parabolic subgroups) of induced representations, as representations induced by Jacquet modules of inducing representations. Therefore, the most important requirement of the Hopf algebra structure, the Hopf axiom (12.1), can be viewed as an algebraization of the Geometric lemma in this case.

The Hopf algebra R is commutative, but not cocommutative.

By construction, the structure of a Hopf algebra contains in itself a number of the most important basic properties of representation theory, applied to general linear groups: induction in stages and induction from associate parabolic subgroups, transitivity of Jacquet modules and the Geometric lemma. There is one more remarkable property of R: it is a **polynomial algebra** over D (this follows from the fact that tempered induction for general linear groups is irreducible, and from basic properties of the Langlands classification).

Remark. *Although the definition of the Hopf algebra R involves only Jacquet modules for maximal parabolic subgroups, the transitivity of Jacquet modules enables one to compute other Jacquet modules using the Hopf algebra structure of R. For example, let $\rho_1, \dots, \rho_l$ be irreducible cuspidal representations of general linear groups. Using the Hopf algebra structure of R, one can prove by simple induction the following fact: if σ is an irreducible cuspidal subquotient of a Jacquet module of $\rho_1 \times \cdots \times \rho_l$ with respect to a standard parabolic subgroup, then there exists a permutation p of $\{1, \dots, l\}$ such that $\sigma \cong \rho_{p(1)} \otimes \cdots \otimes \rho_{p(l)}$; further, each $\rho_{p(1)} \otimes \cdots \otimes \rho_{p(l)}$ with p as above, shows up as a subquotient of the Jacquet module with respect to some standard parabolic subgroup.*

Later on, we shall show how to extend the above algebraic approach to other classical groups. Before that, we shall describe some very basic facts from the Bernstein-Zelevinsky theory.

13. Square integrable representations of general linear groups

The set of all equivalence classes of irreducible cuspidal representations of all $GL(n, F)$, $n \geq 1$, will be denoted by

$$\mathcal{C}.$$

A set of the form $\{\rho, \nu\rho, \dots, \nu^k\rho\}$, where $\rho \in \mathcal{C}, k \in \mathbb{Z}_{\geq 0}$, will be called a **segment in** $\mathcal{C}$, and denoted by $[\rho, \nu^k\rho]$. The set of all such segments will be denoted by

$$\mathcal{S}.$$

The geometric lemma implies that if $\rho \times \rho'$ reduces ($\rho, \rho' \in \mathcal{C}$), then $\rho' \cong \nu^\alpha\rho$ for some $\alpha \in \mathbb{R}$. A fundamental fact, proved by J. Bernstein and A.V. Zelevinsky, using Gelfand-Kazhdan derivatives, is that $\rho \times \rho'$ reduces if and only if $\rho' \cong \nu^{\pm 1}\rho$, i.e., precisely when $\alpha = \pm 1$ (later on, this was proved by F. Shahidi using L-functions).

One can attach to each segment $\Delta = [\rho, \nu^k\rho] = \{\rho, \nu\rho, \dots, \nu^k\rho\} \in \mathcal{S}$ the unique irreducible subrepresentation $\delta(\Delta)$ of

$$\nu^k\rho \times \nu^{k-1}\rho \times \cdots \times \nu\rho \times \rho.$$

The representation $\delta(\Delta)$ is essentially square integrable.

The irreducible representation $\delta(\Delta)$ is characterized by the fact that $\nu^k\rho \otimes \nu^{k-1}\rho \otimes \cdots \otimes \nu\rho \otimes \rho$ is in its Jacquet module (for the corresponding standard Jacquet module). From this, the following simple formula follows easily

$$m^*(\delta([\rho, \nu^k\rho])) = \sum_{i=-1}^{k} \delta([\nu^{i+1}\rho, \nu^k\rho]) \otimes \delta([\rho, \nu^i\rho]), \tag{13.1}$$

which is very useful in applications (we take formally $GL(0,F)$ to be the trivial group; the representation of the trivial group on one-dimensional vector space is denoted by 1; in (13.1) we take formally $\delta(\emptyset) = 1$). The mapping $\Delta \mapsto \delta(\Delta)$ is a bijection from $\mathcal{S}$ onto D (the set of all the equivalence classes of irreducible essentially square integrable representations of general linear groups $GL(n,F)$, $n \geq 1$).

The above classification of J. Bernstein and A.V. Zelevinsky reduced the problem of establishing local Langlands correspondences for general linear groups to the case of cuspidal representations in the following way. For a positive integer n and an irreducible cuspidal representation ρ of a general linear group, let

$$\delta(\rho, n) = \delta([\nu^{-(n-1)/2}\rho, \nu^{(n-1)/2}\rho]).$$

Let E_n be (the unique up to an equivalence) irreducible algebraic representation of $SL(2,\mathbb{C})$ of dimension n. Denote by Φ the local Langlands correspondences for general linear groups. Then

$$\Phi(\delta(\rho, n)) = \Phi(\rho) \otimes E_n.$$

Therefore, it is enough to establish the local Langlands correspondences for irreducible cuspidal representations, which was done later, first by G. Laumon, M. Rapoport and U. Stuhler in positive characteristic ([19]), and after that by M. Harris and R. Taylor ([9]), and also G. Henniart ([13]) in characteristic 0.

We say that $\Delta_1, \Delta_2 \in \mathcal{S}$ are linked, if

$$\Delta_1 \cup \Delta_2 \in \mathcal{S} \backslash \{\Delta_1, \Delta_2\}.$$

The representation $\delta(\Delta_1) \times \delta(\Delta_2)$ reduces if and only if Δ_1 and Δ_2 are linked, and then

$$\{L(\delta(\Delta_1), \delta(\Delta_2)), \delta(\Delta_1 \cup \Delta_2) \times \delta(\Delta_1 \cap \Delta_2)\}$$

is the composition series of $\delta(\Delta_1) \times \delta(\Delta_2)$.

Let $a = (\Delta_1, \ldots, \Delta_l) \in M(\mathcal{S})$. If there are $i, j \in \{1, \ldots, l\}$ such that Δ_i and Δ_j are linked, then replacing the pair Δ_i, Δ_j by the pair

$$\Delta_i \cup \Delta_j, \Delta_i \cap \Delta_j$$

in a, gives a multiset a'. We write $a' \prec a$. Using $\prec$ we generate partial ordering $\leq$ on $M(\mathcal{S})$. Then for $a = (\Delta_1, \ldots, \Delta_l), a' = (\Delta'_1, \ldots, \Delta'_{l'}) \in M(\mathcal{S})$, $L(\delta(\Delta'_1), \ldots, \delta(\Delta'_{l'}))$ is a subquotient of $\delta(\Delta_1) \times \cdots \times \delta(\Delta_l)$ if and only if $a' \leq a$. Further, if a' is minimal among those satisfying $a' \leq a$, then the multiplicity is one.

In particular, if among $\Delta_1, \ldots, \Delta_l$ we do not have linked pairs of segments, then the representation

$$\delta(\Delta_1) \times \cdots \times \delta(\Delta_l)$$

is irreducible. One can find all these results in [46], and much more.

We denote by 1_G the trivial representation (on a one-dimensional space) of a group G. The trivial representation of the trivial group will be denoted simply by 1.

In the Grothendieck group, $\delta([\nu^{-(n-1)/2} 1_{F^\times}, \nu^{(n-1)/2} 1_{F^\times}])$ is an alternated sum of

$$\mathrm{Ind}_{P_\alpha}^{GL(n,F)}(\delta_{P_\alpha}^{-1/2}|_{M_\alpha}), \tag{13.2}$$

where $\alpha = (n_1, \ldots, n_l)$ runs over all ordered partitions of n.

Remark. *The notion of a Steinberg representation comes from the finite field case. In that setting, this representation is an irreducible representation whose dimension is equal to the cardinality of a p-Sylow subgroup, where p is the characteristic of the finite field, and it is alternated sum of $Ind_P^G(1_M)$ where P runs over standard parabolic subgroups. Since the last property holds for $\delta([\nu^{-(n-1)/2} 1_{F^\times}, \nu^{(n-1)/2} 1_{F^\times}])$ (it is an alternated sum of representations (13.2)), this representation is called the Steinberg representation of $GL(n, F)$.*

Our representations $\delta([\rho, \nu^k \rho])$ have similar properties to the Steinberg representations. This is the reason that they are called generalized Steinberg representations. They can also be related to the Steinberg representations using Hecke algebra isomorphisms.

In general, for generalized Steinberg representations all the non-trivial Jacquet modules are irreducible essentially square integrable representations, and the inducing representation is regular (this notion will be explained later). These properties hold only for a small portion of square integrable representations of other classical groups. This is the reason that we shall later call irreducible square integrable representations which satisfy these two properties square integrable representations of Steinberg type.

The definition of the Steinberg representation for general connected reductive group G over a local field is the following:

Definition. *The representation* $Ind_{P_\emptyset}^G(\delta_{P_\emptyset}^{1/2}|_{M_\emptyset})$ *contains a unique square integrable subquotient. This subquotient is called the Steinberg representation of* G *and denoted by* St_G.

14. Other classical groups

In this section we shall recall the definitions of other split classical groups, and introduce notation which will again incorporate some very basic properties of the representation theory of these groups. We shall work with symplectic and split odd orthogonal groups in this paper, but we could work in a similar way with other classical groups (see [24]).

Let $J_n \in GL(n, F)$ be the matrix having entries 1 on the second diagonal, and 0 in the remaining places, and let I_n be the identity matrix in $GL(n, F)$. Denote by ${}^tS$ (resp., ${}^\tau S$) the transposed matrix of S (resp., the transposed matrix of S with respect to the second diagonal). The symplectic group is

$$Sp(2n, F) = \left\{ S \in GL(2n, F);\ {}^tS \begin{bmatrix} 0 & J_n \\ -J_n & 0 \end{bmatrix} S = \begin{bmatrix} 0 & J_n \\ -J_n & 0 \end{bmatrix} \right\},$$

and the split odd-orthogonal group is

$$SO(2n+1, F) = \{S \in SL(2n+1, F);\ {}^\tau S\, S = I_{2n+1}\}.$$

We could also introduce these groups geometrically.

In what follows, sometimes we denote $GL(n, F)$, $Sp(2n, F)$, or $SO(2n+1, F)$, simply by $GL(n)$, $Sp(2n)$, or $SO(2n+1)$, respectively (i.e. if we do not specify which rational points we consider, then it will always mean that we consider F-rational points).

We shall fix a series of symplectic or odd-orthogonal groups, and denote by S_n corresponding group of rank n from the series (i.e. $Sp(2n, F)$ or $SO(2n+1, F)$). We can parameterize the standard parabolic subgroups of S_n by ordered partitions $\alpha = (n_1, \ldots, n_k)$ of integers m, where $0 \leq m \leq n$, in the following way. For $S_n = Sp(2n, F)$ (resp., $S_n = SO(2n+1, F)$) let

$$\alpha' = (n_1, ..., n_k, 2n-2m, n_k, ..., n_1)$$

$$\text{(resp., } \alpha' = (n_1, ..., n_k, 2n+1-2m, n_k, ..., n_1)).$$

Set

$$P_\alpha = P_{\alpha'}^{GL} \cap S_n, \quad M_\alpha = M_{\alpha'}^{GL} \cap S_n, \quad N_\alpha = N_{\alpha'}^{GL} \cap S_n.$$

Now $(g_1, \ldots, g_k, h, \ {}^\tau g_k^{-1}, \ldots, \ {}^\tau g_1^{-1}) \mapsto \text{quasi-diag}(g_1, \ldots, g_k, h)$ is an isomorphism of the product $GL(n_1, F) \times \cdots \times GL(n_k, F) \times S_{n-m}$ onto M_α. We shall use this identification in the sequel.

In particular, for $m \le n$, the bijection $(g, h) \leftrightarrow \text{quasi-diag}(g, h, \ {}^\tau g^{-1})$ gives an identification of $GL(m, F) \times S_{n-m}$ with $M_{(m)}$. For admissible representations π and σ of finite length of $GL(m, F)$ and S_{n-m} respectively, define

$$\pi \rtimes \sigma = \text{Ind}_{P_{(m)}}^{S_n}(\pi \otimes \sigma)$$

Then induction in stages implies

$$(\pi_1 \times \pi_2) \rtimes \sigma \cong \pi_1 \rtimes (\pi_2 \rtimes \sigma).$$

Let

$$R(S) = \underset{n \in \mathbb{Z}_{\ge 0}}{\oplus} \mathcal{R}(S_n).$$

One lifts $\rtimes$ to a biadditive mapping

$$R \times R(S) \to R(S).$$

The lift is also denoted by $\rtimes$. In this way $R(S)$ becomes an R-**module**. In this module we have

$$\pi \rtimes \sigma = \tilde{\pi} \rtimes \sigma,$$

which follows from the fact that parabolically induced representations from associate parabolic subgroups (and representations) have the same composition series. The mapping $\rtimes : R \times R(S) \to R(S)$ factors through $R \otimes R(S)$ by

$$\mu : R \otimes R(S) \to R(S)$$

in a natural way.

For $\pi \in \tilde{S}_n$ define

$$\mu^*(\pi) = \sum_{k=0}^{n} \text{s.s.}\left(r_{M_{(k)}}^{S_n}(\pi)\right) \in \sum_{k=0}^{n} R_k \otimes R(S_{n-k}) \hookrightarrow R \otimes R(S).$$

Extend μ^* to an additive mapping $\mu^* \colon R(S) \to R \otimes R(S)$. We call it **comultiplication** on $R(S)$, and $R(S)$ is a **comodule** over R (coassociativity again follows from the transitivity of Jacquet modules).

Up to now we have observed that $R(S)$ is a module and a comodule over R. It is easy to see that it is not a Hopf module, but it is not far from Hopf module as we shall see soon. Let

$$M^* = (m \otimes 1) \circ (\sim \otimes m^*) \circ s \circ m^* : R \to R \otimes R, \tag{14.1}$$

where 1 denotes the identity mapping (on R), $\sim$ the contragredient mapping and s the transposition mapping $\sum x_i \otimes y_i \mapsto \sum y_i \otimes x_i$. Then

$$\mu^*(\pi \rtimes \sigma) = M^*(\pi) \rtimes \mu^*(\sigma) \tag{14.2}$$

($R \otimes R(S)$ is an $R \otimes R$-module in a natural way). We say that $R(S)$ is an M^*-**Hopf module** over R (in this terminology m^*-Hopf module corresponds to usual Hopf module).

The above formula incorporates the Geometric lemma in the structure of $R(S)$ over R, i.e., algebraizes the Geometric lemma. Again, this is a simple formula, which enables us to get, in a simple way, compositions factors of Jacquet modules of parabolically induced representations of classical groups.

There exists a distinguished Jacquet module in the sense that it requires only the GL-theory to handle it (and that it is non-trivial): let π be a non-trivial subquotient of $\pi' \rtimes \sigma$, where π is a representation of some $GL(p, F)$ and σ is an irreducible cuspidal representation of some S_q. Then we define

$$s_{GL} = r^{S_{p+q}}_{M_{(p)}}(\pi).$$

This will be called the **Jacquet module of π of GL-type**.

Let ρ be an irreducible cuspidal representation of a general linear group and $x, y \in \mathbb{R}$ such that $y - x \in \mathbb{Z}_{\geq 0}$. In what follows, we shall use formula for $M^*\big(\delta([\nu^x\rho, \nu^y\rho])\big)$ several times. One gets the following directly from (13.1) and (14.1):

$$M^*\big(\delta([\nu^x\rho, \nu^y\rho])\big) \tag{14.3}$$

$$= \sum_{i=x-1}^{y} \sum_{j=i}^{y} \delta([\nu^{-i}\tilde{\rho}, \nu^{-x}\tilde{\rho}]) \times \delta([\nu^{j+1}\rho, \nu^{y}\rho]) \otimes \delta([\nu^{i+1}\rho, \nu^{j}\rho]),$$

where $y - i, y - j \in \mathbb{Z}_{\geq 0}$ in the above sums.

Jacquet module methods of determining reducibility of parabolically induced representations

15. Reducibility - irreducibility

Reducibility of parabolically induced representations is closely related to construction of non-cuspidal irreducible square integrable representations. This fact is already shown by the simple example of induction from a maximal parabolic subgroup by a non-unitarizable irreducible cuspidal representation. In general, the induction from a maximal parabolic subgroup by a non-cuspidal non-unitarizable irreducible square integrable representation also quite often gives square integrable subquotient(s). Nevertheless, in the case of general linear groups, this does not happen too often (it happens if and only if corresponding linked segments are disjoint).

We shall explore the idea of constructing non-cuspidal irreducible square integrable representations of classical groups starting from non-cuspidal irreducible square integrable representations of general linear groups. For this, it is necessary to have an efficient method of controlling reducibility of parabolically induced representations when it will be necessary (for square integrable subquotients we must have reducibility of considered parabolically induced representation).

We shall now try to explain how one can have a fairly good control of these reducibilities. Before that, we shall recall briefly of the Casselman's

16. Square integrability criterion

We shall recall of the Casselman square integrability criterion only in the case of classical groups that we consider. We consider $\mathbb{R}^n$ supplied with the standard inner product $((x_1, \ldots, x_n), (y_1, \ldots, y_n)) = \sum_i x_i y_i$. Let

$$\beta_i = (\underbrace{1, 1, \ldots, 1}_{i \text{ times}}, 0, 0, \ldots, 0) \in \mathbb{R}^n, \ 1 \leq i \leq n.$$

Fix a non-cuspidal irreducible admissible representation π of S_n and take $\alpha = (n_1, \ldots, n_\ell)$ with $r_{M_\alpha}^{S_n}(\pi) \neq \{0\}$, such that there exists an irreducible cuspidal subquotient of this Jacquet module. Then all irreducible subquotients are cuspidal. Denote such a subquotient by σ. We can write σ as $\rho_1 \otimes \rho_2 \otimes \cdots \otimes \rho_\ell \otimes \rho$ where $\rho_i \in GL(n_i, F)\check{}$, $\rho \in \tilde{S}_{n-(n_1+\cdots+n_\ell)}$. Denote $m = n_1 + \cdots + n_\ell$. Define

$$e_*(\sigma) = (\underbrace{e(\rho_1), \ldots, e(\rho_1)}_{n_1 \text{ times}}, \ldots, \underbrace{e(\rho_\ell), \ldots, e(\rho_\ell)}_{n_\ell \text{ times}}, \underbrace{0, \ldots, 0}_{n-m \text{ times}})$$

If π is square integrable, then

$$(e_*(\sigma), \beta_{n_1}) > 0, (e_*(\sigma), \beta_{n_1+n_2}) > 0, \cdots, (e_*(\sigma), \beta_{n_1+n_2+\cdots+n_\ell}) > 0.$$

The converse also holds: if the above inequalities hold for any α and σ as above, then π is square integrable (if instead of strict inequalities > 0 hold the weaker conditions ≥ 0 in all the above relations, then this is a criterion for π to be tempered).

17. Cuspidal reducibilities

Fix irreducible unitarizable cuspidal representations ρ and σ of $GL(p, F)$ and S_q respectively (irreducible cuspidal σ is automatically unitarizable). Now section 11. implies that if for some $\alpha \in \mathbb{R}$, $\nu^\alpha \rho \rtimes \sigma$ reduces then $\rho \cong \tilde{\rho}$ (representations satisfying this condition will be called **selfdual**). Conversely, considering complementary series, one gets that if $\rho \cong \tilde{\rho}$, then $\nu^\alpha \rho \rtimes \sigma$ reduces for some $\alpha \geq 0$. Induction from associate parabolic subgroups (from section 7.) implies that $\nu^\alpha \rho \rtimes \sigma$ reduces if and only if $\nu^{-\alpha} \rho \rtimes \sigma$ reduces. Therefore, regarding reducibility, it is enough to study only the non-negative exponents.

Determining the reducibility points is a very important and very hard problem. **Fundamental results** in that direction are the following:

(1) For selfdual ρ there is exactly one $\alpha \geq 0$ for which $\nu^\alpha \rho \rtimes \sigma$ reduces ([33]). This point will be denoted by

$$\alpha_{\rho,\sigma}.$$

(2) If ρ is selfdual and σ is **generic**, then

$$\alpha_{\rho,\sigma} \in \{0, 1/2, 1\}$$

([31], [32]).
In a particular case, when σ is the trivial representation of the trivial group (i.e. we are inducing from the Siegel parabolic subgroup), $\alpha_{\rho,1}$ is equal to 1 if and only if we work with symplectic groups and if $\rho = 1_{F^\times}$ (i.e. the induction is in $Sp(2, F) = SL(2, F)$) ([32], [22]).

(3) For $p > 1$ and ρ a selfdual representation of $GL(p, F)$ (it is enough that $p \geq 1$ and $\rho \not\cong 1_{F^\times}$), $\rho \rtimes 1$ reduces in $Sp(2p, F)$ if and only if it is irreducible in $SO(2p + 1, F)$ ([32]).

(4) F. Shahidi has proved that for generic σ, we have

$$\alpha_{\rho,\sigma} - \alpha_{\rho,1} \in \mathbb{Z}.$$

In general, some important conjectures from the theory of automorphic forms would imply that for **any** σ the following (conjecture) holds

$$\alpha_{\rho,\sigma} - \alpha_{\rho,1} \in \mathbb{Z}, \tag{BA}$$

which will be called the **basic assumption** (see [20] for more details, together with section 12. of [24]). This conjecture would obviously imply that

$$\alpha_{\rho,\sigma} \in (1/2)\mathbb{Z}$$

for any σ.

Further, for fixed σ, there should exist only finitely many ρ with $\alpha_{\rho,\sigma} \notin \{0, 1/2\}$. There is a natural and precise conjecture of C. Mœglin regarding such ρ's (see [22] and [24], and also remarks in section 34.). This is the reason why reducibilities $\alpha_{\rho,\sigma} \notin \{0, 1/2\}$ can be called exceptional.

Some examples. *Let χ_0 be a character of order 2 of $F^\times$.*

(1) As is well known, for $Sp(2,F) = SL(2,F)$ we have $\alpha_{\chi_0,1} = 0$ and $\alpha_{1_{F^\times},1} = 1$.
(2) For $SO(3,F)$, we have $\alpha_{\chi_0,1} = \alpha_{1_{F^\times},1} = 1/2$.
(3) One can get in a relatively simple way (using Clifford theory) the following fact: Suppose that ρ is selfdual and that the central character ω_ρ of ρ satisfies $\omega_\rho \not\equiv 1$. Then the representation $\rho \rtimes 1$ of $Sp(2p,F)$ reduces, and $\nu^\alpha\rho \rtimes 1$ is irreducible for $\alpha > 0$.
(4) F. Shahidi has proved the following: for $Sp(4,F)$ and ρ selfdual irreducible cuspidal representation of $GL(2,F)$ one has $\alpha_{\rho,1} = 0$ if $\omega_\rho \not\equiv 1$, and $\alpha_{\rho,1} = 1/2$ if $\omega_\rho \equiv 1$ ([31]). For $SO(5,F)$, the situation is reversed ([32]).

18. Regular induced representations

Let σ be an irreducible cuspidal representation of a Levi subgroup M of a parabolic subgroup P of G. We shall say that σ is regular (in G), or that $\mathrm{Ind}_P^G(\sigma)$ is regular, if all (non-zero) Jacquet modules are multiplicity one representations. For this, it is equivalent to require one minimal non-zero Jacquet module to be a multiplicity one representation. Equivalently, this is the same as asking that all the representations $w\sigma$, when w runs over all representatives of $W_M\backslash W_G/W_M$ which normalize M (W_G and W_M denote the Weyl groups of G and M respectively) are nonequivalent.

This technical condition enables one to study composition series of such induced representations easily. Namely, let $\mathrm{Ind}_P^G(\sigma)$ be a regular representation as above, and $P' = M'N'$ a parabolic subgroup such that $r^G_{M'}(\mathrm{Ind}_P^G(\sigma))$ is non-zero. Then

$$\phi : \pi \mapsto \text{composition series of } r^G_{M'}(\pi)$$

can be interpreted as an injection from the partitive set of composition series of $\mathrm{Ind}_P^G(\sigma)$ into the partitive set of composition series of $r^G_{M'}(\mathrm{Ind}_P^G(\sigma))$. Further, $\phi(\pi_1 \cap \pi_2) = \phi(\pi_1) \cap \phi(\pi_2)$, and if $\phi(\pi)$ is irreducible, then π is irreducible.

This can be used to construct some interesting representations in a rather easy way, as we shall see soon.

19. Square integrable representations of Steinberg type

The square integrable representations, which we shall introduce now, are natural generalization of usual Steinberg representation. Their characteristic is, among others, that they have a very simple Jacquet modules, and therefore it is pretty easy to control representations parabolically induced from them.

Below we shall denote Jacquet module

$$r^{S_n}_{M_{(k)}}(\pi)$$

simply by

$$s_{(k)}(\pi).$$

Let ρ and σ be irreducible unitarizable cuspidal representations of groups $GL(p, F)$ and S_q respectively, such that $\nu^\alpha\rho \rtimes \sigma$ reduces for some $\alpha > 0$. Now the fact that

$$\text{s.s.}(s_{(p)}(\nu^\alpha\rho \rtimes \sigma)) = \nu^\alpha\rho \otimes \sigma + \nu^{-\alpha}\rho \otimes \sigma,$$

implies that $\nu^\alpha\rho \rtimes \sigma$ has a unique irreducible subrepresentation. We denote this subrepresentation by $\delta(\nu^\alpha\rho; \sigma)$. We get directly

$$\mu^*(\delta(\nu^\alpha\rho; \sigma)) = 1 \otimes \delta(\nu^\alpha\rho; \sigma) + \nu^\alpha\rho \otimes \sigma.$$

The square integrability criterion implies that $\delta(\nu^\alpha\rho; \sigma)$ is square integrable.

We can continue in the following way. Look at the subquotients

$$\nu^{\alpha+1}\rho \rtimes \delta(\nu^\alpha\rho; \sigma) \text{ and } \delta([\nu^\alpha\rho, \nu^{\alpha+1}\rho]) \rtimes \sigma \text{ in } \nu^{\alpha+1}\rho \times \nu^\alpha\rho \rtimes \sigma.$$

We get the semi simplifications of GL-type Jacquet modules easily using formula (14.2). They are

$$\nu^{\alpha+1}\rho \times \nu^\alpha\rho \otimes \sigma + \nu^{-(\alpha+1)}\rho \times \nu^\alpha\rho \otimes \sigma$$

and

$$\delta([\nu^\alpha\rho, \nu^{\alpha+1}\rho]) \otimes \sigma + \nu^{\alpha+1}\rho \times \nu^{-\alpha}\rho \otimes \sigma + \delta([\nu^{-(\alpha+1)}\rho, \nu^{-\alpha}\rho]) \otimes \sigma$$

respectively (we use also (14.3)). From this we get that there exists a unique irreducible subquotient which has $\delta([\nu^\alpha\rho,\nu^{\alpha+1}\rho])\otimes\sigma$ in its Jacquet module. Denote this subquotient by $\delta([\nu^\alpha\rho,\nu^{\alpha+1}\rho];\sigma)$. It is easy to see that we can define this representation as the unique irreducible subrepresentation of $\delta([\nu^\alpha\rho,\nu^{\alpha+1}\rho])\rtimes\sigma$, and that we have

$$\mu^*(\delta([\nu^\alpha\rho,\nu^{\alpha+1}\rho];\sigma)$$

$$=1\otimes\delta([\nu^\alpha\rho,\nu^{\alpha+1}\rho];\sigma)+\nu^{\alpha+1}\rho\otimes\delta(\nu^\alpha\rho;\sigma)+\delta([\nu^\alpha\rho,\nu^{\alpha+1}\rho])\otimes\sigma.$$

Continuing this procedure we get that the representation

$$\nu^{\alpha+n}\rho\times\nu^{\alpha+n-1}\rho\times\cdots\times\nu^{\alpha+1}\rho\times\nu^\alpha\rho\rtimes\sigma$$

has a unique irreducible subrepresentation, which we denote by

$$\delta([\nu^\alpha\rho,\nu^{\alpha+n}\rho];\sigma)\quad(n\geq 0),$$

and that

$$\mu^*\left(\delta([\nu^\alpha\rho,\nu^{\alpha+n}\rho];\sigma)\right)=\sum_{k=-1}^{n}\delta([\nu^{\alpha+k+1}\rho,\nu^{\alpha+n}\rho])\otimes\delta([\nu^\alpha\rho,\nu^{\alpha+k}\rho];\sigma)$$

(we take $\delta(\emptyset;\sigma)$ in the above formula to be just σ). The representation $\delta([\nu^\alpha\rho,\nu^{\alpha+n}\rho];\sigma)$ is square integrable and we have

$$\delta([\nu^\alpha\rho,\nu^{\alpha+n}\rho];\sigma)\tilde{}\cong\delta([\nu^\alpha\rho,\nu^{\alpha+n}\rho];\tilde{\sigma}).$$

It follows directly that $\delta([\nu^\alpha\rho,\nu^{\alpha+n}\rho];\sigma)$ is a subquotient of $\delta([\nu^\alpha\rho,\nu^{\alpha+n}\rho])\rtimes\sigma$ (actually, a subrepresentation; moreover, the unique irreducible one).

20. A reducibility criterion

In a significant number of cases, reducibility of parabolically induced representations can be determined in the following way. Let $P_0=M_0N_0$ be a parabolic subgroup of G and σ an irreducible representation of M_0. Suppose that π and Π are representations of G of finite length, and that there exists a parabolic subgroup $P=MN$ such that the following three conditions hold

(R1) $\text{Ind}_{P_0}^G(\sigma)\leq\Pi$, $\pi\leq\Pi$;

(R2) $r_M^G(\text{Ind}_{P_0}^G(\sigma))+r_M^G(\pi)\not\leq r_M^G(\Pi)$;

(R3) $r_M^G(\text{Ind}_{P_0}^G(\sigma))\not\leq r_M^G(\pi)$.

Then it follows directly that $\mathrm{Ind}_{P_0}^G(\sigma)$ is reducible (since the Jacquet module functor is exact). In applications of this criterion, one usually chooses π and Π to be parabolically induced representations (from other parabolic subgroups). Moreover, it is easy to get upper and lower bounds on the Jacquet modules of common irreducible subquotients of $\mathrm{Ind}_{P_0}^G(\sigma)$ and π (in a simpler cases they can give exact Jacquet modules).

In what follows we shall use the following convention for $\alpha \in \mathbb{R}$: the representation

$$\nu^\alpha 1_{F^\times}$$

of $F^\times$, will be simply denoted by

$$\nu^\alpha$$

(this will considerably simplify some formulas, making them more transparent).

Example 1. *We shall now prove that*

$$\pi_0 := \nu^{1/2} \rtimes St_{SO(3)} = \nu^{1/2} \rtimes \delta(\nu^{1/2}; 1)$$

reduces. Set

$$\pi = \delta([\nu^{-1/2}, \nu^{1/2}]) \rtimes 1,$$

$$\Pi = \nu^{1/2} \times \nu^{1/2} \rtimes 1.$$

Obviously, (R1) holds for $\mathrm{Ind}_{P_0}^G(\sigma) = \pi_0$. *We easily compute from (14.2)*

$$s_{GL}(\pi_0) = \nu^{1/2} \times \nu^{1/2} \otimes 1 \quad + \quad \nu^{-1/2} \times \nu^{1/2} \otimes 1, \tag{20.1}$$

$$s_{GL}(\pi) = \nu^{1/2} \times \nu^{1/2} \otimes 1 \quad + \quad 2\delta([\nu^{-1/2}, \nu^{1/2}]) \otimes 1, \tag{20.2}$$

$$s_{GL}(\Pi) =$$

$$\nu^{1/2} \times \nu^{1/2} \otimes 1 \,+\, 2\nu^{-1/2} \times \nu^{1/2} \otimes 1 \,+\, \nu^{-1/2} \times \nu^{-1/2} \otimes 1.$$

The multiplicity of $\nu^{1/2} \times \nu^{1/2} \otimes 1$ *in all three above Jacquet modules is 1. Therefore, (R2) holds. Observe that* $L(\nu^{-1/2}, \nu^{1/2}) \otimes 1$ *is a subquotient of (20.1), but not of (20.2). Therefore, (R3) also holds. Thus, we have proved that* π_0 *is reducible.* □

Remark. *The above considerations also easily give the reducibility of*

$$\pi = \delta([\nu^{-1/2}, \nu^{1/2}]) \rtimes 1.$$

Namely, the multiplicity of $\delta([\nu^{-1/2}, \nu^{1/2}]) \otimes 1$ *in (20.1) is 1, and in (20.2) is 2, which implies* $s_{GL}(\pi) \not\leq s_{GL}(\pi_0)$*. Therefore* π *is also reducible.*

Since π *is unitarizable, Frobenius reciprocity implies that each irreducible subquotient of* π *has* $\delta([\nu^{-1/2}, \nu^{1/2}]) \otimes 1$ *in its Jacquet module, which implies that the length of* π *is 2. Denote by* τ_+ *the unique irreducible subquotient of* π *which has* $\nu^{1/2} \times \nu^{1/2} \otimes 1$ *in its Jacquet module, and the other one by* τ_-*. Clearly* $\pi = \tau_+ \oplus \tau_-$*. Then the above discussion implies*

$$s_{GL}(\tau_+) = \delta([\nu^{-1/2}, \nu^{1/2}]) \otimes 1 + \nu^{1/2} \times \nu^{1/2} \otimes 1 \tag{20.3}$$

and

$$s_{GL}(\tau_-) = \delta([\nu^{-1/2}, \nu^{1/2}]) \otimes 1.$$

Observe that $\tau_+ \leq \nu^{1/2} \rtimes \delta(\nu^{1/2}; 1)$ *since* $\nu^{1/2} \times \nu^{1/2} \otimes 1$ *has multiplicity one in both (20.1) and (20.2) (use that the Jacquet functor is exact). It also has multiplicity one in* $s_{GL}(\Pi)$*. Thus*

$$\tau_+ \leq \nu^{-1/2} \rtimes \delta(\nu^{1/2}; 1) \textit{ and } \tau_+ \leq \delta([\nu^{-1/2}, \nu^{1/2}]) \rtimes 1. \tag{20.4}$$

The above representation $\pi = \delta([\nu^{-1/2}, \nu^{1/2}]) \rtimes 1$ is a particular case covered by the following very simple but pretty general criterion (Theorem 13.2 in [39]; it also covers non-unitary situations):

Theorem. *Let* $\rho \in \mathcal{C}$ *be unitarizable, and let* σ *be an irreducible cuspidal representation of* S_q*. If* ρ *is selfdual, suppose that* $\nu^\alpha \rho \rtimes \sigma$ *is irreducible for all* $\alpha \in \mathbb{R} \setminus (1/2)\mathbb{Z}$*. Assume* $\Delta \in \mathcal{S}$ *such that* $\Delta \subset \{\nu^\alpha \rho; \alpha \in \mathbb{R}\}$*. Then*

$$\delta(\Delta) \rtimes \sigma \textit{ reduces} \iff \rho' \rtimes \sigma \textit{ reduces for some } \rho' \in \Delta.$$

Let us illustrate the power of the above result with an example. Let $\Delta = [\chi, \nu^n \chi]$, where $\chi \in (F^\times)\tilde{}$ and $n \geq 0$. Consider odd orthogonal groups in this example. Then $\delta(\Delta) \rtimes 1$ reduces if and only if

$[\chi, \nu^n \chi] \in$

$$\{[\nu^{-n-1/2}\chi_0, \nu^{-1/2}\chi_0], [\nu^{-n+1/2}\chi_0, \nu^{1/2}\chi_0], \dots, [\nu^{1/2}\chi_0, \nu^{n+1/2}\chi_0]\}$$

for some $\chi_0 \in (F^\times)\hat{}$ satisfying $\chi_0^2 \equiv 1$. So, we have a complete answer for the reducibility points of the (not necessarily unitary) degenerate principal series of $SO(2n+3, F)$ which are induced from the Siegel parabolic subgroup.

Example 2. *Let $m \in \mathbb{Z}$, $m \geq 2$. Now we shall prove the reducibility of*

$$\pi_0 := St_{GL(2m)} \rtimes St_{SO(3)} = \delta([\nu^{-(m-1/2)}, \nu^{m-1/2}]) \rtimes \delta(\nu^{1/2}; 1).$$

Set

$$\pi = \delta([\nu^{-(m-1/2)}, \nu^{1/2}]) \rtimes \delta([\nu^{1/2}, \nu^{m-1/2}]; 1),$$

$$\Pi = \delta([\nu^{-(m-1/2)}, \nu^{1/2}]) \times \delta([\nu^{1/2}, \nu^{m-1/2}]) \rtimes 1.$$

Now one sees easily that condition (R1) above is satisfied with $Ind_{P_0}^{G}(\sigma) = \pi_0$. From (14.2) one gets directly that the multiplicity of $\delta([\nu^{1/2}, \nu^{m-1/2}])^2 \times \nu^{1/2} \otimes 1$ in the Jacquet modules of all three above representations is 1. This implies that condition (R2) holds.

From (14.2) (and (14.3)) we get

$$s_{GL}(\pi) = \left(\sum_{i=-m-1/2}^{1/2} \delta([\nu^{-i}, \nu^{m-1/2}]) \times \delta([\nu^{i+1}, \nu^{1/2}]) \right) \times \delta([\nu^{1/2}, \nu^{m-1/2}]) \otimes 1,$$

$$s_{GL}(\pi_0) = \left(\sum_{i=m-1/2}^{-m-1/2} \delta([\nu^{-i}, \nu^{m-1/2}]) \times \delta([\nu^{i+1}, \nu^{m-1/2}]) \right) \times \nu^{1/2} \otimes 1.$$

The multiplicity of $\delta([\nu^{-(m-1/2)}, \nu^{m-1/2}]) \times \nu^{1/2} \otimes 1$ in the Jacquet modules of π and π_0 is 1 and 2 respectively (since for π we can get this subquotient only for $i = -m - 1/2$, and then the multiplicity is 1, and for π_0 we get it for $i = -m - 1/2$ and $m - 1/2$, and the multiplicity is one in both cases). This shows (R3). Therefore, the reducibility of π_0 is proved. □

An interesting question is what happens for $m = 1$. The above strategy does not work in this case. We shall address this question later.

Remarks. *(1) The representation $\pi_0 = St_{GL(2m)} \rtimes 1$ of $SO(2m+1, F)$ is reducible (for $m \geq 1$). One proves this in a similar way as in the Example 2. One takes $\pi = \delta([\nu^{-(m-1/2)}, \nu^{-1/2}]) \rtimes \delta([\nu^{1/2}, \nu^{m-1/2}]; 1)$, $\Pi = \delta([\nu^{-(m-1/2)}, \nu^{-1/2}]) \times \delta([\nu^{1/2}, \nu^{m-1/2}]) \rtimes 1$ and considers the multiplicities of $\delta([\nu^{1/2}, \nu^{m-1/2}])^2 \otimes 1$ and $\delta([\nu^{-(m-1/2)}, \nu^{m-1/2}]) \otimes 1$ in the Jacquet modules.*

The above reducibility also follows from the last theorem.

(2) Let $\rho \not\cong 1_{F^\times}$ be an irreducible unitarizable cuspidal representation of $GL(p, F)$ and let the representation $\delta([\nu^{-(m-1)/2}\rho, \nu^{(m-1)/2}\rho]) \rtimes 1$ of $SO(mp+1, F)$ be reducible. Then

$$\pi_0 := \delta([\nu^{-(m-1)/2}\rho, \nu^{(m-1)/2}\rho]) \rtimes \delta(\nu^{1/2}; 1) \quad \textit{reduces.}$$

This can be proved in the following way. We get directly

$$\delta([\nu^{-(m-1)/2}\rho, \nu^{(m-1)/2}\rho]) \rtimes 1 = \tau_1 \oplus \tau_2,$$

where the τ_i are irreducible and the multiplicity of

$$\delta([\nu^{-(m-1)/2}\rho, \nu^{(m-1)/2}\rho]) \otimes 1$$

in the Jacquet module of τ_i is 1. Now take $\pi = \nu^{1/2} \rtimes \tau_i$ and $\Pi = \delta([\nu^{-(m-1)/2}\rho, \nu^{(m-1)/2}\rho]) \times \nu^{1/2} \rtimes 1$. The multiplicity of

$$\delta([\nu^{-(m-1)/2}\rho, \nu^{(m-1)/2}\rho]) \times \nu^{1/2} \otimes 1$$

in the Jacquet modules of π_0, π and Π is 2, 1 and 2 respectively. This implies the reducibility of π_0.

21. Proving irreducibility

The reducibility of a parabolically induced representation has strong implications for its Jacquet modules (we shall list them below). A way to show the irreducibility of such a representation is to show that these implications can not occur. We discuss this in more detail below.

Let σ be an irreducible representation of a Levi factor M_0 of a parabolic subgroup $P_0 = M_0N_0$. Suppose that $\mathrm{Ind}_{P_0}^G(\sigma)$ reduces. Then in the Grothendieck group we can write $\mathrm{Ind}_{P_0}^G(\sigma) = \pi_1 + \pi_2$, with both $\pi_1 > 0$, $\pi_2 > 0$. For any standard parabolic subgroup P with standard Levi decomposition $P = MN$, let

$$T_{i,P} = r_M^G(\pi_i), \quad i = 1, 2.$$

We can consider $T_{1,P}$ and $T_{2,P}$ as elements of $\mathcal{R}(M)$. Then, the following must hold:

($\mathfrak{R}1$) $T_{i,P} \geq 0$ and $T_{1,P} \neq 0$ if and only if $T_{2,P} \neq 0$;

($\mathfrak{R}2$) $T_{1,P} + T_{2,P} = r_M^G(\mathrm{Ind}_{P_0}^G(\sigma))$ in $\mathcal{R}(M)$;

($\mathfrak{R}3$) $r_{M_2}^{M_1}(T_{i,P_1}) = T_{i,P_2}$ when $P_1 \supset P_2$.

The first property follows from section 10. (see (3) there).

Therefore, if one can show that there is no system of $T_{i,P} \in \mathcal{R}(G)$, $i = 1, 2$, when P is running over a subset of standard parabolic subgroups, which satisfy ($\mathfrak{R}1$) - ($\mathfrak{R}3$), then $\mathrm{Ind}_{P_0}^G(\sigma)$ is irreducible. Actually, in showing irreducibility it is often possible to show the non-existence of such a system for three proper standard parabolic subgroups P, P_1, P_2 satisfying $P \subset P_1, P_2$.

Note that conditions ($\mathfrak{R}1$) - ($\mathfrak{R}3$) are necessary for reducibility. An existence of such a system does not prove reducibility.

In the case that σ is unitarizable, one can have a simpler way of proving irreducibility of $\mathrm{Ind}_{P_0}^G(\sigma)$. For example, suppose that

$$\mathrm{Ind}_{P_0}^G(\sigma) \hookrightarrow \mathrm{Ind}_{P'}^G(\sigma')$$

for some irreducible representation σ' of the standard Levi subgroup M' of a standard parabolic subgroup P' of G. Now, if the multiplicity of σ' in $r_{M'}^G(\mathrm{Ind}_{P_0}^G(\sigma))$ is one (by Frobenius reciprocity it must be ≥ 1), then $\mathrm{Ind}_{P_0}^G(\sigma)$ is irreducible (see Example 1 in the following section, where irreducibility is proved in this way). This follows directly from the exactness of the Jacquet functors. In the other examples in the following section, one can find more subtle ways of proving irreducibility when σ is unitarizable.

22. Some half-integral examples of irreducibility

The examples that we study in this section illustrate methods for proving irreducibility. These examples are interesting for understanding phenomena which show up for general square integrable representations. This is the reason that we consider (unitary) tempered induced representations. We start with a very simple example and proceed with more complicated ones (one can find simple examples of different types in the following section).

Example 1 - Irreducibility of $St_{GL(1)} \rtimes St_{SO(3)}$. *Let*

$$\pi_0 = St_{GL(1)} \rtimes St_{SO(3)} = \nu^0 \rtimes \delta(\nu^{1/2}; 1).$$

Applying (14.2) write

$$\begin{aligned}\mu^*(\pi_0) &= M^*(\nu^0) \rtimes \mu^*(\delta(\nu^{1/2}; 1)) \\ &= (1 \otimes \nu^0 + 2\nu^0 \otimes 1) \rtimes (1 \otimes \delta(\nu^{1/2}; 1) + \nu^{1/2} \otimes 1) \\ &= 1 \otimes \pi_0 + \left[2\nu^0 \otimes \delta(\nu^{1/2}; 1) + \nu^{1/2} \otimes \nu^0 \rtimes 1\right] + 2\nu^0 \times \nu^{1/2} \otimes 1.\end{aligned}$$

Let π be any irreducible subquotient of π_0. Then π is a subrepresentation since π_0 is unitarizable. Now we have

$$\pi \hookrightarrow \nu^0 \rtimes \delta(\nu^{1/2}; 1) \hookrightarrow \nu^0 \times \nu^{1/2} \rtimes 1 \cong \nu^{1/2} \times \nu^0 \rtimes 1.$$

Frobenius reciprocity implies that (irreducible representation) $\nu^{1/2} \otimes \nu^0 \rtimes 1$ is in the Jacquet module of π. Since the multiplicity of $\nu^{1/2} \otimes \nu^0 \rtimes 1$ in the Jacquet module of π_0 is 1, we conclude that the length of π_0 is 1 (since the Jacquet functor is exact), i.e. that π_0 is irreducible. □

Example 2 - Irreducibility of $St_{GL(2m+1)} \rtimes St_{SO(3)}$ for $m \in \mathbb{Z}_{\geq 1}$. *Let*

$$\pi_0 = St_{GL(2m+1)} \rtimes St_{SO(3)} = \delta([\nu^{-m}, \nu^m]) \rtimes \delta(\nu^{1/2}; 1).$$

We shall sketch the proof (it uses a similar idea as the proof of Proposition 4.2 in [39]; see there for more details). Applying (14.2) and (14.3) we get

$$\mu^*(\pi_0) = M^*(\delta([\nu^{-m}, \nu^m])) \rtimes \mu^*(\delta(\nu^{1/2}; 1))$$

$$= \Big(\sum_{i=-m-1}^{m} \sum_{j=i}^{m} \delta([\nu^{-i}, \nu^m]) \times \delta([\nu^{j+1}, \nu^m]) \otimes \delta([\nu^{i+1}, \nu^j]) \Big)$$

$$\rtimes (1 \otimes \delta(\nu^{1/2}; 1) + \nu^{1/2} \otimes 1).$$

Let π be any irreducible subquotient of π_0. Then it is a subrepresentation since π_0 is unitarizable. Using simple facts about irreducibility of principal series in $GL(2, F)$ and $SO(3, F)$, and properties of $\rtimes$ and $\times$ (see sections 12. and 14.) we get

$$\pi \hookrightarrow \pi_0 \hookrightarrow \nu^m \times \nu^{m-1} \times \cdots \times \nu^{-m+1} \times \nu^{-m} \times \nu^{1/2} \rtimes 1$$

$$\cong \nu^m \times \nu^{m-1} \times \cdots \times \nu^{-m+1} \times \nu^{1/2} \times \nu^{-m} \rtimes 1$$

$$\cong \nu^m \times \nu^{m-1} \times \cdots \times \nu^{-m+1} \times \nu^{1/2} \times \nu^{m} \rtimes 1$$

$$\cong \nu^m \times \nu^{m-1} \times \cdots \times \nu^{-m+1} \times \nu^{m} \times \nu^{1/2} \rtimes 1$$

$$\cdots\cdots\cdots\cdots\cdots\cdots\cdots$$

$$\cong \nu^m \times \nu^{m-1} \times \cdots \times \nu^{0} \times \nu^{m} \times \nu^{m-1} \times \cdots \times \nu \times \nu^{1/2} \rtimes 1.$$

Frobenius reciprocity implies that

$$\sigma_1 := \nu^m \otimes \nu^{m-1} \otimes \cdots \otimes \nu^{0} \otimes \nu^{m} \otimes \nu^{m-1} \otimes \cdots \otimes \nu \otimes \nu^{1/2} \otimes 1$$

appears in the Jacquet module of π (as a quotient).

Now we want to see from which terms in $s_{(2m)}(\pi_0)$ we can get σ_1 as a subquotient (by the transitivity of Jacquet modules, σ_1 must be a subquotient of some term of $s_{(2m)}(\pi_0)$ since $\pi \leq \pi_0$). From the above formula for $\mu^*(\pi_0)$ we get

$$s.s.(s_{(2m)}(\pi_0)) = \Big(\sum_{i=-m-1}^{m} \delta([\nu^{-i}, \nu^m]) \times \delta([\nu^{i+1}, \nu^m]) \otimes \delta(\nu^{1/2}; 1) \Big)$$

$$+ \Big(\sum_{i=-m-1}^{m-1} \delta([\nu^{-i}, \nu^m]) \times \delta([\nu^{i+2}, \nu^m]) \times \nu^{1/2} \otimes \nu^{i+1} \rtimes 1 \Big).$$

We can get σ_1 *as a subquotient of a term in* $s.s.(s_{(2m)}(\pi_0))$ *only from the first sum showing up in the formula for* $s.s.(s_{(2m)}(\pi_0))$ *(since irreducible subquotients of the Jacquet module with respect to the standard minimal parabolic subgroup of terms in the second sum have the form*

$$\cdots\cdots\cdots\cdots \otimes \nu^{\pm(i+1)} \otimes 1,$$

with $\pm(i+1) \in \mathbb{Z}$*, which cannot give* σ_1 *since we always have* $\pm(i+1) \neq 1/2$*). Further, since all exponents of the tensor factors of* σ_1 *are* ≥ 0*, we can get* σ_1 *for a subquotient of Jacquet module of* π_0 *only if* $-i \geq 0$ *and* $i+1 \geq 0$ *(use the remark in section 12. and the definition of* $\delta(\Delta)$ *in section 13.). Thus* $i = -1$ *or* 0*. Both choices give the same representation (up to an equivalence), the irreducible representation* $\delta([\nu, \nu^m]) \times \delta([\nu^0, \nu^m]) \otimes \delta(\nu^{1/2}; 1)$*. This representation also must be a subquotient of the Jacquet module of* π*.*

Since

$$\delta([\nu, \nu^m]) \times \delta([\nu^0, \nu^m]) \otimes \delta(\nu^{1/2}; 1) \hookrightarrow \nu^m \times \nu^{m-1} \times \cdots \times \nu^2 \times \nu \times \nu^m \times \nu^{m-1} \times \cdots \times \nu \times \nu^0 \otimes \nu^{1/2} \rtimes 1,$$

Frobenius reciprocity implies that

$$\sigma_2 := \nu^m \otimes \nu^{m-1} \otimes \cdots \otimes \nu^2 \otimes \nu \otimes \nu^m \otimes \nu^{m-1} \otimes \cdots \otimes \nu \otimes \nu^0 \otimes \nu^{1/2} \otimes 1$$

is in the Jacquet module of $\delta([\nu, \nu^m]) \times \delta([\nu^0, \nu^m]) \otimes \delta(\nu^{1/2}; 1)$*, and further also in the Jacquet module of* π*.*

Now we shall analyze from which subquotients of $s_{(2m-1)}(\pi_0)$ *we can get* σ_2 *in the Jacquet module. The formula for* $\mu^*(\pi_0)$ *implies*

$$s.s.(s_{(2m-1)}(\pi_0)) = \Big(\sum_{i=-m-1}^{m-1} \delta([\nu^{-i}, \nu^m]) \times \delta([\nu^{i+2}, \nu^m]) \otimes \nu^{i+1} \rtimes \delta(\nu^{1/2}; 1) \Big)$$

$$+ \Big(\sum_{i=-m-1}^{m-2} \delta([\nu^{-i}, \nu^m]) \times \delta([\nu^{i+3}, \nu^m]) \times \nu^{1/2} \otimes \delta([\nu^{i+1}, \nu^{i+2}]) \rtimes 1 \Big).$$

Again, in a similar way as before, we get that σ_2 *can come only from the first sum. Since the tensor product of* σ_2 *ends with* $\nu^0 \otimes \nu^{1/2} \otimes 1$*, we conclude that* σ_2 *can come only from the term in the first sum corresponding to* $i+1 = 0$*, i.e.* $i = -1$*.*

Therefore, the only term of $s.s.(s_{(2m-1)}(\pi_0))$ *which can give* σ_2 *is*

$$\delta([\nu, \nu^m]) \times \delta([\nu, \nu^m]) \otimes \nu^0 \rtimes \delta(\nu^{1/2}; 1).$$

This representation is irreducible by the previous example (and section 13.), and we have seen that it has multiplicity one in $s_{(2m-1)}(\pi_0)$. *Since* σ_2 *is in the Jacquet module of* π, *this representation is also in the Jacquet module of* π. *Now the multiplicity one property of this representation in* $s_{(2m-1)}(\pi_0)$ *implies that* $\pi = \pi_0$, *i.e. that* π_0 *is irreducible.* □

The following example of irreducibility is not quite as simple to prove, but it is very important since this irreducibility is in a sense singular. Namely, recall that we have proved in section 20. that $St_{GL(2m)} \rtimes St_{SO(3)}$ is reducible for all $m \geq 2$. The case of $m = 1$ which we consider below is therefore a kind of surprise:

Example 3 - Irreducibility of $St_{GL(2)} \rtimes St_{SO(3)}$. *Our aim will now be to prove the irreducibility of*

$$\pi_0 = St_{GL(2)} \rtimes St_{SO(3)} = \delta([\nu^{-1/2}, \nu^{1/2}]) \rtimes \delta(\nu^{1/2}; 1).$$

Suppose that π_0 *reduces. First we get from (14.2) and (14.3))*

$$\mu^*(\pi_0) = 1 \otimes \pi_0 + \tag{22.1}$$

$$\big[2\nu^{1/2} \otimes \nu^{-1/2} \rtimes \delta(\nu^{1/2}; 1) + \nu^{1/2} \otimes \delta([\nu^{-1/2}, \nu^{1/2}]) \rtimes 1\big] +$$

$$\big[2\,\boxed{\delta([\nu^{-1/2}, \nu^{1/2}]) \otimes \delta(\nu^{1/2}; 1)} + 3\,\nu^{1/2} \times \nu^{1/2} \otimes \delta(\nu^{1/2}; 1) + 2\nu^{1/2} \times \nu^{1/2} \otimes 1_{SO(3)}\big]$$

$$+ \big[2\delta([\nu^{-1/2}, \nu^{1/2}]) \times \nu^{1/2} \otimes 1 + \nu^{1/2} \times \nu^{1/2} \times \nu^{1/2} \otimes 1\big].$$

Since π_0 *is unitarizable, each irreducible subquotient of* π_0 *is a subrepresentation. Now Frobenius reciprocity implies that* $\delta([\nu^{-1/2}, \nu^{1/2}]) \otimes \delta(\nu^{1/2}; 1)$ *is in the Jacquet module of each irreducible subquotient of* π_0. *Observe that the Jacquet module of* $\delta([\nu^{-1/2}, \nu^{1/2}]) \otimes \delta(\nu^{1/2}; 1)$ *with respect to the minimal parabolic subgroup is*

$$\nu^{1/2} \otimes \nu^{-1/2} \otimes \nu^{1/2} \otimes 1. \tag{22.2}$$

Note that all the representations in the third and fourth lines of (22.1) are irreducible. Observe that the fourth line in (22.1) has length 3, and that $\nu^{1/2} \times \nu^{1/2} \times \nu^{1/2} \otimes 1$ *has multiplicity one in that line. Denote by* π_1 *the (unique) irreducible subquotient of* π_0 *containing* $\nu^{1/2} \times \nu^{1/2} \times \nu^{1/2} \otimes 1$ *in its Jacquet module. The (semi simplification of the) Jacquet module of* $\nu^{1/2} \times \nu^{1/2} \times \nu^{1/2} \otimes 1$ *with respect to the minimal parabolic subgroup is*

$$6\,\nu^{1/2} \otimes \nu^{1/2} \otimes \nu^{1/2} \otimes 1. \tag{22.3}$$

Recall that π_1 *has (22.2) in its Jacquet module, which does not belong to (22.3). This fact and the reducibility of* π_0 *(together with the transitivity of Jacquet modules) imply*

$$s_{(3)}(\pi_1) = \delta([\nu^{-1/2}, \nu^{1/2}]) \times \nu^{1/2} \otimes 1 + \nu^{1/2} \times \nu^{1/2} \times \nu^{1/2} \otimes 1.$$

Now from the fourth line of (22.1), it follows that π_0 *is a length two representation. Denote the other irreducible subquotient by* π_2*. Then*

$$s_{(3)}(\pi_2) = \delta([\nu^{-1/2}, \nu^{1/2}]) \times \nu^{1/2} \otimes 1.$$

The above two formulas imply

$$s_{(1,1,1)}(\pi_1) = \tag{22.4}$$

$$\nu^{1/2} \otimes \nu^{-1/2} \otimes \nu^{1/2} \otimes 1 + 2\, \nu^{1/2} \otimes \nu^{1/2} \otimes \nu^{-1/2} \otimes 1 + 6\, \nu^{1/2} \otimes \nu^{1/2} \otimes \nu^{1/2} \otimes 1,$$

$$s_{(1,1,1)}(\pi_2) = \nu^{1/2} \otimes \nu^{-1/2} \otimes \nu^{1/2} \otimes 1 + 2\, \nu^{1/2} \otimes \nu^{1/2} \otimes \nu^{-1/2} \otimes 1. \tag{22.5}$$

From the other side, we see from the second line of (22.1) and from (20.4) that

$$3\nu^{1/2} \otimes \tau_+ \leq s_{(1)}(\pi_0).$$

Observe that $\nu^{1/2} \otimes \tau_+ \not\leq s_{(1)}(\pi_2)$ *since*

$$\nu^{1/2} \otimes s_{(1,1)}(\tau_+) = 2\, \nu^{1/2} \otimes \nu^{1/2} \otimes \nu^{1/2} \otimes 1 + \nu^{1/2} \otimes \nu^{1/2} \otimes \nu^{-1/2} \otimes 1$$

and $\nu^{1/2} \otimes \nu^{1/2} \otimes \nu^{1/2} \otimes 1$ *does not show up in (22.5). Now* $\pi_0 = \pi_1 \oplus \pi_2$ *implies* $3\nu^{1/2} \otimes \tau_+ \leq s_{(1)}(\pi_1)$*, which implies* $3\, \nu^{1/2} \otimes s_{(1,1)}(\tau_+) \leq s_{(1,1,1)}(\pi_1)$*. This further implies* $3\, \nu^{1/2} \otimes \nu^{1/2} \otimes \nu^{-1/2} \otimes 1 \leq s_{(1,1,1)}(\pi_1)$*. This clearly contradicts (22.4). This contradiction completes the proof of the irreducibility of* π_0*.* □

Remarks. *(1) Let us sum up what we have proved up to now for*

$$St_{GL(m)} \rtimes St_{SO(3)} = \delta(1_{F^\times}, m) \rtimes \delta(\nu^{1/2}; 1).$$

This representation is

- ***irreducible*** *for all* **odd** m*,*
- ***reducible*** *for all* **even** m ***except*** $m = 2$*.*

(2) One can prove that the representation $\delta([\nu^{-m}, \nu^m]) \rtimes 1$*,* $m \in \mathbb{Z}_{\geq 0}$ *of* $SO(2m+1, F)$ *is irreducible (this follows from Proposition 4.2 in [39]; it can be proved in a similar way as the second example of this section).*

(3) *Suppose that ρ is an irreducible unitarizable cuspidal representation of some general linear group such that $\delta([\nu^{-(m-1)/2}\rho, \nu^{(m-1)/2}\rho]) \rtimes 1$ is irreducible. In a similar way as in the second example of this section one can prove that*

$$\pi_0 := \delta([\nu^{-(m-1)/2}\rho, \nu^{(m-1)/2}\rho]) \rtimes \delta(\nu^{1/2}; 1)$$
$$= \delta([\nu^{-(m-1)/2}\rho, \nu^{(m-1)/2}\rho]) \rtimes St_{SO(3)}$$

is irreducible.

A very nice fact is that the above considerations hold much more generally. They hold whenever we have irreducible unitarizable cuspidal representations ρ and σ of $GL(p, F)$ and S_q respectively, such that $\nu^{1/2}\rho \rtimes \sigma$ reduces, with completely analogous proofs. It is also not important which series of classical groups is in question, and whether ρ and σ are just characters or infinite dimensional representations, etc. Everything is done in $R(S)$ (our case was $p = 1$, $S_q = SO(1) = \{1\}$, $\rho = 1_{F^\times}$ and $\sigma = 1$). For example, in this case

$$\delta([\nu^{-(m-1)/2}\rho, \nu^{(m-1)/2}\rho]) \rtimes \delta(\nu^{1/2}\rho; \sigma)$$

is irreducible for each odd positive integer, and is reducible for each even m with one exception. This exception is 2.

Remark. *Now we take a close look at the following special case (which follows from previous considerations). We shall consider only representations of $SO(2n+1, F)$ in this remark. Let ρ be a selfdual irreducible cuspidal representation of $GL(p, F)$ such that the representation $\nu^{1/2}\rho \rtimes 1$ (of $SO(2p+1, F)$) reduces. Then for any selfdual irreducible cuspidal representation ρ' of $GL(p', F)$ the following holds: for exactly one parity of positive integers we have that*

(1) *$\delta(\rho', m) \rtimes \delta(\nu^{1/2}\rho; 1)$ reduces for all m from that parity, with possible finitely many exceptions;*
(2) *$\delta(\rho', m) \rtimes \delta(\nu^{1/2}\rho; 1)$ is irreducible for all m from the other parity.*

Note that the only exception in (1) shows up for $\rho' \cong \rho$ and $m = 2$. Therefore, we can characterize $\delta(\nu^{1/2}\rho; 1)$ with this exception.

The parity in (1) is determined by ρ': it is odd if $\nu^{1/2}\rho' \rtimes 1$ is irreducible, and even otherwise.

At this point the reader who is more interested in square integrability than in the reducibility questions can skip directly to section 28.

23. Some integral examples of irreducibility (and reducibility)

We shall illustrate how one gets reducibility points in the case of

$$\chi \rtimes St_{Sp(2)} = \chi \rtimes \delta(\nu;1), \quad \chi \in (F^\times)\tilde{\ },$$

in a little bit different way than above.

Using (14.2), we write semi simplifications of Jacquet modules of $\chi \rtimes St_{Sp(2)}$ with respect to the maximal parabolic subgroups in the following way:

$$\begin{array}{ccccc}
\text{s.s.}(s_{(1)}(\chi \rtimes St_{Sp(2)})) & \longleftarrow & \chi \rtimes St_{Sp(2)} & \longrightarrow & \text{s.s.}(s_{(2)}(\chi \rtimes St_{Sp(2)})) \\
\| & & & & \| \\
\begin{array}{c}\chi \otimes St_{Sp(2)} + \chi^{-1} \otimes St_{Sp(2)} \\ + \ \underline{\nu \otimes \chi \rtimes 1}\end{array} & & & & \underline{\chi \times \nu \otimes 1} + \underline{\chi^{-1} \times \nu \otimes 1}.
\end{array}$$

Assume

$$\chi^2 \neq 1_{F^\times} \text{ and } \chi \neq \nu^{\pm 1}, \nu^{\pm 2}.$$

Then all five representations in the above two sums are irreducible. In particular,

$$\nu \otimes \chi \rtimes 1, \ \chi \times \nu \otimes 1 \text{ and } \chi^{-1} \times \nu \otimes 1$$

are irreducible. Now we shall concentrate our attention on the Jacquet modules of these three representations:

$$\begin{array}{ccccc}
\text{——} + \text{——} + \nu \otimes \chi \rtimes 1 & & \chi \times \nu \otimes 1 & + & \chi^{-1} \times \nu \otimes 1 \\
\downarrow & & \downarrow & & \downarrow \\
\text{——} + \text{——} + \nu \otimes \chi \otimes 1 + \nu \otimes \chi^{-1} \otimes 1 & & \begin{array}{c}\chi \otimes \nu \otimes 1 + \\ \underline{\nu \otimes \chi \otimes 1}\end{array} & + & \begin{array}{c}\chi^{-1} \otimes \nu \otimes 1 + \\ \underline{\nu \otimes \chi^{-1} \otimes 1}\end{array}
\end{array}$$

Let π be an irreducible subquotient of $\chi \rtimes St_{Sp(2)}$ which has

$$\nu \otimes \chi \rtimes 1$$

as a subquotient of the Jacquet module with respect to $P_{(1)}$ (clearly, such a π exists). Then π must have

$$\nu \otimes \chi \otimes 1 \text{ and } \nu \otimes \chi^{-1} \otimes 1$$

as subquotients of its Jacquet module with respect to the minimal parabolic subgroup (see the diagram). These characters are the underlined members on the right hand side of the above diagram. Since these underlined members are coming from different subquotients of the Jacquet module with respect to $P_{(2)}$, we see that π must have

$$\chi \times \nu \otimes 1 \text{ and } \chi^{-1} \times \nu \otimes 1$$

in its Jacquet module. Since these two representations form the whole corresponding Jacquet module of $\chi \rtimes St_{Sp(2)}$, the exactness of the Jacquet modules implies $\pi = \chi \rtimes St_{Sp(2)}$. Thus, we have proved that $\chi \rtimes St_{Sp(2)}$ is irreducible in this case.

Now we go to reducibility. The representation $\chi \rtimes St_{Sp(2)}$ is reducible if $\chi = \nu^{\pm 2}$ or $\chi^2 = \nu^0$. We shall explain how one proves the reducibility in the case $\chi = \nu^0$ (the case $\chi = \nu^{\pm 2}$ is easy, since then one immediately gets $St_{Sp(4)}$ as a subquotient). Obviously

$$\nu^0 \rtimes St_{Sp(2)} \leq \nu^0 \times \nu \rtimes 1, \tag{23.1}$$

$$\nu^{1/2} St_{GL(2)} \rtimes 1 = \delta([\nu^0, \nu]) \rtimes 1 \leq \nu^0 \times \nu \rtimes 1, \tag{23.2}$$

and further from (14.2) and the representation theory of $GL(n)$ (actually, of $GL(2)$), it follows that

$$s_{(2)}(\nu^0 \rtimes St_{Sp(2)}) = 2(\nu^0 \times \nu \otimes 1) \nleq$$

$$s_{(2)}(\nu^{1/2} St_{GL(2)} \rtimes 1) = \delta([\nu^0, \nu]) \otimes 1 + \nu^0 \times \nu \otimes 1 + \delta([\nu^{-1}, \nu^0]) \otimes 1, \tag{23.3}$$

$$s_{(2)}(\nu^0 \rtimes St_{Sp(2)})) + s_{(2)}(\nu^{1/2} St_{GL(2)} \rtimes 1) \nleq$$

$$s_{(2)}(\nu^0 \times \nu \rtimes 1) = 2(\nu^0 \times \nu \otimes 1) + 2(\nu^0 \times \nu^{-1} \otimes 1). \tag{23.4}$$

The relations (23.1) – (23.4) and the exactness of the Jacquet modules imply the reducibility of $\nu^0 \rtimes St_{Sp(2)}$. In a similar way, one shows reducibility if $\chi^2 = 1_{F^\times}$ but $\chi \not\equiv 1_{F^\times}$ (here one considers the subquotients $\chi \rtimes St_{Sp(2)}$ and $\nu \rtimes \tau$ of $\chi \times \nu \rtimes 1$, where τ is an irreducible subrepresentation of $\chi \rtimes 1$; here one also needs to consider Jacquet modules with respect to $P_{(1)}$).

In the case of $\chi = \nu^{\pm 1}$ we again get irreducibility, but the proof of this fact does not rely on Jacquet modules of $\nu^{\pm 1} \rtimes St_{Sp(2)}$ only (see the following section).

24. A delicate case

In this section, we consider the remaining case of irreducibility which we have mentioned above (this is a special case of Proposition 5.1 of [39]). In this case we can neither prove reducibility nor irreducibility by the above methods. This is one of a few cases where we have a system as in section 21., but we cannot prove reducibility using the strategy of section 20. The idea is to go to larger groups, and get a contradiction there (with Frobenius reciprocity and/or the theory of R-groups). Namely, if we suppose that some parabolically induced representation of $Sp(4, F)$ reduces, it does not affect only the representation theory of $Sp(4, F)$, but also of all $Sp(6, F)$, $Sp(8, F)$, etc. (for example, if the inducing representation is unitarizable, it affects all the unitary duals).

Such cases, which we call delicate cases, are infrequent in the following sense. For example, in the case of symplectic and odd-orthogonal groups, for solving the question of reducibility of any degenerate principal series induced from any maximal parabolic subgroup (i.e. representation induced by one-dimensional representation), which was settled by C. Jantzen in [14], there are only two such cases: irreducibility of the representations $\nu^{1/2}\psi_0 1_{GL(2)} \rtimes 1$ of $SO(5)$ and of $\nu \rtimes 1_{Sp(2)}$ of $Sp(4)$, where ψ_0 is a character of $GL(2, F)$ whose square is trivial (i.e., a character of order one or two). These cases (actually, much more general cases) were settled in [39] (Propositions 6.3 and 5.1). We shall give below the proof of the second irreducibility in the "dual" situation, since later on our primary interest will be square integrable representations. For the moment, let us observe that if we suppose reducibility of the first representation, a very simple case of the theory of R-groups leads to a contradiction in the representation theory of $SO(9)$, while for the second representation we get contradiction in the representation theory of $Sp(6)$ (this contradiction is related to Frobenius reciprocity).

Let us note that this (sometimes very effective) method of solving a problem by going to larger groups was applied systematically in our work on unitarizability of general linear groups ([34]).

We shall now handle a more general case which implies the irreducibility of the representation $\nu \rtimes St_{Sp(2)}$ of $Sp(4)$:

Proposition. *Let ρ be an irreducible unitarizable cuspidal representation of $GL(p, F)$ and σ an irreducible cuspidal representation of S_q. Assume that $\nu\rho \rtimes \sigma$ reduces (which implies that $\rho \rtimes \sigma$ does not reduce). Then $\nu\rho \rtimes \delta(\nu\rho; \sigma)$ is irreducible.*

Proof. Let

$$\pi_0 = \nu\rho \rtimes \delta(\nu\rho; \sigma).$$

We have

$$\mu^*(\pi_0) = 1 \otimes \pi_0 + [\nu\rho \otimes \nu\rho \rtimes \sigma + \nu\rho \otimes \delta(\nu\rho; \sigma) + \nu^{-1}\rho \otimes \delta(\nu\rho; \sigma)] + [\nu\rho \times \nu\rho \otimes \sigma + \nu^{-1}\rho \times \nu\rho \otimes \sigma].$$

From this formula we see that there exists an irreducible subrepresentation π of π_0 satisfying $\nu\rho \times \nu\rho \otimes \sigma \leq s_{(2p)}(\pi)$. Suppose that $\nu\rho \rtimes \delta(\nu\rho; \sigma)$ reduces. Then since $s_{GL}(\pi_0)$ has length 2, we conclude

$$s_{(2p)}(\pi) = \nu\rho \times \nu\rho \otimes \sigma.$$

Now consider

$$\Pi = \rho \times \nu\rho \times \nu^{-1}\rho \rtimes \sigma.$$

Note that

$$\rho \rtimes \pi \leq \Pi, \quad \delta([\nu^{-1}\rho, \nu\rho]) \rtimes \sigma \leq \Pi \tag{24.1}$$

in the Grothendieck group. We have directly

$$\text{s.s.}(s_{(3p)}(\Pi)) = 4\rho \times \nu\rho \times \nu^{-1}\rho \otimes \sigma + 2\rho \times \nu^{-1}\rho \times \nu^{-1}\rho \otimes \sigma + 2\rho \times \nu\rho \times \nu\rho \otimes \sigma,$$

$$\text{s.s.}(s_{(3p)}(\rho \rtimes \pi)) = 2\rho \times \nu\rho \times \nu\rho \otimes \sigma$$

and

$$\text{s.s.}(s_{(3p)}(\delta([\nu^{-1}\rho, \nu\rho]) \rtimes \sigma)) = 2\delta([\nu^{-1}\rho, \nu\rho]) \otimes \sigma + 2\delta([\rho, \nu\rho]) \times \nu\rho \otimes \sigma.$$

Note that the multiplicity of

$$\nu\rho \times \delta([\rho, \nu\rho]) \otimes \sigma$$

is two in the above three Jacquet modules. Now (24.1) implies that there exists an irreducible representation φ such that

$$\varphi \leq \delta([\nu^{-1}\rho, \nu\rho]) \rtimes \sigma$$

and

$$\varphi \leq \rho \rtimes \pi.$$

Since

$$\delta([\nu^{-1}\rho, \nu\rho]) \rtimes \sigma$$

is unitarizable, each irreducible subquotient of it is actually a subrepresentation. Frobenius reciprocity implies the existence of a non-trivial intertwining

$$s_{(3p)}(\varphi) \twoheadrightarrow \delta([\nu^{-1}\rho, \nu\rho]) \otimes \sigma,$$

which implies

$$\delta([\nu^{-1}\rho, \nu\rho]) \otimes \sigma \leq s_{(3p)}(\varphi).$$

Note that $\varphi \leq \rho \rtimes \pi$ implies

$$\delta([\nu^{-1}\rho, \nu\rho]) \otimes \sigma \leq s_{(3p)}(\rho \rtimes \pi).$$

This contradicts

$$\text{s.s.}(s_{(3p)}(\rho \rtimes \pi)) = 2\rho \times \nu\rho \times \nu\rho \otimes \sigma.$$

This contradiction finishes the proof. □

Now we shall present another (more complicated) delicate case, where in the proof we use a consequence of the theory of R-groups (later on we shall recall this result). This is the reason why we restrict to the characteristic zero case. We shall only state the result here, which is Proposition 6.3 of [39]:

Proposition. *Let ρ and σ be irreducible cuspidal representations of groups $GL(p, F)$ and S_q respectively, such that ρ is selfdual (i.e. $\rho \cong \tilde{\rho}$). Suppose that char $F = 0$ and that both $\rho \rtimes \sigma$ and $\nu\rho \rtimes \sigma$ are irreducible. Then $\delta([\rho, \nu\rho]) \rtimes \sigma$ is also irreducible.*

25. Langlands parameters of irreducible subquotients

As we have already mentioned, one can solve the reducibility question in surprising generality using the above simple methods. Moreover, one can get in many cases also lengths and Langlands parameters of irreducible subquotients at the reducibility points (this is done in [14] for all degenerate principal series representations of groups S_n induced from a maximal parabolic subgroup). We shall not go much into this direction. Rather we shall show with a simple example the very basic idea of the method (C. Jantzen and G. Muić have improved the basic method introduced in [39] very significantly).

First we shall recall the Langlands classification in the case of the groups S_n. Denote by T the set of all equivalence classes of irreducible tempered representations of all S_n, $n \geq 0$. Set

$$D_+ = \{\delta \in D; e(\delta) > 0\}.$$

Let

$$t = (\delta_1, \ldots, \delta_k; \tau) \in M(D_+) \times T.$$

Take a permutation p of $\{1, \ldots, k\}$ such that

$$e(\delta_{p(1)}) \geq \cdots \geq e(\delta_{p(k)}).$$

Then the representation

$$\delta_{p(1)} \times \cdots \times \delta_{p(k)} \rtimes \tau \tag{25.1}$$

has a unique irreducible quotient, which we denote by $L(t)$. The mapping $t \mapsto L(t)$ is a bijection of $M(D_+) \times T$ onto $\cup_{n=0}^{\infty} \widetilde{S}_n$. This is the Langlands classification for the groups S_n.

The multiplicity of $L(t)$ in (25.1) is 1. Further, the representation

$$\tilde{\delta}_{p(1)} \otimes \cdots \otimes \tilde{\delta}_{p(k)} \otimes \tau$$

is a subquotient in the Jacquet module of $L(t)$ (later on, we shall use this fact).

Example. *Consider the following representation*

$$\pi_0 := \nu^{n+1} \rtimes L(\nu^n, \nu^{n-1}, \ldots, \nu; 1).$$

Obviously we have an epimorphism

$$\phi : \nu^{n+1} \times \nu^n \times \cdots \times \nu \rtimes 1 \to \pi_0.$$

Therefore

$$L(\nu^{n+1}, \nu^n, \ldots, \nu; 1)$$

is a (sub)quotient of π_0. *The embedding* $\delta([\nu^n, \nu^{n+1}]) \hookrightarrow \nu^{n+1} \times \nu^n$ *lifts to an embedding*

$$i : \delta([\nu^n, \nu^{n+1}]) \times \nu^{n-1} \times \cdots \times \nu \rtimes 1 \hookrightarrow \nu^{n+1} \times \nu^n \times \cdots \times \nu \rtimes 1.$$

We want to show

$$\phi \circ i \neq 0.$$

This would imply that

$$L(\delta([\nu^n, \nu^{n+1}]), \nu^{n-1}, \ldots, \nu; 1)$$

is a subquotient of π_0 *(note that this would further imply that* π_0 *is reducible).*

Suppose $\phi \circ i = 0$. Then we would have an epimorphism from the quotient of the domain of ϕ by the domain of i onto π_0. This quotient is isomorphic to

$$L(\nu^n, \nu^{n+1}) \times \nu^{n-1} \times \cdots \times \nu \rtimes 1.$$

This implies

$$s_{GL}(\pi_0) \leq s_{GL}(L(\nu^n, \nu^{n+1}) \times \nu^{n-1} \times \cdots \times \nu \rtimes 1).$$

Now (14.2) and Bernstein-Zelevinsky theory implies that this is not possible. Thus, $\phi \circ i \neq 0$.

A simple analysis of Jacquet modules of π_0 would give that the length of π_0 is 2. Thus, in the Grothendieck group we have

$$\pi_0 = \nu^{n+1} \rtimes L(\nu^n, \nu^{n-1}, \ldots, \nu; 1)$$

$$= L(\nu^{n+1}, \nu^n, \ldots, \nu; 1) + L(\delta([\nu^n, \nu^{n+1}]), \nu^{n-1}, \ldots, \nu; 1).$$

One can find in [10] and [11] examples of the study of the internal structure of induced representations with help of Jacquet modules (the semi simplicity of certain induced representations is studied there, from which is obtained non-unitarizability of some (non-induced) representations).

26. An interesting integral tempered irreducibility

Now we shall prove the following proposition (later on, we shall explain why this example is interesting for our considerations).

Proposition. *The representation*

$$St_{GL(3)} \rtimes St_{Sp(2)} = \delta([\nu^{-1}, \nu]) \rtimes \delta(\nu; 1)$$

is irreducible.

Proof. Denote the representation in the proposition by π_0. From (14.3), it follows that

$$M^*(\delta([\nu^{-1}, \nu])) = \sum_{k=-2}^{1} \sum_{l=k}^{1} \delta([\nu^{-k}, \nu]) \times \delta([\nu^{l+1}, \nu]) \otimes \delta([\nu^{k+1}, \nu^l]).$$

Recall $\mu^*(\delta(\nu; 1)) = 1 \otimes \delta(\nu; 1) + \nu \otimes 1$. Now applying (14.2), we get directly

$$\mu^*(\pi_0) = 1 \otimes \pi_0$$

$$+ \left[2(\nu \otimes \delta([\nu^{-1}, \nu^0]) \rtimes \delta(\nu; 1)) + \nu \otimes \delta([\nu^{-1}, \nu]) \rtimes 1 \right] \quad (26.1)$$

$$+\left[2\left(\boxed{\delta([\nu^0,\nu])\otimes\nu^{-1}\rtimes\delta(\nu;1)}\right)+2(\nu\times\nu\otimes\delta([\nu^{-1},\nu^0])\rtimes 1)+\nu\times\nu\otimes\nu^0\rtimes\delta(\nu,1)\right] \tag{26.2}$$

$$+\left[2\left(\boxed{\delta([\nu^{-1},\nu])\otimes\delta(\nu;1)}\right)+2(\delta([\nu^0,\nu])\times\nu\otimes\nu^{-1}\rtimes 1) \right. \tag{26.3}$$

$$\left. +2(\nu\times\delta([\nu^0,\nu])\otimes\delta(\nu;1))+\boxed{\nu\times\nu\times\nu\otimes\nu^0\rtimes 1}\right] \tag{26.4}$$

$$\left[2(\delta([\nu^{-1},\nu])\times\nu\otimes 1)+2\left(\boxed{\delta([\nu^0,\nu])\times\nu\times\nu\otimes 1}\right)\right]. \tag{26.5}$$

Now we analyze the above formulas. First observe that all boxed representations are irreducible (recall that the boxed representation in line (26.2) is irreducible by the first proposition in section 24.). Further, if we have an irreducible subquotient π' of π_0, then unitarizability implies that it is a subrepresentation. Now Frobenius reciprocity implies that it must have the boxed term in line (26.3) in its Jacquet module.

The transitivity of Jacquet modules implies that $s_{(1,1,1,1)}(\pi')$ must have

$$\nu\otimes\nu^0\otimes\nu^{-1}\otimes\nu\otimes 1$$

as a subquotient. Note that this subquotient cannot be in the Jacquet modules of the two non-boxed representations in line (26.2) (all irreducible subquotients of these terms start with $\nu\otimes\nu\otimes\dots$). This implies

$$\delta([\nu^0,\nu])\otimes\nu^{-1}\rtimes\delta(\nu;1)\le s_{(2)}(\pi'),$$

which implies

$$\nu\otimes\nu^0\otimes\nu\otimes\nu\otimes 1\le s_{(1,1,1,1)}(\pi').$$

Now $\nu\otimes\nu^0\otimes\nu\otimes\nu\otimes 1$ is not in the Jacquet module of the first term in line (26.5) (since there is always tensor factor ν^{-1}). This implies that

$$\delta([\nu^0,\nu])\times\nu\times\nu\otimes 1\le s_{(4)}(\pi') \tag{26.6}$$

holds for any irreducible subquotient π' of π_0.

Let π be the unique irreducible subquotient of π_0 which has the boxed term from line (26.4) in its Jacquet module. If we show that $s_{(4)}(\pi)$ contains the boxed term from line (26.5) with multiplicity two, then (26.6) implies $\pi=\pi_0$, i.e., π_0 is irreducible. Suppose that the multiplicity is one. Since

$$\nu\otimes\nu\otimes\nu\otimes\nu^0\otimes 1$$

is not in the Jacquet module of the first representation in line (26.5), the (semi simplification of the) Jacquet module of $\nu \times \nu \times \nu \otimes \nu^0 \rtimes 1$ with respect to the minimal parabolic subgroup, which is equal to $12\ \nu \otimes \nu \otimes \nu \otimes \nu^0 \otimes 1$, must be less then or equal to the Jacquet module of $\delta([\nu^0, \nu]) \times \nu \times \nu \otimes 1$ with respect to the minimal parabolic subgroup, which is impossible (since the last Jacquet module is also a representation of length 12, but has $\nu \otimes \nu^0 \otimes \nu \otimes \nu \otimes 1$ as a subquotient). This contradiction ends the proof of irreducibility.

Remark. *The representation*

$$St_{GL(m)} \rtimes St_{Sp(2)} = \delta(1_{F^\times}, m) \rtimes \delta(\nu; 1)$$

is

- ***irreducible*** *for all* **even** m,
- ***reducible*** *for all* **odd** m <u>***except***</u> $m = 3$.

For $m = 3$ this is proved in the last proposition. One proves the above reducibility claim in a similar way as we proved reducibility in the second example of section 20. The remaining irreducibility claim is proved in a similar way as we proved irreducibility in the second example of section 21.

27. On R-groups

We shall very briefly describe the R-groups that show up for the groups that we consider. Let G be such a group (some S_n), and let σ be an irreducible square integrable representation of the standard Levi subgroup M of a standard parabolic subgroup P. One considers the stabilizer $W(\sigma)$ of σ in the Weyl group of G. Let $W(\sigma)'$ be the subgroup of those $w \in W(\sigma)$ for which the corresponding normalized standard intertwining operator acts on $\mathrm{Ind}_P^G(\sigma)$ as a scalar. Then $W(\sigma)/W(\sigma)'$ is the R-group of σ. Actually, it is isomorphic to a subgroup of W (see [17]). Moreover, for our G one has $r^2 = 1$ for each element r of the R-group ([7]). So the R-group is commutative. Further, the intertwining algebra of $\mathrm{Ind}_P^G(\sigma)$ is isomorphic to the group algebra $\mathbb{C}[R]$ of R. Therefore, $\mathrm{Ind}_P^G(\sigma)$ is a multiplicity one representation. Further, characters of R naturally parameterize irreducible pieces of $\mathrm{Ind}_P^G(\sigma)$ (see the second section of [17]). Since the R-group is a free $\mathbb{Z}/2\mathbb{Z}$-module, characters of the R-group are in bijection with functions from (some fixed) basis of the R-group into $\{\pm 1\}$. Therefore, functions from the basis of the R-group into $\{\pm 1\}$ parameterize irreducible pieces of $\mathrm{Ind}_P^G(\sigma)$.

Now we shall present one consequence, quite interesting in our considerations (which also tells what the R-group is in this case):

Theorem. *Let $\delta_1, \dots, \delta_n \in D_u$, and let τ be an irreducible square integrable representation of S_q. Denote by l the number of nonequivalent representations δ among them having $\delta \rtimes \tau$ reducible. Then $\delta_1 \times \cdots \times \delta_n \rtimes \tau$ is a multiplicity one representation of length 2^l.*

This result was proved by Goldberg in [7] in characteristic 0.

On irreducible square integrable representations and their parameters

28. Introductory remarks on invariants of square integrable representations

(1) Recall that by the Langlands classification of the non-unitary dual, we need to know irreducible tempered representations of Levi subgroups, which are tensor products of such representations of groups $GL(p, F)$ and of S_q. The Bernstein-Zelevinsky theory classifies the irreducible tempered representations of general linear groups modulo cuspidal representations. The previous section tells us that we have reduction of the irreducible tempered representations of groups S_q to the square integrable representations of the same groups and the knowledge when $\delta \rtimes \pi$ reduces, where δ and π are irreducible square integrable representations of $GL(p, F)$ and S_q respectively. Recall that $\delta = \delta(\rho, m)$ for some unitarizable irreducible cuspidal representation ρ of a general linear group and a positive integer m. If ρ is not selfdual, then it follows easily that $\delta(\rho, m) \rtimes \pi$ is irreducible.
It remains to consider the case of selfdual ρ. Assuming that basic (technical) assumption (BA) from section 17. holds (what we shall often do in the rest of the paper), the following general phenomenon (which we have already observed in some examples) holds for such ρ:

for exactly one parity of positive integers the following holds:

- $\delta(\rho, m) \rtimes \pi$ reduces for all m from that parity, with possibly finitely many exceptions;
- $\delta(\rho, m) \rtimes \pi$ is irreducible for all m from the other parity.

We have seen in sections 20. and 22. that this holds for the representation $\pi = \delta(\nu^{1/2}\rho; 1)$ of $SO(2p+1, F)$, and that the only exception is $\delta = \delta(\rho, 2)$, and that this exception characterizes $\delta(\nu^{1/2}\rho; 1)$ (recall that we have chosen ρ such that $\nu^{1/2}\rho \rtimes 1$ reduces as a representation of $SO(2p+1, F)$).

Therefore, for understanding the non-unitary dual, it is crucial to know the parity which satisfies the above properties for the given selfdual ρ, and to know the exceptions. The set of all such exceptions, when ρ runs over all the selfdual irreducible cuspidal representations ρ of general linear groups will be denoted by

$$\mathrm{Jord}(\pi).$$

Actually, to abbreviate notation, we shall simply denote the elements of $\mathrm{Jord}(\pi)$ by (ρ, m) instead of $\delta(\rho, m)$ (recall that (ρ, m) completely determines the representation $\delta(\rho, m)$). The sets $\mathrm{Jord}(\pi)$ are finite (see [22]). The set $\mathrm{Jord}(\pi)$ is called the set of **Jordan blocks** of π. As we have already noted, this invariant is very important from the point of view of representation theory (for the tempered representations and the non-unitary dual). But it is also important for the Galois side, as we shall discuss later.

In the case that $\nu^{1/2}\rho \rtimes 1$ reduces, we have seen that $(\rho, 2) \in \mathrm{Jord}(\delta(\nu^{1/2}\rho; 1))$. Using the estimate (34.1) of C. Mœglin we get.

$$\mathrm{Jord}(\delta(\nu^{1/2}\rho; 1)) = \{(\rho, 2)\}.$$

In a special case, we have

$$\mathrm{Jord}(St_{SO(3)}) = \{(1_{F^\times}, 2)\}.$$

For an irreducible square integrable representation π of S_q and a selfdual irreducible cuspidal representation ρ of a general linear group let

$$\mathrm{Jord}_\rho(\pi) := \{m; (\rho, m) \in \mathrm{Jord}(\pi)\}.$$

This set is called the **Jordan blocks of π along ρ**.

Consider a more general representation

$$\delta([\nu^{1/2}\rho, \nu^{k-1/2}\rho]; \sigma), \quad k \in \mathbb{Z}_{\geq 1},$$

when $\nu^{1/2}\rho \rtimes \sigma$ reduces (σ is an irreducible cuspidal representation of some S_q). In this case, we would get the set of "exceptions"

$$\mathrm{Jord}_\rho(\delta([\nu^{1/2}\rho, \nu^{k-1/2}\rho]; \sigma)) = \{2k\}$$

(this is again a single element set; we shall see examples of bigger exception sets later).

(2) From our previous considerations and estimate (34.2), one sees that for the trivial representation of S_0 in the case of odd orthogonal groups we have

$$\mathrm{Jord}(1_{SO(1))}) = \emptyset,$$

and in the symplectic case we have

$$\mathrm{Jord}(1_{Sp(0)}) = \{(1_{F^\times}, 1)\}$$

(this interesting difference is related to the fact that the dual complex L-group of $SO(1, F)$ is $Sp(0, \mathbb{C})$, and of $Sp(0, F)$ is $SO(1, \mathbb{C})$).

(3) Now we shall show how Jordan blocks help distinguish among representations. In general we have

$$\mathrm{Jord}(St_{SO(2n+1)}) = \{(1_{F^\times}, 2n)\}.$$

In particular,

$$\mathrm{Jord}(St_{SO(5)}) = \{(1_{F^\times}, 4)\}.$$

Take any two different characters χ_1, χ_2 of $F^\times$ satisfying $\chi_1^2 = \chi_2^2 \equiv 1$. Then the following inequalities hold

$$\nu^{1/2}\chi_1 \rtimes \delta(\nu^{1/2}\chi_2; 1) \quad \leq \quad \nu^{1/2}\chi_1 \times \nu^{1/2}\chi_2 \rtimes 1,$$

$$\nu^{1/2}\chi_2 \rtimes \delta(\nu^{1/2}\chi_1; 1) \quad \leq \quad \nu^{1/2}\chi_1 \times \nu^{1/2}\chi_2 \rtimes 1.$$

It is very easy to show that the two representations on the left hand side have exactly one irreducible subquotient in common, which we shall denote by

$$\delta(\nu^{1/2}\chi_1, \nu^{1/2}\chi_2; 1),$$

and that

$$\nu^{1/2}\chi_1 \otimes \nu^{1/2}\chi_2 \otimes 1 + \nu^{1/2}\chi_2 \otimes \nu^{1/2}\chi_1 \otimes 1$$

is its Jacquet module with respect to the minimal parabolic subgroup. Now Casselman's square integrability criterion implies that the representation $\delta(\nu^{1/2}\chi_1, \nu^{1/2}\chi_2; 1)$ is square integrable. Similar calculations as before give

$$\mathrm{Jord}(\delta(\nu^{1/2}\chi_1, \nu^{1/2}\chi_2; 1)) = \{(\chi_1, 2), (\chi_2, 2)\}.$$

This is an example of Jordan blocks with more then one element (more interesting examples of this type will come later). So we see that in this case Jordan blocks distinguish $\delta(\nu^{1/2}\chi_1, \nu^{1/2}\chi_2; 1)$ and $St_{SO(5)}$.
Later on, we shall have examples where Jordan blocks are not enough to distinguishing square integrable representations that we consider. For the moment, let us mention one simple example of this type. One gets such representations when one has nonequivalent irreducible subrepresentations of $\rho|SL(2)$, where ρ is an irreducible cuspidal representation

of $GL(2) = GSp(2)$ (R. Langlands and J.-P. Labesse called these representations L-indistinguishable). More interesting examples will be given later.

(4) From the point of view of the local Langlands conjectures for irreducible square integrable representations (discussed in section 4.), we know that the first parameter attached to an irreducible square integrable representation should be a (discrete) admissible homomorphism of the Weil-Deligne group into the complex dual group. Further, we know that the second parameter of an irreducible square integrable representation should be an irreducible representation of the component group, which is one-dimensional in our case, i.e., a character.

Let us take a closer look, from the point of view of the local Langlands correspondences, at the exceptions that we have obtained and discussed in the last few sections. Look first at $St_{SO(2n+1,F)}$. Recall that by the local Langlands correspondence for odd-orthogonal groups, $St_{SO(2n+1,F)}$ should correspond to an admissible homomorphism

$$W_F \times SL(2,\mathbb{C}) \to Sp(2n,\mathbb{C}).$$

Recall that

$$\mathrm{Jord}(St_{SO(2n+1,F)}) = \{(1_{F^\times}, 2n)\},$$

which means that the only exception is the square integrable representation $\delta(1_{F^\times}, 2n) = St_{GL(2n,F)}$. Since this exception is a representation of $GL(2n, F)$, we can apply the local Langlands correspondence Φ for general linear groups (discussed in section 13) to it. We get

$$\Phi(\delta(1_{F^\times}, 2n)) = 1_{W_F} \otimes E_{2n}.$$

A well known fact states that E_m is:

- symplectic (i.e., respects a non-degenerate symplectic form on $\mathbb{C}^m$) if m is even;
- orthogonal (i.e., respects a non-degenerate orthogonal form on $\mathbb{C}^m$) if m is odd.

This implies that in fact

$$\Phi(\delta(1_{F^\times}, 2n)) : W_F \times SL(2,\mathbb{C}) \to Sp(2n,\mathbb{C}).$$

Therefore, $\Phi(\delta(1_{F^\times}, 2n))$ goes just where the admissible homomorphism corresponding to $St_{SO(2n+1,F)}$ should go.

If we consider $St_{Sp(2n,F)}$, where

$$\mathrm{Jord}(St_{Sp(2n,F)}) = \{(1_{F^\times}, 2n+1)\},$$

then

$$\Phi(\delta(1_{F^\times}, 2n+1)) : W_F \times SL(2,\mathbb{C}) \to SO(2n+1,\mathbb{C}),$$

and we again get that $\Phi(\delta(1_{F^\times}, 2n+1))$ goes just where the admissible homomorphism corresponding to $St_{Sp(2n,F)}$ should go.
Consider the square integrable representation $\delta(\nu^{1/2}\chi_1, \nu^{1/2}\chi_2; 1)$ of $SO(5,F)$ considered in (3). Recall

$$\mathrm{Jord}(\delta(\nu^{1/2}\chi_1, \nu^{1/2}\chi_2; 1)) = \{(\chi_1, 2), (\chi_2, 2)\}.$$

Now

$$\bigoplus_{(\rho,k)\in \mathrm{Jord}(\delta(\nu^{1/2}\chi_1,\nu^{1/2}\chi_2;1))} \Phi(\delta(\rho,k)) =$$

$$\Phi(\chi_1)\otimes E_2 \oplus \Phi(\chi_2)\otimes E_2 : W_F \times SL(2,\mathbb{C}) \to Sp(4,\mathbb{C})$$

($\Phi(\chi_i)$ is here simply obtained by the Artin reciprocity of local class field theory). So, this representation goes precisely where the admissible homomorphism corresponding to $\delta(\nu^{1/2}\chi_1, \nu^{1/2}\chi_2; 1)$ should go. Actually, G. Muić conjectured in section 3. of [26] that in this way one should get natural candidates for the admissible homomorphisms corresponding to the generic square integrable representations of classical groups (see (28.1) below).
Further, considering the cases of Langlands functoriality between classical and general linear groups which are expected to hold, C. Mœglin realized that it is natural to expect that the (discrete) admissible homomorphism corresponding to a general irreducible square integrable representation π of S_n should be

$$\bigoplus_{(\rho,k)\in \mathrm{Jord}(\pi)} \Phi(\delta(\rho,k)) = \bigoplus_{(\rho,k)\in \mathrm{Jord}(\pi)} \Phi(\rho)\otimes E_k \tag{28.1}$$

(see remark (7) below for a few more details regarding this).
Suppose for a moment that (28.1) holds.

(5) Consider the case $S_n = SO(2n+1,F)$. Denote by $e_{(\rho,k)}$ the linear operator on $\mathbb{C}^{2n}$ which acts as multiplication by -1 on $\Phi(\rho)\otimes E_k$, and as identity on all other summands in (28.1). Then the component group is

$$\left\{ \prod_{(\rho,k)\in \mathrm{Jord}(\pi)} e_{(\rho,k)}^{\mu_{(\rho,k)}} ; \mu_{(\rho,k)} \in \mathbb{Z}/2\mathbb{Z} \right\} \Big/ \{\pm \mathrm{Id}_{\mathbb{C}^{2n}}\}.$$

Now characters (i.e., irreducible representations) of this group are in a natural bijection with each of the following sets:

- functions $\mathrm{Jord}(\pi) \to \{\pm 1\}$ which have a prescribed value on one fixed $(\rho, k) \in \mathrm{Jord}(\pi)$;
- functions $\mathrm{Jord}(\pi) \to \{\pm 1\}$ which take the value 1 on even number of times in the case $\mathrm{card}(\mathrm{Jord}(\pi)) \in 2\mathbb{Z}$;
- functions $\{e_{(\rho_i,k_i)} e^{-1}_{(\rho_{i+1},k_{i+1})}; i = 1, \dots, m-1\} \to \{\pm 1\}$, if we enumerate $\mathrm{Jord}(\pi) = \{(\rho_i, k_i); i = 1. \dots, m\}$.

Note that in all three above cases we can forget the Galois side. We can work simply with functions related to $\mathrm{Jord}(\pi)$ (in the last case we can work formally with $(\rho_i, k_i)(\rho_{i+1}, k_{i+1})^{-1}$'s). So, if we expect that (28.1) is the admissible homomorphism corresponding the irreducible square integrable representation π, for the purpose of classifying square integrable representations, we may try to parameterize them by functions related to $\mathrm{Jord}(\pi)$.
Therefore, regarding the second parameter of an irreducible square integrable representation π (i.e. an irreducible representation of the component group, which is a character in our case), we shall not work on the Galois side (except giving Galois side interpretations at some places). We shall relate such a character to a function related to $\mathrm{Jord}(\pi)$, and consider it as an invariant of π. This function, denoted by

$$\epsilon_\pi,$$

will be defined only partially on $\mathrm{Jord}(\pi)$ (for reasons explained below), and will be called the **partially defined function** attached to π. All the non-cuspidal square integrable representations considered up to now can be already distinguished by $\mathrm{Jord}(\pi)$. In the cases of more complicated square integrable representations, we shall need this function for further distinction of that representations.
Let us mention that this function has some relation with the function that parameterizes irreducible tempered pieces in section 27. We shall not go into more detail here, but this may become clear if we know that the construction of square integrable representations is closely related to the question of (reducibility of) tempered induction (see [41] for examples of this very direct connection). In the present paper, we shall see the relation of tempered induction to square integrable representations only in the simplest case.
Note that in the case of general linear groups, tempered induction is irreducible, i.e., parabolic induction carries irreducible tempered representations to irreducible (tempered) ones, and the component groups are trivial (so its irreducible representations can be only trivial).

(6) Since the classification that we present here is modulo cuspidal data (similarly as it was in the case of Bernstein and Zelevinsky for general linear groups; there the reducibility point is always 1), we shall have a third invariant, which is an irreducible cuspidal representation of the classical group that shows up in the construction of a square integrable representation. This invariant is called the **partial cuspidal support**. In our constructions up to now it was the trivial representation 1 of S_0 in the case of Steinberg representation, and σ in the case of the generalized Steinberg representation. In general, the partial cuspidal support of a general irreducible square integrable representation π of S_q is an irreducible cuspidal representation

$$\pi_{\text{cusp}}$$

of some $S_{q'}$, $q' \leq q$, such that there exists a representation ψ of $GL(q-q', F)$, such that

$$\pi \hookrightarrow \psi \rtimes \pi_{\text{cusp}}.$$

Note that π_{cusp} should have its own character $\epsilon_{\pi_{\text{cusp}}}$ which parameterizes it. Therefore, roughly the (complete) character corresponding to π should be determined by both ϵ_π and $\epsilon_{\pi_{\text{cusp}}}$. This partially explains why ϵ_π is only partially defined. The other reason is that $\text{Jord}(\pi)$ does not correspond exactly to the basis of the component group: in the orthogonal case the dual group has non-trivial center, while in the symplectic case one needs to take care of the determinant one condition. Our discussion above is not very precise (for a precise discussion, one should consult Mœglin's papers).

The invariants attached to π will form the triple

$$(\text{Jord}(\pi), \epsilon_\pi, \pi_{\text{cusp}}).$$

Recall that in the local Langlands conjecture we should have only two parameters. We do not have (the analogue of) the partial cuspidal support. Actually, the pair consisting of $\text{Jord}(\pi)$ and the (complete) character corresponding to π should determine π_{cusp}. Therefore, in our triple $\text{Jord}(\pi)$ and ϵ_π need to have "control" of π_{cusp} in an appropriate way (to start with fixed $\text{Jord}(\pi)$ we should have at most finitely many possibilities for π_{cusp}). The condition among $\text{Jord}(\pi)$, ϵ_π and π_{cusp} is a technical condition which we shall discuss later. Triples which satisfy this condition will be called **admissible triples**. C. Mœglin has attached to an irreducible square integrable representation π such an admissible triple, and has shown that the process of attaching triples is injective (assuming a natural technical conjecture (BA), which shows up only in proofs).

(7) The Langlands program started in the 1960's with Langlands' conjectures for $GL(n)$-groups. These conjectures generalize the Artin reciprocity law from class field theory (which corresponds to the case $GL(1)$). R.P. Langlands soon realized that these conjectures should be a very special case of a general principle called functoriality. We shall recall here very briefly only the case of functoriality that interests us. For simplicity, we shall consider only the case of odd orthogonal groups $SO(2n+1,F)$ (a completely analogous situation holds with symplectic groups). Recall ${}^L SO(2n+1,F)^0 = Sp(2n,\mathbb{C})$ and ${}^L GL(2n,F)^0 = GL(2n,\mathbb{C})$. The inclusion

$$i : Sp(2n,\mathbb{C}) \hookrightarrow GL(2n,\mathbb{C})$$

yields the inclusion $\varphi \mapsto i \circ \varphi$ of (discrete) admissible homomorphisms for $SO(2n+1,F)$ into admissible homomorphisms for $GL(2n,F)$:

$$\begin{array}{ccc} Sp(2n,\mathbb{C}) & \xrightarrow{\quad i \quad} & GL(2n,\mathbb{C}) \\ {\scriptstyle\varphi}\uparrow & & {\scriptstyle i\circ\varphi}\uparrow \\ W_F \times SL(2,\mathbb{C}) & = & W_F \times SL(2,\mathbb{C}). \end{array} \qquad (28.2)$$

The existences of the local Langlands correspondences for $SO(2n+1,F)$ and $GL(2n,F)$ (see section 4.) would directly give the lift of the mapping $\varphi \mapsto i \circ \varphi$ to the level of representations and L-packets (this would be an instance of a local Langlands functoriality, as well as a local Langlands correspondence is an instance of it). Very roughly (and little bit oversimplified), the functoriality predicts analogous "natural" lifting on the level of adelic groups over global fields ("natural" means that some basic invariants like L-functions should be respected; see [5] for details). Local and global lifts should be compatible, and on unramified factors the lift should have naturally expected form.

Recall that the local Langlands conjecture for general linear groups is proved. Observe now that the knowledge of the local lift of irreducible representations of $SO(2n+1,F)$ to irreducible representations of $GL(2n,F)$ would imply the knowledge of the local Langlands correspondence for $SO(2n+1,F)$. Further, the knowledge of the global lift from $SO(2n+1)$ to $GL(2n)$ and the requirement that the local and global lifts should be compatible, would imply the knowledge of the action of the local Langlands correspondence for $SO(2n+1)$ on local factors of automorphic representations.

The lifts of generic cuspidal automorphic representations of adelic classical groups over number fields to general linear groups were obtained in [5] using a converse theorem for global representations (regarding being automorphic; this theorem generalizes the classical converse theorems of Hecke and Weil). Therefore, [5] gives what will be admissible homomorphisms corresponding to local factors of generic cuspidal automorphic representations of the adelic group $SO(2n+1)$ (they are also generic). The formula obtained in the case of generic square integrable representations of $SO(2n+1, F)$ is the formula that we mentioned in (4) (and which was conjectured by G. Muić in [26]).
Functoriality is not established in the non-generic case, but some expectations of the modern theory of automorphic forms would imply the formula also holds in this case (see [20]). The above discussion also gives support to the basic assumption (see also [32]).

29. An important simple example of construction of square integrable representations

Take an irreducible unitarizable cuspidal representation ρ of $GL(p, F)$ and an irreducible cuspidal representation σ of S_q such that $\rho \rtimes \sigma$ reduces (then $\rho \cong \tilde{\rho}$ and $\nu^\alpha\rho \rtimes \sigma$ does not reduce for all $\alpha \in \mathbb{R}^\times$). From the Jacquet module $s_{(p)}(\rho \rtimes \sigma)$, one gets that $\rho \rtimes \sigma$ is a sum of two irreducible representations. Further, Frobenius reciprocity implies that $\rho \rtimes \sigma$ is a multiplicity one representation. Write

$$\rho \rtimes \sigma = \tau_1 \oplus \tau_2,$$

where τ_1 and τ_2 are irreducible ($\tau_1 \not\cong \tau_2$). Then it follows directly that

$$\mu^*(\tau_i) = 1 \otimes \tau_i + \rho \otimes \sigma.$$

We shall work below with symplectic groups and take ρ to be a character ψ of order 2 of $F^\times$ and $\sigma = 1_{Sp(0)}$ (=1), but the entire construction holds for general ρ and σ as above (i.e., with reducibility at 0). We shall comment a more general case later.

We consider

$$\delta([\psi, \nu\psi]) \rtimes 1 \hookrightarrow \nu\psi \times \psi \rtimes 1 = \nu\psi \rtimes \tau_1 \oplus \nu\psi \rtimes \tau_2.$$

Observe that

$$\mu^*(\nu\psi \rtimes \tau_i) = 1 \otimes \nu\psi \rtimes \tau_i + \tag{29.1}$$

$$[\nu\psi \otimes \tau_i + \nu^{-1}\psi \otimes \tau_i + \psi \otimes \nu\psi \rtimes 1] +$$

$$[\nu\psi \times \psi \otimes 1 + \nu^{-1}\psi \times \psi \otimes 1].$$

From this, using Frobenius reciprocity, we see that $\nu\psi \rtimes \tau_i$ has a unique irreducible subrepresentation (since the multiplicity of $\nu\psi \otimes \tau_i$ in the above Jacquet module is 1). We denote this subrepresentation by

$$\delta([\psi, \nu\psi]_{\tau_i}; 1).$$

Observe

$$\mu^*(\delta([\psi, \nu\psi]) \rtimes 1) = 1 \otimes \delta([\psi, \nu\psi]) \rtimes 1 + \tag{29.2}$$

$$[\nu\psi \otimes \tau_1 + \nu\psi \otimes \tau_2 + \psi \otimes \nu\psi \rtimes 1] +$$

$$\left[\delta([\psi, \nu\psi]) \otimes 1 + \psi \times \nu\psi \otimes 1 + \delta([\nu^{-1}\psi, \psi]) \otimes 1\right].$$

This implies that both representations $\delta([\psi, \nu\psi]_{\tau_i}; 1)$ are subrepresentations of the representation $\delta([\psi, \nu\psi]) \rtimes 1$ (because of imbedding (29.1) and the fact that the multiplicity of $\nu\psi \otimes \tau_i$ in the above Jacquet module is also 1). From the fact that $\delta([\psi, \nu\psi]_{\tau_i}; 1)$ is in both representations, we conclude

$$s_{(2)}(\delta([\psi, \nu\psi]_{\tau_i}; 1)) \leq \nu\psi \times \psi \otimes 1 + \delta([\nu^{-1}\psi, \psi]) \otimes 1.$$

The second summand on the right hand side, i.e. $\delta([\nu^{-1}\psi, \psi]) \otimes 1$, is in the Jacquet module of the Langlands quotient of $\delta([\psi, \nu\psi]) \rtimes 1$ (see section 25). This implies

$$s_{(2)}(\delta([\psi, \nu\psi]_{\tau_i}; 1)) \leq \nu\psi \times \psi \otimes 1,$$

which yields $s_{(1,1)}(\delta([\psi, \nu\psi]_{\tau_i}; 1)) \leq \nu\psi \otimes \psi \otimes 1 + \psi \otimes \nu\psi \otimes 1$.

From the Casselman criterion we conclude that $\delta([\psi, \nu\psi]_{\tau_i}; 1)$ is tempered. Now we claim that $\nu\psi \times \psi \rtimes 1$ does not have tempered subquotients which are not square integrable. Suppose that it has. Then there exists a proper standard parabolic subgroup $P = MN$ and an irreducible square integrable modulo center representation δ of M such that $\nu\psi \times \psi \rtimes 1$ and $\mathrm{Ind}_P^{Sp(4,F)}(\delta)$ have an irreducible subquotient in common. If P is the minimal parabolic subgroup, then $\mathrm{Ind}_P^{Sp(4,F)}(\delta) = \chi_1 \times \chi_2 \rtimes 1$ for some $\chi_1, \chi_2 \in (F^\times)\hat{}$. Now, section 10. implies that the above two representations cannot have an irreducible subquotient in common. Suppose that P is the Siegel parabolic subgroup. Then from section 10., it follows that δ cannot be cuspidal. This implies $\mathrm{Ind}_P^{Sp(4,F)}(\delta) = (\chi \circ \det) St_{GL(2,F)} \rtimes 1 \leq \nu^{1/2}\chi \times \nu^{-1/2}\chi \rtimes 1$ for some $\chi \in (F^\times)\hat{}$. Now again section 10. implies that the representations $\nu\psi \times \psi \rtimes 1$ and $\nu^{1/2}\chi \times \nu^{-1/2}\chi \rtimes 1$ can not have an irreducible subquotient in common. We handle the remaining proper standard parabolic subgroup similarly.

In this way we have proved that $\delta([\psi, \nu\psi]_{\tau_i}; 1)$ is square integrable. From this it follows that

$$s_{(2)}(\delta([\psi, \nu\psi]_{\tau_i}; 1)) = \delta([\psi, \nu\psi]) \otimes 1,$$

which further implies

$$\mu^* \left(\delta([\psi, \nu\psi]_{\tau_i}; 1)\right) = 1 \otimes \delta([\psi, \nu\psi]_{\tau_i}; 1) + \nu\psi \otimes \tau_i + \delta([\psi, \nu\psi]) \otimes 1.$$

Remark. *The above simple example is important since this is how, in systematic way, we shall get two square integrable representations of bigger groups (in this case of $Sp(4, F)$) from a square integrable representation of a smaller group (in this case from the trivial representation of $Sp(0, F)$).*

The above proof of square integrability is an illustration of the simplest example of a much more general proof of square integrability in [24] (in sections 9. - 11. of that paper; that proof is a crucial part of [24] and it completes the classification of square integrable representations of classical groups modulo cuspidal data).

Now we can analyze representations induced by the above two square integrable representations. We obtain

$$\mathrm{Jord}(\delta([\psi, \nu\psi]_{\tau_i}; 1)) = \{(\psi, 1), (\psi, 3), (1_{F^\times}, 1)\}.$$

So the Jordan blocks do not distinguish these non-isomorphic square integrable representations.

In general, suppose that $\rho \rtimes \sigma$ reduces (ρ and σ are as usual). Then we can write

$$\rho \rtimes \sigma = \tau_1 \oplus \tau_2,$$

for nonequivalent irreducible τ_1 and τ_2. We can continue the above construction: we define

$$\delta([\rho, \nu^k\rho]_{\tau_i}; \sigma)$$

as the unique irreducible subrepresentation of

$$\delta([\nu\rho, \nu^k\rho]) \rtimes \tau_i.$$

Similarly one proves that $\delta([\rho, \nu^k\rho]_{\tau_i}; \sigma)$ are square integrable (one can also determine $\mu^*(\delta([\rho, \nu^k\rho]_{\tau_i}; \sigma))$ - it is a simple formula, similar to that for the Steinberg type representations).

We shall now briefly illustrate

30. A little bit more complicated example of construction of square integrable representations

In this section we consider odd orthogonal groups. From section 20. we know that

$$\delta([\nu^{-1/2},\nu^{1/2}]) \rtimes 1 = \tau_+ \oplus \tau_-.$$

Now

$$\delta([\nu^{-1/2},\nu^{3/2}]) \rtimes 1 \hookrightarrow \nu^{3/2} \times \delta([\nu^{-1/2},\nu^{1/2}]) \rtimes 1 = \nu^{3/2} \rtimes \tau_+ \oplus \nu^{3/2} \rtimes \tau_-.$$

The multiplicity of $\nu^{3/2} \otimes \tau_\pm$ in $\mu^*(\nu^{3/2} \rtimes \tau_\pm)$ (and $\delta([\nu^{-1/2},\nu^{3/2}]) \rtimes 1$) is 1. This implies that $\nu^{3/2} \otimes \tau_\pm$ has a unique irreducible subrepresentation, which will be denoted by

$$\delta([\nu^{-1/2},\nu^{3/2}]_\pm;1).$$

In calculations below we shall write

$$\pi_\pm = \delta([\nu^{-1/2},\nu^{3/2}]_\pm;1).$$

Since the multiplicity of $\nu^{3/2} \otimes \tau_\pm$ in $\mu^*(\delta([\nu^{-1/2},\nu^{3/2}]) \rtimes 1))$ is also 1, both $\pi_\pm$ must be subrepresentations of $\delta([\nu^{-1/2},\nu^{3/2}]) \rtimes 1$. In this way we get two upper bounds for the Jacquet module of $\delta([\nu^{-1/2},\nu^{3/2}]_\pm;1)$. From them, similarly as in the previous section, we first conclude that $\nu^{3/2} \otimes \tau_\pm$ is tempered, and then again similarly as in the previous section, that it must be square integrable. It is not hard to show

$$s_{GL}(\pi_+) = \delta([\nu^{-1/2},\nu^{3/2}]) \otimes 1 + \delta([\nu^{1/2},\nu^{3/2}]) \times \nu^{1/2} \otimes 1,$$

$$s_{GL}(\pi_-) = \delta([\nu^{-1/2},\nu^{3/2}]) \otimes 1.$$

Now we shall recall of a general principle regarding Jordan blocks. From it, we shall easily see what are the Jordan blocks of $\pi_\pm$. Below we shall assume that basic (technical) assumption (BA) from the section 17. holds.

Proposition 2.1 from [24] tells:

Proposition. *Let π' be an irreducible square integrable representation of S_q and let $x, y \in (1/2)\mathbb{Z}$ such that $x-y \in \mathbb{Z}_{\geq 0}$. Let ρ be an irreducible selfdual cuspidal representation of $GL(p)$. We assume that x, $y \in \mathbb{Z}$ if and only if $\nu^t\rho \rtimes 1$ reduces for some $t \in \mathbb{Z}$. Further, suppose that there is an irreducible square integrable representation π embedded in the induced representation*

$$\pi \hookrightarrow \nu^x\rho \times \cdots \times \nu^{x-i+1}\rho \times \cdots \times \nu^y\rho \rtimes \pi'.$$

Then:

(i) If $y > 0$, then $2y-1 \in \mathrm{Jord}_\rho(\pi')$ and

$$\mathrm{Jord}_\rho(\pi) = (\mathrm{Jord}_\rho(\pi') \setminus \{(\rho, 2y-1)\}) \cup \{(\rho, 2x+1)\}.$$

(ii) If $y \leq 0$, then

$$\mathrm{Jord}_\rho(\pi) = \mathrm{Jord}_\rho(\pi') \cup \{(\rho, 2x+1), (\rho, -2y+1)\}.$$

From this directly follows

$$\mathrm{Jord}(\delta([\nu^{-1/2}, \nu^{3/2}]_\pm; 1)) = \{(1_{F^\times}, 2), (1_{F^\times}, 4)\}.$$

Further, all the other Jordan blocks that we have computed up to now can be obtained from the above proposition.

We can also often see elements of the Jordan blocks from Jacquet modules:

Lemma. *Suppose that π is an irreducible square integrable representation of S_q. Let $\nu^x\rho \otimes \tau$ be an irreducible subquotient of a standard Jacquet module of π, where ρ is an irreducible selfdual cuspidal representation of $GL(p, F)$, $x \in \mathbb{R}$, and τ is an irreducible representation of $S_{q'}$. Then*

$$(\rho, 2x+1) \in \mathrm{Jord}(\pi).$$

31. Partially defined function

We shall now explain how one can distinguish the above two representations. This will be done by partially defined function on $\mathrm{Jord}(\pi)$, where π is a square integrable representation. Partially defined function will essentially be a character defined on a subgroup of the free $\mathbb{Z}/2\mathbb{Z}$-module with basis $\mathrm{Jord}(\pi)$ (a character of such a subgroup must take values in $\{\pm 1\}$). Therefore, if this character is defined on the whole $\mathrm{Jord}(\pi)$, it is enough to define the function on $\mathrm{Jord}(\pi)$ (this is the case for reducibilities in $\frac{1}{2} + \mathbb{Z}$). Otherwise, we shall again define it on generators and again talk about a function (with values in $\{\pm 1\}$), rather then about a character.

Now we shall explain how to make a difference between representations π_+ and π_- from the previous section, without going into the internal structure of these representations (recall that we know certain Jacquet modules which happened to be different). Recall

$$s_{GL}(\pi_+) = \delta([\nu^{-1/2}, \nu^{3/2}]) \otimes 1 + \delta([\nu^{1/2}, \nu^{3/2}]) \times \nu^{1/2} \otimes 1.$$

Both representations on the right hand side are irreducible, and have different central characters. Therefore, we have a direct sum. Now the Frobenius reciprocity implies

$$\pi_+ \hookrightarrow \nu^{1/2} \times \delta([\nu^{1/2}, \nu^{3/2}]) \rtimes 1,$$

which easily implies that

$$\pi_+ \hookrightarrow \nu^{1/2} \rtimes \pi' = \nu^{(2-1)/2} \rtimes \pi'$$

for some irreducible representation π' (use the elementary Lemma 3.2 of [24]). From the other side,

$$s_{GL}(\pi_-) = \delta([\nu^{-1/2}, \nu^{3/2}]) \otimes 1$$

implies that always

$$\pi_- \not\hookrightarrow \nu^{1/2} \rtimes \pi' = \nu^{(2-1)/2} \rtimes \pi'.$$

This is the reason that for $\pi \in \{\pi_\pm\}$ can define

$$\epsilon_\pi((1_{F^\times}, 2)) = 1 \iff \pi \hookrightarrow \nu^{(2-1)/2} \rtimes \pi'$$

for some irreducible representation π' (of $SO(5, F)$). Now obviously

$$\epsilon_{\pi_\pm}((1_{F^\times}, 2)) = \pm 1 \tag{31.1}$$

(i.e. $\epsilon_{\pi_+}((1_{F^\times}, 2)) = 1$ and $\epsilon_{\pi_-}((1_{F^\times}, 2)) = -1$). Next question is what is $\epsilon_{\pi_\pm}((1_{F^\times}, 4))$. One can get answer to that question from a general principle covering the case when square integrable representations are obtained in the above way. Before explaining that principle, let us observe that

$$\pi_\pm \hookrightarrow \delta([\nu^{-1/2}, \nu^{3/2}]) \rtimes 1 \hookrightarrow \nu^{3/2} \times \nu^{1/2} \times \nu^{-1/2} \rtimes 1,$$

which implies (again using Lemma 3.2 of [24]) that for some irreducible π' we have

$$\pi_\pm \hookrightarrow \nu^{3/2} \rtimes \pi' = \delta([\nu^{(2+1)/2}, \nu^{(4-1)/2}]) \rtimes \pi'.$$

Now we present definition of the partially defined function attached to a general square integrable representation.

Definition. *Let π be an irreducible square integrable representation of S_q.*

(1) Suppose that $(\rho, a), (\rho, c) \in \mathrm{Jord}(\pi)$ are such that $c < a$, $\mathrm{Jord}_\rho(\pi) \cap \{x; c < x < a\} = \emptyset$. Then we shall denote c by

$$a_-.$$

(2) For $(\rho, b) \in \mathrm{Jord}(\pi)$ the partially defined function may or may not be defined on $(\rho, b) \in \mathrm{Jord}(\pi)$. If it is defined, it takes values in $\{\pm 1\}$. For a, a_- as in (1), $\epsilon_\pi((\rho, a)(\rho, a_-)^{-1})$ is always defined, and it must be equal to

$$\epsilon_\pi((\rho, a))\epsilon_\pi((\rho, a_-))^{-1}$$

if $\epsilon_\pi((\rho, a))$ *and* $\epsilon_\pi((\rho, a_-))$ *are defined. If* $\epsilon_\pi((\rho, a))$ *and* $\epsilon_\pi((\rho, a_-))$ *are not defined, we shall nevertheless write* $\epsilon_\pi((\rho, a)(\rho, a_-)^{-1})$ *formally as* $\epsilon_\pi((\rho, a))\epsilon_\pi((\rho, a_-))^{-1}$. *This is an element of* $\{\pm 1\}$, *and it is 1 if and only if*

$$\pi \hookrightarrow \delta([\nu^{(a_-+1)/2}\rho, \nu^{(a-1)/2}\rho]) \rtimes \pi'.$$

for some irreducible representation π' *of some* $S_{q'}, q' < q$.

(3) *For* $(\rho, a) \in \mathrm{Jord}(\pi)$, $\epsilon_\pi((\rho, a))$ *is always defined if* a *is even. Using (2), we can compute* $\epsilon_\pi((\rho, a))$ *if we know* $\epsilon_\pi((\rho, a_{\min}))$, *where* $a_{\min} =$ *min*$(\mathrm{Jord}_\rho(\pi))$. *One defines*

$$\epsilon_\pi((\rho, a_{\min})) \quad \textit{to be} \quad 1$$

if and only if

$$\pi \hookrightarrow \delta([\nu^{1/2}\rho, \nu^{(a_{\min}-1)/2}\rho]) \rtimes \pi'$$

for some irreducible representation π' *of some* $S_{q'}, q' < q$.

(4) *For* $(\rho, a) \in \mathrm{Jord}(\pi)$, $\epsilon_\pi((\rho, a))$ *is not defined if* a *is odd and if*

$$\mathrm{Jord}_\rho(\pi_{\mathrm{cusp}}) \neq \emptyset.$$

(5) *For* $(\rho, a) \in \mathrm{Jord}(\pi)$, $\epsilon_\pi((\rho, a))$ *is defined if* a *is odd and*

$$\mathrm{Jord}_\rho(\pi_{\mathrm{cusp}}) = \emptyset.$$

Note that in that case $\rho \rtimes \pi_{\mathrm{cusp}}$ *reduces into two non-equivalent irreducible representations. One makes the choice of the irreducible component of* $\rho \rtimes \pi_{\mathrm{cusp}}$ *to which 1 is attached (−1 is attached to the other one), i.e.*

$$\rho \rtimes \pi_{\mathrm{cusp}} = \tau_1 \oplus \tau_{-1}.$$

Then C. Mœglin used normalized intertwining operators to define $\epsilon_\pi((\rho, a))$. *We can define partially defined function in a different way, which we now briefly explain. Using (2), we are able to compute* $\epsilon_\pi((\rho, a))$ *if we know* $\epsilon_\pi((\rho, a_{\max}))$, *where* $a_{\max} =$ *max*$(\mathrm{Jord}_\rho(\pi))$. *We define*

$$\epsilon_\pi((\rho, a_{\max})) \quad \textit{to be} \quad s \in \{\pm 1\}$$

if and only if

$$\pi \hookrightarrow \pi' \times \delta([\nu\rho, \nu^{(a_{\max}-1)/2}\rho]) \rtimes \tau_s$$

for some irreducible representation π' *of a general linear group (see [43] for more details).*

Remarks. *(1) In our case (for square integrable representations defined in section 30.) we have*

$$\epsilon_{\pi_\pm}((1_{F^\times},2))=\epsilon_{\pi_\pm}((1_{F^\times},4))=\pm 1$$

(see (31.1)).

(2) (Add (2) of the previous definition) At this point, it would be interesting to have example when $\epsilon_\pi((\rho,a))\epsilon_\pi((\rho,a_-))^{-1}=-1$. *The first example (in the even parity) not involving cuspidal cases, occurs for* $SO(13)$. *One considers*

$$\delta([\nu^{-1/2},\nu^{3/2}])\rtimes\delta([\nu^{1/2},\nu^{5/2}];1)=\delta([\nu^{-1/2},\nu^{3/2}])\rtimes St_{SO(7)}.$$

This representation has two irreducible subrepresentations. Both are square integrable. Their Jordan blocks are

$$\{(1_{F^\times},2),(1_{F^\times},4),(1_{F^\times},6)\}.$$

One of these subrepresentations contains

$$\delta([\nu^{1/2},\nu^{5/2}])\times\delta([\nu^{1/2},\nu^{3/2}])\times\nu^{1/2}\otimes 1$$

in its Jacquet module. This subrepresentation has partially defined function identically equal to 1. The partially defined function of the other subrepresentation is 1 at $(1_{F^\times},6)$, *and otherwise −1. So 4 and 6 are neighbors in the Jordan blocks along* $1_{F^\times}$, *and partially defined function takes different values on them.*

(3) (Add (3) of the previous definition) We have seen examples (in the even parity) of different $\epsilon_\pi((\rho,a_{\min}))$*'s (the representations* $\pi_\pm$ *from section 30.).*

(4) Look now at the symplectic groups. We have the following simplest example regarding (4) of the above definition.
The representation $\delta([\nu^{-1},\nu^2])\rtimes 1$ *has two irreducible subrepresentations. Denote them by* $\pi_\pm$. *Both are square integrable. We get from the proposition of the section 30. that their Jordan blocks are*

$$\{(1_{F^\times},1),(1_{F^\times},3),(1_{F^\times},5)\}.$$

Their partially defined functions are (up to an order)

$$\epsilon_{\pi_+}((1_{F^\times},3))\epsilon_{\pi_+}((1_{F^\times},1))^{-1}=1,\epsilon_{\pi_+}((1_{F^\times},5))\epsilon_{\pi_+}((1_{F^\times},3))^{-1}=1;$$

$$\epsilon_{\pi_-}((1_{F^\times},3))\epsilon_{\pi_-}((1_{F^\times},1))^{-1}=-1,\epsilon_{\pi_-}((1_{F^\times},5))\epsilon_{\pi_-}((1_{F^\times},3))^{-1}=1.$$

Here, in the full principal series, we have one more irreducible square integrable subquotient (the third one). Now we shall describe it.

The representation $\delta([\nu^0,\nu]) \rtimes \delta([\nu,\nu^2];1)$ has two irreducible subrepresentations. Both are square integrable and one of them is π_+. Denote the remaining one by π_3. Its Jordan blocks are again

$$\{(1_{F^\times},1),(1_{F^\times},3),(1_{F^\times},5)\}.$$

Its partially defined function is

$$\epsilon_{\pi_3}((1_{F^\times},3))\epsilon_{\pi_+}((1_{F^\times},1))^{-1}=1, \epsilon_{\pi_+}((1_{F^\times},5))\epsilon_{\pi_+}((1_{F^\times},3))^{-1}=-1.$$

Up to now, we have constructed 3 representations with the same Jordan blocks. Actually, the classification tells that the representation which would have both values -1, must be strongly positive (see below for the definition of this type of representations).

In the following section, we shall discuss square integrable representations which are the first step in construction of the general square integrable representations from the cuspidal ones.

32. Some examples of strongly positive representations

The criterion of W. Casselman for square integrability tells us that certain sums of exponents must be positive, but all the exponents $e(\rho_i)$ considered in the criterion do not need to be positive. For example, representations in sections 29. and 30. are square integrable, but all their exponents are not positive. From the other side, for the Steinberg representations, not only the sums are positive, but each term of each sum is positive. Such representations, for which all $e(\rho_i)$ in the criterion for square integrability are strictly positive, will be called **strongly positive representations**. They are obviously square integrable, and they are starting building blocks of general square integrable representations.

Examples of such representations are, besides generalized Steinberg representations, irreducible cuspidal representations (which can be viewed as a trivial case of generalized Steinberg representations). Combining several generalized Steinberg representations, one can get a new (just a little bit) more general examples of strongly positive representations (examples of such representations are representations $\delta(\nu^{1/2}\chi_1,\nu^{1/2}\chi_2;1)$ from section 28.; see [40] for more such regular examples).

In general, more characteristic examples of these type of representations have shown up in representation theory of p-adic groups pretty late. To get an idea about these (little bit unusual) representations, it is good to see some characteristic examples.

Before we give the examples, let us explain how one can characterize these representations in terms of admissible triples. To an irreducible square integrable representation π, we have up to now attached $\mathrm{Jord}(\pi)$, π_{cusp} and partially defined function ϵ_π, which may not be completely defined, but the differences $\epsilon_\pi((\rho,a))\epsilon_\pi((\rho,a_-))^{-1}$ are always defined. This is what we need. We shall say that a triple $(\mathrm{Jord}(\pi),\epsilon_\pi,\pi_{\mathrm{cusp}})$ is **alternated** if

$$\epsilon_\pi((\rho,a))\epsilon_\pi((\rho,a_-))^{-1}=-1$$

for all exiting (neighbor) pairs $(\rho,a),(\rho,a_-)$ in $\mathrm{Jord}(\pi)$.

Mœglin has proved in [21] that an irreducible square integrable representation is strongly positive if and only if the attached triple is alternated.

Observe that irreducible cuspidal representations obviously have alternated triples, since they can not be subquotients in representations parabolically induced from any proper parabolic subgroup. Also, generalized Steinberg representations (and their "combinations") have alternated triples for obvious reason: their $\mathrm{Jord}_\rho(\pi)$ has at most one element (so the condition being alternated is automatically satisfied).

Really interesting examples here are the cases where condition of being alternated is not satisfied for such obvious reasons. For this, we need to have a reducibility bigger then 1. Therefore, the involved representation can not be generic. For a long time, we did not know such examples (we knew examples of non-generic representations, but we did not know reducibility points). This explains why such representations has shown relatively lately in the representation theory.

C. Mœglin has constructed an example of (non-generic) irreducible cuspidal representation σ of $Sp(8,F)$ such that $\nu^3\rtimes\sigma$ reduces. Now the theorem from section 20. implies that $\mathrm{Jord}_{1_{F^\times}}(\sigma)$ is $\{1,3,5\}$.

Remarks. *(1) C. Mœglin has explained us how she got the example, using the Howe correspondence (Waldspurger published in 1980'es an example [44] of such use of Howe correspondence for determining reducibility points for $GSp(4)$). Later on, we shall also describe examples of reducibilities >1 obtained by M. Reeder.*

(2) As we noted already, Jordan blocks of σ (as above) along the trivial character of $F^\times$ consist exactly of 1, 3 and 5. Since the sum is 9, we conclude from the estimate (34.2) (which we shall introduce later), that these are all the Jordan blocks of σ. Therefore, this is the representation missing in (4) of Remarks in section 31.

Now we can explain how should look like all the other cuspidal reducibilities of an irreducible selfdual cuspidal representation ρ of a general *linear group, with σ. The above fact, describing Jordan blocks of σ, tells that each other (integral and half-integral) reducibility is either 0 or 1/2. Moreover, basic assumption states that it should hold*

$$\alpha_{\rho,\sigma} - \alpha_{\rho,1} \in \mathbb{Z}.$$

Therefore, for other ρ's, $\alpha_{\rho,\sigma}$ is completely determined by $\alpha_{\rho,1}$ (so it depends only on ρ, not on σ).

(3) Suppose that the cuspidal reducibility $\alpha_{\rho,1}$ is

$$1/2.$$

Then for any irreducible cuspidal representation σ' the reducibility should be some exponent in $1/2 + \mathbb{Z}_{\geq 0}$. Therefore, the Jordan blocks along ρ should be even, i.e. of the form

$$(\rho, 2k).$$

Recall that even dimensional algebraic representations of $SL(2,\mathbb{C})$ are symplectic (while odd-dimensional are orthogonal). In the local Langlands correspondence for $Sp(2n, F)$ we need to get orthogonal representation, which should be of the form

$$\varphi(\rho) \otimes E_{2k},$$

where E_l denotes the l-dimensional irreducible algebraic representation of $SL(2,\mathbb{C})$. Therefore, in the local Langlands correspondence for general linear groups $\varphi(\rho)$ should be symplectic representation.
Because of that, if a representation $\nu^{1/2}\rho \rtimes 1$ of the symplectic group reduces, we say that ρ is a symplectic representation.
For a similar reason, if a representation $\nu^{1/2}\rho \rtimes 1$ of the symplectic group is irreducible, we say that ρ is an orthogonal representation.
F. Shahidi has shown that for selfdual irreducible cuspidal representation ρ of a general linear group, nonequivalent to $1_{F^\times}$, the reducibility in symplectic and orthogonal case of $\rho \rtimes 1$ is always in $\{0, 1/2\}$, and that it is irreducible in one case if and only if it is reducible in the other one (recall $\rho \not\cong 1_{F^\times}$). So orthogonality excludes being symplectic and conversely.

(4) The basic assumption and the theorem from section 20 imply that if (ρ, k) is a Jordan block of an irreducible cuspidal representation of a classical group, where $k \geq 3$, then $(\rho, k-2)$ is always a Jordan block of the same (cuspidal) representation.

Let us return to Mœglin's example, where $\nu^3 \rtimes \sigma$ reduces. We obtain a series of generalized Steinberg representations

$$\delta([\nu^3, \nu^k]; \sigma)$$

for $k \geq 3$ (we can formally say that $k = 2$ produces σ itself). All their Jacquet modules are essentially square integrable. For example, Siegel-like Jacquet module is

$$\delta([\nu^3, \nu^k]) \otimes \sigma.$$

Here $\mathrm{Jord}_{1_{F^\times}}$ is $\{1, 3, 2k+1\}$.

In particular, $\mathrm{Jord}_{1_{F^\times}}(\delta(\nu^3; \sigma)) = \{1, 3, 7\}$.

Looking now at the regular representation $\nu^2 \times \nu^3 \rtimes \sigma$ and its subrepresentations

$$\nu^2 \rtimes \delta(\nu^3; \sigma) \quad \text{and} \quad L(\nu^2, \nu^3) \rtimes \sigma,$$

we directly get that there exists a unique common subrepresentation π. This subrepresentation satisfies

$$\mu^*(\pi) = 1 \otimes \pi + \nu^2 \otimes \delta(\nu^3; \sigma) + L(\nu^2, \nu^3) \otimes \sigma.$$

From this follows that π is square integrable. Further analysis, using the proposition from section 30., gives

$$\mathrm{Jord}_{1_{F^\times}}(\pi) = \{1, 5, 7\},$$

and from the Jacquet modules follows that ϵ_π on the (difference of the) pair 5, 7 takes value -1 (otherwise, we would have a representation of the form $\nu^3 \otimes \tau$ in that Jacquet module, which is impossible). This representation will be denoted by

$$\delta([\nu^2, \nu^3]; \sigma).$$

Analogously, one can define representation

$$\delta([\nu, \nu^3]; \sigma)$$

which has

$$L(\nu, \nu^2, \nu^3) \otimes \sigma$$

in its Jacquet module, and for which the Jordan blocks along $1_{F^\times}$ are $\{3, 5, 7\}$.

"Combining" $\nu^4 \rtimes \delta([\nu, \nu^3]; \sigma)$ and $L([\nu, \nu^2]) \rtimes \delta([\nu^3, \nu^4]; \sigma)$ we get an irreducible (square integrable) subrepresentation, denoted by

$$\delta([\nu, \nu^4]; \sigma),$$

which has

$$L(\nu, \nu^2, \delta([\nu^3, \nu^4])) \otimes \sigma$$

in its Jacquet module, and for which Jordan blocks along $1_{F^\times}$ are $\{3, 5, 9\}$.

A more complicated task is to show that $\nu^3 \rtimes \delta([\nu, \nu^4]; \sigma)$ has an irreducible square integrable subquotient. Its Jordan blocks along $1_{F^\times}$ are $\{3, 7, 9\}$.

All above representations are strongly positive.

At this point, we have enough different examples to be able to get the idea of basic phenomena which show up in classification modulo cuspidal data of general square integrable representations. Instead of giving all details in the general case, which is pretty complicated, we shall describe the classification here on some interesting examples.

First we shall give a reduction to the

33. Cuspidal lines

Let ρ be an irreducible unitarizable cuspidal representation of a general linear group, and let

$$\mathrm{Irr}^{GL}_{\nu^{\mathbb{R}}\rho}$$

be the set of all irreducible subquotients of representations of the form

$$\nu^{e_1}\rho \times \cdots \times \nu^{e_k}\rho, \quad e_i \in \mathbb{R},\ k \geq 1.$$

Then each irreducible representation π of a general linear group can be decomposed

$$\pi \cong \underset{i=1}{\overset{l}{\times}} \pi(\rho_i)$$

where $\pi(\rho_i) \in \mathrm{Irr}^{GL}_{\nu^{\mathbb{R}}\rho_i}$, for some finite set of non-equivalent irreducible unitarizable cuspidal representations ρ_i of general linear groups. The above decomposition is unique up to a permutation. This decomposition, established by A.V. Zelevinsky, enables one to work in cuspidal lines $\mathrm{Irr}^{GL}_{\nu^{\mathbb{R}}\rho}$. Work with irreducible representations in particular cuspidal line is essentially independent of the corresponding cuspidal representation ρ.

We can make similar decomposition for classical groups S_n, but one needs to involve not only parabolic induction, but also Jacquet modules. We shall not go into details here, which appear to be rather elementary in the setting related to the square integrability (see [42]). This decomposition was done in full generality by C. Jantzen ([15]).

The main work which one needs to do in construction of irreducible square integrable representations for groups S_n is in the cuspidal lines. Therefore, we shall concentrate to a cuspidal line

$$\mathrm{Irr}^S_{\nu^{\mathbb{R}}\rho,\sigma},$$

which is the set of all irreducible subquotients of representations of the form

$$\nu^{e_1}\rho \times \cdots \times \nu^{e_k}\rho \rtimes \sigma, \quad e_i \in \mathbb{R},\ k \geq 0$$

(ρ will be assumed to be cuspidal and selfdual; this is enough to get all the square integrable representations). Square integrable classes in $\mathrm{Irr}^S_{\nu^{\mathbb{R}}\rho,\sigma}$ will be denoted by

$$\mathcal{D}(\rho;\sigma).$$

Similarly as in the case of general linear groups (where the situation in cuspidal lines was independent of ρ), here the construction depends on the number

$$\alpha_{\rho,\sigma},$$

but essentially not on ρ and σ.

Now we shall go back to

34. General strongly positive representations

Instead of giving a complete description of irreducible strongly positive representations , we shall follow the example

$$\alpha_{\rho,\sigma} = 3.$$

Note that in all examples of strongly positive square integrable representations in section 32., Jordan blocks had 3 elements. Actually, this holds in general. To any set $\{2k_1+1, 2k_2+1, 2k_3+1\}$ of 3 odd positive integers (we assume $0 \leq k_1 < k_2 < k_3$), one attaches a representation: the unique irreducible subrepresentation of

$$\Pi := \delta([\nu\rho, \nu^{k_1}\rho]) \times \delta([\nu^2\rho, \nu^{k_2}\rho]) \times \delta([\nu^3\rho, \nu^{k_3}\rho]) \rtimes \sigma \qquad (34.1)$$

(observe that if we chose $k_i = i-1$, then factors corresponding to $j \leq i$ do not show up effectively). This subrepresentation is strongly positive, and the Jordan blocks along ρ are $\{2k_1+1, 2k_2+1, 2k_3+1\}$. In this way one gets all strongly positive representations in $\mathcal{D}(\rho;\sigma)$. Now we can easily get previous examples specifying particular choices of k_i's.

The description for any other positive integral reducibility $\alpha_{\rho,\sigma}$ is analogous to the above one (in the product Π we have $\alpha_{\rho,\sigma}$ GL-factors, instead of

3). Note that for $\alpha_{\rho,\sigma} = 1$, each strongly positive representation in $\mathcal{D}(\rho;\sigma)$ is a generalized Steinberg representations.

For $\alpha_{\rho,\sigma} = 0$, σ is the only strongly positive representation in $\mathcal{D}(\rho;\sigma)$.

Now we shall explain situation in the case of half-integral reducibility.

Similar as above, the generalized Steinberg representations are the only strongly positive representations in $\mathcal{D}(\rho;\sigma)$ for $\alpha_{\rho,\sigma} = 1/2$.

Consider now the case $\alpha_{\rho,\sigma} = 5/2$ (this reducibility shows up in an example of M. Reeder, in the case of irreducible degenerate unipotent cuspidal representations; see the remarks below). The theorem in section 20. implies that the Jordan blocks of σ along ρ are

$$\mathrm{Jord}_\rho = \{2, 4\}.$$

One has generalized Steinberg representations $\delta([\nu^{5/2}\rho, \nu^{5/2+k}];\sigma)$, $k \geq -1$ (which are strongly positive), for which Jordan blocks along ρ consist of $\{2, 2k+6\}$. In particular, for $\delta(\nu^{5/2}\rho;\sigma)$ Jordan blocks along ρ consists of $\{2, 6\}$. From the definition of the partially defined function, it is clear that its value is -1 on $(\rho, 2)$.

Similarly as above, we have a unique irreducible (square integrable) subrepresentation of

$$\nu^{3/2}\rho \rtimes \delta(\nu^{5/2};\sigma),$$

denoted by

$$\delta([\nu^{3/2}\rho, \nu^{5/2}];\sigma).$$

Here Jordan blocks along ρ consists of $\{4, 6\}$. From the definition of the partially defined function, it is clear that its value is -1 on $(\rho, 4)$.

Continuing with $\nu^{1/2}\rho \rtimes \delta([\nu^{3/2}\rho, \nu^{5/2}];\sigma)$, and its unique irreducible subrepresentation denoted by $\delta([\nu^{1/2}\rho, \nu^{5/2}];\sigma)$ (which is square integrable), we would get that Jordan blocks along ρ are $\{2, 4, 6\}$. From the definition of the partially defined function, it is clear that its value is 1 on $(\rho, 2)$.

Now we can say how strongly positive representations in $\mathcal{D}(\rho;\sigma)$ look like for this reducibility (i.e. at 5/2). Here Jordan blocks along ρ can have 2 or 3 elements.

Consider first the case when the Jordan blocks along ρ have two elements $\{2k_1, 2k_2\}$, $0 < k_1 < k_2$. Then corresponding strongly positive representation is the unique irreducible subrepresentation of

$$\delta([\nu^{3/2}\rho, \nu^{(2k_1-1)/2}]) \times \delta([\nu^{5/2}\rho, \nu^{(2k_2-1)/2}]) \rtimes \sigma.$$

In this case, the partially defined function takes the value -1 on $(\rho, 2k_1)$.

Consider now the case when the Jordan blocks along ρ have three elements $\{2k_1, 2k_2, 2k_3\}$, $0 < k_1 < k_2 < k_3$. Then corresponding strongly positive representation is unique irreducible subrepresentation of

$$\delta([\nu^{1/2}\rho, \nu^{(2k_1-1)/2}]) \times \delta([\nu^{3/2}\rho, \nu^{(2k_2-1)/2}]) \times \delta([\nu^{5/2}\rho, \nu^{(2k_3-1)/2}]) \rtimes \sigma.$$

In this case partially defined function is 1 on $(\rho, 2k_1)$.

This is general way how one constructs strongly positive representations in the half-integral case. In general, for a half-integral reducibility $\alpha_{\rho,\sigma}$, the Jordan blocks will have $\alpha_{\rho,\sigma} \pm 1/2$ elements. The corresponding partially defined function is 1 on the minimal element if and only if the Jordan blocks have $\alpha_{\rho,\sigma} + 1/2$ elements.

"Control" of π (or Jord(π) and ϵ_π) over π_{cusp} roughly corresponds to the condition related to the Jordan blocks of a strongly positive representation coming with π, and from the Jordan blocks of π_{cusp} along each ρ (roughly, in the case of integral reducibility they should have cardinality equal to the value of the reducibility point, while in the half-integral case the cardinality needs to be equal to the value of the reducibility point $\pm 1/2$; in the half-integral case there is additional condition on partially defined function depending on sign of $1/2$). We shall not go here in more details (which one can find in [21], or [24], or [42]).

Remarks. *(1) M. Reeder proved that there exists the following reducibility: For $k \geq 1$, denote $n = k^2 + k + 1$. He has shown, using computation of Plancherel measures, that for certain irreducible unipotent degenerate cuspidal representations $\theta(k)$, the induced representation*

$$\nu^{k+1/2} \rtimes \theta(k)$$

reduces. Further, $\theta(1)$ is a representation of $SO(5, F)$, $\theta(2)$ is a representation of $SO(13, F)$, $\theta(3)$ is a representation of $SO(26, F)$.

(2) C. Mœglin has proved in [22] that

$$\sum_{(\rho,k)\in Jord(\pi)} k d_\rho \leq 2n \ (resp. \ \leq 2n+1) \tag{34.2}$$

holds if π is a square integrable representation of $SO(2n+1)$ (resp. $Sp(2n)$), where

$$d_\rho$$

is defined with requirement that ρ is a representation of $GL(d_\rho, F)$. It is expected that in (34.2) holds the equality.

(3) *For a generic square integrable representation π, Theorem 8.1 of [5] implies the equality in (34.1). Note that equality in (34.2) is equivalent to the fact that the expected admissible homomorphism (28.1) attached to π goes to the dual group of expected rank.*

(4) *Suppose that σ is an irreducible cuspidal representation of $SO(2n+1, F)$ (resp. $Sp(2n, F)$). Let X be the set of all non-equivalent selfdual irreducible cuspidal representations ρ of general linear groups $GL(d_\rho, F)$ such that*

$$\alpha_{\rho,\sigma} \geq 1$$

(and $\alpha_{\rho,\sigma} \in (1/2)\mathbb{Z}$). Then the theorem in section 20. and (34.2) imply

$$\sum_{\rho\in X,\, \alpha_{\rho,\sigma}\in\mathbb{Z}} \alpha^2_{\rho,\sigma} d_\rho + \sum_{\rho\in X,\, \alpha_{\rho,\sigma}\in(1/2)\mathbb{Z}} (\alpha^2_{\rho,\sigma} - 1/4) d_\rho \leq 2n \ (resp.\ 2n+1).$$

For generic (and cuspidal) σ, we have

$$\sum_{\rho\in X} d_\rho = 2n \ (resp.\ 2n+1).$$

35. The general step

We shall explain a general principle which is used to construct all the square integrable representations, starting from the strongly positive ones.

Theorem. *Let π be an irreducible square integrable representation of S_q, ρ an irreducible cuspidal selfdual representation of $GL(p, F)$, and $a, a_- \in \mathbb{Z}_{\geq 1}$ such that $a - a_- \in 2\mathbb{Z}_{\geq 1}$. We shall assume that (ρ, a) satisfies the parity condition: a is even $\iff$ $\nu^{1/2}\rho \rtimes 1_{S_0}$ reduces (in the series of groups that we consider). Further we assume*

$$\text{Jord}_\rho(\pi) \cap \{x; a_- \leq x \leq a\} = \emptyset.$$

Then the representation

$$\delta([\nu^{-(a_- -1)/2}\rho, \nu^{(a-1)/2}\rho]) \rtimes \pi \tag{35.1}$$

contains exactly two irreducible subrepresentations. They are not equivalent.

Further, we can decompose

$$\delta([\nu^{-(a_- -1)/2}\rho, \nu^{(a_- -1)/2}\rho]) \rtimes \pi = T_1 \oplus T_{-1}$$

as a sum of two nonequivalent irreducible (tempered) representations. Then

$$\delta([\nu^{(a_- -1)/2+1}\rho, \nu^{(a-1)/2}\rho]) \rtimes T_\eta$$

contains a unique irreducible subrepresentation, which we denote by π_η. *These irreducible representations are exactly the irreducible subrepresentations of (35.1).*

Representations π_η *are square integrable,*

$$\text{Jord}(\pi_\eta) = \text{Jord}(\pi) \cup \{(\rho, a), (\rho, a_-)\},$$

and

$$\epsilon_{\pi_\eta}(\rho, a) = \epsilon_{\pi_\eta}(\rho, a_-).$$

Further, ϵ_{π_η} *restricted to* $\text{Jord}(\pi)$ *is just* ϵ_π.

Proofs of square integrability in sections 29. and 30. illustrate on simple examples the proof of above theorem (proof in general case is much more complicated).

Theorem. *Supposing that the basic assumption holds, one gets each irreducible square integrable representation in finitely many steps as described in the previous theorem, starting from an irreducible strongly positive square integrable representation. Admissible triples give one to one parameterization of irreducible square integrable representations.*

For a more precise statement and more details, one should consult [24].

We have already noted that this classification is modulo irreducible cuspidal representations and cuspidal reducibilities (i.e. it reduces irreducible square integrable representations to the irreducible cuspidal representations ρ and σ and cuspidal reducibilities $\alpha_{\rho,\sigma}$). As we have already noted, we can also expect that it will give a reduction in establishing local Langlands correspondence for classical groups to the cuspidal case: the admissible homomorphism corresponding to an irreducible square integrable representation π should be

$$\bigoplus_{(\rho,k)\in\text{Jord}(\pi)} \Phi(\rho) \otimes E_k,$$

where Φ in the above formula denotes the local Langlands correspondence for general linear groups.

From the other side, if we know which Jordan blocks are attached to an irreducible cuspidal representation σ of a classical group (which would determine the admissible homomorphism of the Weil-Deligne group corresponding to σ), then the reducibility point would be:

- $\alpha_{\rho,\sigma} = \frac{\max(\text{Jord}_\rho(\sigma))+1}{2}$ if $\text{Jord}_\rho(\sigma) \neq \emptyset$;
- $\alpha_{\rho,\sigma} = \alpha_{\rho,1}$ if $\text{Jord}_\rho(\sigma) = \emptyset$

(the second equality follows from the basic assumption).

To know the reducibility $\alpha_{\rho,1}$, it is enough to know if the local Langlands correspondence carries ρ into symplectic or orthogonal representation (see Remarks in section 32.; if an irreducible cuspidal representation goes to the symplectic group, then in the second case reducibility should be for symplectic series of groups at 1/2 and for the orthogonal series of groups at 0; if it goes to the orthogonal group, we should have converse situation).

References

1. Arthur, J. *Unipotent automorphic representations: conjectures*, Astérisque 171-172 (1989), 13 - 71.
2. Bernstein, J., *All reductive p-adic groups are tame*, Functional Anal. Appl. 8 (1974), 3-6.
3. Bernstein, J. and Zelevinsky, A.V., *Induced representations of reductive p-adic groups I* , Ann. Sci. École Norm Sup. 10 (1977), 441-472.
4. Casselman, W., *Introduction to the theory of admissible representations of p-adic reductive groups*, preprint (http://www.math.ubc.ca/ cass/research/pdf/p-adic-book.pdf).
5. Cogdell, J., Kim, H.H., Piatetski-Shapiro, I.I. and Shahidi, F., *Functoriality for the classical groups*, Publ. Math. IHES. 99 (2004), 163–233.
6. Gan, W.T. and Takeda, S., *The local Langlands conjecture for Sp(4)* , Int. Math. Res. Not. 2010, No. 15, 2987-3038.
7. Goldberg, D., *Reducibility of induced representations for* $Sp(2n)$ *and* $SO(n)$, Amer. J. Math. 116 (1994), 1101-1151.
8. Gustafson, R., *The degenerate principal series for* $Sp(2n)$, Mem. Amer. Math. Society 248 (1981), 1-81.
9. Harris, M. and Taylor, R., On the geometry and cohomology of some simple Shimura varieties , Princeton University Press, Annals of Math. Studies 151, 2001.
10. Hanzer, M. and Jantzen, C. *A method of proving non-unitarity of representations of p-adic groups II* , preprint.
11. Hanzer, M. and Tadić, M., *A method of proving non-unitarity of representations of p-adic groups I*, Math. Z. 266 (2010), 799-816.
12. Heiermann, V., *Décomposition spectrale et représentations spéciales d'un groupe réductif p-adique*, J. Inst. Math. Jussieu 3 (2004), 327–395.
13. Henniart, G., *Une preuve simple des conjectures de Langlands pour GL(n) sur un corps p-adique*, Invent. Math. 139 (2000), 439-455.
14. Jantzen, C., *Degenerate principal series for symplectic and odd-orthogonal groups*, Mem. Amer. Math. Society 590 (1996), 1-100.
15. Jantzen, C., *On supports of induced representations for symplectic and odd-orthogonal groups*, Amer. J. Math. 119 (1997), 1213-1262.
16. Jiang, D. and Soudry, D. *Local converse theorem for SO(2n+1) and applications*, Ann. of Math. 157(2003), 743–806.
17. Keys, C.D., *L-indistinguishability and R-groups for quasi-split groups:*

unitary groups in even dimension, Ann. Sci. École Norm Sup. 20 (1987), 31-64.

18. Kudla, S.S. and Rallis, S., *Ramified degenerate principal series representations for $Sp(n)$*, Israel J. Math. 78 (1992), 209-256.
19. Laumon, G., Rapoport, M. and Stuhler, U., *P-elliptic sheaves and the Langlands correspondence*, Invent. Math. 113 (1993), 217-338 .
20. Mœglin, C., *Normalisation des opérateurs d'entrelacement et réductibilité des induites de cuspidales; le cas des groupes classiques p-adiques*, Annals of Math. 151 (2000), 817-847.
21. Mœglin, C., *Sur la classification des séries discrètes des groupes classiques p-adiques: paramètres de Langlands et exhaustivité*, J. Eur. Math. Soc. 4 (2002), 143–200.
22. Mœglin, C., *Points de réducibilité pour les induites de cuspidales*, J. Algebra 268 (2003), 81–117.
23. Mœglin, C., *Classification des séries discrètes pour certains groupes classiques p-adiques*, in book Harmonic analysis, group representations, automorphic forms and invariant theory - In honour of Roger E. Howe, World Scientific Publishing Co., Singapore, 2007, 209–245.
24. Mœglin, C. and Tadić, M., *Construction of discrete series for classical p-adic groups*, J. Amer. Math. Soc. 15 (2002), 715-786.
25. Muić, G., *On generic irreducible representations of $Sp(n, F)$ and $SO(2n+1, F)$*, Glasnik Mat. 33(53) (1998), 19-31.
26. Muić, G., *Some results on square integrable representations; Irreducibility of standard representations*, Intern. Math. Research Notices No. 14 (1998), 705-726.
27. Muić, G., *Reducibility of Generalized Principal Series*, Canad. J. Math. 57 (2005), 616-647.
28. Muić, G., *Reducibility of Standard Representations*, Pacific J. Math. 222 (2005), 133–168.
29. Muić, G., *Construction of the Steinberg Type Representations for Reductive p-adic Groups*, Math. Z. 253 (2006), 635-652.
30. Reeder, M., *Hecke algebras and harmonic analysis on p-adic groups*, Amer. J. Math. 119 (1997), 225–249.
31. Shahidi, F., *A proof of Langlands conjecture on Plancherel measures; complementary series for p-adic groups*, Ann. of Math. 132 (1990), 273-330.
32. Shahidi, F., *Twisted endoscopy and reducibility of induced representations for p-adic groups*, Duke Math. J. 66 (1992), 1-41.
33. Silberger, A., *Special representations of reductive p-adic groups are not integrable*, Ann. of Math. 111 (1980), 571-587.
34. Tadić, M., *Classification of unitary representations in irreducible representations of general linear group (non-archimedean case)*, Ann. Sci. École Norm. Sup. 19 (1986), 335-382
35. Tadić, M., *Induced representations of $GL(n, A)$ for p-adic division algebras A*, J. reine angew. Math. 405 (1990), 48-77.
36. Tadić, M., *Representations of p-adic symplectic groups*, Compositio

Math. 90 (1994), 123-181.

37. Tadić, M., *Structure arising from induction and Jacquet modules of representations of classical p-adic groups* , J. Algebra 177 (1995), 1-33.
38. Tadić, M., *An external approach to unitary representations*, Bulletin Amer. Math. Soc. 28 (1993), 215-252.
39. Tadić, M., *On reducibility of parabolic induction*, Israel J. Math. 107 (1998), 29-91.
40. Tadić, M., *On regular square integrable representations of p-adic groups*, Amer. J. Math. 120 (1998), 159-210.
41. Tadić, M., *A family of square integrable representations of classical p-adic groups in the case of general half-integral reducibilities*, Glasnik Mat. 37(57) (2002), 21–57.
42. Tadić, M., *On classification of some classes of irreducible representations of classical groups*, in book Representations of real and p-adic groups, Singapore University Press and World Scientific, Singapore, 2004, 95-162.
43. Tadić, M., *On invariants of discrete series representations of classical p-adic groups*, Manuscripta Math. 136 (2011), 417-435.
44. Waldspurger, J.-L., *Un exercise sur $GSp(4, F)$ et les représentations de Weil*, Bull. Soc. Math. France 115 (1987), 35-69.
45. Waldspurger, J.-L., *La formule de Plancherel pour les groupes p-adiques*, J. Inst. Math. Jussieu 2 (2003), 235–333.
46. Zelevinsky, A. V., *Induced representations of reductive p-adic groups II. On irreducible representations of $GL(n)$*, Ann. Sci. École Norm. Sup. 13 (1980), 165-210.

A survey on Voronoï's theorem

Takao Watanabe

Department of Mathematics, Graduate School of Science, Osaka University,
Toyonaka, Osaka 560-0001, Japan
E-mail: twatanabe@math.sci.osaka-u.ac.jp

Dedicated to Professor Takayuki Oda on his 60th birthday

In the last half of 20th century, various generalizations of Hermite's constant and Voronoï's theorem were studied by many authors. In this paper, we give an account of a recent development concerning Voronoï's theorem.

Keywords: Hermite constant; eutactic form; extreme form; perfect form

Let V_n be the vector space of real $n \times n$ symmetric matrices and P_n the open cone of positive definite symmetric matrices in V_n. By $m_1(a)$, we denote the arithmetical minimum $\inf_{x\in\mathbf{Z}^n\setminus\{0\}} {}^t xax$ of $a \in P_n$. The Hermite invariant is the positive valued function γ on P_n defined by $\gamma(a) = m_1(a)/\det(a)^{1/n}$. Its maximum γ_n is called Hermite's constant. The determination of γ_n is one of main problems in lattice sphere packings or the arithmetic theory of quadratic forms. Voronoï's fundamental theorem [62] gives a characterization of local maxima of γ, i.e., which can be stated that γ attains a local maximum on $a \in P_n$ if and only if a is perfect and eutactic. In the last half of 20th century, various generalizations of Hermite's constant and Voronoï's theorem were studied by many authors. In this paper, we give an account of a recent development concerning Voronoï's theorem.

Contents

Notation. Throughout this paper, V_n denotes the vector space of real $n \times n$ symmetric matrices, P_n the open cone of positive definite symmetric matrices in V_n and P_n^{semi} the closure of P_n in V_n. The vector space V_n is equipped with the inner product $\langle a_1, a_2 \rangle = \mathrm{tr}(a_1 a_2)$ for $a_1, a_2 \in V_n$. The unimodular group $GL_n(\mathbf{Z})$ acts on V_n by $(a, g) \mapsto {}^t gag$ for $a \in V_n$ and $g \in GL_n(\mathbf{Z})$. In general, for a given ring R, the set of all $m \times n$ matrices with coefficients in R is denoted by $M_{m,n}(R)$. We write $M_n(R)$ for $M_{n,n}(R)$ and R^n for $M_{n,1}(R)$. The unit group of the matrix ring $M_n(R)$ is denoted by $GL_n(R)$. The identity matrix in $GL_n(R)$ is denoted by I_n.

A Euclidean space $\mathbf{R}^n$ is equipped with the inner product $(x, y) = {}^t xy$. For $a \in M_n(\mathbf{R})$, $\|a\|$ denotes the operator norm of a, i.e.,

$$\|a\| = \sup_{x \in \mathbf{R}^n \setminus \{0\}} \left(\frac{(ax, ax)}{(x, x)} \right)^{1/2}.$$

For a constant $c \in \mathbf{R}$, $\mathbf{R}_{>c}$ and $\mathbf{R}_{\geq c}$ stand for the open interval $(c, +\infty)$ and the closed interval $[c, +\infty)$, respectively.

1. Type one functions and Voronoï's theorem

There are several methods to prove Voronoï's theorem [62, Théorème 17], e.g., [6], [51], [54], see also [28, §29], [29, §39], [40, §3.4] and [60, §3.1.7]. Convexity of the domain P_n and the concavity of both functions m_1 and $\det^{1/n}$ play key roles in some proofs. Poor and Yuen [47] investigated a family of such kind functions as m_1 and $\det^{1/n}$. This family is called type one functions. Every type one function ϕ is completely characterized by the corresponding semikernel $K_1(\phi)$. In this section, we first discuss type one functions and semikernels, and then formulate Voronoï type theorem in terms of type one functions. The semikernel $K_1(m_1)$ associated with m_1 is called the Ryshkov polyhedron. In the second half of this section, we investigate a description of faces of $K_1(m_1)$. This gives a well-known geometric interpretation of perfection.

1.1. *Type one functions and semikernels*

Definition 1.1. A function $\phi : P_n^{\rm semi} \to \mathbf{R}_{\geq 0}$ is called a type one function if ϕ satisfies the following conditions:
(TO$_1$) $\phi(\theta a) = \theta\phi(a)$ for all $a \in P_n^{\rm semi}$ and $\theta \geq 0$,
(TO$_2$) $\phi(a_1 + a_2) \geq \phi(a_1) + \phi(a_2)$ for all $a_1, a_2 \in P_n^{\rm semi}$,
(TO$_3$) $\phi(a) > 0$ for all $a \in P_n$.
A type one function ϕ is called a type one class function if $\phi({}^tgag) = \phi(a)$ holds for all $a \in P_n^{\rm semi}$ and $g \in GL_n(\mathbf{Z})$.

Example 1.1. The trace tr and the smallest eigenvalue λ_1 are type one functions, but not type one class functions. The reduced determinant $\det^{1/n}$ and the arithmetical minimum

$$m_1(a) = \inf_{x \in \mathbf{Z} \setminus \{0\}} {}^txax$$

are type one class functions.

For a type one function ϕ, the dual type one function $\phi^\circ : P_n^{\rm semi} \to \mathbf{R}_{\geq 0}$ is defined to be

$$\phi^\circ(a) = \inf_{b \in P_n} \frac{\langle a, b\rangle}{\phi(b)}.$$

If ϕ is a type one class function, then so is ϕ°. The dual type one class function of m_1 is denoted by w_1, which is called the dyadic trace. The dual type one class function of $\det^{1/n}$ is $n \det^{1/n}$.

Any type one function is continuous on P_n, but not necessarily continuous on $P_n^{\rm semi}$. For example, w_1 is not continuous on $P_n^{\rm semi}$; however w_1 is upper semicontinuous on P_n^{semi}. Here a type one function ϕ is said to be upper semicontinuous at $a \in P_n^{\mathrm{semi}}$ if

$$\phi(a) = \limsup_{b\to a} \phi(b) = \lim_{\epsilon\downarrow 0}(\sup\{\phi(b) \;:\; \|a-b\| \leq \epsilon,\; b \in P_n^{\mathrm{semi}}\}) .$$

In general, the dual ϕ° of an arbitrary type one function ϕ is necessarily upper semicontinuous on P_n^{semi} ([58, Corollary 2.7]).

Definition 1.2. Let K be a convex subset of P_n^{semi} such that $0 \notin K$, $\mathbf{R}_{\geq 1} \cdot K = K$ and $\mathbf{R}_{>0} \cdot K \supset P_n$.
(1) K is called a kernel if K is closed in P_n^{semi}.
(2) K is called a semikernel if the following three conditions are satisfied:
(SK$_1$) $K \cap (P_n \cup \{0\})$ is closed in $P_n \cup \{0\}$,
(SK$_2$) $\{\,\theta \geq 0 \mid \theta a \in K\,\}$ is closed in $[0,\infty)$ for any $a \in K$,
(SK$_3$) $a + b \subset K$ for all $a \in K$ and $b \in P_n^{\mathrm{semi}}$.

It is easy to see that a kernel is a semikernel. The dual $K^\sqcup$ of a semikernel K is defined to be

$$K^\sqcup = \{\, a \in V_n \;:\; \langle a, b\rangle \geq 1 \text{ for all } b \in K\,\}.$$

This $K^\sqcup$ is a kernel.

There is a natural correspondence between type one functions and semikernels. For a type one function ϕ, we set

$$K_1(\phi) = \{\, a \in P_n^{\mathrm{semi}} \;:\; \phi(a) \geq 1\,\}.$$

Conversely, for a semikernel K, define the function $\psi(K,\cdot) : P_n^{\mathrm{semi}} \to \mathbf{R}_{\geq 0}$ by

$$\psi(K,a) = \max\left(\{\,\theta > 0 \;:\; a \in \theta\cdot K\,\} \cup \{0\}\right).$$

The existence of this maximum follows from the condition (SK$_2$). The following results were proved in [58, §1]

Proposition 1.1. *The correspondence $\phi \mapsto K_1(\phi)$ gives a bijection between the set of type one functions (resp. upper semicontinuous type one functions) and the set of semikernels (resp. kernels). For any type one function ϕ and any semikernel K, one has*

$$\psi(K_1(\phi),\cdot) = \phi\,, \qquad K_1(\psi(K,\cdot)) = K$$

and moreover

$$\psi(K,\cdot)^\circ = \psi(K^\sqcup,\cdot).$$

Proposition 1.2. *For any type one function ϕ, we have*

$$\begin{cases} \phi^{\circ\circ}(a) = \phi(a) & \text{if } a \in P_n \\ \phi^{\circ\circ}(a) \geq \phi(a) & \text{if } a \in P_n^{\text{semi}} \setminus P_n. \end{cases}$$

If ϕ is upper semicontinuous on P_n^{semi}, then $\phi^{\circ\circ} = \phi$ on P_n^{semi}.

1.2. *Voronoï's theorem of m_1/ϕ*

Voronoï's theorem characterizes local maxima of the Hermite invariant $F_{\det^{1/n}} = m_1/\det^{1/n}$. A point $a \in P_n$ is said to be extreme (resp. strict extreme) if $F_{\det^{1/n}}$ attains a local maximum (resp. a strict local maximum) on a up to the multiplication by an element of $\mathbf{R}_{>0}$. Indeed, we do not need to distinguish between extreme points and strictly extreme points since any extreme point is strictly extreme ([40, Theorem 3.4.5]). For $a \in P_n$, $S(a)$ denotes the set of minimal integral vectors of a, i.e.,

$$S(a) = \{\, x \in \mathbf{Z}^n \setminus \{0\} \;:\; {}^t x a x = m_1(a) \,\}.$$

For any $y \in \mathbf{R}^n$, φ_y denotes the linear form $v \mapsto {}^t y v y$ on V_n.

Definition 1.3. Let $a \in P_n$. We fix an element $b \in GL_n(\mathbf{R})$ such that $a = {}^t b b$. An element a is said to be perfect if the linear forms φ_{bx} $(x \in S(a))$ span the dual space V_n^* of V_n. An element a is said to be eutactic (resp. weakly eutactic) if there exist $\rho_x \in \mathbf{R}_{>0}$ (resp. $\rho_x \in \mathbf{R}$), $x \in S(a)$, such that

$$\mathrm{tr} = \sum_{x \in S(a)} \rho_x \varphi_{bx}. \tag{1}$$

We note that these definitions of perfection, eutaxy and weakly eutaxy are independent of a choice of b. It follows from definition that a is perfect if and only if $\{x\,{}^t x \;:\; x \in S(a)\}$ spans V_n. If tr is represented as (1), then we have

$$\mathrm{tr} = \sum_{x \in S(a)} \rho_x \varphi_{hbx}$$

for any orthogonal matrix h. The coefficients ρ_x are independent of h.

Any perfect element a is uniquely determined by $m_1(a)$ and $S(a)$, i.e., a is a unique solution of the system of linear equations in the unknown $v = {}^t v$: $\langle v, x\,{}^t x\rangle = m_1(a)$, $x \in S(a)$. If $m_1(a) \in \mathbf{Q}$, then its solution is contained in $V_n \cap M_n(\mathbf{Q})$ by Cramer's formula. This is none other than the rationality of perfect elements ([36, p.252, 5°]).

Theorem 1.1 (Korkine–Zorotareff). *If $a \in P_n$ is perfect and $m_1(a) \in \mathbf{Q}$, then $a \in P_n \cap M_n(\mathbf{Q})$.*

Voronoï's theorem [62, Théorème 17] is stated as follows.

Theorem 1.2 (Voronoï). *$a \in P_n$ is extreme if and only if a is perfect and eutactic.*

We fix a type one function ϕ. It is natural to ask whether the same kind of Voronoï's theorem holds for the function $F_\phi = m_1/\phi$ on P_n. An element $a \in P_n$ is said to be ϕ-extreme (resp. strictly ϕ-extreme) if F_ϕ attains a local maximum (resp. a strictly local maximum) on a up to the multiplication by an element of $\mathbf{R}_{>0}$. Assume ϕ is differentiable on P_n. Then

$$(\partial \log \phi)_b(v) = \lim_{t \to 0} \frac{\log \phi({}^t b(\mathrm{I}_n + tv)b) - \log \phi({}^t bb)}{t}$$

exists for $b \in GL_n(\mathbb{R})$ and $v \in V_n$. We define ϕ-eutaxy as follows:

Definition 1.4. Let $a \in P_n$, and fix an element $b \in GL_n(\mathbf{R})$ such that $a = {}^t bb$. An element a is said to be ϕ-eutactic if there exist $\rho_x > 0$ ($x \in S(a)$) such that $(\partial \log \phi)_b = \sum_{x \in S(a)} \rho_x \varphi_{bx}$.

In a similar fashion as eutaxy, this definition is independent of a choice of b. If $\phi = \det^{1/n}$, then $(\partial \log \phi)_b = \mathrm{tr}$, and hence $\det^{1/n}$-eutaxy is the same as Definition 1.3.

It follows from (TO_1) and (TO_2) that ϕ is log-concave, i.e,

$$\log \phi((1-\theta)a_1 + \theta a_2)) \geq (1-\theta) \log \phi(a_1) + \theta \log \phi(a_2)$$

holds for all $a_1, a_2 \in P_n$ and $0 < \theta < 1$. We say ϕ is strictly log-concave if this inequality is strict for $a_1 \neq a_2$.

In [58, §2], Voronoï's theorem is generalized as follows.

Theorem 1.3. *Let ϕ be a strictly log-concave and differentiable type one function. Then, $a \in P_n$ is ϕ-extreme if and only if a is perfect and ϕ-eutactic. Moreover, any ϕ-extreme point is strictly ϕ-extreme.*

The line of the proof of Theorem 1.3 is the same as Barnes' [6] and Martinet's [40, §3.4] proof of Voronoï's theorem. We give an outline of the proof. We use the following two lemmas: the first is the same as [40, Lemmas 3.4.2 and 3.4.3] and the second is a generalization of [40, Lemma 3.4.4].

Lemma 1.1. *Let $a \in P_n$, and fix an element $b \in GL_n(\mathbb{R})$ such that $a = {}^t bb$.*
(1) *There exists a neighborhood $\mathcal{U}$ of I_n in $GL_n(\mathbb{R})$ such that $S({}^t b {}^t uub) \subset$*

$S(a)$ for any $u \in \mathcal{U}$.
(2) There exists a neighborhood $\mathcal{V}$ of 0 in V_n such that

$$m_1({}^t b(\mathrm{I}_n + v)b) = m_1(a) \iff \min_{x \in S(a)} \varphi_{bx}(v) = 0$$

for any $v \in \mathcal{V}$.

Lemma 1.2. *Let ϕ be a strictly log-concave and differentiable type one function. Let $a \in P_n$, and fix an element $b \in GL_n(\mathbb{R})$ such that $a = {}^t b b$.*
(1) There exists a neighborhood $\mathcal{V} \subset V_n$ of 0 such that either $v = 0$ or $\phi({}^t b(\mathrm{I}_n + v)b) < \phi(a)$ holds for any $v \in \mathcal{V}$ with $(\partial \log \phi)_b(v) \leq 0$ and $\mathrm{I}_n + v \in P_n$.
(2) Let $\mathcal{C}$ be a closed cone in V_n such that $(\partial \log \phi)_b(v) > 0$ for all $v \in \mathcal{C} \backslash \{0\}$. Then there exists $\alpha > 0$ such that $\phi({}^t b(\mathrm{I}_n + v)b) > \phi(a)$ holds for any $v \in \mathcal{C}$ with $0 < \|v\| < \alpha$.

We set $\mathcal{D}_a = \{v \in V_n \ : \ \min_{x \in S(a)} \varphi_{bx}(v) \geq 0 \ \text{ and } \ (\partial \log \phi)_b(v) \leq 0\}$. By these lemmas, we obtain the following generalization of Korkine and Zolotareff's equivalent condition (cf. [40, Theorem 3.4.5]).

Lemma 1.3. *Let ϕ be a strictly log-concave and differentiable type one function. Then $a \in P_n$ is ϕ-extreme if and only if $\mathcal{D}_a = \{0\}$. Any ϕ-extreme point is strictly ϕ-extreme.*

Lemma 1.3 leads us to Theorem 1.3 as follows: Let $a \in P_n$ be perfect and ϕ-eutactic. Fix an element $b \in GL_n(\mathbb{R})$ such that $a = {}^t b b$. For $v \in \mathcal{D}_a$, ϕ-eutaxy concludes $\varphi_{bx}(v) = 0$ for all $x \in S(a)$, and then $v = 0$ by perfection. Thus $\mathcal{D}_a = \{0\}$ and a is ϕ-extreme. Conversely, let a be ϕ-extreme. If $\varphi_{bx}(v) = 0$ for all $x \in S(a)$, then either v or $-v$ is contained in $\mathcal{D}_a$. Since $\mathcal{D}_a = \{0\}$, we have $v = 0$. This implies that a is perfect. The linear forms $-(\partial \log \phi)_b$ and φ_{bx}, $x \in S(a)$, satisfy

$$\begin{aligned} &\{v \in V_n \ : \ \min_{x \in S(a)} \varphi_{bx}(v) \geq 0 \ \text{ and } \ -(\partial \log \phi)_b(v) \geq 0\} \\ &= \bigcap_{x \in S(a)} \mathrm{Ker}(\varphi_{bx}) \cap \mathrm{Ker}(-(\partial \log \phi)_b) = \{0\}\,. \end{aligned}$$

Then, by Stiemke's theorem, a must be ϕ-eutactic. Here Stiemke's theorem asserts that, for a family of linear forms $\varphi_1, \cdots, \varphi_r$ on $\mathbf{R}^N$, there exists $\rho_1, \cdots, \rho_r \in \mathbf{R}_{>0}$ such that $\rho_1 \varphi_1 + \cdots + \rho_r \varphi_r = 0$ if and only if $\{v \in \mathbf{R}^N \ : \ \min_{1 \leq i \leq r} \varphi_i(v) \geq 0\} = \mathrm{Ker}(\varphi_1) \cap \cdots \cap \mathrm{Ker}(\varphi_r)$.

1.3. *Geometric characterizations of perfect forms*

The kernel $K_1(m_1)$ is called the Ryshkov polyhedron. Ryshkov [54], [56, Chapter III] closely investigated polyhedral geometric structure of $K_1(m_1)$ and its dual $K_1(m_1)^{\sqcup}$. Since m_1 equals zero on the boundary of P_n^{semi}, the Ryshkov polyhedron $K_1(m_1)$ is contained in P_n. For an integral vector $x \in \mathbf{Z}^n \setminus \{0\}$ and a constant $\lambda \in \mathbf{R}$, $H_{x,\lambda}^+$ denotes the affine half-space $\{a \in V_n \ : \ \langle a, x^t x\rangle \geq \lambda\}$ in V_n. Then $K_1(m_1)$ is the intersection of affine half-spaces $H_{x,1}^+$, $(x \in \mathbf{Z}^n \setminus \{0\})$. It is known that $K_1(m_1)$ is a locally finite polyhedron, i.e., the intersection of $K_1(m_1)$ and an arbitrary polytope is a polytope, (see e.g., [28, Proposition 29.5], [60, Theorem 3.1]). In particular, $K_1(m_1) \cap \{a \in V_n \ : \ \mathrm{tr}(a) \leq \lambda\}$ is a polytope for any sufficiently large constant $\lambda > 0$. We denote by $\partial K_1(m_1)$ the boundary of $K_1(m_1)$. In what follows, we give a description of faces of $K_1(m_1)$.

Lemma 1.4. *Let $a_1, \cdots, a_r \in \partial K_1(m_1)$ and S be a non-empty finite subset of $\mathbf{Z}^n \setminus \{0\}$ such that $S \subset S(a_i)$ for $i = 1, \cdots, r$. Then, for any $\lambda_1, \cdots, \lambda_r \in \mathbf{R}_{\geq 0}$ with $\lambda_1 + \cdots + \lambda_r = 1$, one has $\lambda_1 a_1 + \cdots + \lambda_r a_r \in \partial K_1(m_1)$ and $S \subset S(\lambda_1 a_1 + \cdots + \lambda_r a_r)$.*

Proof. Since $K_1(m_1)$ is convex, $\lambda_1 a_1 + \cdots + \lambda_r a_r$ is contained in $K_1(m_1)$. If $x \in S$, then

$$\langle \sum_{i=1}^{r} \lambda_i a_i, x^t x\rangle = \sum_{i=1}^{r} \lambda_i m_1(a_i) = \sum_{i=1}^{r} \lambda_i = 1\,.$$

This means $m_1(\lambda_1 a_1 + \cdots + \lambda_r a_r) = 1$ and $S \subset S(\lambda_1 a_1 + \cdots + \lambda_r a_r)$. □

For a non-empty finite subset $S \subset \mathbf{Z}^n \setminus \{0\}$, define the subset $\mathcal{F}_S$ of $\partial K_1(m_1)$ as

$$\mathcal{F}_S = \{a \in \partial K_1(m_1) \ : \ S \subset S(a)\}\,.$$

We denote by $\mathcal{H}_S$ the affine subspace of V_n generated by $\mathcal{F}_S$, i.e.,

$$\mathcal{H}_S = \{\lambda_1 a_1 + \cdots + \lambda_r a_r \ : \ 1 \leq r \in \mathbf{Z},\ a_i \in \mathcal{F}_S,\ \lambda_i \in \mathbf{R},\ \lambda_1 + \cdots + \lambda_r = 1\}$$

if $\mathcal{F}_S \neq \emptyset$, or $\mathcal{H}_S = \emptyset$ if $\mathcal{F}_S = \emptyset$. Since S is non-empty, $\mathcal{H}_S$ is a proper affine subspace of V_n.

Lemma 1.5. *One has $\mathcal{F}_S = \partial K_1(m_1) \cap \mathcal{H}_S$. In particular, $\mathcal{F}_S$ is a face of $K_1(m_1)$ if $\mathcal{F}_S \neq \emptyset$.*

Proof. We assume $\mathcal{F}_S \neq \emptyset$ and fix an $a_0 \in \mathcal{F}_S$. Let $r = \dim \mathcal{H}_S$. There exist r elements $a_1, \cdots, a_r \in \mathcal{F}_S$ such that $\{a_1 - a_0, \cdots, a_r - a_0\}$ is a basis of the subspace $\{a - a_0 \ : \ a \in \mathcal{H}_S\}$. Any element $b \in \partial K_1(m_1) \cap \mathcal{H}_S$ is represented as

$$b = a_0 + \lambda_1(a_1 - a_0) + \cdots + \lambda_r(a_r - a_0), \qquad \lambda_1, \cdots, \lambda_r \in \mathbf{R}.$$

Since $S \subset S(a_i)$ for $i = 0, 1, \cdots, r$, we have $\langle a_i - a_0, x^t x\rangle = 0$ for all $x \in S$, and hence $\langle b, x^t x\rangle = \langle a_0, x^t x\rangle = 1$ for all $x \in S$. This means $S \subset S(b)$. Therefore, $\partial K_1(m_1) \cap \mathcal{H}_S$ is a subset of $\mathcal{F}_S$. □

Lemma 1.6. *Any face of $K_1(m_1)$ is of the form $\mathcal{F}_S$ for some non-empty finite subset $S \subset \mathbf{Z}^n \setminus \{0\}$.*

Proof. Let $\mathcal{F}$ be a face of $K_1(m_1)$ of dimension r. First, we assume $\mathcal{F}$ is a facet, i.e., $r = \dim V_n - 1$. There exist $r+1$ elements $a_0, a_1, \cdots, a_r \in \mathcal{F}$ such that $a_1 - a_0, \cdots, a_r - a_0$ are linearly independent. We fix constants $\lambda_0, \lambda_1, \cdots, \lambda_r$ such that $\lambda_0 + \lambda_1 + \cdots + \lambda_r = 1$ and $0 < \lambda_i < 1$ for $i = 0, 1, \cdots, r$. Since $\mathcal{F}$ is convex, the element $a = \lambda_0 a_0 + \lambda_1 a_1 + \cdots + \lambda_r a_r$ is also contained in $\mathcal{F}$. For any $x \in S(a)$, we have

$$1 = m_1(a) = \langle a, x^t x\rangle = \sum_{i=0}^{r} \lambda_i \langle a_i, x^t x\rangle \geq \sum_{i=0}^{r} \lambda_i = 1 .$$

Thus, $\langle a_i, x^t x\rangle$ equals $m_1(a_i) = 1$ for all $i = 0, 1, \cdots, r$. This implies $S(a) \subset S(a_i)$ for all $i = 0, 1, \cdots, r$. Let S be the intersection of $S(a_i)$, $i = 0, 1, \cdots, r$. Since $S \subset S(a)$ is obvious, one has $S = S(a)$. By definition, the face $\mathcal{F}_S$ contains $a_0, a_1, \cdots, a_r$. Therefore, $\partial K_1(m_1) \cap \mathcal{H}_S = \mathcal{F}_S$ contains $\mathcal{F}$. Since $\mathcal{H}_S$ is a proper affine subspace and $\mathcal{F}$ is a facet, we obtain $\mathcal{F}_S = \mathcal{F}$.

In general case, $\mathcal{F}$ is an intersection of finite number of facets, say $\mathcal{F}_{S_1}, \cdots, \mathcal{F}_{S_k}$. By definition, we have

$$\mathcal{F} = \bigcap_{i=1}^{k} \mathcal{F}_{S_i} = \mathcal{F}_{S_1 \cup \cdots \cup S_k} .$$

□

We denote by $\partial^0 K_1(m_1)$ the set of all vertices ($=$ 0 dimensional faces) of $K_1(m_1)$. The next theorem is well-known.

Theorem 1.4. *For $a \in \partial K_1(m_1)$, the following three conditions are equivalent each other.*
(1) a is perfect.
(2) $a \in \partial^0 K_1(m_1)$.

(3) *There exists a neighbourhood* $\mathcal{O}$ *of* a *in* P_n *such that* $S(b) \subsetneqq S(a)$ *for any* $b \in \mathcal{O} \setminus \mathbf{R}_{>0}a$.

Proof. First we show the contraposition of (1) $\Longrightarrow$ (2). Let $a \in \partial K_1(m_1) \setminus \partial^0 K_1(m_1)$. Then, there exist $a_1, a_2 \in \partial K_1(m_1)$ and $0 < \lambda_0 < 1$ such that $a = \lambda_0 a_1 + (1 - \lambda_0) a_2$. Both $S(a_1)$ and $S(a_2)$ contain $S(a)$. Assume a is perfect. By Lemma 1.4, $S(\lambda a_1 + (1-\lambda) a_2)$ also contains $S(a)$ for all positive $\lambda < 1$. Therefore, $\lambda a_1 + (1 - \lambda) a_2$ is perfect. This contradicts Theorem 1.1.

Next we show (2) $\Longrightarrow$ (3). Let $a \in \partial^0 K_1(m_1)$. By Lemma 1.6, there exists a finite subset $S \subset \mathbf{Z}^n \setminus \{0\}$ such that $\{a\} = \mathcal{F}_S$. By Lemma 1.1, there is a neighbourhood $\mathcal{O}$ of a in P_n such that $S(b) \subset S(a)$ for all $b \in \mathcal{O}$. If $b \in \partial K_1(m_1) \cap \mathcal{O}$ and $S(b) = S(a)$, then we have $b \in \mathcal{F}_S$, and hence $b = a$. This $\mathcal{O}$ satisfies (3).

We show the contraposition of (3) $\Longrightarrow$ (1). Let $a \in \partial K_1(m_1)$ be a non-perfect point. Then, there exists $c \in V_n \setminus \{0\}$ such that $\langle c, x^t x \rangle = 0$ for all $x \in S(a)$. If $\epsilon > 0$ is sufficiently small, then $a + \epsilon c$ is contained in P_n and $S(a + \epsilon c)$ is a subset of $S(a)$. Since $\langle a + \epsilon c, x^t x \rangle = 1$ for all $x \in S(a)$, we have $m_1(a + \epsilon c) = 1$ and $S(a + \epsilon c) = S(a)$. This means that a does not satisfy the condition (3). $\square$

Corollary 1.1. *The set of all perfect elements of* P_n *coincides with* $\mathbf{R}_{>0} \cdot \partial^0 K_1(m_1)$.

In the rest of this section, we show that $K_1(m_1)$ is the convex hull of $\partial^0 K_1(m_1)$ in V_n. For $a \in \partial^0 K_1(m_1)$, we set

$$\mathcal{C}_a = \{b \in V_n \ : \ \langle b, x^t x \rangle \geq 0 \ \text{ for all } x \in S(a)\} = \bigcup_{x \in S(a)} H^+_{x^t x, 0} \, ,$$

which is a polyhedral cone in V_n of finite faces. For a non-zero $b \in \mathcal{C}_a$, the ray $\mathbf{R}_{\geq 0} \cdot b$ is called an extreme ray of $\mathcal{C}_a$ if for any $b_1, b_2 \in \mathcal{C}_a$, whenever $b = (b_1 + b_2)/2$, we must have $b_1, b_2 \in \mathbf{R}_{\geq 0} \cdot b$.

Lemma 1.7. *Let* $a \in \partial^0 K_1(m_1)$. *If* $\mathbf{R}_{\geq 0} \cdot b$ *is an extreme ray of* $\mathcal{C}_a$, *then* $b \notin P_n^{\text{semi}}$.

Proof. We prove that $b \in P_n^{\text{semi}}$ leads us to a contradiction. Since a is perfect, the set $\{x^t x \ : \ x \in S(a)\}$ spans V_n. We set

$$S' = \{x \in S(a) \ : \ \langle b, x^t x \rangle = 0\}$$

and

$$W = \{c \in V_n \ : \ \langle c, x^t x \rangle = 0 \ \text{ for all } x \in S'\} \, .$$

Since $b \neq 0$, S' is non-empty and W is a subspace of V_n containing the line $\mathbf{R} \cdot b$.

First we assume $\dim W \geq 2$. There is a $c \in W$ such that b and c are linearly independent. If we assume $b \in P_n^{\rm semi}$, then we have $\langle b, x^t x\rangle > 0$ for all $x \in S(a) \setminus S'$. Thus, for sufficiently small $\lambda > 0$, we have $\langle b \pm \lambda c, x^t x\rangle > 0$ for all $x \in S(a) \setminus S'$. From $\langle b \pm \lambda c, x^t x\rangle = 0$ for all $x \in S'$, it follows $b \pm \lambda c \in \mathcal{C}_a$. Then one has

$$b = \frac{1}{2}(b + \lambda c) + \frac{1}{2}(b - \lambda c)$$

and $b \pm \lambda c \notin \mathbf{R}_{\geq 0} \cdot b$. This is a contradiction.

Next we assume $\dim W = 1$, i.e., $W = \mathbf{R} \cdot b$. Let $N = \dim V_n$. Since the subspace spanned by $\{x^t x \,:\, x \in S'\}$ is the orthogonal complement of W, there are $N-1$ linearly independent vectors $x_1{}^t x_1, \cdots, x_{N-1}{}^t x_{N-1}$ in $\{x^t x \,:\, x \in S'\}$. By the perfection of a, there exists $x_N \in S(a) \setminus S'$ such that $x_1{}^t x_1, \cdots, x_{N-1}{}^t x_{N-1}, x_N{}^t x_N$ are linearly independent. If we assume $b \in P_n^{\rm semi}$, then there is a square root $\sqrt{b} \in P_n^{\rm semi}$ such that $(\sqrt{b})^2 = b$. For each $i = 1, \cdots, N-1$, one has

$$0 = \langle b, x_i{}^t x_i\rangle = {}^t(\sqrt{b}x_i)(\sqrt{b}x_i)\,,$$

i.e., $x_1, \cdots, x_{N-1}$ are contained in the nullspace of $\sqrt{b}$. Thus there is a non-zero $y \in \mathbf{R}^n$ such that ${}^t y x_i = 0$ for $i = 1, \cdots, N-1$. We choose a non-zero $z \in \mathbf{R}^n$ which is orthogonal to x_N. Then the non-zero symmetric matrix $(y^t z + z^t y)/2 \in V_n$ is orthogonal to $x_1{}^t x_1, \cdots, x_N{}^t x_N$. This contradicts that $x_1{}^t x_1, \cdots, x_N{}^t x_N$ spans V_n. □

Proposition 1.3. *Let L be an edge (= one dimensional face) of $K_1(m_1)$. Then there are $a_1, a_2 \in \partial^0 K_1(m_1)$ such that $L = \{\lambda a_1 + (1-\lambda)a_2 \,:\, 0 \leq \lambda \leq 1\}$.*

Proof. For a sufficiently large $\theta > 0$, we set

$$K_1(m_1)_\theta = K_1(m_1) \cap \{a \in V_n \,:\, \langle a, \mathrm{I}_n\rangle \leq \theta\} \quad \text{and} \quad L_\theta = L \cap K_1(m_1)_\theta\,.$$

Since L_θ is an edge of the polytope $K_1(m_1)_\theta$, there are vertices a_1, a_1' of $K_1(m_1)_\theta$ such that L_θ is the line joining a_1 and a_1'. Since L_θ is not contained in the affine hyperplane $\{a \in V_n \,:\, \langle a, \mathrm{I}_n\rangle = \theta\}$, at least one of a_1 and a_1' must be a vertex of $K_1(m_1)$. Let $a_1 \in \partial^0 K_1(m_1)$ and $b \in V_n$ be a direction of L. Thus, any point of L is of the form $a_1 + \lambda b$ for some $\lambda \geq 0$.

We show $\mathbf{R}_{\geq 0} \cdot b$ is an extreme ray of $\mathcal{C}_{a_1}$. There is an open interval $(0, \lambda_0)$ such that $a_1 + \lambda b \in L$ for all $\lambda \in (0, \lambda_0)$. Since $a_1 + \lambda b \in \partial K_1(m_1)$

for $\lambda \in (0, \lambda_0)$, we have $m_1(a_1 + \lambda b) = 1$ and

$$1 \leq \langle a_1 + \lambda b, x^t x \rangle = 1 + \lambda \langle b, x^t x \rangle$$

for all $x \in S(a_1)$. This means $b \in \mathcal{C}_{a_1}$. If $\mathbf{R}_{\geq 0} \cdot b$ is not an extreme ray of $\mathcal{C}_{a_1}$, then there are $b_1, b_2 \in \mathcal{C}_{a_1} \setminus \mathbf{R}_{\geq 0} \cdot b$ such that $b = (b_1 + b_2)/2$. For $i = 1, 2$ and a sufficiently small $\lambda > 0$, we have $a_1 + \lambda b_i \in P_n$ and $S(a_1 + \lambda b_i) \subset S(a_1)$. From $b_i \in \mathcal{C}_{a_1}$, it follows that for $x \in S(a_1 + \lambda b_i)$,

$$m_1(a_1 + \lambda b_i) = \langle a_1 + \lambda b_i, x^t x \rangle = 1 + \lambda \langle b_i, x^t x \rangle \geq 1 \, .$$

Namely, both $a_1 + \lambda b_1$ and $a_1 + \lambda b_2$ are contained in $K_1(m_1) \setminus L$ and $a_1 + \lambda b$ is the middle point of $a_1 + \lambda b_1$ and $a_1 + \lambda b_2$. This is impossible since L is an edge of $K_1(m_1)$. Therefore $\mathbf{R}_{\geq 0} \cdot b$ must be an extreme ray of $\mathcal{C}_a$.

Since $b \notin P_n^{\mathrm{semi}}$ by Lemma 1.7, the value $\lambda_1 = \sup\{\lambda \geq 0 \, : \, a_0 + \lambda b \in K_1(m_1)\}$ is finite. Thus L is written as $L = \{a_1 + \lambda b \, : \, 0 \leq \lambda \leq \lambda_1\}$. Finally we show the point $a_2 = a_1 + \lambda_1 b$ is a vertex of $K_1(m_1)$. If $a_2 \notin \partial^0 K_1(m_1)$, then there are $c_1, c_2 \in \partial K_1(m_1)$ such that $a_2 = (c_1 + c_2)/2$ and $\lambda c_1 + (1 - \lambda) c_2 \in \partial K_1(m_1)$ for $0 \leq \lambda \leq 1$. In this case, the triangle of vertices a_1, c_1 and c_2 is contained in $\partial K_1(m_1)$. This contradicts that L is an edge of $K_1(m_1)$. □

Corollary 1.2. *The Ryshkov domain $K_1(m_1)$ is the convex hull of $\partial^0 K_1(m_1)$.*

Proof. We fix an arbitrary $a \in K_1(m_1)$. If $\theta > \mathrm{tr}(a)$, then $a \in K_1(m_1)_\theta$. Let $\{b_1, \cdots, b_r\}$ be the set of all vertices of the polytope $K_1(m_1)_\theta$. Since $K_1(m_1)_\theta$ is the convex hull of $\{b_1, \cdots, b_r\}$, a is of the form $\lambda_1 b_1 + \cdots + \lambda_r b_r$ with $\lambda_1 + \cdots + \lambda_r = 1$ and $\lambda_i \geq 0$, $i = 1, \cdots, r$. Each b_i is either a vertex of $K_1(m_1)$ or the intersection of an edge of $K_1(m_1)$ and the affine hyperplane $\{c \in V_n \, : \, \mathrm{tr}(c) = \theta\}$. In any case b_i is a point on an edge of $K_1(m_1)$. By Proposition 1.3, all b_i are contained in the convex hull of $\partial^0 K_1(m_1)$, and hence a is also contained in the convex hull of $\partial^0 K_1(m_1)$. □

Proposition 1.3 is regarded as the dual statement of [56, Theorem 12.1]. From Proposition 1.3, it follows that the convex cone $K_1(m_1)$ does not have any extreme direction. Thus, Corollary 1.2 is a consequence of more general theorem [53, Theorem 18.5].

All subsets $K_1(m_1)$, $\partial K_1(m_1)$ and $\partial^0 K_1(m_1)$ of P_n are invariant by the action of $GL_n(\mathbf{Z})$. The finiteness of $\partial^0 K_1(m_1)/GL_n(\mathbf{Z})$ is due to Voronoï [62, §7 Théorèm]).

Theorem 1.5 (Voronoï). *The cardinality of $\partial^0 K_1(m_1)/GL_n(\mathbf{Z})$ is finite.*

Proof. We follows the argument of [60, Theorem 3.4]. Let $a \in \partial^0 K_1(m_1)$. By the reduction theory of Hermite or Minkowski, there exists an equivalent $a' \in aGL_n(\mathbf{Z})$ such that $\lambda_1\lambda_2 \cdots \lambda_n \leq c_n \det a'$, where λ_i denotes the i-th diagonal component of a' and c_n is the constant depending only on n. Since a' is perfect, there are n linearly independent minimal vectors $x_1, \cdots, x_n$ in $S(a')$. Since $m_1(a') = m_1(a) = 1$, Hadamard's inequality leads us to $\det a' \leq \langle a', x_1{}^t x_1\rangle \cdots \langle a', x_n{}^t x_n\rangle = 1$. Therefore, we have

$$\mathrm{tr}(a') = \lambda_1 + \cdots + \lambda_n \leq n\lambda_1 \cdots \lambda_n \leq nc_n$$

because of $1 = m_1(a') \leq \lambda_i$ for $i = 1, \cdots, n$. This shows that any perfect element of $\partial^0 K_1(m_1)$ is $GL_n(\mathbf{Z})$-equivalent to a vertex of polytope $K_1(m_1) \cap \{a \in V_n \ : \ \mathrm{tr}(a) \leq nc_n\}$. $\square$

More generally, it is known that the set of all faces of $\partial K_1(m_1)$ has only finitely many $GL_n(\mathbf{Z})$-orbits ([55, Theorem 5'], see Theorem 3.2 below). The actual value of the cardinality $\rho_n = \sharp(\partial^0 K_1(m_1)/GL_n(\mathbf{Z}))$ is known up to $n = 8$ (cf. [59, §3.1]): one has $\rho_2 = \rho_3 = 1, \rho_4 = 2, \rho_5 = 3, \rho_6 = 7, \rho_7 = 33, \rho_8 = 10916$.

We note that the face $\mathcal{F}_S$ may not necessarily be compact in general. It is known that the non-empty face $\mathcal{F}_S$ is compact if and only if S spans $\mathbf{R}^n$ ([55, Theorem 1], see also [40, Remark 9.1.12]). If $S(a)$ spans $\mathbf{R}^n$, then $a \in P_n$ is said to be well-rounded. Any perfect point is obviously well-rounded. Any weakly eutactic point is also well-rounded ([14, Théorème 2.3]). The finiteness and the algebraicity of weakly eutactic classes are verified by Bergé and Martinet [14, Théorèmes 3.5 and 4.1].

Theorem 1.6 (Bergé and Martinet). *Let $\partial^{we} K_1(m_1)$ be the set of all weakly eutactic points in $\partial K_1(m_1)$. Then the cardinality of $\partial^{we} K_1(m_1)/GL_n(\mathbf{Z})$ is finite. Any $a \in \partial^{we} K_1(m_1)$ is contained in $GL_n(\overline{\mathbf{Q}})$, where $\overline{\mathbf{Q}}$ stands for the algebraic closure of $\mathbf{Q}$.*

Let $\partial^{wr} K_1(m_1)$ be the set of all well-rounded points in $\partial K_1(m_1)$. The quotient $\partial^{wr} K_1(m_1)/GL_n(\mathbf{Z})$ is compact (cf. [40, Proposition 9.1.6]).

1.4. *Hermite like constants*

Let ϕ be a type one class function and S_p be a complete set of representatives for $\partial^0 K_1(m_1)/GL_n(\mathbf{Z})$. From $P_n \subset \mathbf{R}_{>0} \cdot K_1(m_1)$, it follows

$$\sup_{a \in P_n} F_\phi(a) = \sup_{a \in K_1(m_1)} F_\phi(a) = \sup_{a \in K_1(m_1)} \frac{1}{\phi(a)}.$$

By Corollary 1.2, any $a \in K_1(m_1)$ is represented as

$$a = \lambda_1 a_1 + \cdots + \lambda_r a_r$$

by some $a_1, \cdots, a_r \in \partial^0 K_1(m_1)$ and $\lambda_1, \cdots, \lambda_r \in \mathbf{R}_{\geq 0}$ with $\lambda_1 + \cdots + \lambda_r = 1$. Then, since $\phi(a) \geq \min\{\phi(a_1), \cdots, \phi(a_r)\}$, one has

$$\sup_{a \in K_1(m_1)} \frac{1}{\phi(a)} = \sup_{a \in \partial^0 K_1(m_1)} \frac{1}{\phi(a)} = \max_{a \in S_p} \frac{1}{\phi(a)} .$$

Therefore, the Hermite like constant

$$\delta_\phi = \max_{a \in P_n} F_\phi(a)$$

of F_ϕ is well-defined. In the case of $\phi = \det^{1/n}$, δ_ϕ coincides with the Hermite constant γ_n.

Let ϕ° be the dual type one class function of ϕ. By definition, the inequality $m_1(a) \leq \delta_{\phi^\circ} \phi^\circ(a)$ holds for all $a \in P_n$. By passing to the dual, one has $\phi^{\circ\circ}(a) \leq \delta_{\phi^\circ} w_1(a)$ for $a \in P_n$, and by Proposition 1.2,

$$\sup_{a \in P_n} \frac{\phi(a)}{w_1(a)} \leq \delta_{\phi^\circ} .$$

Thus, we can define the dual constant

$$\widehat{\delta}_\phi = \sup_{a \in P_n} \frac{\phi(a)}{w_1(a)} .$$

Indeed, we can show $\widehat{\delta}_\phi = \delta_{\phi^\text{o}}$ for any type one class function ϕ. In particular, this gives

$$\gamma_n = \delta_{\det^{1/n}} = \widehat{\delta}_{(\det^{1/n})^\circ} = n \sup_{a \in P_n} \frac{\det(a)^{1/n}}{w_1(a)} .$$

We write ξ_ϕ for the product $\delta_\phi \cdot \widehat{\delta}_\phi$. This satisfies the invariance $\xi_{\phi^\circ} = \xi_{C\phi} = \xi_\phi$ for any constant $C > 0$. For example, $\xi_{w_1} = \xi_{m_1} = \delta_{w_1}$ and $\xi_{\det^{1/n}} = \gamma_n^2/n$. By definition, we have the following:

Proposition 1.4. *The inequality $\xi_{w_1} \leq \xi_\phi$ holds for any type one class function ϕ.*

See [58, Propositions 3.2 and 3.3] for details.

2. Rankin's constant and Voronoï's theorem

Rankin [52] defined the constant $\gamma_{n,k}$ as a generalization of Hermite's constant, and proved Rankin's inequality among $\gamma_{n,k}$. About 40 years later, Coulangeon [22] formulated Voronoï's theorem of this case in terms of k-perfection and k-eutaxy. It is an open problem to find a geometric characterization of k-perfect forms. Bergé and Martinet [13] introduced the constant $\gamma'_{n,k}$ and proved several inequalities among $\gamma_{n,k}$ and $\gamma'_{n,k}$. In this section, we will survey Voronoï's theorem for $\gamma_{n,k}$ and $\gamma'_{n,1}$.

2.1. *Rankin's constant*

Let k be a positive integer with $1 \leq k \leq n-1$. We denote by $M^*_{n,k}(\mathbf{Z})$ the subset $\{X = (x_1, \cdots, x_k) \in M_{n,k}(\mathbf{Z}) \ : \ x_1 \wedge x_2 \wedge \cdots \wedge x_k \neq 0\}$ of $M_{n,k}(\mathbf{Z})$. The unimodular group $GL_k(\mathbf{Z})$ (resp. $GL_n(\mathbf{Z})$) acts on $M^*_{n,k}(\mathbf{Z})$ by right (resp. left) multiplications. For $X \in M^*_{n,k}(\mathbf{Z})$, define the function $D_X : P_n^{\mathrm{semi}} \longrightarrow \mathbf{R}_{\geq 0}$ by

$$D_X(a) = \det({}^t X a X)^{1/k}$$

for $a \in P_n^{\mathrm{semi}}$. It is obvious that D_X is a type one function. The function $m_k : P_n^{\mathrm{semi}} \longrightarrow \mathbf{R}_{\geq 0}$ defined by

$$m_k(a) = \inf_{X \in M^*_{n,k}(\mathbf{Z})} D_X(a)$$

is a type one class function. This is regarded as a generalization of the arithmetical minimum function m_1. It is obvious that $m_k(a) > 0$ if $a \in P_n$, and $m_k(a) = 0$ otherwise. Rankin [52] defined the constant

$$\gamma_{n,k} = \left(\max_{a \in P_n} \frac{m_k(a)}{\det(a)^{1/n}} \right)^k,$$

and then proved the inequality

$$\gamma_{n,k} \leq \gamma_{j,k} (\gamma_{n,j})^{k/j}$$

for $1 \leq k < j \leq n-1$ as a generalization of Mordell's inequality. By using this inequality, Rankin determined the value $\gamma_{4,2} = 3/2$. See **2.4** for other explicit values of $\gamma_{n,k}$.

2.2. *Voronoï's theorem of $m_k / \det^{1/n}$*

A point $a \in P_n$ is said to be k-extreme (strictly k-extreme) if $m_k / \det^{1/n}$ attains a local maximum (resp. a strict local maximum) on a up to the multiplication by an element of $\mathbf{R}_{>0}$. A Voronoï type characterization of k-extreme points was studied by Coulangeon [22]. The subset

$$S_k^*(a) = \{X \in M_{n,k}^*(\mathbf{Z}) \; : \; D_X(a) = m_k(a)\}$$

corresponding to $a \in P_n$ plays a key role. Since $D_{Xh}(a) = D_X(a)$ for all $h \in GL_k(\mathbf{Z})$, the set $S_k^*(a)$ is invariant by the action of $GL_k(\mathbf{Z})$, and hence the quotient $S_k(a) = S_k^*(a)/GL_k(\mathbf{Z})$ exists. We write $[X]$ for the element $X \cdot GL_k(\mathbf{Z})$ in $S_k(a)$. The following was proved in [22, Proposition 2.7].

Proposition 2.1 (Coulangeon). *The cardinality of $S_k(a)$ is finite.*

We recall the notion of k-perfection and k-eutaxy. For each $i = 1, 2, \cdots, k$, we define the map $*_i : V_n \times M_{n,k}(\mathbf{R}) \longrightarrow M_{n,k}(\mathbf{R})$ by

$$v *_i X = (x_1, \cdots, x_{i-1}, vx_i, x_{i+1}, \cdots, x_k)$$

for $v \in V_n$ and $X = (x_1, \cdots, x_k) \in M_{n,k}(\mathbf{R})$. Note that $*_i$ is linear in X but not in v. Then, for each $X \in M_{n,k}^*(\mathbf{Z})$, the linear map $\varphi_X : V_n \longrightarrow \mathbf{R}$ is defined by

$$\varphi_X(v) = \sum_{i=1}^{k} \det({}^tX \cdot (v *_i X)) \, .$$

It is obvious that φ_X depends only on the class $[X] = X \cdot GL_k(\mathbf{Z})$. Another definition of φ_X is given by

$$\varphi_X(v) = \det({}^tX \cdot X) \cdot < p_X, v > ,$$

where p_X denotes the matrix representation of the orthogonal projection from $\mathbf{R}^n$ onto the subspace spanned by $\{x_1, \cdots, x_k\}$.

Definition 2.1. Let $a \in P_n$. We fix an element $b \in GL_n(\mathbf{R})$ such that $a = {}^tbb$. An element a is said to be k-perfect if $\{\varphi_{bX}\}_{[X] \in S_k(a)}$ spans the dual space V_n^* of V_n. An element a is said to be k-eutactic if there exist $\rho_X > 0$ ($[X] \in S_k(a)$) such that

$$\mathrm{tr} = \sum_{[X] \in S_k(a)} \rho_X \varphi_{bX} .$$

These definitions of k-perfection and k-eutaxy do not depend on a choice of b. Now the main theorem of [22] is stated as follows:

Theorem 2.1 (Coulangeon). *A point $a \in P_n$ is k-extreme if and only if a is k-perfect and k-eutactic. Any k-extreme point is strictly k-extreme.*

The line of the proof of Theorem 2.1 is parallel to that of Theorem 1.3. Namely, the following sufficient and necessary condition for k-extremeness is shown: $a = {}^t bb \in P_n$ is k-extreme if and only if the set

$$\{v \in V_n \ : \ \min_{X \in S_k(a)} \varphi_{bX}(v) \geq 0 \ \text{ and } \ \mathrm{tr}(v) \leq 0\}$$

is reduced to $\{0\}$ ([22, Théorème 3.2.2]). Theorem 2.1 follows from this and Stiemke's theorem.

The finiteness of k-perfect points was proved in [22, Théorèm 4.5].

Theorem 2.2 (Coulangeon). *The number of k-perfect points in P_n modulo $\mathbf{R}_{>0}GL_n(\mathbf{Z})$ is finite.*

Example 2.1. Let $L \subset \mathbf{R}^n$ be a full lattice, which means a $\mathbf{Z}$-module of rank n. The dual lattice L^* of L is defined by $L^* = \{y \in \mathbf{R}^n : {}^t xy \in \mathbf{Z} \ \text{ for all } x \in L\}$. If $x_1, \cdots, x_n$ is a basis of L, then we denote by $[L]$ the class of the Gram matrix $({}^t x_i x_j)_{1 \leq i,j \leq n}$ in $\mathbf{R}_{>0} \backslash P_n / GL_n(\mathbf{Z})$. In dimension 4, there are at least 5 inequivalent 2-perfect points, i.e., $[A_4], [A_4^*], [D_4], [W_4], [W_4^*]$ ([22, §5.1]). Here we use standard notations of root lattices. By W_4, we denote the Watson lattice of rank 4, i.e.,

$$[W_4] = \begin{pmatrix} 3 & -1 & -1 & -1 \\ -1 & 3 & -1 & -1 \\ -1 & -1 & 3 & -1 \\ -1 & -1 & -1 & 3 \end{pmatrix} \quad \bmod \mathbf{R}_{>0}GL_4(\mathbf{Z})\,.$$

The maximum of $m_2/\det^{1/4}$ is attained on $[D_4]$. In general $n \geq 4$, the class of any irreducible root lattice of rank n is 2-extreme and k-eutactic for all $k < n$ ([22, Théorème 5.1.1]).

When $k = 1$, Theorem 1.1 shows that the Hermite constant $\gamma_n = \gamma_{n,1}$ is an algebraic number. The algebraicity of k-perfect points and $\gamma_{n,k}$ for $k \geq 2$ was verified by Bavard [9, Théorèm 2.2], [11, §1.5]. It is based on the following general result [11, Lemme 1.11].

Lemma 2.1. *Let $W \subset \mathbf{R}^N$ be an algebraic subset defined by polynomials with coefficients in $\overline{\mathbf{Q}} \cap \mathbf{R}$. Then the set of isolated points in W is a finite subset contained in $W \cap (\overline{\mathbf{Q}} \cap \mathbf{R})^N$.*

If a is k-perfect, then the set

$$W = \{(b, \lambda) \in V_n \times \mathbf{R} \ : \ D_X(b) - \lambda = 0 \ \text{ for all } X \in S_k^*(a)\}$$

satisfies the assumption of Lemma 2.1. Since k-perfect points are isolated in W ([11, Proposition 1.8]), we obtain

Theorem 2.3 (Bavard). *Any k-perfect point is contained in $P_n \cap M_n(\overline{\mathbf{Q}})$. In particular, $\gamma_{n,k}$ is an algebraic number.*

2.3. *Some problems on k-perfect forms*

Since m_k is a continuous type one function and vanishes on the boundary of $P_n^{\rm semi}$, $K_1(m_k)$ is a kernel contained in P_n. As we have seen in §1.3, $K_1(m_1)$ is a locally finite polyhedral convex cone and perfect points are characterized as vertices of $K_1(m_1)$. When $k \geq 2$, $K_1(m_k)$ is not polyhedral, and we have the following problem.

Problem 2.1. *Determine locations of k-perfect points in $K_1(m_k)$.*

Theorem 1.4 (3) gives another characterization of perfect points in P_n. On variations of the set $S_k(a)$, the next is elementary ([22, Lemme 2.9]).

Proposition 2.2 (Coulangeon). *For $a \in P_n$, there is a neighborhood $\mathcal{U}$ of a in P_n such that $S_k(a') \subset S_k(a)$ for $a' \in \mathcal{U}$.*

We say that $S_k(a)$ is locally maximal if there is a neighborhood $\mathcal{U}$ of a in P_n such that $S_k(a') \subsetneqq S_k(a)$ for all $a' \in \mathcal{U} \setminus \mathbf{R}_{>0}a$. Let $P_n^{(k)}$ be the set of $a \in P_n$ such that $S_k(a)$ is locally maximal. If $k=1$, $P_n^{(1)}$ coincides with the set of all perfect points in P_n.

Problem 2.2. *Does $P_n^{(k)}$ coincide with the set of all k-perfect points in P_n for any $k \geq 2$?*

The cardinality $\sharp(S(a)) = 2\sharp(S_1(a))$ is called the kissing number of a. Determination of the maximum $\max_{a\in P_n} 2\sharp(S_1(a))$ is known as the lattice kissing number problem [20, Chapter 1, §2]. The actual value of $\max_{a\in P_n} \sharp(S_1(a))$ is known for $1 \leq n \leq 9$ and $n = 24$ (cf. [67]). One can prove the estimate $\max_{a\in P_n} \sharp(S_1(a)) \leq 2^n - 1$ for all n as follows: Let $a \in P_n$ and $x, y \in S(a)$. If $y - x = 2z \in 2\mathbf{Z}^n$, then $m_1(a) = {}^t(x+2z)a(x+2z) = {}^txax$, and hence ${}^tzaz = -{}^txaz$. From ${}^txax \leq {}^t(x+z)a(x+z)$, it follows ${}^tzaz \leq 0$, i.e., $z = 0$. This means that the natural map $\mathbf{Z}^n \longrightarrow \mathbf{Z}^n/2\mathbf{Z}^n$ is injective on $S_1(a)$. This proof is due to Voronoï [62, p.107, Lemme], see also [43, §31, p.80] for more general result. When $n \geq 10$, Watson [66, Theorem 1] proved $\max_{a\in P_n} \sharp(S_1(a)) \leq 2^{n-2} + 8$. We have a similar problem for $k \geq 2$.

Problem 2.3. *Bound the maximum* $\max_{a\in P_n} \sharp(S_k(a))$.

If ϕ is a type one function, then one can ask about Voronoï's theorem for m_k/ϕ.

Problem 2.4. *Prove Voronoï type theorem for m_k/ϕ when $k \geq 2$.*

Let w_k be the dual type one class function of m_k. When $k \geq 2$, it is not trivial that the Hermite–Rankin like constant of m_k/ϕ exists for a given type one class function ϕ. We set

$$\delta_{\phi,k} = \sup_{a \in P_n} \frac{m_k(a)}{\phi(a)}, \qquad \widehat{\delta}_{\phi,k} = \sup_{a \in P_n} \frac{\phi(a)}{w_k(a)}$$

for a type one class function ϕ.

Problem 2.5. *When are both $\delta_{\phi,k}$ and $\widehat{\delta}_{\phi,k}$ finite?*

It is easy to see that $\delta_{\phi,k} = \widehat{\delta}_{\phi^\circ,k}$ provided that both $\delta_{\phi,k}$ and $\widehat{\delta}_{\phi^\circ,k}$ are finite.

2.4. *The Bergé–Martinet constant*

The constant

$$\gamma'_{n,k} = \left(\sup_{a \in P_n} m_k(a) m_k(a^{-1}) \right)^{k/2}$$

was first defined by Bergé and Martinet. In [13], they proved several inequalities among $\gamma_{n,k}$ and $\gamma'_{n,k}$.

Theorem 2.4 (Bergé and Martinet). *One has the following:*
(1) $\gamma'_{n,k} \leq \gamma_{n,k} \leq (\gamma_n)^k$ *for* $1 \leq k \leq n-1$.
(2) $(\gamma_{n,k})^n \leq (\gamma_{n-k,k})^{n-k} (\gamma'_{n,k})^{2k}$ *for* $1 \leq k \leq n/2$.
(3) $\gamma'_{n,2k} \leq (\gamma'_{n-k,k})^2$ *for* $1 \leq k \leq n/2$.
(4) $(\gamma_{n,k})^{n-2k} \leq (\gamma_{n-k,k})^{n-k}$ *for* $1 \leq k \leq n-1$.
(5) $\gamma_{n,n/2} = \gamma'_{n,n/2}$ *if n is even.*

When $k = 1$, an analog of Voronoï's theorem holds for the Bergé–Martinet invariant $F_{\mathrm{BM}}(a) = \sqrt{m_1(a) m_1(a^{-1})}$. A point $a \in P_n$ is said to be dual-extreme (strictly dual extreme) if F_{BM} attains a local maximum (resp. a strict local maximum) on a up to the multiplication by an element of $\mathbf{R}_{>0}$. To define the dual-perfection and the dual-eutaxy, we use the same notation as in **1.2**.

Definition 2.2. Let $a \in P_n$. We fix an element $b \in GL_n(\mathbf{R})$ such that $a = {}^t b b$. An element a is said to be dual-perfect if $\{\varphi_{bx}\}_{x \in S(a)} \cup \{\varphi_{{}^t b^{-1} y}\}_{y \in S(a^{-1})}$

spans the dual space V_n^* of V_n. An element a is said to be dual-eutactic if there exist $\rho_x > 0$ $(x \in S(a))$ and $\rho_y > 0$ $(y \in S(a^{-1}))$ such that

$$\sum_{x \in S(a)} \rho_x \varphi_{bx} = \sum_{y \in S(a^{-1})} \rho_y \varphi_{{}^t b^{-1} y}.$$

Then one has:

Theorem 2.5 (Bergé and Martinet). *A point $a \in P_n$ is dual-extreme if and only if a is dual-perfect and dual-eutactic. Any dual extreme point is a strict dual extreme.*

As noticed in [40, p.99], the number of dual-perfect points in P_n modulo $\mathbf{R}_{>0} GL_n(\mathbf{Z})$ is infinite in general. In [12], Bergé proved the following:

Theorem 2.6 (Bergé). *The number of dual-extreme points in P_n modulo $\mathbf{R}_{>0} GL_n(\mathbf{Z})$ is finite. If $a \in P_n$ is dual extreme, then there exists $\lambda \in \mathbf{R}_{>0}$ such that $\lambda a \in GL_n(\overline{\mathbf{Q}})$. In particular , $\gamma'_n = \gamma'_{n,1}$ is an algebraic number.*

As to the explicit value of γ'_n, $\gamma'_8 = \gamma_8 = 2$ immediately follows from the self-duality of the E_8-lattice. Bergé and Martinet determined the values $\gamma'_2 = 2/\sqrt{3}$, $\gamma'_3 = \sqrt{3/2}$ and $\gamma'_4 = \sqrt{2}$. In [49], Poor and Yuen proved the inequality

$$\frac{n}{(\gamma_n)^2} \leq \inf_{(a,b) \in P_n \times P_n} \frac{\langle a, b \rangle}{m_1(a) m_1(b)} \leq \frac{n}{(\gamma'_n)^2} \tag{2}$$

and, by using this, they determined the following values.

Theorem 2.7 (Poor and Yuen). $\gamma'_5 = \sqrt{2}$, $\gamma'_6 = \sqrt{8/3}$ *and* $\gamma'_7 = \sqrt{3}$.

All known values of γ'_n satisfy $(\gamma'_n)^2 \in \mathbf{Q}$. The following problem is due to Martinet [40, Questions 3.8.12].

Problem 2.6. *Is $(\gamma'_n)^2$ rational for all n?*

Applying Theorem 2.4 to the explicit values of γ'_5, γ'_7, $\gamma_{4,2}$ and γ_n for $n = 2, \cdots, 8$, we have

Theorem 2.8. $\gamma_{6,2} = 3^{2/3}$, $\gamma'_{6,2} = 2$, $\gamma_{8,2} = \gamma'_{8,2} = 3$ *and* $\gamma_{8,3} = \gamma'_{8,3} = \gamma_{8,4} = \gamma'_{8,4} = 4$.

See [57, §2] for details. Barnes and Cohn [7] also proved the first part of the inequality (2). Since one has

$$\frac{1}{\xi_{w_1}} = \inf_{(a,b) \in P_n \times P_n} \frac{\langle a, b \rangle}{m_1(a) m_1(b)},$$

the first part of (2) is a special case of Proposition 1.4. Furthermore, the inequality (2) is generalized to $\gamma_{n,k}$ and $\gamma'_{n,k}$ as follows:

$$\frac{n}{(\gamma_{n,k})^{2/k}} \leq \inf_{(a,b)\in P_n\times P_n} \frac{\langle a,b\rangle}{m_k(a)m_k(b)} \leq \frac{n}{(\gamma'_{n,k})^{2/k}},$$

(see [57, Theorem 1]). To find an analog of Voronoï's theorem for the k-th Bergé–Martinet invariant $F_{\mathrm{BM}}^{(k)}(a) = (m_k(a)m_k(a^{-1}))^{k/2}$ ([40, Problem 10.6.10]) solved by Bavard, see Example 3.7 below. See [26] for other Hermite like invariants.

3. Generalizations of Voronoï's theorem

There are several directions to generalize Voronoï's theorem. A natural generalization of the domain P_n was considered by Koecher [35] and Ash [3]. An extension of the geometric framework was developed by Bavard [9], [11]. A change of a base field from $\mathbf{Q}$ to an algebraic number field was studied by several authors [22], [32], [38], [45], [46]. In this section, we will survey these theories. We do not exhaust all of generalizations. See e.g., [40, Chaprter 13] for other variations.

3.1. *Voronoï's theorem of packing functions on symmetric cones*

A generalization of the domain P_n is given by the notion of symmetric cones. Let Ω be an open convex cone in the Euclidean space $\mathbf{R}^N$. The open dual cone Ω^* of Ω is defined to be

$$\Omega^* = \{a \in \mathbf{R}^N \ : \ (a,b) > 0 \ \text{ for all } b \in \overline{\Omega}\setminus\{0\}\},$$

where $\overline{\Omega}$ denotes the closure of Ω in $\mathbf{R}^N$. If $\Omega = \Omega^*$ holds, then Ω is called a self-dual cone. We denote by G_Ω the stabilizer of Ω in $GL_N(\mathbf{R})$, i.e.,

$$G_\Omega = \{g \in GL_N(\mathbf{R}) \ : \ g\Omega = \Omega\}.$$

If G_Ω acts transitively on Ω, then Ω is said to be homogeneous. By a symmetric cone, we mean a self-dual homogeneous cone. See the textbook [25] for details of symmetric cones.

We fix a symmetric cone Ω. Let G_Ω° be the connected component of the identity in G_Ω. Then G_Ω° also acts transitively on Ω. We denote by K_a the stabilizer of $a \in \Omega$ in G_Ω°. There exists a point $\boldsymbol{e} \in \Omega$ such that $K_{\boldsymbol{e}} = G_\Omega^\circ \cap O_N(\mathbf{R})$. The group $K_{\boldsymbol{e}}$ is connected and gives a maximal compact subgroup of G_Ω°. Thus Ω is identified with the Riemannian symmetric space $G_\Omega^\circ/K_{\boldsymbol{e}}$.

Let $L \subset \mathbf{R}^N$ be a lattice of rank N which contains e. Now we define the packing function $F_{(\Omega,L)} : \Omega \longrightarrow \mathbf{R}_{>0}$, of which we study local maxima. First, the characteristic function φ_Ω of Ω is defined by

$$\varphi_\Omega(a) = \int_\Omega e^{-(a,b)} db\,, \qquad (a \in \Omega)\,.$$

The Lebesgue measure db is normalized so that $\varphi_\Omega(\boldsymbol{e}) = 1$. The defining integral is uniformly convergent on any compact subset in Ω. It follows from the definition that $\varphi_\Omega(ga) = |\det g|^{-1}\varphi_\Omega(a)$ for all $g \in G_\Omega$ and $a \in \Omega$. Next, the minimum function m_L on Ω is defined by

$$m_L(a) = \min\{(a,b) \ : \ b \in (L \setminus \{0\}) \cap \overline{\Omega}\}\,.$$

Since Ω is self-dual, the value $m_L(a)$ is positive. Then the packing function $F_{\Omega,L}$ is defined by

$$F_{(\Omega,L)}(a) = m_L(a)^N \varphi_\Omega(a)\,.$$

A point $a \in \Omega$ is said to be extreme if $F_{(\Omega.L)}$ attains a local maximum on a up to the multiplication by an element of $\mathbf{R}_{>0}$.

To state Ash's definition of eutaxy, we need a Jordan algebra structure of $\mathbf{R}^N$ induced from Ω. Let $\mathfrak{g}$ be the Lie algebra of G_Ω°, i.e.,

$$\mathfrak{g} = \{X \in M_N(\mathbf{R}) \ : \ \exp(X) \in G_\Omega^\circ\}\,.$$

Since Ω is a symmetric cone, $\mathfrak{g}$ is invariant by the transpose $X \mapsto {}^tX$. We set $\mathfrak{g}_\pm = \{X \in \mathfrak{g} \ : \ {}^tX = \pm X\}$. Then $\mathfrak{g}_-$ coincides with the Lie algebra of $K_{\boldsymbol{e}}$. Moreover, the map $\psi : \mathfrak{g}_+ \longrightarrow \mathbf{R}^N$ defined by $\psi(X) = X\boldsymbol{e}$ gives a linear isomorphism. We define the binary product $* : \mathbf{R}^N \times \mathbf{R}^N \longrightarrow \mathbf{R}^N$ by

$$a * b = \psi^{-1}(a)b\,.$$

This product satisfies

(J_1) $a * b = b * a$
(J_2) $a * (a^2 * b) = a^2 * (a * b)$, where a^2 means $a * a$
(J_3) $\boldsymbol{e} * a = a * \boldsymbol{e} = a$
(J_4) $(a * c, b) = (a, c * b)$

for all $a, b, c \in \mathbf{R}^N$. Namely, $*$ gives $\mathbf{R}^N$ a formally real Jordan algebra structure with the identity $\boldsymbol{e}$. We denote by J_Ω this formally real Jordan algebra. For $a \in J_\Omega$, the subalgebra $\mathbf{R}[a]$ of J_Ω generated by a and $\boldsymbol{e}$ is an associative algebra. An element a is said to be invertible if there exists an element $b \in \mathbf{R}[a]$ such that $a * b = \boldsymbol{e}$. This b is unique and is denoted by

a^{-1}. Let $J_\Omega^\times$ be the subset of all invertible elements in J_Ω. Then Ω coincides with the connected component of $J_\Omega^\times$ which contains e.

We assume $L \otimes_{\mathbf{Z}} \mathbf{Q}$ gives a $\mathbf{Q}$-structure of J_Ω. For $a \in \Omega$, we set

$$S_{(\Omega,L)}(a) = \{b \in (L \setminus \{0\}) \cap \overline{\Omega} \ : \ (a,b) = m_L(a)\}\,.$$

Definition 3.1. Let $a \in \Omega$ and a^{-1} be the inverse of a in the Jordan algebra J_Ω. A point a is said to be perfect if $S_{(\Omega.L)}(a)$ spanns $\mathbf{R}^N$. A point a is said to be eutactic if there exist $\lambda_b \in \mathbf{R}_{>0}$, $b \in S_{(\Omega,L)}$, such that

$$a^{-1} = \sum_{b \in S_{(\Omega,L)}} \lambda_b b\,.$$

Let $g \in \exp(\mathfrak{g}_+)$. The Taylor expansion of $1/\varphi_\Omega$ at the point $a = ge$ is given by

$$\frac{1}{\varphi_\Omega(a+v)} = \det g \cdot \{1 + (a^{-1}, v) + \frac{1}{2}((a^{-1}, v)^2 - (g^{-1}v, g^{-1}v)) + O((v,v)^{3/2})\}$$

for $v \in \mathbf{R}^N$ ([3, Corollary to Proposition 3]). By using this formula, Ash proved that the function $1/F_{(\Omega,L)}$ is a topological Morse function on $\Omega/\mathbf{R}_{>0}$. Voronoï's theorem of $F_{(\Omega,L)}$ follows from this fact.

Theorem 3.1 (Ash). *A point $a \in \Omega$ is extreme if and only if a is perfect and eutactic.*

Let $K_1(m_L) = \{a \in \overline{\Omega} \ : \ (a,b) \geq 1 \text{ for all } b \in (L \setminus \{0\}) \cap \overline{\Omega}\}$, which is a polyhedral cone and is regarded as a generalization of the Ryshkov domain $K_1(m_1)$. Any perfect point $a \in \Omega$ of $m_L(a) = 1$ is a vertex of $K_1(m_L)$. We have the following finiteness:

Theorem 3.2 (Ash). *The discrete group $\Gamma = \{g \in G_\Omega^\circ \ : \ {}^t gL = L\}$ of G_Ω° acts on $K_1(m_L)$. The set of faces of $K_1(m_L)$ has only finitely many Γ-orbits. In particular, the number of perfect points in Ω modulo $\mathbf{R}_{>0}\Gamma$ is finite. Moreover, the number of eutactic points in Ω modulo $\mathbf{R}_{>0}\Gamma$ is finite.*

It is proved in [3, Theorem 2] that a point $a \in \Omega$ is eutactic if and only if a is critical non-degenerate for $F_{(\Omega,L)}$. The finiteness of eutactic points is derived from the finiteness of Γ-orbits of critical points of $F_{(\Omega,L)}$.

Example 3.1. The cone P_n of positive definite symmetric matrices is a symmetric cone in V_n. In this case, e is chosen as the identity matrix I_n and the product $*$ is defined by

$$a * b = \frac{1}{2}(ab + ba)$$

for $a, b \in V_n$. When $L = \{v \in M_n(\mathbf{Q}) \cap V_n \,:\, 2v \in M_n(\mathbf{Z}),\ v_{11}, \cdots, v_{nn} \in \mathbf{Z}\}$, the packing function $F_{(P_n,L)}$ is equal to $(m_1/\det^{1/n})^{n(n+1)/2}$. Ash's definition of perfection and eutaxy is equivalent to Definition 1.3 ([3, Corollary to Proposition 2]).

Example 3.2. Let B be the non degenerate bilinear form on $\mathbf{R}^n$ defined by

$$B(x,y) = x_1y_1 - x_2y_2 - \cdots - x_ny_n$$

for $x = {}^t(x_1, \cdots, x_n)$ and $y = {}^t(y_1, \cdots, y_n) \in \mathbf{R}^n$. Then the Lorentz cone $\Omega_n = \{x \in \mathbf{R}^n \,:\, B(x,x) > 0,\ x_1 > 0\}$ is a symmetric cone in $\mathbf{R}^n$. We choose $\boldsymbol{e}$ as the unit vector ${}^t(1, 0, \cdots, 0) \in \mathbf{R}^n$. Let $\{\boldsymbol{e}\}^\perp$ be the orthogonal complement of $\boldsymbol{e}$ with respect to the usual inner product $(\cdot,\cdot)$ of $\mathbf{R}^n$. The product of the Jordan algebra J_{Ω_n} is defined by

$$(\lambda\boldsymbol{e} + u) * (\lambda'\boldsymbol{e} + u') = (\lambda\lambda' - B(u,u'))\boldsymbol{e} + \lambda u' + \lambda' u$$

for $\lambda, \lambda' \in \mathbf{R}$ and $u, u' \in \{\boldsymbol{e}\}^\perp$. The packing function $F_{(\Omega_n,\mathbf{Z}^n)}$ is given as

$$F_{(\Omega_n,\mathbf{Z}^n)}(a) = \frac{m_{\mathbf{Z}^n}(a)^n}{B(a,a)^{n/2}}.$$

Since $G^\circ_{\Omega_n} = \mathbf{R}_{>0} \cdot SO_0(1, n-1)$ acts transitively on Ω_n, we have

$$F_{(\Omega_n,\mathbf{Z}^n)}(\lambda g\boldsymbol{e}) = \frac{m_{\mathbf{Z}^n}(g\boldsymbol{e})^n}{B(g\boldsymbol{e}, g\boldsymbol{e})^{n/2}} = m_{\mathbf{Z}^n}(g\boldsymbol{e})^n$$

for $\lambda \in \mathbf{R}_{>0}$ and $g \in SO_0(1, n-1)$, and moreover,

$$m_{\mathbf{Z}^n}(g\boldsymbol{e}) = \min_{x \in (\mathbf{Z}^n \setminus \{0\}) \cap \overline{\Omega}_n} (\boldsymbol{e}, {}^t g x) = \min_{x \in (\mathbf{Z}^n \setminus \{0\}) \cap \overline{\Omega}_n} \left(\frac{({}^t g x, {}^t g x) + B(x,x)}{2} \right)^{1/2}.$$

Therefore, we have

$$\max_{a \in \Omega_n} F_{(\Omega_n,\mathbf{Z}^n)}(a) = \max_{[g] \in K_e \backslash SO_0(1,n-1)/\Gamma} \ \min_{x \in (\mathbf{Z}^n \setminus \{0\}) \cap \overline{\Omega}_n} \left(\frac{(gx, gx) + B(x,x)}{2} \right)^{n/2},$$

where $\Gamma = SO_0(1, n-1) \cap SL_n(\mathbf{Z})$.

Problem 3.1. *Let Ω be an arbitrary symmetric cone. Replacing P_n and P_n^{semi} with Ω and $\overline{\Omega}$, respectively, in Definitions 1.1 and 1.2, we can define type one functions on $\overline{\Omega}$ and semikernels in $\overline{\Omega}$. For example, both $\varphi_\Omega^{-1/N}$ and m_L are continuously extended to type one functions on $\overline{\Omega}$. Can Proposition 1.1 and Theorem 1.3 be generalized to this setting?*

As stated in Theorem 3.2, the number of eutactic classes in $\mathbf{R}_{>0}\backslash\Omega/\Gamma$ is finite. When $\Omega = P_n$, Ash verified a "mass formula with signs" of eutactic classes ([4], see also [40, Theorem 9.5.3]). For $a \in \partial K_1(m_1)$, we set $\Gamma_a = \{g \in SL_n(\mathbf{Z}) \ : \ {}^t gag = a\}$. Let $\mathcal{F}_{S(a)}$ be the face of $K_1(m_1)$ defined in **1.3**, i.e., $\mathcal{F}_{S(a)} = \{b \in \partial K_1(m_1) \ : \ S(a) \subset S(b)\}$.

Theorem 3.3 (Ash). *The set $\partial^e K_1(m_1)$ of eutactic points in $\partial K_1(m_1)$ satisfies*

$$\sum_{[a]\in\partial^e K_1(m_1)/GL_n(\mathbf{Z})} \frac{(-1)^{\dim \mathcal{F}_{S(a)}}}{\sharp\Gamma_a} = \chi(SL_n(\mathbf{Z})) = \begin{cases} -1/12 & (n=2) \\ 0 & (n \geq 3) \end{cases},$$

where $\chi(SL_n(\mathbf{Z}))$ stands for the Euler characteristic of $SL_n(\mathbf{Z})$.

The actual value of $\epsilon_n = \sharp(\partial^e K_1(m_1)/GL_n(\mathbf{Z}))$ is known up to $n = 5$ (cf. [8]): one has $\epsilon_2 = 2, \epsilon_3 = 5, \epsilon_4 = 16, \epsilon_5 = 118$.

3.2. *Bavard's theory*

Let V be a Riemannian manifold and Γ a discrete subgroup of the isometry group of V. Let C be a set endowed with a right action of Γ. We consider a family of C^1-functions $f_s : V \longrightarrow \mathbf{R}$ parameterized by $s \in C$. We assume the following two conditions:

(B_1) $f_s \circ \gamma = f_{s\gamma}$ for all $s \in C$ and $\gamma \in \Gamma$.
(B_2) The cardinality of the subset $\{s \in C \ : \ f_s(v) \leq \lambda\}$ of C is finite for any $v \in V$ and $\lambda \in \mathbf{R}$.

Each f_s is called a length function on V. We write $\mathcal{E}$ for the quadruplet $(V, \Gamma, C, \{f_s\})$. What we do is to characterize local maxima of the function $F_{\mathcal{E}}(v) = \min_{s\in C} f_s(v)$ in $v \in V$. A point $v \in V$ is said to be extreme (resp. strictly extreme) if v attains a local maximum (resp. a strictly local maximum) of $F_{\mathcal{E}}$.

For a given $v \in V$, $T_v V$ stands for the tangent space of V at v and $X_s(v)$ stands for the gradient vector of f_s at v. By the condition (B_2), $S_{\mathcal{E}}(v) = \{s \in C \ : \ f_s(v) = F_{\mathcal{E}}(v)\}$ is a finite subset of C. Let $\mathrm{Conv}(v)$ be the convex hull of $\{X_s(v)\}_{s\in S_{\mathcal{E}}(v)}$ in $T_v V$ and $\mathrm{Aff}(v)$ the affine subspace spanned by $\{X_s(v)\}_{s\in S_{\mathcal{E}}(v)}$ in $T_v V$.

Definition 3.2. A point $v \in V$ is said to be perfect if $T_v V = \mathrm{Aff}(v)$ holds. A point $v \in V$ is said to be eutactic if the origin $0 \in T_v V$ is contained in the interior of $\mathrm{Conv}(v)$.

We need the following condition for $\mathcal{E}$.

(C) For any $v \in V$, any subset $S' \subset S_{\mathcal{E}}(v)$ and any non-zero vector $X \in T_v V$ orthogonal to $\{X_s(v)\}_{s \in S'}$, there exists a C^1-curve $c : [0, \epsilon) \longrightarrow V$ for a sufficiently small $\epsilon > 0$ such that $c(0) = v$, $c'(0) = X$ and $f_s(v) < f_s(c(t))$ for all $t \in (0, \epsilon)$ and $s \in S'$.

Now Bavard's theorem is stated as:

Theorem 3.4 (Bavard). *Assume $\mathcal{E}$ satisfies the condition* (C). *Then any extreme point in V is strictly extreme, and a point $v \in V$ is extreme if and only if v is perfect and eutactic.*

A function $f : V \longrightarrow \mathbf{R}$ is said to be convex if f is convex on any geodesic line on V, i.e.,

$$f(\ell(\lambda\alpha + (1 - \lambda)\beta)) \leq \lambda f(\ell(\alpha)) + (1 - \lambda) f(\ell(\beta))$$

holds for any geodesic $\ell : [0, \epsilon) \longrightarrow V$, $\alpha, \beta \in (0, \epsilon)$, $\alpha \neq \beta$, and $0 < \lambda < 1$. If this inequality is strict, then f is said to be strictly convex. It is proved that $\{f_s\}_{s \in C}$ satisfies the condition (C) if f_s is strictly convex for all $s \in C$.

Theorem 3.5 (Bavard). *Assume f_s is convex for all $s \in C$. Then a point $v \in V$ is strictly extreme if and only if v is perfect and eutactic.*

Example 3.3. We consider the subset $P_n^1 = \{a \in P_n : \det a = 1\}$ of P_n, which is identified with the Riemannian symmetric space $SL_n(\mathbf{R})/SO_n(\mathbf{R})$. For $x \in \mathbf{Z}^n \setminus \{0\}$, define the length function $f_x : P_n^1 \longrightarrow \mathbf{R}$ by $f_x(a) = {}^t x a x$. The family $\{f_x\}_{x \in \mathbf{Z}^n \setminus \{0\}}$ satisfies (B_1) for $\Gamma = SL_n(\mathbf{Z})$, ($\mathrm{B}_2$) and the condition (C) ([9, Example 1]). Thus one can apply Theorem 3.4 to $\mathcal{E} = (P_n^1, SL_n(\mathbf{Z}), \mathbf{Z}^n \setminus \{0\}, \{f_x\})$. Since the definition of perfection and eutaxy of Definition 3.2 is equivalent to that of Definition 1.3, this case verifies Voronoï's theorem. The length function f_x is convex on P_n^1 for all $x \in \mathbf{Z}^n \setminus \{0\}$.

Example 3.4. Let G be a connected Lie subgroup of $SL_n(\mathbf{R})$ and $G \cdot \mathrm{I}_n$ be the G-orbit of the identity matrix I_n in P_n^1, i.e., $G \cdot \mathrm{I}_n = \{{}^t g g : g \in G\}$. Assume G is invariant by the transpose $g \mapsto {}^t g$. Then $G \cdot \mathrm{I}_n$ is totally geodesic, and hence the restriction $f_x|_{G \cdot \mathrm{I}_n}$ of the length function f_x to $G \cdot \mathrm{I}_n$ is convex for all $x \in \mathbf{Z}^n \setminus \{0\}$. Thus one can apply Theorem 3.5 to $\mathcal{E} = (G \cdot \mathrm{I}_n, G \cap SL_n(\mathbf{Z}), \mathbf{Z}^n \setminus \{0\}, \{f_x|_{G \cdot \mathrm{I}_n}\})$.

Example 3.5. Assume n is even. Let G be the symplectic group, i.e.,

$$G = Sp_n(\mathbf{R}) = \left\{ g \in SL_n(\mathbf{R}) \ : \ {}^t g \begin{pmatrix} 0 & -\mathrm{I}_{n/2} \\ \mathrm{I}_{n/2} & 0 \end{pmatrix} g = \begin{pmatrix} 0 & -\mathrm{I}_{n/2} \\ \mathrm{I}_{n/2} & 0 \end{pmatrix} \right\}.$$

In this case, the family $\{f_x|_{G\cdot \mathrm{I}_n}\}_{x\in\mathbf{Z}^n\setminus\{0\}}$ satisfies the condition (C). This is a particular case of more general family [1, Lemme 3.3]. See [1, Théorème 3.1] for other symmetric spaces of classical type.

Example 3.6. For $X \in M_{n,k}^*(\mathbf{Z})$, define the length function $f_X : P_n^1 \longrightarrow \mathbf{R}$ by $f_X(a) = \det({}^tXaX)$, i.e., $f_X = D_X^k$. Then the family $\{f_X\}_{X\in M_{n,k}^*(\mathbf{Z})}$ satisfies the condition (C) ([9, Proposition 2.8]). Theorem 2.1 is verified again by Theorem 3.4 specialized to $\mathcal{E} = (P_n^1, SL_n(\mathbf{Z}), M_{n,k}^*(\mathbf{Z}), \{f_X\})$.

Example 3.7. We define the subset $C_{n,k}$ of $M_{n,k}(\mathbf{Z}) \times M_{n,k}(\mathbf{Z})$ by

$$C_{n,k} = \{(X,0),(0,Y) \ : \ X, Y \in M_{n,k}^*(\mathbf{Z})\}.$$

This set is stable by the action of $SL_n(\mathbf{Z})$: $(X,Y)g = ({}^tgX, g^{-1}Y)$. For $(X,Y) \in C_{n,k}$, define the length function $f_{(X,Y)} : P_n \longrightarrow \mathbf{R}$ by $f_{(X,Y)}(a) = D_X(a)^k + D_Y(a^{-1})^k$ for $a \in P_n$. Then the quadruplet $\mathcal{E} = (P_n, SL_n(\mathbf{Z}), C_{n,k}, \{f_{(X,Y)}\})$ satisfies the condition (C) [11, Théorèm 5]. Bavard proved the set of extreme points of k-th Bergé–Martinet invariant $F_{\mathrm{BM}}^{(k)}$ coincides with that of $F_{\mathcal{E}}$ ([11, Proposition 2.21]). Thus Voronoï's theorem for $F_{\mathrm{BM}}^{(k)}$ results in that of $F_{\mathcal{E}}$.

In some cases, the finiteness and the algebraicity of perfect points were also proved by Bavard [11, Corollaire 2.12 et Théorème 1]. We explain the simplest case of Bavard's result. Let f be one of the real number field $\mathbf{R}$, the complex number field $\mathbf{C}$ or the Hamilton quaternion field $\mathbf{H}$. For an $n \times n$ matrix $(\lambda_{ij}) \in M_n(\mathsf{f})$ with entries in f, we write $(\lambda_{ij})^*$ for ${}^t(\overline{\lambda}_{ij})$, where $\lambda \mapsto \overline{\lambda}$ stands for the main involution of f. The set $P_n^1(\mathsf{f}) = \{g^*g : g \in SL_n(\mathsf{f})\}$ is a Riemannian symmetric space. We fix a subring o_f in f as $\mathsf{o}_\mathsf{f} = \mathbf{Z}$ if $\mathsf{f} = \mathbf{R}$, $\mathsf{o}_\mathsf{f} = \mathbf{Z}[\sqrt{-1}]$ if $\mathsf{f} = \mathbf{C}$ and $\mathsf{o}_\mathsf{f} = \mathbf{Z}[\mathbf{i},\mathbf{j},\mathbf{k}]$ if $\mathsf{f} = \mathbf{H}$, where $\{1,\mathbf{i},\mathbf{j},\mathbf{k}\}$ denotes the usual quaternion basis of $\mathbf{H}$. For $x \in \mathsf{o}_\mathsf{f}^n \setminus \{0\}$, define the length function $f_x : P_n^1(\mathsf{f}) \longrightarrow \mathbf{R}$ by $f_x(a) = x^*ax$. Then $\mathcal{E} = (P_n^1(\mathsf{f}), SL_n(\mathsf{o}_\mathsf{f}), \mathsf{o}_\mathsf{f}^n \setminus \{0\}, \{f_x\})$ satisfies the condition (C) ([1, Corollaire 3.1]).

Theorem 3.6 (Bavard). *For $\mathcal{E} = (P_n^1(\mathsf{f}), SL_n(\mathsf{o}_\mathsf{f}), \mathsf{o}_\mathsf{f}^n \setminus \{0\}, \{f_x\})$, the number of perfect points in $P_n^1(\mathsf{f})$ modulo $SL_n(\mathsf{o}_\mathsf{f})$ is finite. Any perfect point in $P_n^1(\mathsf{f})$ is algebraic over $\mathbf{Q}$, i.e., which is contained in $M_n(\mathsf{o}_\mathsf{f} \otimes_\mathbf{Z} \overline{\mathbf{Q}})$.*

More generally, Bavard proved such result for some totally geodesic subvarieties in $P_n^1(\mathsf{f})$. However the algebraicity of $\gamma'_{n,k}$ for $k \geq 2$ is still unknown.

In connection with Ash's mass formula (Theorem 3.3), we append Bavard's mass formula [10, Théorème 1]. We recall $\partial^{wr} K_1(m_1)$ denotes the set of all well-rounded points in $\partial K_1(m_1)$. If $a \in \partial^{wr} K_1(m_1)$, then $\mathcal{F}_{S(a)}$ is a compact face. The family $\{\mathcal{F}_{S(a)}\}_{a \in \partial^{wr} K_1(m_1)}$ of compact faces has only finitely many $SL_n(\mathbf{Z})$-orbits by Theorem 3.2. Let $\{\mathcal{F}_1, \cdots, \mathcal{F}_r\}$ be the complete set of representatives of $SL_n(\mathbf{Z})$-orbits in $\{\mathcal{F}_{S(a)}\}_{a \in \partial^{wr} K_1(m_1)}$.

Theorem 3.7 (Bavard). *$\{\mathcal{F}_1, \cdots, \mathcal{F}_r\}$ satisfies*

$$\sum_{i=1}^{r} \frac{(-1)^{\dim \mathcal{F}_i}}{\sharp \Gamma_i} = \chi(SL_n(\mathbf{Z})) = \begin{cases} -1/12 & (n = 2) \\ 0 & (n \geq 3) \end{cases},$$

where Γ_i stands for the stabilizer of $\mathcal{F}_i$ in $SL_n(\mathbf{Z})$.

For a further study of this mass formula, see [15].

3.3. *Voronoï's theorem over an algebraic number field I*

There are two methods of an extension of the base field. One is the additive generalization (35], [38], [45], [46]) and another is the multiplicative generalization ([22], [32]). Both methods give the original Voronoï's theorem if the base field is $\mathbf{Q}$. We first explain the additive generalization.

Let k be an algebraic number field of degree r and $\mathsf{o_k}$ the ring of integers of k. The set of all infinite (resp. real and imaginary) places of k is denoted by p_∞ (resp. p_1 and p_2). Let k_σ be the completion of k at $\sigma \in \mathsf{p}_\infty$, i.e., $\mathsf{k}_\sigma = \mathbf{R}$ if $\sigma \in \mathsf{p}_1$ and $\mathsf{k}_\sigma = \mathbf{C}$ if $\sigma \in \mathsf{p}_2$. We use the étale $\mathbf{R}$-algebra $\mathsf{k}_\mathbf{R} = \mathsf{k} \otimes_\mathbf{Q} \mathbf{R}$, which is identified with $\prod_{\sigma \in \mathsf{p}_\infty} \mathsf{k}_\sigma$. For $\boldsymbol{x} = (x_\sigma) \in \mathsf{k}_\mathbf{R}$, the conjugate $\overline{\boldsymbol{x}}$ of $\boldsymbol{x}$ is defined to be $\overline{\boldsymbol{x}} = (\overline{x}_\sigma)$, where $\overline{x}_\sigma$ is the complex conjugate of x_σ. The trace and the norm of $\mathsf{k}_\mathbf{R}$ are defined as

$$\mathrm{Tr}_{\mathsf{k}_\mathbf{R}}(\boldsymbol{x}) = \sum_{\sigma \in \mathsf{p}_\infty} \mathrm{Tr}_{\mathsf{k}_\sigma/\mathbf{R}}(x_\sigma), \qquad \mathrm{Nr}_{\mathsf{k}_\mathbf{R}}(\boldsymbol{x}) = \prod_{\sigma \in \mathsf{p}_\infty} \mathrm{Nr}_{\mathsf{k}_\sigma/\mathbf{R}}(x_\sigma)$$

for $\boldsymbol{x} = (x_\sigma) \in \mathsf{k}_\mathbf{R}$.

Let $\mathsf{k}_\mathbf{R}^n = \mathsf{k}^n \otimes_\mathbf{Q} \mathbf{R}$ be the $\mathsf{k}_\mathbf{R}$-module of rank n. An element of $\mathsf{k}_\mathbf{R}^n$ is denoted by a column vector $X = {}^t(\boldsymbol{x}_1, \cdots, \boldsymbol{x}_n)$ with $\boldsymbol{x}_i \in \mathsf{k}_\mathbf{R}$, $i = 1, \cdots, n$. The group consisting of $\mathsf{k}_\mathbf{R}$-linear automorphisms of $\mathsf{k}_\mathbf{R}^n$ is denoted by $GL_n(\mathsf{k}_\mathbf{R})$, which is identified with $\prod_{\sigma \in \mathsf{p}_\infty} GL_n(\mathsf{k}_\sigma)$. As an $\mathbf{R}$-vector space, $\mathsf{k}_\mathbf{R}^n$ is equipped with the inner product

$$(X, Y) = \mathrm{Tr}_{\mathsf{k}_\mathbf{R}}({}^t\overline{X}Y) = \mathrm{Tr}_{\mathsf{k}_\mathbf{R}}(\overline{\boldsymbol{x}}_1 \boldsymbol{y}_1 + \cdots + \overline{\boldsymbol{x}}_n \boldsymbol{y}_n),$$

for $X = {}^t(\boldsymbol{x}_1, \cdots, \boldsymbol{x}_n), Y = {}^t(\boldsymbol{y}_1, \cdots, \boldsymbol{y}_n) \in \mathsf{k}_{\mathbf{R}}^n$. The group of isometries

$$O_n(\mathsf{k}_{\mathbf{R}}) = \{g \in GL_n(\mathsf{k}_{\mathbf{R}}) \ : \ (gX, gY) = (X, Y) \ \text{ for all } X, Y \in \mathsf{k}_{\mathbf{R}}^n\}$$

is a maximal compact subgroup of $GL_n(\mathsf{k}_{\mathbf{R}})$. We define the subsets $V_n(\mathsf{k}_{\mathbf{R}})$ and $P_n(\mathsf{k}_{\mathbf{R}})$ of $M_n(\mathsf{k}_{\mathbf{R}})$ as follows:

$$V_n(\mathsf{k}_{\mathbf{R}}) = \{a \in M_n(\mathsf{k}_{\mathbf{R}}) \ : \ (aX, Y) = (X, aY) \ \text{ for all } X, Y \in \mathsf{k}_{\mathbf{R}}^n\},$$
$$P_n(\mathsf{k}_{\mathbf{R}}) = \{a \in V_n(\mathsf{k}_{\mathbf{R}}) \ : \ (aX, X) > 0 \ \text{ for all } X \in \mathsf{k}_{\mathbf{R}}^n \setminus \{0\}\}.$$

The set $V_n(\mathsf{k}_{\mathbf{R}})$ is an $\mathbf{R}$-subspace of $M_n(\mathsf{k}_{\mathbf{R}})$ of dimension $n(n+1)\sharp\mathsf{p}_1/2 + n^2\sharp\mathsf{p}_2$, and $P_n(\mathsf{k}_{\mathbf{R}})$ is a symmetric cone in $V_n(\mathsf{k}_{\mathbf{R}})$. Let $V_n(\mathsf{k}_{\mathbf{R}})^*$ be the dual space of $V_n(\mathsf{k}_{\mathbf{R}})$ as an $\mathbf{R}$-vector space. The trace $\mathrm{Tr}_{M_n(\mathsf{k}_{\mathbf{R}})} \in V_n(\mathsf{k}_{\mathbf{R}})^*$ is defined to be the composition of the matrix trace tr and $\mathrm{Tr}_{\mathsf{k}_{\mathbf{R}}}$, i.e, $\mathrm{Tr}_{M_n(\mathsf{k}_{\mathbf{R}})} = \mathrm{Tr}_{\mathsf{k}_{\mathbf{R}}} \circ \mathrm{tr}$. For $X \in \mathsf{k}_{\mathbf{R}}^n$, define the linear form $\varphi_X \in V_n(\mathsf{k}_{\mathbf{R}})^*$ by $\varphi_X(a) = (aX, X)$ for $a \in V_n(\mathsf{k}_{\mathbf{R}})$.

An o_k-submodule Λ in $\mathsf{k}_{\mathbf{R}}^n$ is called an o_k-lattice if Λ is discrete and $\Lambda \otimes_{\mathbf{Z}} \mathbf{R} = \mathsf{k}_{\mathbf{R}}^n$. Any projective o_k-module in k^n of rank n is regarded as an o_k-lattice in $\mathsf{k}_{\mathbf{R}}^n$ by the natural inclusion $\mathsf{k}^n \subset \mathsf{k}_{\mathbf{R}}^n$. Conversely, for any o_k-lattice Λ in $\mathsf{k}_{\mathbf{R}}^n$, there exists $g \in GL_n(\mathsf{k}_{\mathbf{R}})$ such that $g^{-1}\Lambda$ is a projective o_k-module in k^n (see e.g., [37, Lemma 3.2]). Thus, by Steinitz's theorem, any o_k-lattice is isomorphic with an o_k-module of the form $\mathsf{o}_\mathsf{k}^{n-1} \oplus \mathsf{q}$, where q is an ideal of o_k. Let $\mathsf{q}_1 = \mathsf{o}_\mathsf{k}, \mathsf{q}_2, \cdots, \mathsf{q}_h$ be a complete system of representatives of the ideal class group of k. If $\mathcal{H}_i$ denotes the $GL_n(\mathsf{k}_{\mathbf{R}})$-orbit of the o_k-lattice $\mathsf{o}_\mathsf{k}^{n-1} \oplus \mathsf{q}_i$, then the set $\mathcal{H}(\mathsf{k}_{\mathbf{R}}^n)$ of all o_k-lattices in $\mathsf{k}_{\mathbf{R}}^n$ is given by the disjoint union of $\mathcal{H}_1, \cdots, \mathcal{H}_h$:

$$\mathcal{H}(\mathsf{k}_{\mathbf{R}}^n) = \bigsqcup_{i=1}^{h} \mathcal{H}_i .$$

Each component $\mathcal{H}_i$ is identified with $GL_n(\mathsf{k}_{\mathbf{R}})/GL(\mathsf{o}_\mathsf{k}^{n-1} \oplus \mathsf{q}_i)$, where $GL(\mathsf{o}_\mathsf{k}^{n-1} \oplus \mathsf{q}_i)$ denotes the stabilizer of $\mathsf{o}_\mathsf{k}^{n-1} \oplus \mathsf{q}_i$ in $GL_n(\mathsf{k}_{\mathbf{R}})$.

For $\Lambda \in \mathcal{H}(\mathsf{k}_{\mathbf{R}}^n)$, the minimum $m_+(\Lambda)$ and the discriminant $\mathrm{disc}(\Lambda)$ of Λ are defined to be

$$m_+(\Lambda) = \min_{X \in \Lambda \setminus \{0\}} (X, X), \qquad \mathrm{disc}(\Lambda) = \left(\frac{\omega(\mathsf{k}_{\mathbf{R}}^n/\Lambda)}{\omega(\mathsf{k}_{\mathbf{R}}^n/\mathsf{o}_\mathsf{k}^n)} \right)^2,$$

where ω denotes an invariant measure on $\mathsf{k}_{\mathbf{R}}^n$. We denote by $S_+(\Lambda)$ the set of shortest vectors in Λ, i.e.,

$$S_+(\Lambda) = \{X \in \Lambda \ : \ (X, X) = m_+(\Lambda)\}.$$

As an analog of the Hermite invariant, we consider the function $F_+ : \mathcal{H}(\mathsf{k}_{\mathbf{R}}^n) \longrightarrow \mathbf{R}_{>0}$ defined by

$$F_+(\Lambda) = \frac{m_+(\Lambda)}{\mathrm{disc}(\Lambda)^{1/(rn)}} .$$

Obviously, F_+ depends only on the similar isometry class $\mathbf{R}^\times O_n(\mathsf{k}_{\mathbf{R}})\Lambda$ of Λ, i.e., F_+ is a function on $\mathbf{R}^\times O_n(\mathsf{k}_{\mathbf{R}})\backslash\mathcal{H}(\mathsf{k}_{\mathbf{R}}^n)$. An o_k-lattice $\Lambda \in \mathcal{H}(\mathsf{k}_{\mathbf{R}}^n)$ is said to be extreme if F_+ attains a local maximum on $\mathbf{R}^\times O_n(\mathsf{k}_{\mathbf{R}})\Lambda$.

Definition 3.3. An o_k-lattice $\Lambda \in \mathcal{H}(\mathsf{k}_{\mathbf{R}}^n)$ is said to be perfect if $\{\varphi_X\}_{X\in S_+(\Lambda)}$ spanns $V_n(\mathsf{k}_{\mathbf{R}})^*$, and Λ is said to be eutactic if there are $\rho_X \in \mathbf{R}_{>0}$, $X \in S_+(\Lambda)$, such that

$$\mathrm{Tr}_{M_n(\mathsf{k}_{\mathbf{R}})} = \sum_{X\in S_+(\Lambda)} \rho_X \varphi_X .$$

Leibak [38] proved a weak version of Voronoï's theorem for F_+ restricted to the component $\mathcal{H}_1$ of free o_k-lattices. Leibak's definition of eutaxy is weaker than that of Definition 3.3. Okuda and Yano [45] found a suitable definition of eutaxy to complete Leibak's result.

Theorem 3.8 (Leibak, Okuda and Yano). *An o_k-lattice $\Lambda \in \mathcal{H}(\mathsf{k}_{\mathbf{R}}^n)$ is extreme if and only if Λ is perfect and eutactic.*

By Humbert's reduction theory and Cramer's formula, one has the following finiteness and algebraicity of perfect o_k-lattices.

Theorem 3.9 (Okuda and Yano). *The number of similar isometry classes of perfect o_k-lattices in $\mathcal{H}(\mathsf{k}_{\mathbf{R}}^n)$ is finite. Let $\Lambda \in \mathcal{H}(\mathsf{k}_{\mathbf{R}}^n)$ be a perfect o_k-lattice with $m_+(\Lambda) = 1$. If $g \in GL_n(\mathsf{k}_{\mathbf{R}})$ such that $g^{-1}\Lambda \subset \mathsf{k}^n$, then ${}^t\overline{g}_\sigma g_\sigma \in M_n(\mathsf{k}')$ for all $\sigma \in \mathsf{p}_\infty$, where k' is the Galois closure of k over $\mathbf{Q}$.*

In the case that k is a real quadratic field, a classification of some perfect o_k-lattices of small rank was given by Ong [46] and Leibak [39].

Koecher studied the function F_+ in connection with the reduction theory of $P_n(\mathsf{k}_{\mathbf{R}})$ and proved the finiteness of the number of similar isometry classes of perfect o_k-lattices in $\mathcal{H}_1$ ([35, §9, 10]). The bound $\sharp(S_+(\Lambda)) \leq 2(2^{rn} - 1)$ for all $\Lambda \in \mathcal{H}(\mathsf{k}_{\mathbf{R}}^n)$ is proved by the similar way as in **2.3** (see [35, Lemma 12]).

3.4. *Voronoï's theorem over an algebraic number field II*

We use the same notation as in the previous section. For $a \in P_n(\mathsf{k}_{\mathbf{R}})$, the multiplicative minimum $m_*(a)$ and the discriminant $\mathrm{disc}(a)$ are defined to be

$$m_*(a) = \min_{X \in \mathsf{o}_\mathsf{k}^n \setminus \{0\}} \mathrm{Nr}_{\mathsf{k}_\mathbf{R}}({}^t\overline{X}aX)\,, \qquad \mathrm{disc}(a) = \mathrm{Nr}_{\mathsf{k}_\mathbf{R}}(\det a)\,.$$

We denote by $S_*(a)$ the set of minimal integral vectors, i.e.,

$$S_*(a) = \{X \in \mathsf{o}_\mathsf{k}^n \ : \ \mathrm{Nr}_{\mathsf{k}_\mathbf{R}}({}^t\overline{X}aX) = m_*(a)\}\,.$$

The unit group $\mathsf{o}_\mathsf{k}^\times$ acts on $S_*(a)$ by multiplication. The set $S_*(a)/\mathsf{o}_\mathsf{k}^\times$ of classes is finite ([32, Lemma 1]). Define the function $F_* : P_n(\mathsf{k}_\mathbf{R}) \longrightarrow \mathbf{R}_{>0}$ by

$$F_*(a) = \frac{m_*(a)}{\mathrm{disc}(a)^{1/n}}\,.$$

The group $GL_n(\mathsf{k}_\mathbf{R})$ acts transitively on $P_n(\mathsf{k}_\mathbf{R})$ by $(a,g) \mapsto {}^t\overline{g}ag$ for $a \in P_n(\mathsf{k}_\mathbf{R})$ and $g \in GL_n(\mathsf{k}_\mathbf{R})$. By definition, F_* is invariant by the action of the discrete subgroup $GL_n(\mathsf{o}_\mathsf{k}) \subset GL_n(\mathsf{k}_\mathbf{R})$, i.e., one has

$$F_*({}^t\overline{g}ag) = F_*(a)$$

for all $a \in P_n(\mathsf{k}_\mathbf{R})$ and $g \in GL_n(\mathsf{o}_\mathsf{k})$. Moreover, if we set

$$\mathsf{k}_\mathbf{R}^+ = \{x = (x_\sigma)_{\sigma \in \mathsf{p}_\infty} \in \mathsf{k}_\mathbf{R}^\times \ : \ x_\sigma \in \mathbf{R}_{>0} \ \text{ for all } \sigma \in \mathsf{p}_\infty\}\,,$$

then $F_*(xa) = F_*(a)$ holds for all $a \in P_n(\mathsf{k}_\mathbf{R})$ and $x \in \mathsf{k}_\mathbf{R}^+$. Therefore, F_* is considered as a function on $\mathsf{k}_\mathbf{R}^+ \backslash P_n(\mathsf{k}_\mathbf{R})/GL_n(\mathsf{o}_\mathsf{k})$. An element $a \in P_n(\mathsf{k}_\mathbf{R})$ is said to be extreme if F_* attains a local maximum on the class $\mathsf{k}_\mathbf{R}^+ aGL_n(\mathsf{o}_\mathsf{k})$.

Let $a \in P_n(\mathsf{k}_\mathbf{R})$ and $X \in \mathsf{k}^n \setminus \{0\}$. Then ${}^t\overline{X}aX$ is invertible in $\mathsf{k}_\mathbf{R}^\times$. The map $b \mapsto (bX, ({}^t\overline{X}aX)^{-1}X)$ on $V_n(\mathsf{k}_\mathbf{R})$ defines an $\mathbf{R}$-linear form. We write φ_X^a for this linear form. For $x = (x_\sigma) \in \mathsf{k}_\mathbf{R}^+$, we set $\log(x) = (\log(x_\sigma)) \in \mathsf{k}_\mathbf{R}$. From the definition, it follows $\varphi_X^a(\log(x)a) = \mathrm{Tr}_{\mathsf{k}_\mathbf{R}}(\log(x))$ for $x \in \mathsf{k}_\mathbf{R}^+$. We denote by $\mathsf{k}_\mathbf{R}^1$ the subset of $x \in \mathsf{k}_\mathbf{R}^+$ such that $\mathrm{Tr}_{\mathsf{k}_\mathbf{R}}(\log(x)) = 0$, i.e.,

$$\mathsf{k}_\mathbf{R}^1 = \{x = (x_\sigma)_{\sigma \in \mathsf{p}_\infty} \in \mathsf{k}_\mathbf{R}^+ \ : \ \mathrm{Nr}_{\mathsf{k}_\mathbf{R}}(x) = 1\}\,.$$

Thus φ_X^a is null on the $(\sharp(\mathsf{p}_1)+\sharp(\mathsf{p}_2)-1)$-dimensional subspace $\log(\mathsf{k}_\mathbf{R}^1)\cdot a \subset V_n(\mathsf{k}_\mathbf{R})$.

Definition 3.4. An element $a \in P_n(\mathsf{k}_\mathbf{R})$ is said to be perfect if $\{\varphi_X^a\}_{[X] \in S_*(a)/\mathsf{o}_\mathsf{k}^\times}$ spanns the dual space $(V_n(\mathsf{k}_\mathbf{R})/\log(\mathsf{k}_\mathbf{R}^1)\cdot a)^*$, and a is said to be eutactic if there are $\rho_X \in \mathbf{R}_{>0}$, $[X] \in S_*(a)/\mathsf{o}_\mathsf{k}^\times$, such that the linear form $b \mapsto \mathrm{Tr}_{M_n(\mathsf{k}_\mathbf{R})}(a^{-1}b)$ on $V_n(\mathsf{k}_\mathbf{R})$ is represented as

$$(b \mapsto \mathrm{Tr}_{M_n(\mathsf{k}_\mathbf{R})}(a^{-1}b)) = \sum_{[X] \in S_*(a)/\mathsf{o}_\mathsf{k}^\times} \rho_X \varphi_X^a\,.$$

This definition is due to Coulangeon. Icaza [32, Proposition 3] first proved a weak version of Voronoï's theorem for F_*, and later Coulangeon [23] completed a full version.

Theorem 3.10 (Icaza, Coulangeon). *An element $a \in P_n(\mathsf{k}_{\mathbf{R}})$ is extreme if and only if a is perfect and eutactic.*

The finiteness and the algebraicity of perfect points were also proved by Coulangeon [23, Proposition 4.1].

Theorem 3.11 (Coulangeon). *The number of perfect elements in $P_n(\mathsf{k}_{\mathbf{R}})$ modulo $\mathsf{k}_{\mathbf{R}}^+ GL_n(\mathsf{o}_\mathsf{k})$ is finite. If $a = (a_\sigma) \in P_n(\mathsf{k}_{\mathbf{R}})$ is perfect, then there exists $x = (x_\sigma) \in \mathsf{k}_{\mathbf{R}}^+$ such that $x_\sigma a_\sigma \in M_n(\overline{\mathbf{Q}})$ for all $\sigma \in \mathsf{p}_\infty$.*

Theorem 3.10 is also verified by Bavard's theory. We set $P_n(\mathsf{k}_{\mathbf{R}})^1 = \{a \in P_n(\mathsf{k}_{\mathbf{R}}) : \mathrm{disc}(a) = 1\}$, and for $X \in \mathsf{o}_\mathsf{k}^n \setminus \{0\}$ define the length function $f_X : P_n(\mathsf{k}_{\mathbf{R}})^1 \longrightarrow \mathbf{R}$ by $f_X(a) = \log(\mathrm{Nr}_{\mathsf{k}_{\mathbf{R}}}({}^t\overline{X}aX))$. Let $\mathcal{E} = (P_n(\mathsf{k}_{\mathbf{R}})^1, SL_n(\mathsf{o}_\mathsf{k}), \mathsf{o}_\mathsf{k}^n \setminus \{0\}, \{f_X\})$. Then Bavard [11, Proposition 2.22] proved that the definition of perfection and eutaxy in Definition 3.4 coincides with that in Definition 3.2 for $\mathcal{E}$, and $\mathcal{E}$ satisfies the condition (C). In contrast to Theorem 3.2, the number of classes in $\mathsf{k}_{\mathbf{R}}^+ \backslash P_n(\mathsf{k}_{\mathbf{R}})/GL_n(\mathsf{o}_\mathsf{k})$ of eutactic elements is not finite in general if $\mathsf{k} \neq \mathbf{Q}$, see [23, p.162]. To resolve this problem, Bavard introduced the notion of non-degenerate points in his framework ([11, Définition 1.5]), and proved that the number of classes of non-degenerate eutactic elements is finite ([11, Proposition 2.24]).

If k is an imaginary quadratic field, then F_* is essentially the same as F_+ restricted to $\mathcal{H}_1$. More precisely, we have

$$4F_*({}^t\overline{g}g) = F_+(g\mathsf{o}_\mathsf{k}^n)^2$$

for all $g \in GL_n(\mathsf{k}_{\mathbf{R}}) = GL_n(\mathbf{C})$. In particular, the number of $S_*(a)/\mathsf{o}_\mathsf{k}^\times$ is bounded by $2(4^n - 1)/\sharp(\mathsf{o}_\mathsf{k}^\times)$. In general, any estimate of the number of $S_*(a)/\mathsf{o}_\mathsf{k}^\times$ is unknown.

Problem 3.2. *Bound the maximum* $\max_{a \in P_n(\mathsf{k}_{\mathbf{R}})} \sharp(S_*(a)/\mathsf{o}_\mathsf{k}^\times)$.

The assertion of Theorem 3.10 is true even if the free o_k-lattice o_k^n in the definitions of m_* and $S_*(a)$ is replaced with a general o_k-lattice $\Lambda \in \mathcal{H}(\mathsf{k}_{\mathbf{R}}^n)$. This was verified by Meyer [41, Théorème 3.21] in more general setting.

4. Generalized Hermite constants of flag varieties

A generalization of Hermite's constant to algebraic groups was studied in [63] and [64]. The main problem in this theory is to formulate and verify

Voronoï type theorems. This problem was completely solved by Meyer [41], [42] in the case of GL_n. Some inner forms of GL_n were studied in [24]. It is likely that Bavard's theory applies to many cases, e.g., see Example 4.2 below. However, to approach adelic Voronoï theorems for the generalized Hermite constants involving positive characteristic cases, we will need a suitable definition of perfection for our adelic setting.

4.1. *Generalized Hermite constants*

Let G be a connected affine algebraic group defined over $\mathbf{Q}$. For any $\mathbf{Q}$-algebra A, $G(A)$ stands for the group of A-rational points of G. In particular, $G(\mathbf{A})$ denotes the adele group of G. Let $\mathbf{X}^*_{\mathbf{Q}}(G)$ be the module of $\mathbf{Q}$-rational characters of G. We denote by $G(\mathbf{A})^1$ the subgroup $\{g \in G(\mathbf{A}) \ : \ |\chi(g)|_{\mathbf{A}} = 1 \quad \text{for all } \chi \in \mathbf{X}^*_{\mathbf{Q}}(G)\}$, where $|\cdot|_{\mathbf{A}}$ denotes the usual idele norm of the idele group of $\mathbf{Q}$. By the product formula of the idele norm, $G(\mathbf{Q})$ is contained in $G(\mathbf{A})^1$.

In the following, let G be a connected reductive algebraic group defined over $\mathbf{Q}$. We fix a minimal $\mathbf{Q}$-parabolic subgroup P of G and a Levi subgroup M_P of P. The maximal central $\mathbf{Q}$-split torus Z_P of M_P is a maximal $\mathbf{Q}$-split torus of G. We choose a maximal $\mathbf{Q}$-parabolic subgroup Q of G and its Levi subgroup M_Q such that $P \subset Q$ and $M_P \subset M_Q$. Let Z_G be the maximal central $\mathbf{Q}$-split torus of G. Since Q is maximal, the module $\mathbf{X}^*_{\mathbf{Q}}(M_Q/Z_G)$ is of rank one, and hence there is a unique generator $\widehat{\alpha}_Q$ of $\mathbf{X}^*_{\mathbf{Q}}(M_Q/Z_G)$ such that the restriction of $\widehat{\alpha}_Q$ to Z_P/Z_G is a positive scalar multiple of a positive simple root with respect to (P, Z_P). Let U_Q be the unipotent radical of Q, and let K be a maximal compact subgroup of $G(\mathbf{A})$ such that $G(\mathbf{A}) = P(\mathbf{A})K$. Then the height function $H_Q : G(\mathbf{A}) \longrightarrow \mathbf{R}_{>0}$ is defined by

$$H_Q(umh) = |\widehat{\alpha}_Q(m)|_{\mathbf{A}}^{-1}$$

for $u \in U_Q(\mathbf{A})$, $m \in M_Q(\mathbf{A})$ and $h \in K$. Indeed H_Q is a function on the space $Z_G(\mathbf{A})Q(\mathbf{A})^1\backslash G(\mathbf{A}) = Q(\mathbf{A})^1\backslash G(\mathbf{A})^1$. Define the function F_Q : $G(\mathbf{A}) \longrightarrow \mathbf{R}_{>0}$ by

$$F_Q(g) = \min_{[v] \in Q(\mathbf{Q})\backslash G(\mathbf{Q})} H_Q(vg)\,.$$

The generalized Hermite constant γ_Q of $Q\backslash G$ is defined to be the maximum

$$\gamma_Q = \max_{[g] \in Z_G(\mathbf{A})G(\mathbf{Q})\backslash G(\mathbf{A})/K} F_Q(g) = \max_{[g] \in G(\mathbf{Q})\backslash G(\mathbf{A})^1/K} F_Q(g)\,.$$

We assume the following two conditions for G and Q:

(C$_1$) $G(\mathbf{A}) = G(\mathbf{Q})G(\mathbf{R})K$.
(C$_2$) $G(\mathbf{Q}) = Q(\mathbf{Q})G(\mathbf{Z})$, where $G(\mathbf{Z}) = G(\mathbf{Q}) \cap G(\mathbf{R})K$.

The condition (C$_1$) means that G is of class number one. By [17, Proposition 7.5], the condition (C$_2$) is satisfied if M_Q is of class number one. Then γ_Q is represented as

$$\gamma_Q = \max_{[g]\in G(\mathbf{Z})\backslash G(\mathbf{R})/K_\infty} F_Q(g) = \max_{[g]\in G(\mathbf{Z})\backslash G(\mathbf{R})/K_\infty} \min_{\gamma\in G(\mathbf{Z})} H_Q^\infty(\gamma g)\,,$$

where K_∞ and H_Q^∞ denote the infinite components of K and H_Q, respectively.

Example 4.1. For $k = 1, \cdots, n-1$, let

$$R_k(\mathbf{Q}) = \left\{ \begin{pmatrix} a & b \\ 0 & d \end{pmatrix} \ : \ a \in GL_k(\mathbf{Q}),\ b \in M_{k,n-k}(\mathbf{Q}),\ d \in GL_{n-k}(\mathbf{Q}) \right\}.$$

Then R_k is a maximal $\mathbf{Q}$-parabolic subgroup of GL_n. It is well-known that GL_n and R_k satisfy both conditions (C$_1$) and (C$_2$). The character $\widehat{\alpha}_{R_k}$ is given by

$$\widehat{\alpha}_{R_k}\left(\begin{pmatrix} a & 0 \\ 0 & d \end{pmatrix}\right) = (\det a)^{(n-k)/\gcd(k,n-k)}(\det d)^{-k/\gcd(k,n-k)}\,.$$

For $\gamma \in GL_n(\mathbf{Z})$, X_γ denotes the n by k matrix consisting of the first k-colums of γ. It is an easy exercise to prove that

$$H_{R_k}^\infty(\gamma g)^{\gcd(k,n-k)/n} = \det({}^tX_{\gamma^{-1}}{}^tg^{-1}g^{-1}X_{\gamma^{-1}})^{1/2}$$

holds for any $\gamma \in GL_n(\mathbf{Z})$ and $g \in SL_n(\mathbf{R})$. From this relation, it follows that $(\gamma_{R_k})^{2\gcd(k,n-k)/n}$ equals the Rankin constant $\gamma_{n,k}$. Thus, Coulangeon's result in **2.2** is interpreted as Voronoï's theorem of the function F_{R_k}.

Example 4.2. Let B be a non degenerate bilinear form on $\mathbf{Q}^n$ defined by

$$B(x, y) = x_1y_1 - x_2y_2 - \cdots - x_ny_n$$

for $x = {}^t(x_1, \cdots, x_n)$ and $y = {}^t(y_1, \cdots, y_n) \in \mathbf{Q}^n$. We assume $n \geq 3$ and put $\boldsymbol{e}_{11} = {}^t(1, 1, 0, \cdots, 0) \in \mathbf{Q}^n$. Let $\mathcal{N}_B(\mathbf{Q})$ be the set of all non-zero isotropic vectors in $\mathbf{Q}^n$ with respect to B. The special orthogonal group $SO_B(\mathbf{Q})$ of B transitively acts on $\mathcal{N}_B(\mathbf{Q})$, i.e., one has $\mathcal{N}_B(\mathbf{Q}) = SO_B(\mathbf{Q})\boldsymbol{e}_{11}$. Let $P(\mathbf{Q})$ be the stabilizer of the isotropic line $\mathbf{Q}\boldsymbol{e}_{11}$ in $SO_B(\mathbf{Q})$. Then P is a unique proper $\mathbf{Q}$-parabolic subgroup of the algebraic group SO_B up to $SO_B(\mathbf{Q})$-conjugates. For any finite prime p, K_p denotes the stabilizer of the $\mathbf{Z}_p$-lattice $\mathbf{Z}_p^n$ in $SO_B(\mathbf{Q}_p)$. Since $\mathbf{Z}_p^n$ is a unimodular maximal lattice with respect to B, K_p is a maximal compact subgroup of $SO_B(\mathbf{Q}_p)$. At the infinite place ∞, the stabilizer K_∞ of the vector $\boldsymbol{e} = {}^t(1, 0, \cdots, 0)$ in $SO_B(\mathbf{R})$ gives a maximal compact subgroup of $SO_B(\mathbf{R})$. The intersection $SO_B(\mathbf{Q}) \cap (SO_B(\mathbf{R}) \times \prod_{p<\infty} K_p)$ is the stabilizer $SO_B(\mathbf{Z})$ of the lattice $\mathbf{Z}^n$ in $SO_B(\mathbf{Q})$. The adele group $SO_B(\mathbf{A})$ has the Iwasawa decomposition: $SO_B(\mathbf{A}) = P(\mathbf{A})(\prod_{p\leq\infty} K_p)$. The height function $H_P : SO_B(\mathbf{A}) \longrightarrow \mathbf{R}_{>0}$ is given by

$$H_P(g) = \|g^{-1}\boldsymbol{e}_{11}\|_{\mathbf{A}^n} = \prod_{p\leq\infty} \|g_p^{-1}\boldsymbol{e}_{11}\|_p$$

for $g = (g_p)_{p\leq\infty} \in SO_B(\mathbf{A})$. Here, the local height $\|\cdot\|_p$ is defined by

$$\|x\|_p = \begin{cases} \sqrt{x_1^2 + \cdots + x_n^2} & (p = \infty) \\ \max(|x_1|_p, \cdots, |x_n|_p) & (p < \infty) \end{cases}$$

for $x = {}^t(x_1, \cdots, x_n) \in \mathbf{Q}_p^n$. If $g = umh$ with $u \in U_P(\mathbf{A})$, $m \in M_P(\mathbf{A})$ and $h \in \prod_{p\leq\infty} K_p$, then $H_P(g)$ equals the idele norm of the first component of the vector $m^{-1}\boldsymbol{e}_{11}$. Since the class number of the indefinite lattice $(B, \mathbf{Z}^n)$ equals one, SO_B satisfies the condition (C$_1$), and hence the generalized Hermite constant γ_P is written as

$$\gamma_P = \max_{[g]\in SO_B(\mathbf{Z})\backslash SO_B(\mathbf{R})/K_\infty} \min_{[v]\in P(\mathbf{Q})\backslash SO_B(\mathbf{Q})} \|g^{-1}v^{-1}\boldsymbol{e}_{11}\|_{\mathbf{A}^n} .$$

Let $\mathcal{N}_B^*(\mathbf{Z})$ denote the set of primitive vectors in $\mathcal{N}_B(\mathbf{Q}) \cap \mathbf{Z}^n$, i.e.,

$$\mathcal{N}_B^*(\mathbf{Z}) = \{x \in \mathbf{Z}^n \setminus \{0\} \ : \ B(x, x) = 0 \text{ and } \gcd(x_1, \cdots, x_n) = 1\} .$$

From $SO_B(\mathbf{Q})\boldsymbol{e}_{11} = \mathcal{N}_B(\mathbf{Q}) = \mathbf{Q}^\times \cdot \mathcal{N}_B^*(\mathbf{Z})$, it follows that

$$\min_{[v]\in P(\mathbf{Q})\backslash SO_B(\mathbf{Q})} \|g^{-1}v^{-1}\boldsymbol{e}_{11}\|_{\mathbf{A}^n} = \min_{x\in\mathcal{N}_B^*(\mathbf{Z})} \|g^{-1}x\|_{\mathbf{A}^n} .$$

Since x is a primitive isotropic vector and $g \in SO_B(\mathbf{R})$, we have

$$\|g^{-1}x\|_{\mathbf{A}^n} = \|g^{-1}x\|_\infty \times \prod_{p<\infty} \|x\|_p = \|g^{-1}x\|_\infty ,$$

and hence

$$\gamma_P = \max_{[g]\in K_\infty\backslash SO_B(\mathbf{R})/SO_B(\mathbf{Z})} \min_{x\in\mathcal{N}_B^*(\mathbf{Z})} \|gx\|_\infty .$$

We may compare this with Example 3.2. The group $SO_0(1, n-1)$ is the identity connected component of $SO_B(\mathbf{R})$. Since $K_\infty\backslash SO_B(\mathbf{R})/SO_B(\mathbf{Z}) = K_e\backslash SO_0(1, n-1)/\Gamma$ and $\mathcal{N}_B^*(\mathbf{Z})$ equals the subset of primitive vectors in $\mathbf{Z}^n \setminus \{0\} \cap \partial\overline{\Omega}_n$, one has

$$2^{n/2} \max_{a\in\Omega_n} F_{(\Omega_n, \mathbf{Z}^n)}(a) \leq (\gamma_P)^n . \tag{3}$$

If $n = 3$, then $\gamma_P = \sqrt{\gamma_2} = \sqrt{2/\sqrt{3}}$ since SO_B is isomorphic with PGL_2 over $\mathbf{Q}$. In this case, the equality of (3) holds. In general n, Birch and Davenport's theorem [16, Theorem A] gives $\gamma_P \leq (\sqrt{2n}\gamma_{n-1})^{(n-1)/2}$. It is unknown whether the equality $2^{n/2} \max_{a\in\Omega_n} F_{(\Omega_n,\mathbf{Z}^n)}(a) = (\gamma_P)^n$ holds for all n or not. Bavard's theory applies to this example. By using the same notation as in Example 3.4, we can choose $\mathcal{E} = (SO_0(1, n-1) \cdot \mathrm{I}_n, \Gamma, \mathcal{N}_B^*(\mathbf{Z}), \{f_x|_{SO_0(1,n-1)}\})$ as a quadruplet in question (cf. [9, Proposition 2.6]).

We should mention two remarks. Let G and P be the same as above, i.e., G is a connected reductive algebraic group defined over $\mathbf{Q}$ and P a minimal $\mathbf{Q}$-parabolic subgroup. First, let R be a $\mathbf{Q}$-parabolic subgroup of G such that $P \subset R$. Then the generalized Hermite constant γ_R is defined even if R is not maximal. However, in this case, there is no canonical height function on $R(\mathbf{A})^1\backslash G(\mathbf{A})^1$. As a result, there are infinitely many multiplicatively independent generalized Hermite constants of $R\backslash G$. To define a height function on $R(\mathbf{A})^1\backslash G(\mathbf{A})^1$, we choose a $\mathbf{Q}$-rational embedding of the projective variety $R\backslash G$ into a projective space. Such an embedding is constructed by a strongly $\mathbf{Q}$-rational irreducible representation $\pi : G \longrightarrow GL_N$ of G. By strongly $\mathbf{Q}$-rational, we mean that π is a $\mathbf{Q}$-rational morphism and the highest weight line l_π in $\overline{\mathbf{Q}}^N$ of π is defined over $\mathbf{Q}$. Assume $\pi(R)$ is the stabilizer of l_π in $\pi(G)$. Then the map $g \mapsto \pi(g^{-1})l_\pi$ gives rise to a $\mathbf{Q}$-rational embedding of $R\backslash G$ into the projective space $\mathbf{P}^{N-1}$. If R_1 denotes the maximal parabolic subgroup of GL_N defined in a similar fashion as in Example 4.1, then $\mathbf{P}^{N-1}$ is identified with $R_1\backslash GL_N$, and the height H_{R_1} is defined on $R_1(\mathbf{A})^1\backslash GL_N(\mathbf{A})^1$. The composition of π and H_{R_1} gives a height function $H_{R,\pi}$ on $R(\mathbf{A})^1\backslash G(\mathbf{A})^1$. Then the generalized Hermite constant $\gamma_{R,\pi}$ is defined to be

$$\gamma_{R,\pi} = \max_{[g]\in R(\mathbf{A})^1\backslash G(\mathbf{A})^1} F_{R,\pi}(g)\,, \quad \text{where } F_{R,\pi}(g) = \min_{[v]\in R(\mathbf{Q})\backslash G(\mathbf{Q})} H_{R,\pi}(vg)\,.$$

Second remark is a base change from $\mathbf{Q}$ to a number field k. The adelic definition of the generalized Hermite constants is immediately extended to reductive groups defined over k. Thus one can define $\gamma_{R,\pi}(\mathsf{k})$ for a connected reductive group G defined over k, a parabolic k-subgroup $R \subset G$ and a strongly k-rational representation π. In this notation, the constant $\gamma_{R,\pi}(\mathbf{Q})$ means $\gamma_{R,\pi}$. We write $\gamma_n(\mathsf{k})$ (resp. $\gamma_n(\mathsf{k})_1$ and $\gamma_{n,k}(\mathsf{k})$) for the generalized Hermite constant of $R_1\backslash GL_n$ (resp. $(R_1 \cap SL_n)\backslash SL_n$ and $R_k\backslash GL_n$) defined over k. When k is a totally real number field, $\gamma_2(\mathsf{k})_1$ was implicitly occurred in Cohn's paper [19, §5, §9]. Newman [44, Chapter XI] defined $\gamma_n(\mathsf{k})_1$ for imaginary quadratic fields k. For a general number field k, $\gamma_n(\mathsf{k})_1$ was studied by Icaza [32] in terms of Humbert forms, i.e., $\gamma_n(\mathsf{k})_1$ is relating with the function F_* as follows:

$$(\gamma_n(\mathsf{k})_1)^{2n} = \max_{[a]\in \mathsf{k}_{\mathbf{R}}^{+}\backslash P_n(\mathsf{k}_{\mathbf{R}})/GL_n(\mathfrak{o}_{\mathsf{k}})} F_*(a)\,,$$

(cf. [65, §3]). As a generalization of Rankin's constant, Thunder [61] defined $\gamma_{n,k}(\mathsf{k})$ by using twisted heights on Grassmann varieties. The above definition of $\gamma_{R,\pi}(\mathsf{k})$ was given in [63]. Meyer [41] investigated $\gamma_{R,\pi}(\mathsf{k})$ when $G = GL_n$ and π is a Schur module realized as a polynomial representation, and established Voronoï's theorem for $F_{R,\pi}$ by using Bavard's theory. Among other results in [41], we present here only the algebraicity of $\gamma_{R,\pi}(\mathsf{k})$:

Theorem 4.1 (Meyer). *Let k be an arbitrary algebraic number field. When $G = GL_n$, all $\gamma_{R,\pi}(\mathsf{k})$ are algebraic numbers.*

In the case that k is a function field of one variable over a finite field, $\gamma_Q(\mathsf{k})$ was studied in [64].

4.2. *Generalized Hermite constants of Sp_n*

Let $n = 2m$ be an even integer. We consider a symplectic group

$$G(\mathbf{Q}) = Sp_n(\mathbf{Q}) = \left\{ g \in GL_{2m}(\mathbf{Q}) \;:\; {}^t g \begin{pmatrix} 0 & -\mathrm{I}_m \\ \mathrm{I}_m & 0 \end{pmatrix} g = \begin{pmatrix} 0 & -\mathrm{I}_m \\ \mathrm{I}_m & 0 \end{pmatrix} \right\}.$$

For $1 \le k \le m$, Q_k denotes the maximal parabolic subgroup of G given as follows:

$$Q_k(\mathbf{Q}) = U_k(\mathbf{Q}) L_k(\mathbf{Q})\,,$$

$$L_k(\mathbf{Q}) = \left\{ \delta(a,b) = \begin{pmatrix} a & 0 & 0 & 0 \\ 0 & b_{11} & 0 & b_{12} \\ 0 & 0 & {}^t a^{-1} & 0 \\ 0 & b_{21} & 0 & b_{22} \end{pmatrix} \;:\; \begin{array}{l} a \in GL_k(\mathbf{Q}) \\ b = (b_{ij}) \in Sp_{2(m-k)}(\mathbf{Q}) \end{array} \right\},$$

$$U_k(\mathbf{Q}) = \left\{ \begin{pmatrix} \mathrm{I}_k & * & * & * \\ 0 & \mathrm{I}_{m-k} & * & 0 \\ 0 & 0 & \mathrm{I}_k & 0 \\ 0 & 0 & * & \mathrm{I}_{m-k} \end{pmatrix} \in G(\mathbf{Q}) \right\}.$$

The module of $\mathbf{Q}$-rational characters $\mathbf{X}^*_{\mathbf{Q}}(L_k)$ of L_k is a free $\mathbf{Z}$-module of rank 1 and its base is given by

$$\widehat{\alpha}_{Q_k}(\delta(a,b)) = \det a\,.$$

We fix a good maximal compact subgroup K of $G(\mathbf{A})$ so that $G(\mathbf{A})$ has the Iwasawa decomposition $G(\mathbf{A}) = Q_k(\mathbf{A})K$. Since G and Q_k satisfy both conditions (C_1) and (C_2), one has

$$\gamma_{Q_k} = \max_{[g]\in G(\mathbf{Z})\backslash G(\mathbf{R})/K_\infty} \min_{\gamma\in G(\mathbf{Z})} H^\infty_{Q_k}(\gamma g)\,. \tag{4}$$

We restrict ourselves to the case $k = m$. An element of $L_m(\mathbf{A})$ is denoted by

$$\delta(a) = \begin{pmatrix} a & 0 \\ 0 & {}^t a^{-1} \end{pmatrix}, \qquad (a \in GL_m(\mathbf{A})).$$

By definition, we have

$$H^\infty_{Q_m}(u\delta(a)h) = |\det a|^{-1}, \qquad (u \in U_m(\mathbf{R}),\ \delta(a) \in L_m(\mathbf{R}),\ h \in K_\infty).$$

Let

$$\mathsf{H}_m = \{Z \in M_m(\mathbf{C}) \ :\ \mathrm{Re}\, Z \in V_m,\ \mathrm{Im}\, Z \in P_m\}$$

be the Siegel upper half space. The group $G(\mathbf{R})$ acts on H_m by

$$g\langle Z\rangle = (aZ+b)(cZ+d)^{-1}, \qquad (g = \begin{pmatrix} a & b \\ c & d \end{pmatrix} \in G(\mathbf{R}),\ Z \in \mathsf{H}_m).$$

Since it is possible to choose the maximal compact subgroup K_∞ as the stabilizer of $Z_0 = \sqrt{-1}\mathrm{I}_m \in \mathsf{H}_m$ in $G(\mathbf{R})$, we have $\mathrm{Im}\,\{(u\delta(a)h)\langle Z_0\rangle\} = a\,{}^t a$, and hence

$$H^\infty_{Q_m}(g) = (\det \mathrm{Im}\,\{g\langle Z_0\rangle\})^{-1/2}$$

for any $g \in G(\mathbf{R})$. Combining this with (4), we get

$$\gamma_{Q_m} = \max_{[g]\in G(\mathbf{Z})\backslash G(\mathbf{R})/K_\infty} \min_{\gamma\in G(\mathbf{Z})} (\det \mathrm{Im}\,\{\gamma g\langle Z_0\rangle\})^{-1/2}\,.$$

Since $g\langle Z_0\rangle$ runs over a fundamental domain of $G(\mathbf{Z})\backslash\mathsf{H}_m$, we have

$$\gamma_{Q_m} = \frac{1}{\displaystyle\min_{[Z]\in G(\mathbf{Z})\backslash \mathsf{H}_m} \max_{\gamma\in G(\mathbf{Z})} (\det \mathrm{Im}\,\{\gamma\langle Z\rangle\})^{1/2}}\,.$$

Note that

$$\det \mathrm{Im}\,\{\gamma\langle Z\rangle\} = |\det(cZ+d)|^{-2} \det \mathrm{Im}\, Z \ \text{ for } \ \gamma = \begin{pmatrix} a & b \\ c & d \end{pmatrix} \in G(\mathbf{Z}).$$

Siegel's fundamental domain S_m of $G(\mathbf{Z})\backslash\mathsf{H}_m$ is given as follows:

$$\mathsf{S}_m = \left\{ Z \in \mathsf{H}_m \ :\ \begin{array}{l} \bullet\ |\det(cZ+d)| \geq 1 \text{ for all } \begin{pmatrix} * & * \\ c & d \end{pmatrix} \in G(\mathbf{Z}) \\ \bullet\ \mathrm{Im}\, Z \in \mathsf{M}_m, \quad |\mathrm{Re}\, Z_{ij}| \leq 1/2 \text{ for all } i, j \end{array} \right\},$$

where M_m denotes Minkowski's domain:

$$\left\{ Y \in P_m \ : \ \begin{array}{l} \bullet\ {}^t x Y x \geq Y_{ii} \text{ for all } x \in \mathbf{Z}^m \text{ with } \gcd(x_i, \cdots, x_m) = 1 \\ \bullet\ Y_{j,j+1} > 0, \ \ i = 1, \cdots, m, \ j = 1, \cdots, m-1 \end{array} \right\}.$$

From

$$Z \in \mathsf{S}_m \implies \max_{\gamma \in G(\mathbf{Z})} \det \operatorname{Im} \{\gamma\langle Z \rangle\} = \det \operatorname{Im} Z,$$

it follows that

$$\gamma_{Q_m} = \frac{1}{\min\limits_{Z \in \mathsf{S}_m} (\det \operatorname{Im} Z)^{1/2}},$$

namely

$$\min_{Z \in \mathsf{S}_m} \det \operatorname{Im} Z = \frac{1}{(\gamma_{Q_m})^2}.$$

If $m = 1$, we have $\min_{Z \in \mathsf{S}_1} \det \operatorname{Im} Z = \sqrt{3}/2$. Recently, Kawamura [33] determined the actual value of $\min_{Z \in \mathsf{S}_2} \det \operatorname{Im} Z$ by using Gottschling's description of S_2.

Theorem 4.2 (Kawamura). *One has* $\min_{Z \in \mathsf{S}_2} \det \operatorname{Im} Z = 2/3$, *and hence* $\gamma_{Q_2} = \sqrt{3/2}$. *This minimum is attained only when* $Z = Z_8$ *or* $-\overline{Z_8}$, *where*

$$Z_8 = \frac{1}{3}\begin{pmatrix} 1 & -1 \\ -1 & 1 \end{pmatrix} + \sqrt{-1}\frac{\sqrt{2}}{3}\begin{pmatrix} 2 & 1 \\ 1 & 2 \end{pmatrix}.$$

The boundary $\partial\mathsf{S}_2$ of S_2 is described by 28 polynomials in 6 real variables. Hayata, Oda and Yatouga [30] computed 0-dimensional cells of $\partial\mathsf{S}_2$. There are at least 180 0-dimensional cells of $\partial\mathsf{S}_2$. The points Z_8 and $-\overline{Z_8}$ are contained in Hayata's list.

For general $n = 2m$, we have the following bound by [63, Example 3]:

$$\min_{Z \in \mathsf{S}_m} \det \operatorname{Im} Z \leq \left\{ \frac{1}{m+1} \cdot \frac{\prod\limits_{j=1}^{[\frac{m-1}{2}]} \xi(2j+1)}{\prod\limits_{j=1}^{m} \xi(m+j)} \right\}^{\frac{2}{m+1}},$$

where $\xi(s)$ denotes the zeta function $\pi^{-s/2}\Gamma(s/2)\zeta(s)$.

Problem 4.1. *Give a good lower bound of* $\min_{Z \in \mathsf{S}_m} \det \operatorname{Im} Z$.

Problem 4.2. *Formulate a Voronoï type theorem for the function* $Z \mapsto \det \operatorname{Im} Z$ *on* S_m.

We finish this example by some observation. Let

$$R_m(\mathbf{Q}) = \left\{ \begin{pmatrix} a & b \\ 0 & d \end{pmatrix} \ : \ a, d \in GL_m(\mathbf{Q}),\ b \in M_m(\mathbf{Q}) \right\}.$$

be a maximal parabolic subgroup of GL_n. We have $\gamma_{R_m} = \gamma_{n,m}$ as seen in Example 4.1. It is obvious that $Sp_n(\mathbf{A}) \subset GL_n(\mathbf{A})$, $Sp_n(\mathbf{A}) \cap R_m(\mathbf{A}) = Q_m(\mathbf{A})$ and $H_{R_m}(g) = H_{Q_m}(g)^2$ for $g \in Sp_n(\mathbf{R})$. Since $Q_m(\mathbf{Q})\backslash Sp_n(\mathbf{Q}) \subset R_m(\mathbf{Q})\backslash GL_n(\mathbf{Q})$, the inequality

$$F_{R_m}(g) \leq F_{Q_m}(g)^2 \leq (\gamma_{Q_m})^2$$

holds for any $g \in Sp_n(\mathbf{A})$. If the maximum of F_{R_m} is attained on a point in $Sp_n(\mathbf{A})$, $\gamma_{n,m}$ would be bounded by $(\gamma_{Q_m})^2$. This happens when $m = 1$ or 2 and in fact we have $\gamma_{2,1} = (\gamma_{Q_1})^2$ and $\gamma_{4,2} = (\gamma_{Q_2})^2$.

4.3. *Variation of the set of minimal points*

We return to the general setting of **4.1**. We write Y_Q and X_Q for $Q(\mathbf{A})^1\backslash G(\mathbf{A})^1$ and $Q(\mathbf{Q})\backslash G(\mathbf{Q})$, respectively. For $g \in G(\mathbf{A})^1$, the set $S_Q(g)$ of minimal points in X_Q is defined as

$$S_Q(g) = \{[x] \in X_Q \ : \ H_Q(xg) = F_Q(g)\}.$$

By Northcott's theorem, $S_Q(g)$ is a finite set. We prove the following:

Proposition 4.1. *For* $g \in G(\mathbf{A})^1$, *there is a neighbourhood* $\mathcal{U}$ *of* g *in* $G(\mathbf{A})^1$ *such that* $S_Q(g') \subset S_Q(g)$ *for all* $g' \in \mathcal{U}$.

Proof. We define the operator norm $\|g\|_Q$ of $g \in G(\mathbf{A})^1$ by

$$\|g\|_Q = \sup_{[y] \in Y_Q} \frac{H_Q(yg)}{H_Q(y)} = \max_{h \in K} H_Q(hg).$$

It is easy to see the following properties:

- $\|g_1 g_2\|_Q \leq \|g_1\|_Q \|g_2\|_Q$ for all $g_1, g_2 \in G(\mathbf{A})^1$.
- $\|h_1 g h_2\|_Q = \|g\|_Q$ for all $g \in G(\mathbf{A})^1$ and $h_1, h_2 \in K$.
- $\|h\|_Q = 1$ for all $h \in K$.
- $g \mapsto \|g\|_Q$ is continuous on $G(\mathbf{A})^1$.

We fix a $g \in G(\mathbf{A})^1$ and put $C = \min_{[x]\in X_Q\setminus S_Q(g)} H_Q(xg)$. Then $F_Q(g) < C$ and we can take a constant δ so that $1 < \delta < C/F_Q(g)$. Since $g \mapsto F_Q(g)$ is continuous on $G(\mathbf{A})^1$, the set

$$\mathcal{U} = \left\{ u \in G(\mathbf{A})^1 \ : \ \|u^{-1}\|_Q < \frac{C}{\delta F_Q(g)}, \ \frac{F_Q(gu)}{F_Q(g)} < \delta \right\}$$

is a neighbourhood of the identity in $G(\mathbf{A})^1$. Let $u \in \mathcal{U}$ and $[x] \in S_Q(gu)$. We have

$$\|u^{-1}\|_Q^{-1} H_Q(xg) \leq H_Q(xgu) = F_Q(gu)\,.$$

If $[x] \notin S_Q(g)$, then $C \leq H_Q(xg)$ and

$$\delta F_Q(g) \leq \delta F_Q(g) H_Q(xg)/C < \|u^{-1}\|_Q^{-1} H_Q(xg) \leq F_Q(gu) < \delta F_Q(g)\,.$$

This is a contradiction. Therefore, we have $[x] \in S_Q(g)$, and hence $S_Q(gu) \subset S_Q(g)$ for any $u \in \mathcal{U}$. □

This proposition is a generalization of Proposition 2.2. As a consequence, one can define the local maximality of $S_Q(g)$. This leads us to similar problems as in **2.3**, e.g.,

Problem 4.3. *If $g \in G(\mathbf{A})^1$ is an extreme point of F_Q, is $S_Q(g)$ locally maximal?*

Problem 4.4. *Bound $\sharp S_Q(g)$ by a constant independent of g.*

References

1. H. Akrout, *Théorèm de Voronoï dans les espaces symétriques*, Canad. J. Math. **54** (2002), 449–467.
2. A. Ash, D. Mumford, M. Rappaport and Y. Tai, *Smooth Compactification of Locally Symmetric Varieties*, Lie Groups: History, Frontiers and Applications, vol. 4, Math. Sci. Press, Brookline, Massachusetts, 1975.
3. A. Ash, *On eutactic forms*, Can. J. Math. **29** (1977), 1040–1054.
4. A. Ash, *On the existence of eutactic forms*, Bull. London Math. Soc. **12** (1980), 192–196.
5. R. Baeza, R. Coulangeon, M. I. Icaza and M. O'Ryan, *Hermite's constant for quadratic number fields*, Experiment. Math. **16** (2007), no.4, 455–462.
6. E. S. Barnes, *On a theorem of Voronoi*, Proc. Cambridge Philos. Soc. **53** (1957), 537–539.
7. E. S. Barnes and M. J. Cohn, *On the inner product of positive quadratic forms*, J. Lond. Math. Soc. (2) **12**(1976), 32–36.
8. C. Batut, *Classification of quintic eutactic forms*, Math. Comp. **70** (2001), 395–417.

9. C. Bavard, *Systole et invariant d'Hermite*, J. reine angew. Math. **482** (1997), 93–120.
10. C. Bavard, *Classes minimales de re'seaux et re'tractions ge'ome'triques e'quivariantes dans les espaces syme'triques*, J. London Math. Soc. (2) **64** (2001), 275–286.
11. C. Bavard, *Théorie de Voronoï géométrique. Propriétés de finitude pour les familes de résaux et analgues*, Bull. Soc Math. France **133** (2005), 205–257.
12. A.-M. Bergé, *Minimal vectors of pairs of dual lattices*, J. Number Theory **52** (1995), 284–298.
13. A.-M. Bergé and J. Martinet, *Sur un problém de dualité lié aux sphères en géométrié des nombres*, J. Number Theory **32** (1989), 14–42.
14. A.-M. Bergé and J. Martinet, *Sur la classification des réseaux eutactiques*, J. London Math. Soc. (2) **53** (1996), 417–432.
15. A.-M. Bergé and J. Martinet, *On weakly eutactic forms*, J. Lond. Math. Soc. (2) **75** (2007), 187–198.
16. B. J. Birch and H. Davenport, Quadtratic equations in several variables, Proc. Cambridge Phil. Soc. **54** (1958), 135–138.
17. A. Borel, *Some finiteness properties of adele groups over number fields*, Publ. Math. Inst. Hautes Etud. Sci. **16** (1963), 5–30.
18. J. M. Borwein and A. S. Lewis, *Convex Analysis and Nonlinear Optimization: Theory and Examples*, second edition, CMS Books in Mathematics, vol. 3, Springer-Verlag, New York, 2006.
19. H. Cohn, *On the shape of the fundamental domain of the Hilbert modular group*, Proc. Symp. Pure Math. **VIII**, Amer. Math. Soc. (1965), 190–202.
20. J. H. Conway and N. J. A. Sloane, *Sphere Packings, Lattices and Groups*, Grund. Math. Wiss. Vol. 290, Springer-Verlag 1999.
21. R. Coulangeon, M. I. Icaza and M. O'Ryan, *Lenstra's constant and extreme forms in number fields*, Experiment. Math. **16** (2007), no.4, 455–462.
22. R. Coulangeon, *Réseaux k-extremes*, Proc. London Math. Soc **73** (1996), no.3, 555–574.
23. R. Coulangeon, *Voronoï theory over algebraic number fields*, in *Réseaux euclidens, designs sphériques et formes modulaires*, Monogr. Enseign. Math. vol. 37, 2001, 147–162.
24. R. Coulangeon and T. Watanabe, *Hermite constant and Voronoi theory over a quaternion skew field*, Osaka J. Math. **43** (2006), 517 - 556.
25. J. Faraut and A. Korányi, *Analysis on Symmetric Cones*, Oxford Mathematical Monographs, Clarendon Press, 1994.
26. M. Gindraux, *Lamination et antilamination des réseaux euclidiens*, J. Th. Nombres Bordeaux **21** (2009), 535–557.
27. E. Gottschling, *Explizite Bestimmung der Randflächen des Fundamentalbereiches der Modulgruppe zweiten Grades*, Math. Ann. **138** (1959), 103–124.
28. P. M. Gruber, *Convex and Discrete Geomety*, Grundlehren Math. Wiss., vol. 336, Springer-Verlag, 2007.
29. P. M. Gruber and C. G. Lekkerkerker, *Geometry of Numbers*, North-Holland 1987.
30. T. Hayata, T. Oda and T. Yatougo, *Zero cells of the Siegel-Gottschling fun-*

damental domain of degree 2, preprint, (2011).

31. P. Humbert, *Théorie de la réduction des formes quadratiques définies positives dans un corps algébrique K fini*, Comment. Math. Helv. **12** (1939/40), 263–306.
32. M. I. Icaza, *Hermite constant and extreme forms for algebraic number fields*, J. London Math. Sco. **55** (1997), 11–22.
33. T. Kawamura, *The minimum of the determinant in the fundamental region of the modular group of degree 2*, preprint, 2009.
34. Y. Kitaoka, *Arithmetic of Quadratic Forms*, Cambridge, 1993.
35. M. Koecher, *Beiträge zu einer Reductionstheorie in Positivitätsbereichen. I*, Math. Ann. **141** (1960), 384 – 432.
36. A. Korkine and G. Zolotareff, *Sur les formes quadratiques positives*, Math. Ann. **11** (1877), 242–292.
37. M. Laca, N. Larsen and S. Neshveyev, *On Bost-Connes types systems for number fields*, J. Number Theory **129** (2009), 325–338.
38. A. Leibak, *On additive generalization of Voronoï's thoery to algebraic number fields*, Proc. Estonian Acad. Sci. Phys. Math. **54** (2005), no.4, 195–211.
39. A. Leibak, *The complete enumeration of binary perfect forms over the algebraic number field* $\mathbf{Q}(\sqrt{6})$, Proc. Estonian Acad. Sci. Phys. Math. **54** (2005), no.4, 212–234.
40. J. Martinet, *Perfect Lattices in Euclidean Spaces*, Grundlehren Math. Wiss., vol. 327, Springer-Verlag, 2003.
41. B. Meyer, *Constantes d'Hermite et théorie de Voronoï*, Thèse, L'Université Bordeaux 1, 2008.
42. B. Meyer, *Generalized Hermite constants, Voronoi theory and heights on flag varieties*, Bull. Soc. Math. Fr. **137** (2009), 127–158.
43. H. Minkowski, *Geometrie der Zahlen*, Druck und Verlag von B. G. Teubner, 1910.
44. M. Newman, Integral Matrices, Academic Press, Pure and Applied Math. Vol.45, 1972.
45. K. Okuda and S. Yano, *A generalization of Voronoï's theorem to algebraic lattices*, J. Th. Nombres Bordeaux **22** (2010), 727–740.
46. H. E. Ong, *Perfect quadratic forms over real-quadratic number fields*, Geom. Dedicata **20** (1986), no.1, 51–77.
47. C. Poor and D. S. Yuen, *Linear dependence among Siegel Modular forms*, Math. Ann. **318**(2000), no.2, 205–234.
48. C. Poor and D. S. Yuen, *The extreme core*, Abh. Math. Sem. Univ. Hamburg **75**(2005), 51–75.
49. C. Poor and D. S. Yuen, *The Bergé-Martinet constant and slopes of Siegel cusp forms*, Bull. Lond. Math. Soc. **38**(2006), no.6, 913–924.
50. M. E. Pohst and M. Wagner, *On the computation of Hermite–Humbert constants for real quadratic number fields*, J. Théor. Nombres Bordeaux **17** (2005), no. 3, 905–920.
51. M. Kneser, *Two remarks on extreme forms*, Canad. J. Math. **7** (1955), 145–149.
52. R. A. Rankin, *On positive definite quadratic forms*, J. London Math. Soc. **28**

(1953), 309–314.
53. R. T. Rockafellar, *Convex Analysis*, Princeton Mathematical Series, vol. 28, Princeton Univ. Press, 1970.
54. S. S. Ryshkov, *The polyhedron $\mu(m)$ and some extremal problems of the geomety of numbers*, Soviet Math. Dokl. **11** (1970), 1240–1244.
55. S. S. Ryshkov, *On the problem of the determination of the perfect quadratic forms in many variables*, Proc. Steklov Math. **3** (1979), 233–259.
56. S. S. Ryshkov and E. P. Baranovskii, *Classical methods in the theory of lattice packings*, Russian Math. Surveys **34** (1979), 1–68.
57. K. Sawatani, T. Watanabe and K. Okuda, *A note on the Hermite–Rankin constant*, J. Th. Nombres Bordeaux **22** (2010), 209–217.
58. K. Sawatani and T. Watanabe, *Type one functions and Voronoï's theorem*, J. Convex Analysis **18** (2011), no. 2, 341–351.
59. A. Schürmann, *From positive definite quadratic forms to lattice sphere packings and coverings*, lecture notes, 2006, `http://fma2.math.uni-magdeburg.de/~achill/publications.html`.
60. A. Schürmann, *Computational Geometry of Positive Definite Quadratic Forms*, University Lecture Series 48, Amer. Math. Soc., 2008.
61. J. L. Thunder, *Higher dimensional analogues of Hermite's constant*, Michigan Math. J. **45** (1998), 301–314.
62. G. Voronoï, *Nouvelles applications des paramétres coninus á la théorie des formes quadratiques, premier mémoire. Sur quelques propri—'etés des formes quadratiques positives parfaites*, J. reine angew. Math. **133** (1908), 97–178.
63. T. Watanabe, *On an analog of Hermite's constant*, J. Lie Theory **10** (2000), 33–52.
64. T. Watanabe, *Fundamental Hermite constants of linear algebraic groups*, J. Math. Soc. Japan **55** (2003), 1061–1080.
65. T. Watanabe, *A survey and a complement of fundamental Hermite constants*, Contemporary Math. Vol. **344**, Amer. Math. Soc. (2004), 339–350.
66. G. L. Watson, *The number of minimum points of a positive quadratic form*, Dissertationes Math. Rozprawy Mat. **84** (1971) 42 pp.
67. C. Zong, *The kissing numbers of convex bodies—a brief survey*, Bull. London Math. Soc. **30** (1998), 1–10.